Date Due

9 SEP 1978

D1273674

Mechanics, waves, and thermal physics

R. L. ARMSTRONG / J. D. KING
UNIVERSITY OF TORONTO

Mechanics, waves,

and thermal physics

PRENTICE-HALL, INC.

ENGLEWOOD CLIFFS, NEW JERSEY

Mechanics, waves, and thermal physics

R. L. Armstrong / J. D. King

© 1970 by Prentice-Hall, Inc., Englewood Cliffs, N.J.
All rights reserved. No part of this book may be reproduced in any
form or by any means without permission in writing from the publisher.

Current printing (last digit): 10 9 8 7 6 5 4 3 2 1

13-571810-4

Library of Congress Catalog Card Number 78-76874

Printed in the United States of America

PRENTICE-HALL INTERNATIONAL, INC.   *London*
PRENTICE-HALL OF AUSTRALIA, PTY. LTD.   *Sydney*
PRENTICE-HALL OF CANADA, LTD.   *Toronto*
PRENTICE-HALL OF INDIA PRIVATE LTD.   *New Delhi*
PRENTICE-HALL OF JAPAN, INC.   *Tokyo*

JOINT UNIVERSITY LIBRARIES NASHVILLE TENN.

Science
QC
21
.2
.A74

m

663345

Preface

The course of study presented in this book has evolved from the authors' experience in teaching the first year of a two-year sequence of general physics courses at the University of Toronto. The course was designed to satisfy a number of specific requirements. We wished to present a program which represented a more unified approach to physics than is traditionally encountered in introductory texts. The material had to contain a comprehensive treatment of some of the basic ideas of physics and yet contain topics of current interest to the students. Since many of the students would eventually major in chemistry or biology, material was often selected with their needs in mind. We also felt that it was essential to stimulate the brightest students and yet to provide a palatable course for the average ones. Rather more material is presented than can normally be covered in any single year, thus allowing an instructor a certain amount of freedom to construct his own course. All of the material has been tested in the classroom.

We have adopted the approach of beginning with a study of the dynamics of single particles, progressing to systems of two and then a few particles, and finally treating systems of large numbers of particles first by the kinetic theory and then by introducing quantum and statistical ideas.

The first section of the course is concerned with an in-depth treatment of the kinematics of single particles. Both the classical and relativistic descriptions of motion are presented. It is our observation that students welcome the early introduction of the special theory of relativity. It is a topic about which they have heard a little, and one which is very different from anything they have had in high school. They find it both exciting and perplexing. Since relativistic kinematics can be studied without the use of calculus, the early study of this topic delays the necessity of intro-ducing calculus into the physics course until the students have attained some proficiency in it in the usual calculus co-requisite course.

In the following dynamics section, the notion of fundamental forces is stressed, conservation laws and their importance to a physicist are emphasized, and a general introduction to the oscillations of a single particle is given. The section concludes with a chapter on relativistic dynamics.

Planetary motion is discussed as an example of a two-particle system. At present students are quite interested in satellite dynamics and seem to enjoy this material. The internal motions of molecules are next discussed in terms of specific molecular models. This material is particularly relevant to the study of chemistry and biology.

Since a knowledge of wave motion is essential to the understanding of some of the material in the latter portion of the text, an introduction to the properties of waves and to the principle of superposition is given.

The kinetic theory description of the behavior of gases is developed. The predictions of the kinetic theory are compared with the behavior of real gases and modifications to the theory suggested. Kinetic theory is then applied to the calculation of specific heats and found to fail for diatomic gases. After elementary quantum and statistical ideas are introduced, a quantum statistical approach is shown to account successfully for the specific heats of diatomic gases. This approach is then extended to a consideration of the static and dynamic properties of crystalline solids. The transition from the microscopic to the macroscopic world is completed with a discussion of thermodynamics.

Detailed mathematical developments of important results are separated from the main body of the text; they are included for those who are interested.

The problems at the end of each chapter are arranged according to the sections, the more difficult ones being designated by a star (★). Some of the problems contain data with associated experimental errors. Such data may be used to obtain experimental errors associated with the deduced quantities, using the procedures outlined in an appendix on experimental measurement.

This course may be logically followed by a number of other courses. In particular, a course in electricity, magnetism, optics, and some selected topics in modern physics would complete a reasonable two-course package.

The four years of development that are represented by this book were for the most part enjoyable. The experience of trying new material on eager young students is refreshing and illuminating. We hope that you may benefit from our experience.

R. L. ARMSTRONG
J. D. KING

Contents

APPENDICES

Mechanics, waves, and thermal physics

The beginning 1

1.1 THE SCIENTIFIC METHOD

Any science begins with observations, but observations are only a beginning, not an end in themselves. In an attempt to explain the observations, the scientist uses his experience, his imagination, and his intuition to formulate theories, usually in terms of mathematical models. These theories are very often of an evolutionary nature; they extend a line of thought farther in a direction in which it was already going. For example, the discovery of the basic laws of electricity and magnetism was followed by James Clerk Maxwell's consolidation of the basic laws into a comprehensive set of four equations; these equations predicted the existence of electromagnetic waves, which were subsequently produced in the laboratory by Heinrich Hertz. Then light was shown to be an electromagnetic wave, and its behavior was found to be described by the theory. The theory was thus extended, and ultimately all observations of electric and magnetic phenomena became describable within the framework of the basic theory.

Occasionally, however, it becomes necessary to abandon a traditional mode of thinking and to strike out in a new direction in order to describe adequately a particular observation or set of observations. The introduction of the quantum principle by Max Planck and the formulation of the special theory of relativity by Albert Einstein are two well-known examples of entirely fresh approaches to the explanation of phenomena.

The formulation of such theories is more than just a convenient way of summarizing a body of experimental observations; it provides the scientist with a means of predicting the results of experiments not yet executed or allows him to see relations between already familiar phenomena. That is, prediction and interpretation play an equal role in the formulation of a scientific theory. In either case the deduced theories are exposed to further experimental tests, which often lead to modifications or even reformulations of the theories.

We may think of physics very generally as that branch of science that deals with the nature of the material world and its phenomena; more specifically, physics deals with such basic problems as the structure of space and time, the nature of the most elementary particles from which matter is composed, the forces responsible for the motions of particles and agglomerations of particles, and the nature of radiation and its interaction with matter. In his first attempt to understand the behavior of his universe, man naturally studied the visible objects around him. These were things he could readily take hold of, whose motion he could easily follow. Soon, however, his observations went outward to larger objects—the moon, the sun, the stars. By the time Isaac Newton had formulated his laws of motion and universal gravitation, which accounted most satisfactorily for the motions of the largest planets down to the smallest visible objects, men were beginning to be concerned with the basic structure of visible matter.

To do this, better tools were needed in order to reveal what could not be seen with the naked eye. As these were developed (the microscope, the electroscope, cathode ray tubes, etc.) the atomic nature of matter was slowly discerned. All matter was found to be composed of more or less complex agglomerations of a relative handful of atoms that form the basic elements. Later, men discovered that the nucleus of a given atom could undergo changes that would transform it into the nucleus of a different atom. These transformations, known under the various names of radioactivity, nuclear reactions, fission, and fusion, have enabled men to develop new sources of energy, some controlled as in nuclear reactors, others uncontrolled as in nuclear (fission) and hydrogen (fusion) bombs.

As more information became available concerning the atomic structure of nature, more and more of the large scale phenomena readily visible to man became explainable from an atomic point of view. For example, most substances exist in three distinct phases—solid, liquid, and vapor. The elastic properties of solids were found to be describable in terms of the behavior of atoms packed into a regular structure or lattice; in the liquid phase the atoms are released from their relatively fixed places in a regular lattice and are free to move about throughout the volume of the liquid, but are constrained from moving very far away from the other atoms in the liquid; in the vapor phase, the atoms are free to move about relatively independently of each other. Such a picture, with an appropriate mathematical formulation, has been reasonably successful in explaining many of the properties of solids, liquids, and gases, although there is much work yet to be done, especially with respect to the liquid state.

As a second example, we shall consider the constitution of the nearest star, our sun. By analysis with spectroscopic techniques of the light emitted from the sun, it is known that the interior of the sun consists mostly of hydrogen. As the nature of fusion processes was investigated, it became apparent that the enormous energy output of the sun could be explained on the basis of a series of nuclear reactions that result in the disappearance of four hydrogen nuclei (protons) and the appearance of the nucleus of a helium atom, which is almost, but not quite, four times as heavy as a proton. The small loss in mass summed over countless numbers of such reactions accounts for the very large energy radiated by the sun.

Because of the importance of the atomic structure of matter in most of present-day scientific research, it is only reasonable to develop our knowledge of physics from an atomic basis from the outset, gradually working our way from one-, to two-, to many-particle systems and finally relating the large-scale properties of ordinary matter to its microscopic structure. The atomic picture has relevance not only to the physicist but also to the chemist and the biologist, who find it increasingly necessary to invoke the atomic picture of nature in explaining their observations.

As a result of the enormous volume of scientific knowledge, it has been found expedient to divide science into several disciplines—physics, geology, astronomy, chemistry, and biology. Physics, as outlined in the first few paragraphs of this section, is the fundamental science.

Geology deals with the macroscopic structure and behavior of the earth; astronomy deals with the structure and origin of the solar system, the Galaxy, and the remainder of the universe. Both of these sciences are closely related to physics and draw substantially from the theoretical and technological advances of physics.

Chemistry is primarily concerned with studies of the structure of molecular systems, the synthesis of compounds, the rates of chemical reactions, and the manner in which these reactions develop. Physics and chemistry merge in the realm of physical chemistry, where the experimental techniques of physics are applied to the study of simple molecular systems. The interaction between physics and chemistry has always been substantial. As an example, the atomic theory developed by physicists was substantiated to a large extent by experiments in chemistry.

Biology deals with the study of living material and life processes. The union of biology and chemistry, called biochemistry, deals with the analysis and synthesis of substances formed in biological systems; the union of biology and physics, called biophysics, applies the experimental knowledge and techniques of physics to the problem of biology.

1.2 MEASUREMENT AND UNITS

All physical quantities are ultimately defined in terms of four "fundamental" quantities: length, time, mass, and charge.

The fundamental unit of length in the MKS system of units is the **meter**. It was originally defined to be one ten-millionth of one-quarter of the earth's circumference. This is neither a convenient nor an accurate standard and was replaced, by international agreement, by the distance between two scratches on a platinum-iridium bar kept in a carefully controlled environment at Sèvres, France. All secondary standards in turn had to be calibrated with respect to this distance. This remained the standard until 1960, when, again by international agreement, the meter was redefined in terms of the wavelength of a particular color of light emitted by a certain isotope of the gas krypton. In particular, a length of one meter is taken to be equal to 1,650,763.73 wavelengths of the orange light emitted by the krypton isotope of mass number 86. This standard is far more precisely defined than the previous one. It is also much more convenient, since this standard can be set up in any laboratory in the world.

The fundamental unit of time, the **second**, was originally defined in terms of the rotational period of the earth. When careful measurements are made it is found that the earth's rotation is not exactly periodic and is therefore not acceptable as the basis of a standard of time. A new standard of time was adopted in 1966 when the second was redefined to be 1/31,556,925.9747 of the tropical year 1900 A.D. This redefined second corresponds to 9,192,631,770 cycles of vibration of the cesium atom of mass number 133 when it is in one particular state of vibration (the language required to specify the state of vibration comes from the theory of atomic structure). Again, this is a very convenient standard that can be set up in any laboratory.

The standard unit of mass, the **kilogram**, was chosen to have a mass approximately equal to that of a cube of water, one-tenth of a meter on a side. The actual standard is a piece of platinum alloy kept in a carefully controlled environment at Sèvres, France. No suitable atomic standard has yet been found to replace it. A detailed discussion of the concept of mass is given in Section 7.2.

In this text, we shall not be dealing with electrical phenomena to any great extent, and therefore we shall not give a precise definition of charge.

Other physical quantities will have dimensions that are derived from these fundamental quantities. Some derived quantities are obtained by definition (for example, speed is defined as rate of change of distance with time), whereas others are derived from relationships (for example, force is equal to the rate of change of momentum, where momentum is the product of mass and speed). One can denote mass by the letter M, length by L, time by T, and charge by Q. The dimensions of other physical quantities are represented by appropriate combinations of M, L, T, and Q. Table 1.1 lists the dimensions of some common physical quantities that we shall use in the text.

We shall be encountering numbers that are very large as well as numbers that are very small throughout the text. In order to avoid writing

Table 1.1   Dimensions of physical quantities

Quantity	Dimension	Unit (MKS)
length	L	meter (m)
time	T	second (sec)
mass	M	kilogram (kg)
charge	Q	coulomb (C)
speed	LT^{-1}	meter $\cdot$ sec^{-1} (m $\cdot$ sec^{-1})
acceleration	LT^{-2}	meter $\cdot$ sec^{-2} (m $\cdot$ sec^{-2})
force	MLT^{-2}	newton (N)
work and energy	ML^2T^{-2}	joule (J)
current	QT^{-1}	ampere (A)
magnetic field (intensity)	$MT^{-1}Q^{-1}$	weber $\cdot$ m^{-2} (Wb $\cdot$ m^{-2} or T) or tesla
frequency	T^{-1}	hertz (Hz) $\equiv$ cycles $\cdot$ sec^{-1}

down numbers such as 12,300,000 and 0.0000067, we shall use powers of ten and write 1.23×10^7 and 6.7×10^{-6}, respectively. In addition, we shall often find it convenient to use prefixes in front of a unit to indicate powers of ten. The standard list of prefixes and their abbreviations is given in Table 1.2.

Table 1.2   Prefix names of multiples and submultiples of units

Factor by which unit is multiplied	Prefix	Symbol	Factor by which unit is multiplied	Prefix	Symbol
10^{12}	tera	T	10^{-2}	centi	c
10^9	giga	G	10^{-3}	milli	m
10^6	mega	M	10^{-6}	micro	μ
10^3	kilo	k	10^{-9}	nano	n
10^2	hecto	h	10^{-12}	pico	p
10	deka	da	10^{-15}	femto	f
10^{-1}	deci	d	10^{-18}	atto	a

Vectors 2

2.1 INTRODUCTION

Vector analysis is an almost indispensable aid to the study of physics. It provides a concise notation for the writing of the equations that describe mathematically the physical behavior of physical systems. Vectors may also be very helpful to us in the formation of mental pictures of physical or geometrical ideas. For these reasons, vector analysis has proven to be most useful in the development and appreciation of the fundamental laws of physics.

2.2 DEFINITION OF A VECTOR

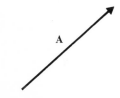

Fig. 2.1. The graphical representation of a vector.

A **vector** is a quantity that has both a magnitude and a direction. For example, the statement that Montreal is 175 miles due north of New York contains both a magnitude and a direction. This information can be described by means of a vector. A vector may be represented graphically as an arrow whose direction defines the direction of the vector and whose length indicates the magnitude of the vector, as illustrated in Fig. 2.1. The symbol **A** stands for the quantity under consideration. For example, in Fig. 2.2 the vector **S** represents a displacement of 3.6 km from O to P in the northeast direction. The tail end O of the vector **S** is called the **origin** of the vector, whereas the end point P is known as the **terminus** of the vector. It is quite common to refer to the origin and terminus points of a vector when representing a vector mathematically. That is, in Fig. 2.2 the vector **S** could also be written as the vector **OP**; the forms are equivalent, and both are in common use. The **magnitude** of the vector **A** is written as $|\mathbf{A}|$ or A; again both forms are commonly used.

In many cases, the position of the origin and terminus points of a vector are not important; we shall feel free to move a vector about on a graph, but we shall be careful not to change its magnitude or direction. If the positions of the ends are important, this information will be added when the vector is used.

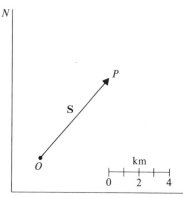

Fig. 2.2. The graphical representation of a displacement vector in the northeast direction.

A **scalar** is a quantity that has magnitude but no direction; it is simply a number. For example, time is a scalar quantity. That is, your watch ticks at the same rate independent of the direction in which you are facing.

2.3 VECTOR ALGEBRA

The algebraic operations of addition, subtraction, and multiplication can be extended to provide an algebra of vectors, with suitable definitions. The operation of division, however, appears to serve no useful purpose in a vector algebra, and its use is not extended to vectors. We shall deal with addition and subtraction of vectors in this section and with multiplication of vectors in Section 2.5.

We note first that two vectors **A** and **B** that have the same direction and magnitude are defined to be **equal**. Therefore, in Fig. 2.3 we may write **A** = **B**. Figure 2.4 shows two vectors **A** and −**A** that have equal magnitude but opposite directions. It is apparent that successive applications of the vectors **A** and −**A** will have a net zero or null effect. Mathematically, we write the vector equation

$$\mathbf{A} + (-\mathbf{A}) = 0.$$

The addition and subtraction of two arbitrary vectors **A** and **B** is performed as shown in Fig. 2.5. The vector **B** in Fig. 2.5(a) is translated parallel to itself, so that the origin of **B** coincides with the terminus of **A**; the vector **A** + **B** joins the origin of **A** to the terminus of **B**. Subtraction of the vector **B** and the vector **A** is shown in Fig. 2.5(b). In subtraction, the vector −**B** is added to the vector **A** to yield the sum **A** + (−**B**) = **A** − **B**; the resulting sum is, by definition, the difference between the vectors **A** and **B**.

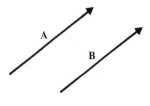

Fig. 2.3. The vectors **A** and **B** have the same direction, the same magnitude, and **A** = **B**.

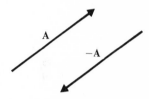

Fig. 2.4. **A** and −**A** are equal but opposite vectors.

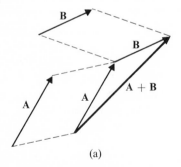 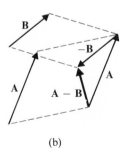

(a)   (b)

Fig. 2.5. (a) Vectors **A** and **B** are added to yield the vector **A** + **B**. (b) Vector **B** is subtracted from vector **A** to yield the vector **A** − **B**.

Example. A prospector walks two miles due north from his campsite and then three miles northwest and there finds some interesting ore samples. In what direction should he head, and how far must he travel, in order to return to his campsite in a straight line?

Graphical solution. In Fig. 2.6 we take O to be the prospector's campsite. We mark off a length $OP = 2$ miles to the north and a length $PQ = 3$ miles to the north-

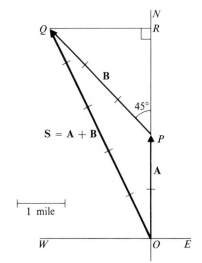

Fig. 2.6. Graphical solution to the example.

west. The length OQ is the distance he must walk and is measured to be 4.6 miles (approximately). The angle OQR is measured to be 63.0° (by protractor). Note that the vector **OQ** = **S** = **A** + **B** is the **displacement** of the prospector from his campsite when he finds the ore samples and is the sum of the two displacements **OP** = **A** and **PQ** = **B**. The prospector must walk 4.6 miles in a direction 63.0° south of east to return directly to his campsite.

Mathematical solution. From triangle OPQ we have, by the cosine law of trigonometry,

$$S^2 = A^2 + B^2 - 2AB \cos \angle OPQ$$
$$= 2^2 + 3^2 - 2(2)(3) \cos 135°$$
$$= 4 + 9 - 12(-0.707)$$
$$= 13 + 8.48$$
$$= 21.48 \text{ (miles)}^2;$$

therefore,

$$S = 4.63 \text{ miles.}$$

By the sine law of trigonometry,

$$\frac{B}{\sin \angle POQ} = \frac{S}{\sin \angle OPQ}$$

$$\frac{3}{\sin \angle POQ} = \frac{4.63}{\sin 135°}$$

$$\sin \angle POQ = \frac{3(0.707)}{4.63}$$

$$= 0.458.$$

Therefore,

$$\angle POQ = 27.3°.$$

Therefore, the vector **OQ** has a magnitude of 4.63 miles. The direction south of east in which the prospector must walk is given by the angle OQR, where

$$\angle OQR = 90.0° - \angle ROQ = 90.0° - 27.3° = 62.7°.$$

2.4   COMPONENTS OF A VECTOR

Any vector **A** in two dimensions may be positioned with its initial point at the origin of a Cartesian (rectangular) coordinate system. It is evident from Fig. 2.7 that

$$\mathbf{A} = \mathbf{A}_x + \mathbf{A}_y,$$

where $\mathbf{A}_x$ and $\mathbf{A}_y$ are the **component vectors** of $\mathbf{A}$ in the x and y directions, respectively. A_x and A_y are the coordinate points of the terminus of the vector $\mathbf{A}$, as well as being the magnitudes of the vectors $\mathbf{A}_x$ and $\mathbf{A}_y$, respectively. A_x and A_y are called the **components** of $\mathbf{A}$ in the x and y directions, respectively. The term "component" is often used incorrectly when the component vectors are being referred to; the reader should be careful to be consistent in using these two terms. The definitions of component vectors and components may be readily extended to a three-dimensional rectangular coordinate system.

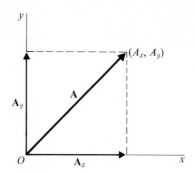

Fig. 2.7. The component vectors of $\mathbf{A}$ along Cartesian coordinate axes are $\mathbf{A}_x$ and $\mathbf{A}_y$.

The use of component vectors is not restricted to rectangular coordinate systems. Any set of vectors that sum to give the vector in question may be regarded as component vectors. However, component vectors along Cartesian axes are by far the most useful component vectors.

Example. Referring to the example in Section 2.3, determine how far north and how far west of his campsite the prospector was when he made his find.

Solution. The displacement of the prospector from his campsite when he made his find is given by the vector **OQ** in Fig. 2.6. We require the components of this vector. The north component is $OR = OP + PR$ in Fig. 2.8, and the west component is OT. Now,

$$PR = PQ \sin 45°$$

$$= 3(0.707)$$

$$= 2.12 \text{ miles},$$

and

$$OR = 2 + 2.12$$

$$= 4.14 \text{ miles}.$$

The angle OQR was determined to be 62.7°, and OQ was 4.63 miles. Since $\angle TOQ = \angle OQR = 62.7°$, we have

$$OT = OQ \cos 62.7°$$

$$= 4.63(0.459)$$

$$= 2.12 \text{ miles}.$$

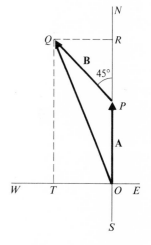

Fig. 2.8. The north and west components of the displacement vector **OQ** are OR and OT, respectively.

Therefore, the northward component of his displacement was 4.14 miles, and the westward component was 2.12 miles.

2.5   VECTOR MULTIPLICATION

At first thought, it might appear that the concept of vector multiplication is not very meaningful. However, there are two combinations of vectors that resemble algebraic multiplication in form and are of use in the description of physical phenomena.

The **dot or scalar product** of two vectors **A** and **B** is defined as the product of the magnitudes of **A** and **B** and the cosine of the angle between them (see Fig. 2.9). The scalar product is denoted by **A** · **B** (read **A** dot **B**) and has the value

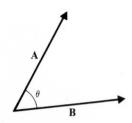

Fig. 2.9. The scalar product of vectors **A** and **B** is given by **A** · **B** = $AB \cos\theta$.

$$\mathbf{A} \cdot \mathbf{B} = AB \cos \theta,$$

where the angle θ is given by $0 \leqslant \theta \leqslant \pi$. Note particularly that **A** · **B** is a scalar and not a vector.

Example. A man exerts a steady force **F** at an angle θ to the horizontal on a heavy box via a rope and succeeds in moving the box a distance s along the floor (see Fig. 2.10). Find an expression for the work done by the man in moving the box,

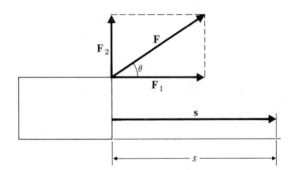

Fig. 2.10. A force **F** acting at an angle θ to the horizontal succeeds in moving a box a distance s along the floor.

given that work is defined as the product of a force times the distance through which it acts.

Solution. The force **F** has component forces $\mathbf{F}_1$ in the direction of motion of the box and $\mathbf{F}_2$ perpendicular to the direction of motion. It is clear that only the component force $\mathbf{F}_1$ produces movement of the box, and the work done, according to our definition, will be $F_1 s$. However, since F_1 is a component of **F** we have

$$F_1 = F \cos \theta,$$

and the work done can be written as

$$(F \cos \theta)s = Fs \cos \theta.$$

But s is the magnitude of the vector **s**, which represents the displacement of the box. Therefore, the work done is

$$Fs \cos \theta = \mathbf{F} \cdot \mathbf{s},$$

where **F** is the applied force and **s** is the resulting displacement.

The following laws for the scalar multiplication of vectors are valid:
1. $\mathbf{A} \cdot \mathbf{B} = \mathbf{B} \cdot \mathbf{A}$.   (Commutative law)
2. $\mathbf{A} \cdot (\mathbf{B} + \mathbf{C}) = \mathbf{A} \cdot \mathbf{B} + \mathbf{A} \cdot \mathbf{C}$.   (Distributive law)
3. $m(\mathbf{A} \cdot \mathbf{B}) = (m\mathbf{A} \cdot \mathbf{B}) = \mathbf{A} \cdot (m\mathbf{B}) = (\mathbf{A} \cdot \mathbf{B})m$,   where m is a scalar.
4. If $\mathbf{A} \cdot \mathbf{B} = 0$, then either $\mathbf{A} = \mathbf{B} = 0$, or **A** and **B** are perpendicular to each other.

The **cross or vector product** of the two vectors **A** and **B** is defined to be a vector whose magnitude is the product of the magnitudes of **A** and **B** and the sine of the angle between them and whose direction is perpendicular to the plane containing the vectors **A** and **B**. In symbols,

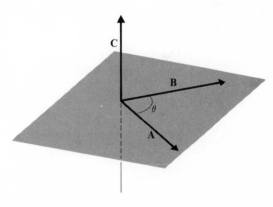

$$\mathbf{C} = \mathbf{A} \times \mathbf{B}$$

$$C = AB \sin \theta, \qquad 0 \leqslant \theta \leqslant \pi.$$

The direction of **C** is as shown in Fig. 2.11 and is chosen so that the vectors **A**, **B**, and **C** form a **right-handed system**.

Fig. 2.11. The vector **C** is the vector product of vectors **A** and **B**.

In general, three vectors **A**, **B**, and **C**, which have coincident initial points and are not all in the same plane, form a right-handed system if a right-threaded screw rotated from **A** to **B** through an angle θ less than 180° will advance in the direction of **C**, as shown in Fig. 2.12.

Example. Show that the area of a parallelogram with sides **A** and **B** is $|\mathbf{A} \times \mathbf{B}|$.

Solution. From Fig. 2.13 we see that the area of the parallelogram is the same as the area of the rectangle with sides h and $|\mathbf{B}|$. But $h = |\mathbf{A}| \sin \theta$, so that the area of the parallelogram is

$$A = h|\mathbf{B}|$$

$$= |\mathbf{A}||\mathbf{B}| \sin \theta$$

$$\equiv AB \sin \theta$$

$$= |\mathbf{A} \times \mathbf{B}|.$$

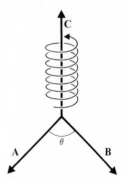

Fig. 2.12. The vectors **A**, **B**, and **C** form a right-handed system.

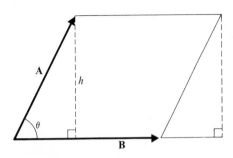

Fig. 2.13. The area of a parallelogram is $|\mathbf{A} \times \mathbf{B}|$.

The following laws for the cross product of vectors are valid:

1. $\mathbf{A} \times \mathbf{B} = -\mathbf{B} \times \mathbf{A}$.

   (Commutative law not valid)

2. $\mathbf{A} \times (\mathbf{B} + \mathbf{C}) = \mathbf{A} \times \mathbf{B} + \mathbf{A} \times \mathbf{C}$

   (Distributive law)

3. $m(\mathbf{A} \times \mathbf{B}) = (m\mathbf{A}) \times \mathbf{B} = \mathbf{A} \times (m\mathbf{B}) = (\mathbf{A} \times \mathbf{B})m$, where m is a scalar.

4. If $\mathbf{A} \times \mathbf{B} = 0$, either $\mathbf{A}$ or $\mathbf{B} = 0$, or $\mathbf{A}$ and $\mathbf{B}$ are parallel.

Finally, we should note that a scalar times a scalar is a scalar, whereas a scalar times a vector is a vector; we shall have many examples of this in the following chapters.

2.6 UNIT VECTORS

A **unit vector** is a vector having unit magnitude. Any vector $\mathbf{A}$ can be represented as a unit vector $\mathbf{a}$ in the direction of $\mathbf{A}$ multiplied by A, the magnitude of $\mathbf{A}$. In symbols

$$\mathbf{A} = A\mathbf{a}.$$

Of special importance are the unit vectors $\mathbf{i}, \mathbf{j}$, and $\mathbf{k}$, which have the directions of the positive x-, y-, and z-coordinate axes, respectively, of a three-dimensional rectangular coordinate system, as shown in Fig. 2.14.

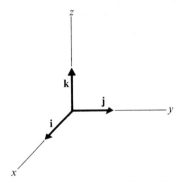

Fig. 2.14. The unit vectors $\mathbf{i}, \mathbf{j}$, and $\mathbf{k}$.

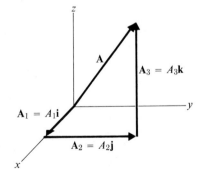

Fig. 2.15. The components of a vector $\mathbf{A}$ in terms of unit vectors.

Any vector $\mathbf{A}$ in three dimensions can be written as the sum of three component vectors $\mathbf{A}_1, \mathbf{A}_2$, and $\mathbf{A}_3$ having the directions of the x, y, and z axes, respectively, as indicated in Fig. 2.15. In symbols

$$\mathbf{A} = \mathbf{A}_1 + \mathbf{A}_2 + \mathbf{A}_3.$$

However, we may also write

$$\mathbf{A}_1 = A_1\mathbf{i}, \qquad \mathbf{A}_2 = A_2\mathbf{j}, \qquad \mathbf{A}_3 = A_3\mathbf{k}.$$

Therefore,

$$\mathbf{A} = A_1\mathbf{i} + A_2\mathbf{j} + A_3\mathbf{k}.$$

Since $\mathbf{i}, \mathbf{j}$, and $\mathbf{k}$ are mutually perpendicular vectors, we have the following identities:

$$\mathbf{i} \cdot \mathbf{i} = \mathbf{j} \cdot \mathbf{j} = \mathbf{k} \cdot \mathbf{k} = 1$$

$$\mathbf{i} \cdot \mathbf{j} = \mathbf{j} \cdot \mathbf{k} = \mathbf{k} \cdot \mathbf{i} = 0$$

$$\mathbf{i} \times \mathbf{i} = \mathbf{j} \times \mathbf{j} = \mathbf{k} \times \mathbf{k} = 0$$

$$\mathbf{i} \times \mathbf{j} = \mathbf{k}, \mathbf{j} \times \mathbf{k} = \mathbf{i}, \mathbf{k} \times \mathbf{i} = \mathbf{j}.$$

Example. Show that the magnitude of the vector

$$\mathbf{A} = A_1\mathbf{i} + A_2\mathbf{j} + A_3\mathbf{k} \quad \text{is} \quad (A_1^2 + A_2^2 + A_3^2)^{1/2}.$$

Solution.

$$A = (A^2)^{1/2}$$

$$= (AA)^{1/2}$$

$$= (\mathbf{A} \cdot \mathbf{A})^{1/2}$$

$$= [(A_1\mathbf{i} + A_2\mathbf{j} + A_3\mathbf{k}) \cdot (A_1\mathbf{i} + A_2\mathbf{j} + A_3\mathbf{k})]^{1/2}$$

$$= (A_1^2 + A_2^2 + A_3^2)^{1/2}.$$

PROBLEMS

1. Make a list of physical quantities that behave
   (a) as scalars.
   (b) as vectors.

2. Represent graphically
   (a) a displacement of 12 miles 20° west of south.
   (b) a displacement of 7 miles 20° south of east.
   (c) a displacement of 3 miles 30° north of west.

3. An airplane flies 200 miles due south, then 100 miles southwest. Represent these displacements graphically and determine the resultant displacement
   (a) graphically.
   (b) analytically.

4. Show that
   (a) $\mathbf{A} + \mathbf{B} = \mathbf{B} + \mathbf{A}$.
   (b) $\mathbf{A} + (\mathbf{B} + \mathbf{C}) = (\mathbf{A} + \mathbf{B}) + \mathbf{C}$.

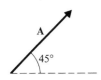

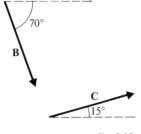

Fig. 2.16.

5. Given the vectors **A**, **B**, and **C** as shown in Fig. 2.16, where $A = 5$, $B = 3$, and $C = 4$ units, construct
(a) $2\mathbf{A} + \mathbf{B} - 3\mathbf{C}$.
(b) $\mathbf{A} - 2\mathbf{B} + 3\mathbf{C}/2$.

6. An airplane moves due west at a speed of 360 mph relative to the ground in the absence of wind. It then enters a region of the atmosphere in which a constant northerly wind of 350 mph is blowing. In what direction must the airplane be pointed in order that it continues to travel due west? What is its speed relative to the ground?

7. Prove that
(a) $\mathbf{A} \cdot \mathbf{B} = \mathbf{B} \cdot \mathbf{A}$.
(b) $\mathbf{A} \cdot (\mathbf{B} + \mathbf{C}) = \mathbf{A} \cdot \mathbf{B} + \mathbf{A} \cdot \mathbf{C}$.

*8. Use the vector dot product to prove the law of cosines for a plane triangle.

9. Show that $\mathbf{A} \times \mathbf{B} = -\mathbf{B} \times \mathbf{A}$.

10. If $\mathbf{A} \times \mathbf{B} = 0$ and if **A** and **B** are not zero, show that **A** is parallel to **B**.

11. Show that $\mathbf{A} \cdot (\mathbf{B} \times \mathbf{C})$ is in absolute value equal to the volume of a parallelepiped with sides **A**, **B**, and **C**.

12. Evaluate $|\mathbf{A}|$, where
$$\mathbf{A} = 3\mathbf{i} + 4\mathbf{j} + 5\mathbf{k}.$$

13. Given that $\mathbf{r}_1 = \mathbf{i} - 2\mathbf{j} - 2\mathbf{k}$
$$\mathbf{r}_2 = -2\mathbf{i} + 4\mathbf{j} + 3\mathbf{k}$$
$$\mathbf{r}_3 = -3\mathbf{i} + 2\mathbf{j} - \mathbf{k},$$
find the magnitudes of
(a) $\mathbf{r}_1$.
(b) $\mathbf{r}_1 + \mathbf{r}_2 + \mathbf{r}_3$.
(c) $5\mathbf{r}_1 + 3\mathbf{r}_2 - \mathbf{r}_3$.

14. The following forces (in newtons) act on a particle P:
$$\mathbf{F}_1 = 4\mathbf{i} - 3\mathbf{j} - 2\mathbf{k}; \qquad \mathbf{F}_3 = -5\mathbf{i} + \mathbf{j} + 3\mathbf{k};$$
$$\mathbf{F}_2 = \mathbf{i} - 2\mathbf{j} + 4\mathbf{k}; \qquad \mathbf{F}_4 = 2\mathbf{i} + 3\mathbf{j} - 5\mathbf{k}.$$
(a) Find the resultant (sum) of the forces.
(b) Find the magnitude of the resultant force.

15. Show that
(a) $\mathbf{i} \cdot \mathbf{i} = \mathbf{j} \cdot \mathbf{j} = \mathbf{k} \cdot \mathbf{k} = 1$.
(b) $\mathbf{i} \cdot \mathbf{j} = \mathbf{j} \cdot \mathbf{k} = \mathbf{k} \cdot \mathbf{i} = 0$.

16. Evaluate
(a) $\mathbf{j} \cdot (2\mathbf{i} - 3\mathbf{j} + \mathbf{k})$.
(b) $(3\mathbf{i} + \mathbf{k}) \cdot (2\mathbf{i} - \mathbf{j})$.

*17. Find the angle between $\mathbf{A} = 6\mathbf{i} - 3\mathbf{j} + 2\mathbf{k}$ and $\mathbf{B} = 2\mathbf{i} + 2\mathbf{j} - \mathbf{k}$.

*18. Find the angles that the vector $\mathbf{A} = 5\mathbf{i} - 7\mathbf{j} + 3\mathbf{k}$ makes with the coordinate axes.

*19. If $\mathbf{A} = \mathbf{i} + 4\mathbf{j} - 2\mathbf{k}$ and $\mathbf{B} = 2\mathbf{i} - 3\mathbf{j} - \mathbf{k}$, find
(a) $\mathbf{A} \times \mathbf{B}$.
(b) $\mathbf{B} \times \mathbf{A}$.

*20. Evaluate $(2\mathbf{i} - 3\mathbf{j}) \cdot [(\mathbf{i} + \mathbf{j} - \mathbf{k}) \times (3\mathbf{i} - \mathbf{k})]$.

3 Introduction to classical kinematics

3.1 INTRODUCTION

The study of the motions of particles through space and time is central to physics. To know the complete motion of a particle, it is necessary to know the particle's position at each instant of time. The description of the motion by mathematical methods is called **kinematics**; a study of the causes of the motion is called **dynamics**. The term **mechanics** embraces both kinematics and dynamics. In this chapter we shall consider only a few introductory ideas in kinematics involving straight-line motion; vector notation will not be necessary as direction along a line can be indicated simply by a plus or minus sign. We shall leave until Chapter 6 a fuller description of kinematics from a more general point of view.

3.2 STRAIGHT-LINE KINEMATICS

We shall introduce the basic kinematical concepts by considering a single particle confined to move in one dimension. The position of the particle at any time is then specified in terms of a single position coordinate with respect to an arbitrarily selected origin. In Fig. 3.1 a particle is shown at point P; the position coordinate or **displacement** of the particle from the origin is represented by x. A complete description of the motion of the

Fig. 3.1. The particle at P is displaced from the origin by a distance x.

particle requires a knowledge of the displacement x at every instant of time. If the particle is displaced to the right of the origin O, the position coordinate will be specified by a positive number; if the particle is to the left of O, by a negative number. We can represent the motion of the particle by means of a **displacement-time graph**. A typical graph might be that shown in Fig. 3.2. At $t = 0$ the particle is 2 m to the right of O. After 3 sec have elapsed, the particle has moved to a position 3 m to the right of O. For the next 5 sec the particle moves to the left, until after 8 sec it is 1 m to the left of O. This graph contains all of the information about the particle's motion.

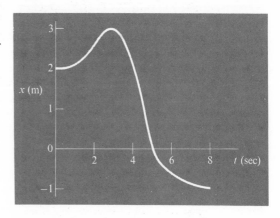

Fig. 3.2. A displacement-time graph.

3.3 SPEED AND VELOCITY

The concept of speed is a familiar one. When you drive along in a car the speedometer registers a certain number. The higher the number, the less time it takes to travel a given distance, and the greater the car's speed. Now we wish to introduce a rather more precise meaning to the concept of speed. Let us consider an expanded portion of Fig. 3.2, such as that shown in Fig. 3.3. At time t_1 the particle's displacement is x_1; at time t_2 it is x_2. The change in displacement is $x_2 - x_1 = \Delta x$ and occurs in a time interval $t_2 - t_1 = \Delta t$. The **average speed** during the interval is defined as

$$\bar{v} = \frac{x_2 - x_1}{t_2 - t_1} = \frac{\Delta x}{\Delta t}$$

and is just the slope of the segment joining the points (x_2, t_2) and (x_1, t_1). Now suppose that point (x_2, t_2) is moved nearer and nearer to the point (x_1, t_1) until the two points coalesce. In this limiting situation the line through the points (x_2, t_2) and (x_1, t_1) becomes a tangent to the curve at the point (x_1, t_1) and the slope of this tangent becomes the **instantaneous** speed at the point (x_1, t_1). We define the instantaneous speed of the particle as the time rate of change of the particle's position at the particular time. We express this result mathematically as

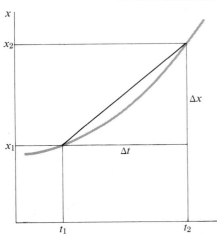

Fig. 3.3. An expanded portion of the curve of Fig. 3.2.

$$v = \lim_{\Delta t \to 0} \left(\frac{\Delta x}{\Delta t} \right) \Bigg|_{x_1, t_1}$$

[read as: v equals the limit, as t approaches zero, of the ratio of Δx to Δt evaluated at the point (x_1, t_1)]. That is, the instantaneous speed at the point (x_1, t_1) is defined as the limiting value of the ratio of the finite change in displacement to the finite change in time as the time interval becomes vanishingly small. As Δt becomes vanishingly small so does Δx, but the ratio $\Delta x / \Delta t$ is determined by the slope of the tangent to the displacement-time curve at the point (x_1, t_1). We define the quantity

$$\frac{dx}{dt} \Bigg|_{x_1, t_1}$$

by

$$\frac{dx}{dt} \Bigg|_{x_1, t_1} = \lim_{\Delta t \to 0} \left(\frac{\Delta x}{\Delta t} \right) \Bigg|_{x_1, t_1}$$

and speak of the **derivative** of x with respect to t evaluated at the point (x_1, t_1). We interpret dx as an **infinitesimal increment** in x and dt as an infinitesimal increment in t. Such derivatives may be evaluated by the methods of differential calculus.

If we have obtained in some way a displacement-time graph for a particle's motion, we might derive the corresponding **speed-time graph** by evaluating the slope of the tangent to the curve at a number of points, plotting the slopes obtained as a function of time, and drawing a smooth curve through them. In this manner Fig. 3.4 was obtained from Fig. 3.2. At $t = 0$, 3, and 8 sec the tangent to the curve in Fig. 3.2 is horizontal, so that the speed is zero at those times. For t between 0 and 3 sec the tangents to the curve have positive slope with the maximum slope occurring for $t = 2$ sec. In this time interval the particle is moving to the right, speeding up for the first 2 sec and slowing down for the last 1 sec. For t between 3 and 8 sec the tangents to the curve have negative slope with the maximum slope occurring for $t = 4.5$ sec. In this time interval the particle is moving to the left, speeding up for the first 1.5 sec and slowing down for the last 3.5 sec. (Actually, Fig. 3.4 is a **velocity-time graph** and not a speed-time graph, since we have introduced directions by using positive numbers to indicate motion to the right and negative numbers to indicate motion to the left.)

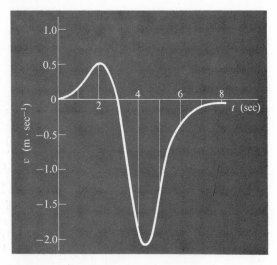

Fig. 3.4. A velocity-time graph obtained by evaluating the tangent to the displacement-time curve of Fig. 3.2 at a number of points on the curve.

Alternatively, we could have found a mathematical expression relating x and t and derived the variation of v with t by the methods of differential calculus. Symbolically, we indicate that x and v are functions of t by writing

$$x = x(t)$$

and

$$v(t) = \frac{dx(t)}{dt}.$$

This approach will be discussed further in Chapter 6.

3.4 ACCELERATION

The concept of **acceleration** is also a familiar one. If you start a car from rest with the accelerator pedal depressed, the speedometer reading increases from zero. The faster the reading increases, the greater the car's acceleration. Now we wish to introduce a rather more precise meaning to the concept of acceleration. To see precisely what this definition means we consider a velocity-time graph in an analogous manner to that just discussed for a displacement-time graph. First we define the **average acceleration** for the time interval $\Delta t = t_2 - t_1$, during which the speed changes from v_1 to v_2, as

$$\bar{a} = \frac{v_2 - v_1}{t_2 - t_1} = \frac{\Delta v}{\Delta t}.$$

We define the **instantaneous acceleration** of the particle as the time rate of change of the particle's speed at a particular time. That is, we define the instantaneous acceleration at the point (v_1, t_1) as the limiting value of the ratio of the finite change in speed (or velocity) to the finite change in time, as the time interval becomes vanishingly small. Using mathematical notation,

$$a = \lim_{\Delta t \to 0} \left(\frac{\Delta v}{\Delta t}\right)\Bigg|_{v_1,t_1} = \frac{dv}{dt}\Bigg|_{v_1,t_1},$$

where $(dv/dt)|_{v_1,t_1}$ is the derivative of v with respect to t evaluated at the point (v_1, t_1).

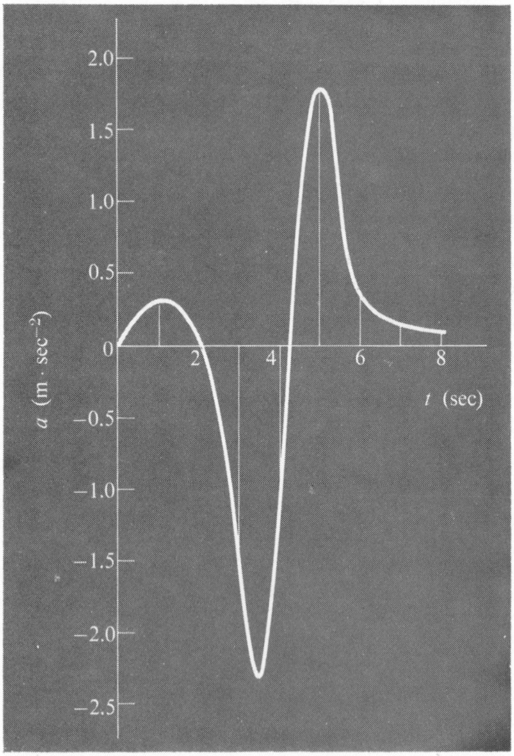

Once we have obtained a velocity-time graph for a particle's motion, we can derive the corresponding **acceleration-time graph** by evaluating the slope of the tangent to the curve at a number of points, plotting the points obtained, and drawing a smooth curve through them. In this manner Fig. 3.5 was obtained from Fig. 3.4. At $t = 2$ and 4.5 sec the tangent to the curve in Fig. 3.4 is horizontal, so that the acceleration is zero at those times. For t between 0 and 2 sec or between 4.5 and 8 sec the tangents to the curve have positive slope. During the first of these intervals the speed increases as the particle moves to the right; during the second interval the speed decreases as the particle moves to the left. For t between 2 and 4.5 sec the tangents have negative slope, and the acceleration is negative, reducing the particle's speed to the right to zero and then giving it an ever increasing speed to the left.

Alternatively, we could have found a mathematical expression relating v and t and derived the variation of a with t by the methods of differential calculus. Symbolically, if

$$v = v(t),$$

Fig. 3.5. The acceleration-time graph obtained from Fig. 3.4.

then

$$a(t) = \frac{dv(t)}{dt}.$$

3.5   DISPLACEMENT, VELOCITY, AND ACCELERATION AS AREAS UNDER CURVES

We wish to emphasize that each of the three curves, displacement-time, velocity-time, and acceleration-time, contains similar information. A knowledge of any one of them permits a determination of the shapes of the other two. So far, we have seen how to go from a displacement-time curve to a velocity-time curve, and thence to an acceleration-time curve. Now we consider the converse problem. Suppose we know the velocity-time curve for a particle's motion. How do we construct the displacement-time curve? First consider a particle traveling with constant speed v_0 for a time t. The instantaneous speed at any time and the average speed during the motion are then equal, so that the distance s_0 traveled in time t_0 is given by

$$s_0 = v_0 t.$$

Figure 3.6 is a velocity-time plot of this particle's motion. (It is also a speed-time plot, since the velocity is constant in direction.) The distance traveled is represented by the shaded area if we start from the origin.

Now let us consider a more complicated motion, in which a particle's velocity varies with time in the manner shown in Fig. 3.7. We can approximate the actual motion by assuming that the particle moves at appropriate

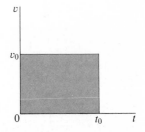

Fig. 3.6. A velocity (speed)-time graph for a particle moving at constant speed.

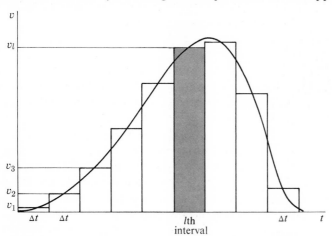

Fig. 3.7. A velocity (speed)-time graph for a variable speed motion.

constant speeds for N equal intervals of time Δt, as shown in Fig. 3.7. The distance Δs_l traveled in the lth time interval Δt is approximately $v_l \Delta t$, where v_l is the constant speed appropriate to that interval and Δs_l is equal to the area of the shaded rectangle. We write

$$\Delta s_l \simeq v_l \Delta t.$$

The total distance s_0 traveled is approximately equal to the sum of the distance traveled during each of the N intervals. That is,

$$s_0 \simeq \sum_{l=1}^{N} \Delta s_l = \sum_{l=1}^{N} v_l \Delta t,$$

which is just the sum of the areas of the N rectangles in Fig. 3.7 ($\sum_{l=1}^{N} \Delta s_l$ means the sum $\Delta s_1 + \Delta s_2 + \cdots + \Delta s_N$). The larger the number of time intervals the smaller Δt, since

$$\Delta t = \frac{t_0}{N},$$

where t_0 is the time taken for the particle to travel the distance s_0. The larger the value of N, the more nearly the approximate calculated value of the distance traveled will approach the actual one. Finally, as the number of intervals becomes infinite (so that the width of the intervals becomes vanishingly small), the calculated value will be equal to the actual distance traveled. In mathematical notation

$$s = \lim_{\Delta t \to 0} \sum_{l=1}^{N} v_l \Delta t.$$

This quantity is just the area under the actual curve in Fig. 3.7. We define the quantity $\int_0^{t_0} v \, dt$ by

$$\int_0^{t_0} v \, dt = \lim_{\Delta t \to 0} \sum_{l=1}^{N} v_l \Delta t$$

and speak of the **integral** of v with respect to t over the range of t values from 0 to t_0. Such integrals may be evaluated by the methods of integral calculus.

If we know the velocity-time graph for a particle's motion, we might derive the corresponding displacement-time graph by evaluating the areas under the curve for a number of consecutive time intervals, plotting the results, and drawing a smooth curve through them. In this manner Fig. 3.9 was obtained from Fig. 3.8. If we take the displacement at time zero to be zero, then during the first three time intervals, when the particle is moving to the right, the increments in displacement are given by the areas labeled (1), (2), and (3) in Fig. 3.8. In Fig. 3.9, area (1) is plotted at $t = 1$ sec, area (1) plus area (2) is plotted at $t = 2$ sec, and area (1) plus area (2) plus area (3) is plotted at $t = 3$ sec, since these three quantities give the displacements at these three times. During the remaining five time intervals the particle's velocity is negative, and the particle is moving to the left. The increments in displacement are now negative and given by the areas labeled (4), (5), (6), (7), and (8) in Fig. 3.8; the corresponding displacements are plotted in Fig. 3.9. From a comparison of Fig. 3.9 and Fig. 3.2 we see that, had we taken the initial displacement to be 2 m to the right of the origin, the two figures would have been identical. Note in particular that to determine a displacement-time graph from a velocity-time graph, one requires a knowledge of the displacement at some instant of time in order to put the origin at the right point on the graph.

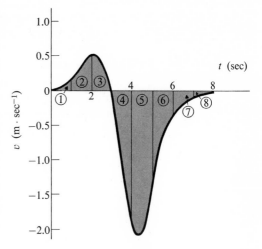

Fig. 3.8. The velocity-time graph for a particular motion.

Alternatively, we could have found a mathematical expression relating v and t and then by methods of integral calculus derived the variation of s with t. Symbolically, if

$$v = v(t),$$

then

$$s(t) = \int v(t)\, dt + \text{constant}.$$

A knowledge of the displacement at a certain time allows a determination of the constant. Examples of this approach will be discussed in Chapter 6.

In an entirely analogous manner one can derive a velocity-time graph. In the calculus notation, if the acceleration is known as a function of the time

$$a = a(t),$$

then the speed as a function of the time is given by

$$v(t) = \int a(t)\, dt + \text{constant}.$$

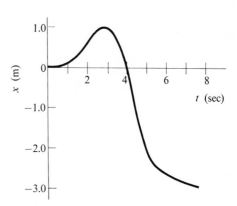

Fig. 3.9. The distance-time graph obtained from Fig. 3.8.

3.6   LINEAR MOTION AT CONSTANT ACCELERATION

Finally, we shall consider motion at constant acceleration. An example in nature that approximates this type of motion is the free fall of objects near the earth's surface under the influence of the earth's gravitational attraction. For example, an apple falling from a tree to the ground moves with an approximately constant acceleration. The acceleration-time graph for such motion is shown in Fig. 3.10. The speed-time graph

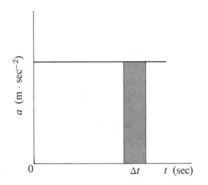

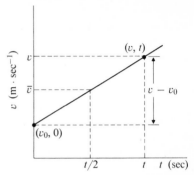

Fig. 3.10. The acceleration-time graph for a particle moving with a constant acceleration.

Fig. 3.11. The velocity-time graph obtained from Fig. 3.10.

follows immediately, since for equal increments in time Δt the increment in speed is always $a\,\Delta t$. If the speed at time $t = 0$ is v_0, then the speed-time graph is, as shown in Fig. 3.11, a straight line having slope a and an intercept v_0 on the v axis. The equation of the line is

$$v = v_0 + at.$$

Since the speed increases linearly with time, the average speed over the time interval from 0 to time t is just the speed at time $t/2$ and, from Fig. 3.11, is

$$\bar{v} = \frac{v + v_0}{2}.$$

The displacement after time t assuming $s = 0$ at $t = 0$ will be

$$s = \bar{v}t = \left(\frac{v + v_0}{2}\right)t.$$

From these equations one can derive two further useful equations relating displacement, speed, acceleration, and time. They are

$$v^2 = v_0^2 + 2as$$

$$s = v_0 t + \tfrac{1}{2}at^2.$$

The proof is left as an exercise for the reader. Using these equations one can solve any problem concerned with motion at constant acceleration.

Example. An electron initially traveling at a speed of $5 \times 10^7 \, \text{m} \cdot \text{sec}^{-1}$ is subjected at time $t = 0$ to an acceleration of $1.8 \times 10^{15} \, \text{m} \cdot \text{sec}^{-2}$ in a direction opposite to its direction of motion.
(a) At what time does the electron come to a stop?
(b) Determine the displacement of the electron from its initial position at times $t_1 = 2 \times 10^{-8}$ sec and $t_2 = 7 \times 10^{-8}$ sec.

Solution.
(a) Since the acceleration is in a direction opposite to the direction of motion of the electron at time $t = 0$, we write

$$v = v_0 - at$$

for the velocity as a function of time. The electron comes to a stop when $v = 0$ or when

$$t = \frac{v_0}{a} = \frac{5 \times 10^7}{1.8 \times 10^{15}} = 2.8 \times 10^{-8} \text{ sec.}$$

The electron comes to a stop after 2.8×10^{-8} sec.
(b) We write

$$s = v_0 t - \tfrac{1}{2}at^2$$

for the displacement as a function of time. For $t_1 = 2 \times 10^{-8}$ sec,

$$s = (5 \times 10^7)(2 \times 10^{-8}) - \tfrac{1}{2}(1.8 \times 10^{15})(2 \times 10^{-8})^2$$

$$= 1 - 0.36$$

$$= 0.64 \text{ m.}$$

For $t_2 = 7 \times 10^{-8}$ sec,

$$s = (5 \times 10^7)(7 \times 10^{-8}) - \tfrac{1}{2}(1.8 \times 10^{15})(7 \times 10^{-8})^2$$

$$= 3.5 - 4.41$$

$$= -0.91 \text{ m.}$$

After 2×10^{-8} sec the electron is 0.64 m from its initial position in the direction of its initial motion; after 7×10^{-8} sec it is 0.91 m from its initial position in a direction opposite to its initial motion.

PROBLEMS

1. The displacement-time graph for a "sticky" molecule initially traveling toward a wall is shown in Fig. 3.12.
   (a) Derive a speed-time graph for the molecule.
   (b) Derive an acceleration-time graph for the molecule.
   (c) Describe the motion of the molecule.

2. The acceleration-time graph for a particle moving in an inhomogeneous electric field is shown below in Fig. 3.13.
   (a) Construct a speed-time graph for the particle, assuming that at $t = 0$ the particle is at rest.
   (b) Construct a displacement-time graph for the particle, assuming that at $t = 0$ the displacement is zero.

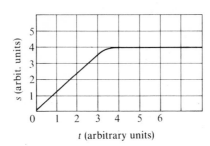

Fig. 3.12   Fig. 3.13

3. The speed-time graph of a particle is shown in Fig. 3.14.
   (a) During what portion of the motion is the particle traveling with constant acceleration?
   (b) Calculate the displacement of the particle during the first 3 sec.
   (c) Calculate the particle's acceleration at $t = 5$ sec and $t = 6$ sec.

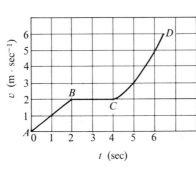

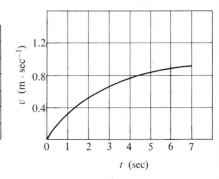

Fig. 3.14   Fig. 3.15

4. The speed-time graph for a particle falling from rest in a retarding medium is shown in Fig. 3.15.
   (a) Comment on the shape of this curve.
   (b) Derive the acceleration-time graph for the particle.
   (c) Derive the displacement-time graph.

5. The velocity-time graph for two colliding particles is shown in Fig. 3.16.
   (a) Comment on the motion of the two particles.
   (b) Derive a displacement-time graph for each of the particles.
   (c) Derive an acceleration-time graph for each of the particles.

6. A particle falling under the influence of gravity near the earth's surface is subject to a constant acceleration of $9.8 \text{ m} \cdot \text{sec}^{-2}$. If the particle falls from rest, calculate
  (a) the time required for the particle to fall through 10 m.
  (b) the speed of the particle after falling through 10 m.

7. Write down the four basic equations relating s, v, t, and a for constant a for the case $s = s_0$ when $t = 0$.

8. The speed-time graph for an object is as shown in Fig. 3.17. How far does the object travel in the first 6 min? What is the average acceleration of the object during the 6-min interval?

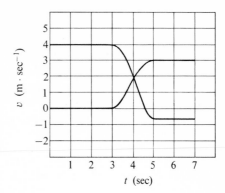

Fig. 3.16

*9. From the displacement-time graph shown in Fig. 3.18, construct
  (a) a speed-time graph.
  (b) an acceleration-time graph.

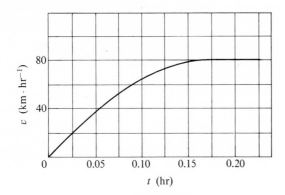

Fig. 3.17

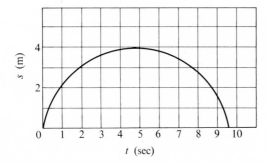

Fig. 3.18

10. From the acceleration-time graph shown in Fig. 3.19, construct
   (a) a speed-time graph.
   (b) a displacement-time graph.

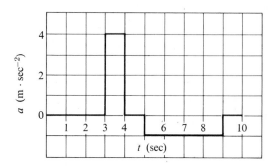

Fig. 3.19

11. From the velocity-time graph shown in Fig. 3.20, construct
   (a) an acceleration-time graph.
   (b) a displacement-time graph.

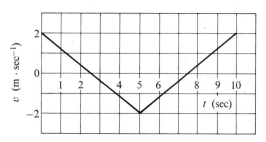

Fig. 3.20

12. An object travels $50\,\text{m}\cdot\text{sec}^{-1}$ for 300 sec, $25\,\text{m}\cdot\text{sec}^{-1}$ for 200 sec, and $60\,\text{m}\cdot\text{sec}^{-1}$ for 700 sec. What is the average speed of the object during the 1200 sec?

*13. Figure 3.21 is taken from a multiple-flash photograph of an object moving from left to right taken at $\frac{1}{20}$-sec intervals.
   (a) Deduce a displacement-time graph.
   (b) From the graph in (a) construct a speed-time graph.
   (c) Suggest an algebraic equation to describe the graph obtained in (b).

*14. You are in an elevator that is accelerating upward at $2.00\,\text{m}\cdot\text{sec}^{-2}$. When the elevator has an upward speed of $4\,\text{m}\cdot\text{sec}^{-1}$, you drop your key from your hand, which is 1.5 m above the elevator floor. Find
   (a) how long before the key strikes the floor.
   (b) the position of the key relative to its starting point at the instant it strikes the floor.

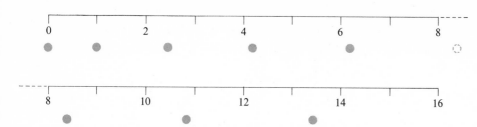

Fig. 3.21

15. Table 3.1 gives the instantaneous speed of an object at 2-sec intervals. Plot the speed-time graph and use the graph to answer the following questions.
   (a) What is the speed of the object at $t_1 = 5.2$ sec and $t_2 = 9.6$ sec?
   (b) What is the average acceleration during the interval $t_2 - t_1$?
   (c) How far did the object travel during the interval $t_2 - t_1$?

Table 3.1

Time (sec)	Speed $(m \cdot sec^{-1})$
0.0	0.0
2.0	2.4
4.0	4.8
6.0	7.2
8.0	9.6
10.0	12.0
12.0	14.4

*16. A man runs at his maximum speed to catch a commuter train. When he is still 24 m from the door, the train begins to accelerate uniformly at $0.75\ m \cdot sec^{-2}$. With the aid of a graph deduce what his speed must be in order that he is just able to catch the train.

*17. A slow freight train is moving at $50\ km \cdot hr^{-1}$. A fast commuter train suddenly appears from behind traveling at $v\ km \cdot hr^{-1}$ on the same track. The commuter engineer applies his brakes knowing that at the speed he is traveling it will require 2 km to stop. If a subsequent crash is to be avoided, what is the maximum value of v? Assume the trains are 0.5 km apart when the brakes are applied.

18. A car travels 180 km at an average speed of $40\ km \cdot hr^{-1}$. A second car begins the trip 1 hr after the first, but the two cars arrive at their destination at exactly the same time. What was the average speed of the second car? Illustrate your solution by means of a graph.

*19. The speed-time graph for an object is given in Fig. 3.22. What is the particle's speed at $t = 2.0$ units of time? From the graph obtain an equation (in a simple form) to describe the dependence of the object's speed on time.

20. Plot a speed-time graph of a particle traveling at a uniform speed of $v_0\ m \cdot sec^{-1}$. On the same graph plot the speed-time graph for a second particle that starts from rest and uniformly accelerates at $a\ m \cdot sec^{-2}$ to a final speed $v_0\ m \cdot sec^{-1}$.
   (a) From your graph deduce the relation between the distances covered by the two particles during the time in which the second one is accelerating.
   (b) Prove your result algebraically.

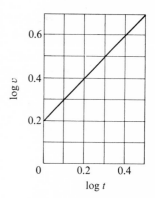

Fig. 3.22

4 Frames of reference in uniform relative motion

4.1 INTRODUCTION

We live in a world filled with motion. All about us we see objects in motion and become aware of the manner of their motion at an early age. The ideas that we form are carried forward into our later years. Although we do not ordinarily have the opportunity to observe objects in very rapid motion, since this requires special equipment much faster of response than our eyes, we have no reason to expect that an extrapolation of our low-speed observations will not be valid. Therefore, we are very surprised to learn and often reluctant to accept the fact that rapidly moving objects behave quite differently than we expect from observing slowly moving objects. We are going to explore that difference in this chapter and in the next.

4.2 FRAMES OF REFERENCE

All measurements in space are made with respect to some reference point. For example, the position of a runner on a track in a 100 m sprint may be specified by stating that the runner is 15 m from the starting line, the distance being measured as shown in Fig. 4.1. The starting line serves as the reference point or **origin** for the measurement. The distance 15 m locates the runner with respect to the origin and is called the **coordinate** of the runner. Since only one coordinate is necessary to denote the position of the runner, the track forms a **one-dimensional coordinate system**.

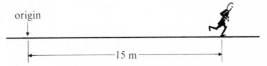

Fig. **4.1.** A one-dimensional motion.

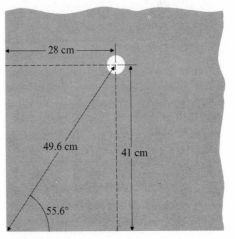

The position of an object, such as a coin on a table, may be given by stating that the center of the coin lies 28 cm from a given edge of the table and 41 cm from a second edge of the table that meets the first edge at an angle of 90°, both distances being measured along a line perpendicular to the edge in question, as shown in Fig. 4.2. One corner of the table serves as the origin, and the two edges that meet at this corner serve as coordinate axes. The tabletop serves as a **two-dimensional rectangular coordinate system**.

Again, we may specify the position of a light in a lecture room by stating the perpendicular distance from a given point on the light to each of three mutually perpendicular sides (walls, floor,

Fig. **4.2.** Two coordinates are necessary to specify the position of an object on a table.

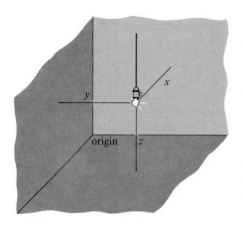

Fig. 4.3. Three coordinates are necessary to specify the position of an object in a room.

or ceiling) of the room. The origin is chosen to be one corner of the room, and the coordinate axes are the lines of intersection of the three sides of the room from which the measurements are made, as shown in Fig. 4.3. These axes are the axes of a **three-dimensional rectangular coordinate system**.

The position of the object with respect to the origin is independent of the coordinate system attached to that origin. For example, in Fig. 4.2 the position of the center of the coin may also be given as 49.6 cm from the corner of the table, with the line joining the coin to the corner making an angle of 55.6° with one edge, as shown in the diagram; **these coordinates are polar coordinates**. The choice of an origin from which to make a measurement establishes a **frame of reference for the measurement**.

A close look at the physical universe shows us that all objects in space are in motion. The earth rotates on its axis and revolves around the sun; in addition the whole solar system is part of a gigantic, rotating, disc-shaped galaxy of stars, which is also moving through space. Thus, all points in space attached to moving objects are in motion and provide frames of reference that are in motion. Therefore, all frames of reference are either accelerated or are moving with uniform velocity with respect to an observer (except for those frames in which the observer is stationary). If a particle on which no observable forces are acting is not accelerated when observed in a certain frame of reference, then that frame of reference is said to be **inertial**; it follows that inertial reference frames are not accelerated.

However, we should note that the concept of an inertial frame is not so simple as it would appear, since the absence of forces can be detected only by the absence of acceleration unless all sources of force are known. Unless we can convince ourselves in a given case that we know all the forces acting on a particle, we are not strictly able to decide whether or not the particle is in an inertial frame of reference. Fortunately, we do know that all known fundamental forces decrease very rapidly with distance from the source of the force, enabling us to find many situations where we would expect only negligible forces to be present. Therefore only negligible acceleration on a particle would result, and we could quite reasonably talk about an inertial frame of reference. The fixed stars, for example, are so far away that we can neglect the gravitational forces acting between them and us, and we may use them to specify an inertial frame.

The earth is a rotating object such that all points on its surface except the north and south poles are rotating about the axis of rotation and consequently are subjected to an acceleration towards the axis of rotation.

An object subjected to no force will travel in a straight line according to Newton's first law (Chapter 7). Therefore, a rotating object must be subjected to an external force and is thus accelerated according to Newton's second law (Section 7.2). We shall see in Section 6.5 that this acceleration is directed towards the axis of rotation. It is known as the **centripetal** (center-seeking) **acceleration**.

In addition, all points on the earth's surface, including the poles, are subjected to an acceleration due to the motion of the earth about the sun. No point on the earth's surface is suitable for the establishment of an inertial frame of reference in the strictest terms. However, the centripetal acceleration due to the earth's rotation is only 3.4×10^{-2} m·sec^{-2} at the equator, where it has its maximum value. The acceleration due to gravity is 9.81 m·sec^{-2}, which is 280 times greater than the centripetal acceleration, and the effect of rotation on the motion of an object is not usually noticeable. Some effects of the earth's rotation are discussed in Section 10.4. The acceleration due to the motion of the earth about the sun is much smaller yet, being only about 7×10^{-3} m·sec^{-2}. It is often a good approximation to consider a frame of reference attached to a point on the earth's surface to be inertial because of the smallness of accelerations due to rotation.

4.3 THE PRINCIPLE OF RELATIVITY

Physicists believe that the laws governing the motions of objects are the same for two observers moving relative to one another at a constant velocity. This means that if a set of self-contained experiments is conducted in a train moving smoothly over the earth's surface at a constant velocity relative to the surface, the results of the experiments would be expected to be identical to the results obtained in a similar set of experiments conducted in a laboratory on the earth's surface. That is, playing Ping-Pong on such a train would be no different from playing Ping-Pong in your home. Any other behavior would be contrary to our own experiences. The belief that physical laws should be independent of the motion of the frame of reference in which they are observed was first enunciated by Isaac Newton, who stated, "The motions of bodies included in a given space are the same among themselves, whether that space is at rest or moving forward in a straight line at constant speed." This statement is known as **the principle of relativity**.

A discussion of relativity is a discussion of the relation between measurements made in frames of reference in relative motion. For the present we shall be concerned only with relative motion at constant velocity. We shall find that a satisfactory description of such relative motion is contained in Albert Einstein's **special theory of relativity**. This theory is well supported by experimental data. Einstein also put forward a **general theory of relativity**, which concerns itself with relative motion in accelerated frames of reference. This latter theory has been and still is a subject of much controversy among physicists because its predictions are very difficult to put to experimental test.

4.4   THE GALILEAN TRANSFORMATION

Consider a baseball game in which a batter lays down a bunt in front of home plate. The catcher fields the ball and throws it to first base past the runner, who is heading in the same direction as the throw. A fast runner might be traveling at a top speed of $9 \text{ m} \cdot \sec^{-1}$, and a good throw by the catcher could send the ball toward first base at $30 \text{ m} \cdot \sec^{-1}$ (see Fig. 4.4). It is the relative speed of the ball and runner that is important, however, since it is a question of whether the ball, traveling at $30 \text{ m} \cdot \sec^{-1}$, will arrive at first base before the runner, who has a head start but is only traveling at $9 \text{ m} \cdot \sec^{-1}$. Our intuition and experience tell us that the ball will be catching up to the runner at a speed of $30 - 9 = 21 \text{ m} \cdot \sec^{-1}$. In general, if the speed of the ball were v_b and the speed of the runner v_r, the ball would be traveling at a speed $v_b - v_r$ with respect to the runner.

Fig. 4.4. The speed of a ball relative to a runner is not the same as the speed of the ball relative to the ground.

We can put this observation into a form more suitable for the discussion of physical concepts. Let us consider two frames of reference R and R' with rectangular coordinate systems oriented so that the directions of the x and x', y and y', and z and z' axes are the same as indicated in Fig. 4.5; the frame R' is moving at uniform velocity **u** in the x (or x') direction, the velocity being measured with respect to the frame R. The coordinates of a point P will be (x, y, z, t) as measured in R and (x', y', z', t') as measured in R'. We introduce time (t, t') as a coordinate here because we shall find it necessary to specify times in both frames of reference in order to relate the observations of phenomena as seen from different frames of reference.

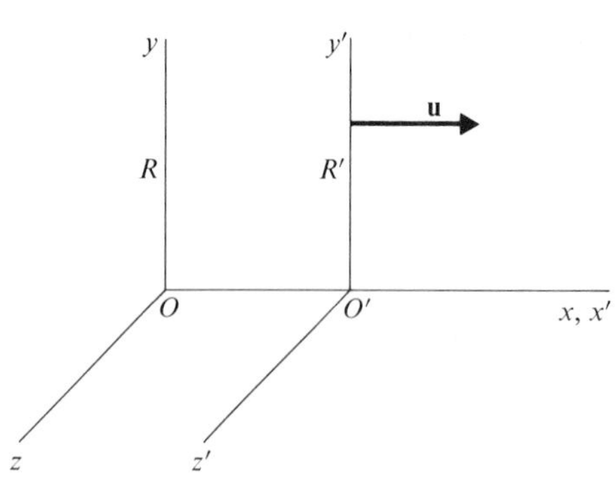

We assume that the origins of the two coordinate systems coincide at time $t = t' = 0$. Since the relative motion is in the x, x' direction, we expect

$$y = y'$$

$$z = z',$$

Fig. 4.5. Coordinate systems attached to reference frames R and R' in uniform relative motion.

independent of the time t or t', because the motion is perpendicular to these coordinate axes. At this stage, we have no reason to suspect that time measurements should be different in different frames of reference, for this is contrary to our normal experience. Therefore, we put

$$t = t'.$$

After a time t, say, the R' system has moved a distance OO' in the x, x' direction, as shown in Fig. 4.6. Since the relative speed is u, the distance OO' is just

$$OO' = ut$$

and

$$x = x' + ut$$

or

$$x' = x - ut.$$

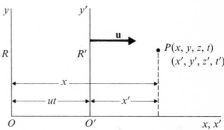

Fig. 4.6. After a time t the R' system has moved a distance ut in the x, x' direction.

(Since the motion is in the x, x' direction only, vector notation is not necessary.)

The set of equations

$$x = x' + ut$$
$$y = y'$$
$$z = z'$$
$$t = t'$$

relates the coordinates of points as measured in the system R to those measured in system R' and is known as the **Galilean transformation**. The inverse transformation is

$$x' = x - ut$$
$$y' = y$$
$$z' = z$$
$$t' = t.$$

This transformation therefore provides a means of correlating sets of measurements made in frames of reference in uniform relative motion.

If the point P is attached to a particle moving in the x, x' direction, we may specify the position of the particle at two different times by (x_1, t_1) and (x_2, t_2) in R, and (x'_1, t'_1) and (x'_2, t'_2) in R'. The average speed of the particle in the x, x' direction as measured in R is

$$\bar{v}_x = \frac{x_2 - x_1}{t_2 - t_1} = \frac{\Delta x}{\Delta t},$$

and the average speed in the x, x' direction as measured in R' is

$$\bar{v}'_x = \frac{x'_2 - x'_1}{t'_2 - t'_1} = \frac{\Delta x'}{\Delta t'}.$$

Substituting into this equation from the Galilean transformation we have

$$\bar{v}'_x = \frac{x_2 - ut_2 - (x_1 - ut_1)}{t_2 - t_1}$$

$$= \frac{(x_2 - x_1) - u(t_2 - t_1)}{t_2 - t_1}$$

$$= \frac{x_2 - x_1}{t_2 - t_1} - u,$$

or

$$\bar{v}'_x = \bar{v}_x - u.$$

This relation may also be written as

$$\frac{\Delta x'}{\Delta t'} = \frac{\Delta x}{\Delta t} - u,$$

which, in the limit $\Delta t \to 0$, $\Delta t' \to 0$, becomes

$$\frac{dx'}{dt'} = \frac{dx}{dt} - u$$

or

$$v'_x = v_x - u,$$

where v'_x and v_x are the x components of the instantaneous velocities as measured in R' and R, respectively. It is apparent that the relation between the y and y' components of the velocity is

$$v'_y = v_y$$

and between the z and z' components is

$$v'_z = v_z,$$

since there is no relative motion in either the y or z directions.

Example. A train travels at a speed of $90 \, \text{km} \cdot \text{hr}^{-1}$ along a straight track. A passenger walks from the rear of the train towards the engine at a speed of

$2 \text{ km} \cdot \text{hr}^{-1}$ relative to the train. What will be the speed of the man relative to the earth as predicted by the Galilean transformation?

Solution. If reference frame R is attached to the earth and reference frame R' to the train, then

$$v'_x = 2 \text{ km} \cdot \text{hr}^{-1}, \qquad u = 90 \text{ km} \cdot \text{hr}^{-1}$$

and

$$v_x = v'_x + u = 90 + 2 = 92 \text{ km} \cdot \text{hr}^{-1}.$$

The relation between accelerations may be found by taking the derivatives of the velocities in the equation

$$v'_x = v_x - u$$

to obtain

$$\frac{dv'_x}{dt} = \frac{dv_x}{dt} - \frac{du}{dt}.$$

However,

$$\frac{du}{dt} = 0,$$

since u is a constant; therefore,

$$\frac{dv'_x}{dt} = \frac{dv_x}{dt}.$$

Since

$$t' = t,$$

we have

$$dt' = dt$$

and

$$\frac{dv_x}{dt} = \frac{dv'_x}{dt'}.$$

Therefore,

$$a_x = a'_x,$$

where a_x and a'_x are the accelerations in the R and R' systems, respectively. The corresponding relations for the y and z directions are

$$a_y = a'_y$$

and

$$a_z = a'_z.$$

That is, according to the Galilean transformation, the acceleration of an object is independent of the reference frame from which it is measured, as long as an inertial frame is selected.

The Galilean transformation is a part of that branch of physics known as **Newtonian physics**, after Isaac Newton, who, by enunciating several basic laws (Chapter 7), was able to unify early experimental data on the motion of both terrestrial and planetary objects into one comprehensive scheme. Moreover, his laws stood unchallenged for over two hundred years, during which time great progress was made in understanding the nature of the universe in which we live.

4.5   THE PI-MESON EXPERIMENT: *a Test of the Galilean Transformation*

Suppose that an observer in R' measures the speed of a light signal traveling in the x' direction and arrives at some value c. We have seen in the last section that, according to the Galilean transformation,

$$v'_x = v_x - u$$

or

$$v_x = v'_x + u.$$

Here

$$v'_x = c,$$

and

$$v_x = c + u$$

is the predicted value for the speed of the light signal measured by an observer in R (see Fig. 4.7). Similarly, if the speed of the light signal had been measured to be c in R, we would expect a speed measured in R' of

$$v'_x = c - u.$$

Obviously, if the Galilean transformation is valid, the speed of a light signal should depend upon the relative motion of source and observer.

The speed of light is rather large, about $3.0 \times 10^8 \, \text{m} \cdot \text{sec}^{-1}$. Therefore, it will be necessary to have either a very large value of u, the relative speed of source and observer, or else an experimental arrangement that is very sensitive to extremely

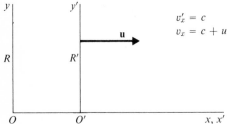

Fig. 4.7. According to the Galilean transformation, the speed of a light signal as seen in R is $c + u$ if the speed as seen in R' is c.

small changes in the speed of light, in order to observe any variation in the speed.

An experiment involving large values of u may be performed with the use of beams of particles produced by very large, high-energy accelerators[1], which are used to investigate the nature of elementary particles. Such an experiment was carried out in 1964 at the CERN proton-synchrotron laboratory in Switzerland. This accelerator produces a beam of high-energy (that is, very fast) protons, the nuclei of hydrogen atoms. The protons are so energetic that when they collide with the nuclei of other atoms in a small piece of material placed in the beam, they can break the nuclei into pieces and even produce hordes of new particles out of the debris of the collisions. One of the new particles that may be produced is called the **neutral pi-meson** (π-meson). The origin of the name of this particle is unimportant; what is important is that the neutral π-mesons in the beam exist for only a very short time (about 10^{-14} sec). After this amount of time they change or decay into **gamma rays** (γ-rays), which are very short wavelength electromagnetic waves[2]. Light is also an electromagnetic wave of a much longer wavelength. All electromagnetic waves must have the same speed according to the theory of electricity and magnetism; this is observed to a very high accuracy for radiation from stationary sources. Measuring the speed of the γ-rays emitted by the moving neutral π-mesons would provide a test for the prediction of the Galilean transformation, which should apply as well to γ-rays as it does to other things in relative motion.

The experimental arrangement is shown in Fig. 4.8. The neutral π-mesons are produced by letting the proton beam strike a target made from beryllium. Since the neutral π-mesons live for such a short time, they all decay before leaving the target. The γ-rays resulting from the decays pass through two magnets that deflect any charged particles out of the beam. The γ-rays are then detected at two positions A and B, and the time taken for the γ-rays to travel the distance AB is measured.

The protons in the synchrotron beam do not come continuously but are produced in short bursts of a few nanoseconds duration (1 nsec = 10^{-9} sec). Each γ-ray detector then records γ-rays for a few nanoseconds only, rather than continuously. Therefore, only the bursts of γ-rays need to be detected, and the time between detection of a given burst at A and at B is easily recorded electronically. The γ-ray counters are set to record γ-rays of energies greater than 6 GeV only. [The GeV is a unit of energy used in high-energy physics. One GeV equals 10^9 eV; **one electron volt** (1 eV) is

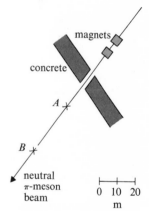

Fig. 4.8. Schematic diagram of the experimental arrangement used to measure the speed of γ-rays emitted by decaying neutral π-mesons.

[1] D. Park, *Contemporary Physics* (New York: Harcourt, Brace & World, Inc., 1964) and R. R. Wilson and R. Littauer, *Accelerators: Machines of Nuclear Physics* (Garden City, N.Y.: Doubleday & Co., Inc., 1960).

[2] A. Beiser, *The Mainstream of Physics* (Reading, Mass.: Addison-Wesley Publishing Co., Inc., 1962), Chapter 16.

the amount of energy that an electron acquires when it is accelerated through a potential difference of 1 V, which is just equivalent to 1.6×10^{-19} joules (J). Therefore, 1 GeV is 1.6×10^{-10} J.]

The distance AB was chosen to be 31.450 ± 0.0015 m for the particular reason that γ-rays traveling at a speed of $c = 2.9979 \times 10^8$ m $\cdot$ sec^{-1} (the speed of light emitted by a stationary source) would travel from A to B in the same time as the time interval between bursts from the machine. Therefore, if the high-energy γ-rays travel at a speed greater than c, any given burst of γ-rays will arrive at B before the next burst arrives at A. However, if the speed of the γ-rays is equal to c, the following burst of γ-rays should arrive at A exactly when the previous burst arrives at B. In the experiment, a difference in arrival time at the two stations was sought. This method of measuring the speed of the γ-rays was chosen because the measurement can be made more accurately than a direct measurement of the time required for the γ-rays to travel from A to B.

The experimental results are shown in Fig. 4.9. It is clear that the γ-rays must have traveled from A to B at very close to the speed c, since the peaks line up very closely. The asymmetric shape of the bursts originates in the manner of producing the proton beam in the synchrotron. A careful analysis of the results shows that the speed of the high-energy γ-rays was $(2.9977 \pm 0.0004) \times 10^8$ m $\cdot$ sec^{-1}. This compares extremely well with the value of c, which is $(2.997925 \pm 0.000004) \times 10^8$ m $\cdot$ sec^{-1}.

What about the speed of the source? Let us assume that almost all of the energy of the proton is transferred to the neutral π-meson. If we use the kinetic energy expression from Newtonian mechanics, namely $E = \frac{1}{2}(mv^2)$ (Section 7.7), where m is 1.23×10^{-28} kg for the neutral π-meson, we find for 6 GeV neutral π-mesons that $v = 4.0 \times 10^9$ m $\cdot$ sec^{-1}. Such a speed for the neutral π-mesons is an order of magnitude greater than the speed of light. We shall see shortly that we shall have to modify our ideas concerning motion at high speeds and that a 6 GeV neutral π-meson actually has a speed of $0.99975\,c$. In any event, **the speed of the γ-rays produced by the moving source is measured to be c to high precision even though the source of the γ-rays is moving at speed c also. This clearly contradicts the prediction of the Galilean transformation.**

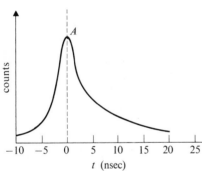

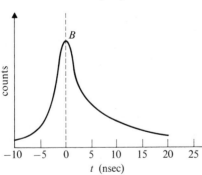

Fig. 4.9. Bursts of γ-rays arrive at detector B after a time very closely equal to the time required for γ-rays traveling from A to B at the speed of light.

4.6   THE POSTULATES OF THE SPECIAL THEORY OF RELATIVITY

The experiment outlined in the previous section indicates that in some situations the classical, or Newtonian, approach to reality breaks down. By the beginning of this century many experiments[3] had been performed that indicated a basic deficiency of classical physics. (The meson experiment was not one of them, of course, since it is very recent.) The experiments produced difficulties in explaining some phenomena of electricity and magnetism. In fact, since light is an electromagnetic wave, we might say that the difficulty was with electromagnetic phenomena in general.

Clerk Maxwell in 1864 succeeded in developing a set of mathematical equations that completely described the phenomena of electricity and magnetism. Also, these equations predicted the existence of electromagnetic waves traveling at very high speed, in fact, at the speed of light. Light was soon shown to be an electromagnetic wave, and other electromagnetic waves were artificially produced (for example, radio waves and x-rays). It was easy to show that the form of the equation describing the behavior of electromagnetic waves is not invariant under a Galilean transformation. That is, the basic equation of electromagnetism does not satisfy the principle of relativity under a Galilean transformation. The speed of the waves is found to change when such a transformation is made. This, however, is contradictory to the experimental evidence that the speed of electromagnetic waves is a constant independent of any relative motion.

Albert Einstein, pondering this situation during the first few years of this century, finally began to suspect that time was the culprit. It was assumed in the Galilean transformation that time is invariant. Perhaps this was the wrong way to proceed. The available experimental evidence indicated a possible constant speed for light waves. Einstein proceeded to assume the speed of light to be constant and to develop the special theory of relativity, first published in 1905, with this assumption as a basic postulate. The other basic assumption was that the principle of relativity should be obeyed. We may state the postulates explicitly in Einstein's own words[4]:

Postulate 1. The laws by which the states of physical systems undergo change are not affected whether these changes of state be referred to the one or the other of two systems of coordinates in uniform translatory motion.

---

[3] Herbert Dingle, *The Special Theory of Relativity*, 4th Edition (London: Methuen & Co. Ltd., 1961), Chapter 2.

[4] A. Einstein, H. A. Lorentz, H. Minkowski, and H. Weyl, *The Principle of Relativity* (New York: Dover Publications, Inc., 1923).

Postulate 2. Any ray of light moves in the "stationary" system of coordinates with the determined velocity c, whether the ray be emitted by a stationary or by a moving body.

4.7 THE LORENTZ TRANSFORMATION

The second postulate indicates the necessity of developing a transformation between coordinate systems in uniform relative motion that will leave the speed of light unchanged. Whatever form this new transformation takes, we should expect it to reduce to the Galilean transformation for the description of motion when relative velocities are small. Such a transformation already existed in 1905, having been developed by Hendrik Lorentz, who was also concerned about the difficulties with electromagnetic phenomena. Lorentz, however, had not realized the full implications of this transformation. Einstein proceeded to develop the transformation from his second postulate in much the same way as we are going to do now.

We consider a short pulse of light spreading out from a point source. From the symmetry of the situation we would expect the wave front (that is, the leading portion of the light signal) to spread out spherically, as shown in Fig. 4.10, and to move with speed c, where c is the constant speed of light. Let us further suppose that a frame of reference R is attached to the light source, which is situated at the origin O of this frame. The light source is stationary in R. Let us imagine a frame of reference R' moving with speed u in the x direction and so situated that the x and x' axes are in the same direction. We further consider that the light pulse is produced at the instant that the origins O and O' coincide. The situation is depicted in Fig. 4.11.

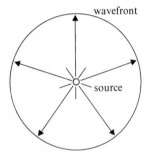

Fig. 4.10. Light spreads out spherically from a point source.

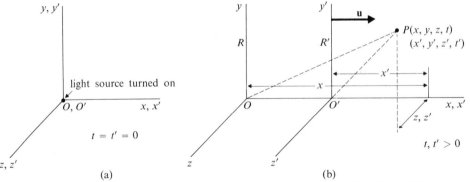

Fig. 4.11. (a) A light source is turned on when the origins of R and R' coincide. (b) Light observed at a point P must travel the path OP in R and the path $O'P$ in R'.

What would be seen at some point P [coordinates (x, y, z, t) in R and coordinates (x', y', z', t') in R'] by an observer at rest in R at this point and by an observer at rest in R' at this point? Obviously, the observer at rest in R will observe a light pulse coming from the origin O of his frame of reference. On the other hand, the observer at rest in R' will see a light pulse coming from O', since O' was coincident with O when the light pulse was produced. We now postulate that the speed of light is independent of the relative motion of source and observer. Therefore, both observers would measure the speed of the light signal to be c.

We take the zero of the time measurements to be the moment of production of the light pulse. Then

$$t = t' = 0$$

when O and O' coincide. The observer in R will detect the light signal at time t where

$$t = \frac{OP}{C} = \frac{(x^2 + y^2 + z^2)^{1/2}}{c}.$$

We have used the Pythagorean relation

$$(OP)^2 = x^2 + y^2 + z^2$$

to obtain this result. The equation for t may be multiplied by c, squared, and rearranged to give

$$x^2 + y^2 + z^2 - c^2 t^2 = 0.$$

The observer in R' will detect the light signal at time t', where

$$t' = \frac{O'P}{C} = \frac{(x'^2 + y'^2 + z'^2)^{1/2}}{c}.$$

Therefore,

$$x'^2 + y'^2 + z'^2 - c^2 t'^2 = 0.$$

Note that we have not assumed that $t = t'$, as is true for the Galilean transformation. We are looking for a transformation between (x, y, z, t) and (x', y', z', t') that, when applied to the equation

$$x^2 + y^2 + z^2 - c^2 t^2 = 0,$$

will produce the equation

$$x'^2 + y'^2 + z'^2 - c^2 t'^2 = 0.$$

Therefore,

$$x^2 + y^2 + z^2 - c^2 t^2 = x'^2 + y'^2 + z'^2 - c^2 t'^2$$

must be satisfied; if this occurs then

$$x^2 + y^2 + z^2 - c^2t^2$$

is **invariant** under the transformation. The invariance of the quantity $x^2 + y^2 + z^2 - c^2t^2$ under a transformation from R to R' means that if a light signal is propagated in all directions with speed c from O at $t = 0$, as observed in R, then a light signal is propagated from O' in all directions at speed c at $t' = 0$, as observed in R'. The required transformation would therefore be in agreement with the first postulate.

What else do we know? It is clear that the origin O' of R' has moved a distance ut in the x direction as measured in R after time t. On the other hand, an observer in R' would say that the origin O of R has moved a distance $-ut'$ in the $-x'$ direction after time t' (see Fig. 4.12). Thus we must have

$$x' = -ut' \qquad \text{when} \quad x = 0,$$

and

$$x = +ut \qquad \text{when} \quad x' = 0.$$

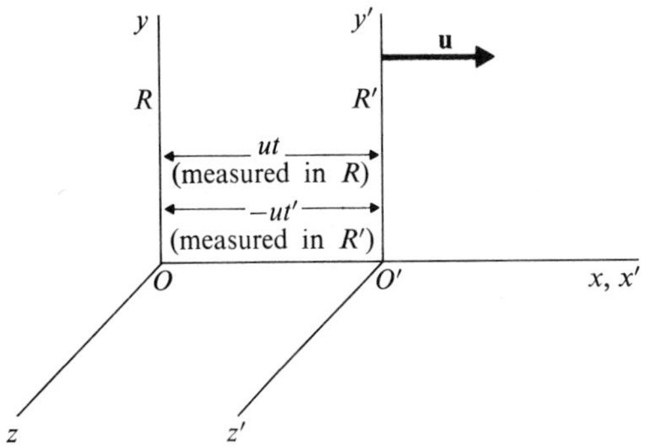

Fig. 4.12. If the origin of R' has moved a distance ut in the x direction as measured in R, then the origin of R has moved a distance ut' in the negative x' direction as measured in R'.

The linear equations

$$x' = \gamma(x - ut)$$

$$x = \gamma'(x' + ut'),$$

where γ and γ' are constants, are the simplest relations involving x, x', t, t', and u that satisfy this requirement; for this reason we look for a transformation satisfying them. Moreover, it is necessary to have linear equations on account of the properties of homogeneity that we attribute to space and time. That is, we expect a one-to-one correspondence between positions and times for a single event as measured by observers in different inertial frames. For this reason, quadratic (or higher order) equations are not satisfactory. Since the motion is in the x, x' direction only, we shall have

$$y' = y$$

$$z' = z$$

at all times.

The constants γ and γ' must be equal as required by the first postulate. The calculations are carried out in the shaded area below. The result is that γ and γ' must have the following form in order to satisfy the requirements for the transformation:

$$\gamma = \gamma' = \frac{1}{\left(1 - \dfrac{u^2}{c^2}\right)^{1/2}}.$$

With this expression for the constants γ and γ', we have the following set of equations to relate the coordinates of points as measured in R to those measured in R':

$$x = \gamma(x' + ut')$$

$$y = y'$$

$$z = z'$$

$$t = \gamma\left(t' + \frac{ux'}{c^2}\right).$$

This is known as the **Lorentz transformation**. The inverse transformation is given by

$$x' = \gamma(x - ut)$$

$$y' = y$$

$$z' = z$$

$$t' = \gamma\left(t - \frac{ux}{c^2}\right).$$

We note that when u is much smaller than c,

$$\frac{u^2}{c^2} \ll 1 \quad \text{and} \quad \gamma \simeq 1.$$

In this limit it is readily seen that the Lorentz and the Galilean transformation become, for all practical purposes, identical.

Derivation of the Lorentz transformation

The equations to be satisfied are

$$x^2 + y^2 + z^2 - c^2t^2 = x'^2 + y'^2 + z'^2 - c^2t'^2 \tag{1}$$

$$x' = \gamma(x - ut) \tag{2}$$

$$x = \gamma'(x' + ut'). \tag{3}$$

From equation (3) we have

$$\frac{x}{\gamma'} = x' + ut'$$

or

$$t' = \frac{1}{u}\left(\frac{x}{\gamma'} - x'\right).$$

Substitution from equation (2) gives

$$t' = \frac{1}{u}\left[\frac{x}{\gamma'} - \gamma(x - ut)\right]$$

$$= \frac{1}{u}\left(\frac{x}{\gamma'} - \gamma x + \gamma ut\right)$$

$$= \gamma\left[t + \frac{x}{u}\left(\frac{1}{\gamma\gamma'} - 1\right)\right].$$

Therefore, equation (1) becomes

$$x^2 + y^2 + z^2 - c^2t^2 - \gamma^2(x - ut)^2 - y'^2 - z'^2$$

$$+ c^2\gamma^2\left[t + \frac{x}{u}\left(\frac{1}{\gamma\gamma'} - 1\right)\right]^2 = 0.$$

Since the motion is in the x, x' direction, we assume

$$y = y' \qquad z = z'$$

$$y^2 = y'^2 \qquad z^2 = z'^2.$$

Squaring the bracketed expressions in the equation and rearranging produces

$$x^2\left[1 - \gamma^2 + \frac{c^2\gamma^2}{u^2}\left(\frac{1}{\gamma\gamma'} - 1\right)^2\right]$$

$$+ xt\left[2\gamma^2u + \frac{2c^2\gamma^2}{u}\left(\frac{1}{\gamma\gamma'} - 1\right)\right]$$

$$+ t^2(c^2\gamma^2 - c^2 - \gamma^2u^2) = 0. \tag{4}$$

Now x is the x coordinate of the point P, and t is the time at which the light pulse reaches P. P could be any arbitrary point, so that x and t could have any value. Therefore, in equation (4) the quantities x^2, xt, and t^2 are in general not zero, and the only way that the equation may be satisfied is for the coefficients of x^2, xt, and t^2 to be each equal to zero. Equating the coefficient of t^2 to zero gives

$$c^2\gamma^2 - c^2 - \gamma^2u^2 = 0$$

$$\gamma^2(c^2 - u^2) = c^2$$

$$\gamma^2 = \frac{1}{1 - \dfrac{u^2}{c^2}}$$

or

$$\gamma = \frac{1}{\left(1 - \dfrac{u^2}{c^2}\right)^{1/2}}.$$

From the coefficient for xt we have

$$2\gamma^2 u + \frac{2c^2\gamma^2}{u}\left(\frac{1}{\gamma\gamma'} - 1\right) = 0$$

$$u^2 + c^2\left(\frac{1}{\gamma\gamma'} - 1\right) = 0$$

$$u^2 = c^2\left(1 - \frac{1}{\gamma\gamma'}\right)$$

$$\frac{1}{\gamma\gamma'} = 1 - \frac{u^2}{c^2}$$

$$\gamma' = \frac{1}{\gamma\left(1 - \dfrac{u^2}{c^2}\right)}$$

$$\gamma' = \gamma = \frac{1}{\left(1 - \dfrac{u^2}{c^2}\right)^{1/2}}.$$

Substituting this expression for γ and γ' into the coefficient for x^2 gives zero, as may easily be checked.

We now have

$$x' = \gamma(x - ut)$$

$$x = \gamma(x' + ut').$$

If we eliminate x between these two equations, we obtain

$$x' = \gamma[\gamma(x' + ut') - ut]$$

$$x' = \gamma^2 x' + \gamma^2 ut' - \gamma ut$$

or

$$\gamma ut = \gamma^2 ut' + x'(\gamma^2 - 1)$$

$$t = \gamma t' + \frac{\gamma x'}{u}\left(1 - \frac{1}{\gamma^2}\right)$$

$$t = \gamma\left[t' + \frac{x'}{u}\left(\frac{u^2}{c^2}\right)\right]$$

$$t = \gamma\left(t' + \frac{ux'}{c^2}\right).$$

In a similar manner we obtain

$$t' = \gamma\left(t - \frac{ux}{c^2}\right).$$

The complete transformation equations are

$$x' = \gamma(x - ut) \qquad x = \gamma(x' + ut')$$

$$y' = y \qquad y = y'$$

$$z' = z \qquad z = z'$$

$$t' = \gamma\left(t - \frac{ux}{c^2}\right) \qquad t = \gamma\left(t' + \frac{ux'}{c^2}\right).$$

PROBLEMS

1. A boy sitting in a train traveling at a speed of $30 \, \text{m} \cdot \text{sec}^{-1}$ throws a ball at a speed of $5 \, \text{m} \cdot \text{sec}^{-1}$. What is the speed of the ball relative to the ground if
(a) the ball is thrown in the direction in which the train is traveling?
(b) the ball is thrown in the direction opposite to the direction of the train?
(c) the ball is thrown at right angles to the direction of the train?

*2. In a typical scene, Laurel and Hardy are on a railway handcar going into a tunnel $\frac{1}{10}$ km long. Just as they enter the tunnel doing $20 \, \text{km} \cdot \text{hr}^{-1}$, an express train goes into the opposite end at $60 \, \text{km} \cdot \text{hr}^{-1}$. What minimum constant acceleration must Laurel and Hardy's handcar have in order not to be hit by the train? (That is, they decelerate to a stop and then accelerate to $60 \, \text{km} \cdot \text{hr}^{-1}$ backwards before the train hits them.) Give the answer in $\text{km} \cdot \text{hr}^{-1} \cdot \text{sec}^{-1}$. Do the calculation twice, first using a frame of reference attached to the earth's surface and again using a frame of reference attached to the train.

*3. A man is rowing upstream in a boat. As he passes under a low bridge his hat is knocked off, and it floats downstream. The man rows for 15 min before he discovers his loss, then turns around and rows downstream to catch his hat. He catches it 1 km below the bridge. How fast is the stream moving? Hint: This problem is most easily solved by changing reference frames.

4. A highly collimated beam of light from a laser[5] is directed from the earth to the moon, a distance of $3.8 \times 10^8 \, \text{m}$. If the laser is then rotated at a rate of 1 radian $\cdot \text{sec}^{-1}$, at what speed does the spot of light from the laser move across the surface of the moon? Comment on your answer.

5. Plot a graph of γ as a function of u/c. Comment on the form of the curve.

6. A particle has coordinates $x = 1 \, \text{m}$, $y = 2 \, \text{m}$, $z = 3 \, \text{m}$ at time $t = 0$ in a certain coordinate system. What are its coordinates in a coordinate system moving

[5] A. L. Schawlow, "Optical Masers," *Scientific American*, June, 1962, and "Advances in Optical Masers," *Scientific American*, June, 1963. Available as *Scientific American Offprints 274* and *294*, respectively (San Francisco: W. H. Freeman and Co., Publishers).

in the x, x' direction at a speed of 3 m · sec^{-1} at times $t = 0, t = 2$ sec, $t = 10$ sec, $t = 100$ sec for

(a) a Galilean transformation?

(b) a Lorentz transformation?

7. Repeat the calculations of Problem 6 for a speed of $u = 0.8c$ for the moving coordinate system.

8. A meter stick is situated at rest in the R' reference system, which is traveling at speed $u = 0.8c$ in the x, x' direction with respect to reference system R. If the ends of the meter stick are at $x_2' = 5$ m and $x_1' = 4$ m in R', what are the co-ordinates of the meter stick in R as a function of time t measured in R according

(a) to the Galilean transformation?

(b) to the Lorentz transformation?

Comment on your answer.

9. A meter stick is at rest in the R' reference frame, which is traveling at speed $u = 0.9c$ in the x, x' direction with respect to reference system R. If the ends of the meter stick are at $(x_2' = 2.71$ m, $y_2' = 1.71$ m) and at $(x_1' = 2.00$ m, $y_1' = 1.00$ m), what are the coordinates of the ends of the meter stick in R as meas-ured when $t = 0$ according

(a) to the Galilean transformation?

(b) to the Lorentz transformation?

What angle does the meter stick make with respect to the x, x' direction in each case? Comment on your answer.

10. Two observers, Clem and Yem, are sitting in reference frames R and R' at positions $x = 5$ m and $x' = 3$ m, respectively. System R' is moving with respect to system R in the x, x' direction at speed $u = 0.6c$. Clem lights a match at time $t = 2$ sec, lights his pipe, and blows out the match at $t = 5$ sec. What are the corresponding times as measured by Yem according

(a) to the Galilean transformation?

(b) to the Lorentz transformation?

Comment on your answer.

11. A particle has coordinates $(2, 0, 0, 2/c)$ in R. What are its coordinates in R' if R' travels with a speed $0.6c$ relative to R along their common x, x' axis?

*12. A particle moves with constant velocity in R' so that its spatial coordinates in meters are $(1, 1, 1)$ at $t' = 0$, and $(1, 0, 3)$ at $t' = 10^{-8}$ sec. If R' is moving in the x, x' direction at speed $u = 0.9c$, what are the spatial and time coordinates of the particle in R when $t' = 10^{-9}$ sec according

(a) to the Galilean transformation?

(b) to the Lorentz transformation?

Determine the magnitude and direction of the velocity vector as observed in R in each case. Comment on the magnitudes.

5 Relativistic kinematics

5.1 INTRODUCTION

A number of rather surprising phenomena are predicted if the Lorentz transformation rather than the Galilean transformation is applied when the relative motion of an observer and his region of observation is involved. These phenomena are observable, however, only when the relative speeds involved approach the speed of light. Therefore it is not surprising that we are not familiar with them from our daily experiences.

The most important fact emerging from consideration of the Lorentz transformation is that space and time are not independent as they appear to be in the realm of Newtonian physics. The time t' at which an event occurs in the R' system depends upon the time t at which the event occurs in the R system, as well as upon the x coordinate of the point at which the event occurs in the R system. The intermixing of space and time is best seen by considering the invariant quantity

$$x^2 + y^2 + z^2 - c^2t^2 = 0.$$

The quantity c^2t^2 has the dimension of (length)^2, as have the quantities x^2, y^2, and z^2. Thus c^2t^2 plays the role of a space-like coordinate. Since c is a constant quantity and t varies, t is often referred to as the **fourth dimension of space**, x, y, and z being the ordinary Cartesian coordinates of the three-dimensional physical space so familiar to us. Of course, this does not mean that space and time are equivalent, but only that we must be careful to consider the time when we are making position measurements and vice versa when we are dealing with relative motion.

We shall consider first the velocity and dimensions of objects observed from frames of reference in relative motion and then go on to discuss the measurement of time.

5.2 RELATIONS BETWEEN VELOCITIES

Suppose that a particle is observed to be in motion relative to two observers, one stationary in system R and the second stationary in system R', where R' is moving with speed u relative to R in the common x, x' direction. The observer in R measures a velocity $\mathbf{v}$ for the particle, whereas the observer in R' measures the velocity $\mathbf{v}'$ (see Fig. 5.1). What is the relation between $\mathbf{v}$ and $\mathbf{v}'$?

Let us suppose that the components of $\mathbf{v}$ are v_x, v_y, and v_z in the x, y, and z directions, respectively, and that the components of $\mathbf{v}'$ are v_x', v_y', and v_z' in the x', y', and z' directions, respectively.

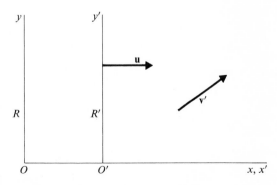

Fig. 5.1. A particle has a velocity $\mathbf{v}'$ as measured in R'. What is the velocity $\mathbf{v}$ as measured in R?

From the Newtonian point of view $\mathbf{v}'$ and $\mathbf{v}$ are related by the vector laws of addition, as illustrated in Fig. 5.2(a). The components of $\mathbf{v}'$ and $\mathbf{v}$ are related by the equations

$$v'_x = v_x - u$$
$$v'_y = v_y$$
$$v'_z = v_z,$$

which are obtained by resolving the vectors into components, as shown in Fig. 5.2(b). They may also be obtained directly from the Galilean transformation.

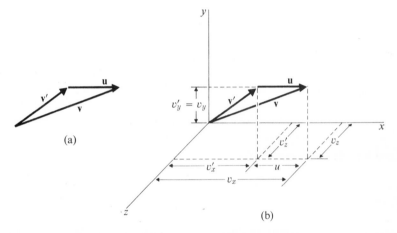

Fig. 5.2. Classical relation between the components of $\mathbf{v}$ and $\mathbf{v}'$.

Let us now derive corresponding equations using the Lorentz transformation. In R' the average speed $\bar{v}'_x$ is defined as

$$\bar{v}'_x = \frac{x'_2 - x'_1}{t'_2 - t'_1}.$$

Substitution from the Lorentz transformation gives

$$\bar{v}'_x = \frac{\gamma(x_2 - ut_2) - \gamma(x_1 - ut_1)}{\gamma\left(t_2 - \dfrac{ux_2}{c^2}\right) - \gamma\left(t_1 - \dfrac{ux_1}{c^2}\right)}$$

$$= \frac{(x_2 - x_1) - u(t_2 - t_1)}{(t_2 - t_1) - \dfrac{u}{c^2}(x_2 - x_1)}.$$

Dividing both numerator and denominator of the right-hand side of the equation by $(t_2 - t_1)$ gives

$$\bar{v}'_x = \frac{\dfrac{x_2 - x_1}{t_2 - t_1} - u}{1 - \dfrac{u}{c^2}\left(\dfrac{x_2 - x_1}{t_2 - t_1}\right)}.$$

But

$$\frac{x_2 - x_1}{t_2 - t_1} = \bar{v}_x,$$

the average speed as observed in R. Therefore,

$$\bar{v}'_x = \frac{\bar{v}_x - u}{1 - \dfrac{u\bar{v}_x}{c^2}}.$$

This is the relation between the average speeds.

The same relation may be shown to hold if instantaneous values of the x components of the velocity are considered. In this case,

$$v'_x = \frac{dx'}{dt'} \quad \text{and} \quad v_x = \frac{dx}{dt},$$

and we obtain the equation

$$v'_x = \frac{v_x - u}{1 - \dfrac{uv_x}{c^2}}$$

for the relation between instantaneous speeds. The derivation is carried out in the shaded area below.

For instantaneous speeds

$$v'_x = \frac{dx'}{dt'} = \frac{dx'}{dt} \cdot \frac{dt}{dt'}.$$

Now

$$\frac{dx'}{dt} = \frac{d}{dt}[\gamma(x - ut)]$$

$$= \gamma\left(\frac{dx}{dt} - u\right)$$

$$= \gamma(v_x - u);$$

$$\frac{dt'}{dt} = \frac{d}{dt}\left[\gamma\left(t - \frac{ux}{c^2}\right)\right]$$

$$= \gamma\left(1 - \frac{u}{c^2}\frac{dx}{dt}\right).$$

Therefore,

$$\frac{dt}{dt'} = \frac{1}{\gamma\left(1 - \dfrac{uv_x}{c^2}\right)}.$$

This gives

$$v'_x = \gamma(v_x - u)\left[\frac{1}{\gamma\left(1 - \dfrac{uv_x}{c^2}\right)}\right]$$

or

$$v'_x = \frac{v_x - u}{1 - \dfrac{uv_x}{c^2}}.$$

The average speed $\bar{v}'_y$ in R' is defined as

$$\bar{v}'_y = \frac{y'_2 - y'_1}{t'_2 - t'_1}.$$

Again, we substitute from the Lorentz transformation to obtain

$$\bar{v}'_y = \frac{y_2 - y_1}{\gamma\left(t_2 - \dfrac{ux_2}{c^2}\right) - \gamma\left(t_1 - \dfrac{ux_1}{c^2}\right)}$$

$$= \frac{y_2 - y_1}{\gamma\left[(t_2 - t_1) - \dfrac{u}{c^2}(x_2 - x_1)\right]}$$

$$= \frac{\bar{v}_y}{\gamma\left(1 - \dfrac{u\bar{v}_x}{c^2}\right)}.$$

The same relation holds between the instantaneous speeds, so that

$$v'_y = \frac{v_y}{\gamma\left(1 - \dfrac{uv_x}{c^2}\right)}.$$

In a similar manner we can obtain

$$v'_z = \frac{v_z}{\gamma\left(1 - \dfrac{uv_x}{c^2}\right)}.$$

Therefore, according to relativistic kinematics, **v** and **v'** are not related by the vector law of addition. Instead the components of $\mathbf{v}'$ and $\mathbf{v}$ are related by

$$v'_x = \frac{v_x - u}{1 - \dfrac{uv_x}{c^2}} \qquad v_x = \frac{v'_x + u}{1 + \dfrac{uv'_x}{c^2}}$$

$$v'_y = \frac{v_y}{\gamma\left(1 - \dfrac{uv_x}{c^2}\right)} \qquad v_y = \frac{v'_y}{\gamma\left(1 + \dfrac{uv'_x}{c^2}\right)}$$

$$v'_z = \frac{v_z}{\gamma\left(1 - \dfrac{uv_x}{c^2}\right)} \qquad v_z = \frac{v'_z}{\gamma\left(1 + \dfrac{uv'_x}{c^2}\right)}.$$

We note especially that the y and z components of the velocity are affected by the transformation even though the relative motion is in the x, x' direction.

Example. In high-energy physics, experiments have been proposed in which beams of protons (the nuclear particles of the hydrogen atom) from two adjoining high-energy accelerators would be brought into the same region of space but traveling in opposite directions. Such an experiment, known as a colliding-beam experiment, is depicted in Fig. 5.3. In a typical experiment each proton beam would contain particles traveling at a speed of $0.950c$ as measured by an observer stationary with respect to the accelerator laboratory. Suppose an observer could be at rest with respect to one proton. What speed of approach would he measure for the second proton?

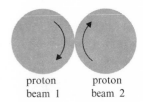

proton   proton
beam 1   beam 2

Fig. 5.3. A schematic diagram of two colliding proton beams.

Solution. We assume that the first proton (proton 1) and the observer are at rest in system R' (see Fig. 5.4). Therefore, the system R' is moving with respect to a system R fixed to a laboratory at a speed of $u = 0.950c$ in the direction of motion of proton 1. In system R, proton 1 is observed to have speed $v_x(1) = 0.950c$ and proton 2 has speed $v_x(2) = -0.950c$. In R', proton 1 has speed $v'_x(1) = 0$. The speed of proton 2 in the R' system would be, **according to Newtonian kinematics,**

$$v'_x(2) = v_x(2) - u$$

$$= -0.950c - 0.950c$$

$$= -1.900c.$$

proton 1
$v_x(1) = 0.95c$
$v'_x(1) = 0$

proton 2
$v_x(2) = -0.95c$
$v'_x(2) = ?$

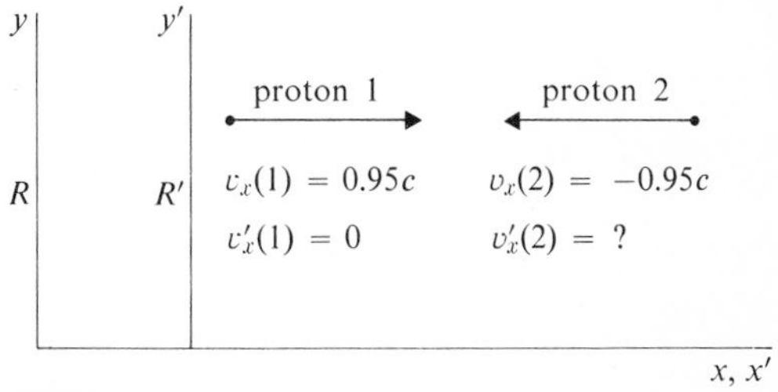

Fig. 5.4. What speed does an observer traveling with proton 1 measure for proton 2?

Therefore, an observer in R' should measure a speed of approach of $1.900c$ for proton 2. **According to relativistic kinematics,**

$$v_x'(2) = \frac{v_x(2) - u}{1 - \dfrac{uv_x(2)}{c^2}}$$

$$= \frac{-0.950c - 0.950c}{1 - \dfrac{0.950c(-0.950c)}{c^2}}$$

$$= -0.998(7)c.$$

In this case, the observer would measure a speed of approach of $0.998(7)c$ for proton 2. This result is very close to, but slightly less than, the speed of light c.

Suppose proton 2 had been a γ-ray traveling in R at speed c. According to the relativistic formula, the speed of approach would be measured as

$$v_x'(2) = \frac{-c - 0.95c}{1 - \dfrac{0.95c(-c)}{c^2}}$$

$$= \frac{-c(1 + 0.95)}{(1 + 0.95)}$$

$$= -c.$$

The speed of approach is still c. This is really just a confirmation of the fact that the Lorentz transformation was derived on the assumption that the speed of light is a constant, for the relativistic formula is derived directly from the Lorentz transformation.

Suppose the speed of the protons had been some speed u much less than c. In this case

$$\frac{u}{c^2} \ll 1$$

and

$$1 - \frac{uv_x}{c^2} \simeq 1.$$

Therefore,

$$v_x'(2) \simeq v_x(2) - u,$$

which is the Newtonian expression. Also

$$\gamma = \frac{1}{\left(1 - \dfrac{u^2}{c^2}\right)^{1/2}} \simeq 1,$$

so that

$$v'_y \simeq v_y$$

$$v'_z \simeq v_z.$$

That is, in the limit of small speeds, the relativistic equations reduce to the Newtonian expressions.

5.3   LENGTH CONTRACTION

Suppose we wish to measure the length of a moving rod. To simplify the discussion, we shall assume that the rod is moving in the direction of its length. We further suppose that the rod is at rest in a system R' and that the x coordinates of the two ends of the rod are x'_1 and x'_2. The situation is illustrated in Fig. 5.5. To measure the length of the moving rod in system R, it is necessary to measure the positions of the ends of the rod at the same time. To achieve this rather difficult task we proceed as follows. We set up a system of observers with synchronized clocks in the system R. Each observer records on his clock the times t_1 and t_2 at which the two ends of the rod pass by him. From these observations we look for two observers at position coordinates x_1 and x_2, one of whom observed the leading end at the same instant of time that the other observed the trailing end. The difference $x_2 - x_1$ in the position coordinates of two such observers gives the length of the moving rod in system R. This situation is illustrated in Fig. 5.6. We shall now consider the problem of the synchronization of the clocks.

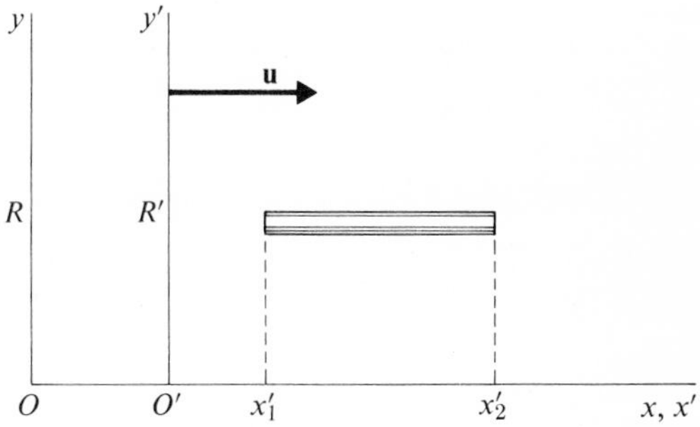

Fig. 5.5. A rod at rest in R' is observed in R where its length is to be measured.

Let us imagine an infinite succession of clocks set up in R along the path of the rod, a few of which are shown in Fig. 5.7. To synchronize the clocks, we send out a light signal from some particular point. Since the position of each clock is known (by reference to its x coordinate), the time taken for the light signal to travel to each clock is known. We now imagine that there is a person stationed at each clock whose task it is to start his clock at the instant that the light signal arrives at the position of the clock. The clock situated at the point where the light signal originates originally reads zero time but is started at the instant that the light signal begins. (Since any light signal has a

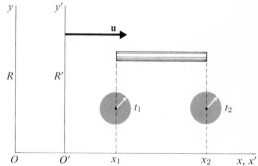

Fig. 5.6. By using properly synchronized clocks we can measure the position of the end of a moving rod at the same instant of time $(t_1 = t_2)$.

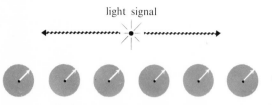

light signal

Fig. 5.7. A few of an infinite succession of clocks set up in R beside the path along which the rod travels.

finite length, in practice all clocks would be started in coincidence with the first portion or "leading edge" of the light signal.) The observers at each of the other clocks set their clocks to a reading equal to the time taken for the light signal to travel the distance from its point of origin to the points in space where their clocks are situated. If a clock is started at the instant the light signal arrives, then the clock is said to be **synchronized** with the clock situated at the point where the light signal originates[1]. This discussion assumes, of course, that all clocks are identical in all respects.

We do not find it difficult to synchronize clocks in our own home, since a light signal travels across an average-sized room in about 5×10^{-7} sec. We just look at another clock and adjust our own while neglecting the time required for the light signal to come to us from the clock. However, if distances are large, or the time intervals we wish to measure are very small, we must be careful to synchronize clocks properly.

According to the Lorentz transformation, the coordinates x_1 and x_2 are related to the coordinates x'_1 and x'_2 by

$$x'_1 = \gamma(x_1 - ut_1), \qquad x'_2 = \gamma(x_2 - ut_2).$$

Therefore, the length of the rod in R' is

$$L' = x'_2 - x'_1 = \gamma(x_2 - x_1) - \gamma u(t_2 - t_1).$$

However, when $t_1 = t_2$, then $x_2 - x_1 = L$, the length of the rod measured in R. Therefore

$$L' = \gamma L = \frac{L}{\left(1 - \dfrac{u^2}{c^2}\right)^{1/2}}$$

or

$$L = L'\left(1 - \frac{u^2}{c^2}\right)^{1/2}.$$

Since u is always less than c,

$$\left(1 - \frac{u^2}{c^2}\right)^{1/2} < 1$$

---

[1] For a more complete discussion of how one might synchronize clocks and use them experimentally, see E. F. Taylor and J. A. Wheeler, *Spacetime Physics* (San Francisco: W. H. Freeman & Co., Publishers, 1966), Chapter 1, Section 4.

at all times, so that L is always less than L'. Therefore, a moving rod, or indeed any moving object, is measured to be **contracted** in its direction of motion by the factor

$$\left(1 - \frac{u^2}{c^2}\right)^{1/2},$$

where u is the speed of the object relative to the observer. No such contraction is observed in the plane perpendicular to the direction of motion.

An observer at rest with respect to the rod is said to measure the **proper length** of the rod. That is, the proper length is defined as the length measured in a coordinate system at rest relative to the rod. It follows that the proper length is the greatest length that may be measured for the rod or for any object. Length contraction is solely a result of the relative motion between source and observer. A meter stick at rest in R has a length of 1 m as measured by an observer in R; similarly, a meter stick at rest in R' has a length of 1 m as measured by an observer at rest in R'. However, if an observer in R measures the meter stick at rest in R', or vice-versa, the contracted length will be obtained. A length measurement is not an absolute quantity but is relative to the observer.

Example. A rocket ship has a length of 30 m as measured by an observer at rest relative to the rocket. What length would an observer on the ground measure for the rocket when it is traveling at a speed of (a) 10^4 m · sec^{-1} (b) 10^7 m · sec^{-1} (c) 10^8 m · sec^{-1} (d) 2×10^8 m · sec^{-1} (e) 2.9×10^8 m · sec^{-1}?

Solution.

(a) The rocket ship has a speed $u = 10^4$ m · sec^{-1}. Therefore,

$$\frac{u}{c} = \frac{10^4}{3 \times 10^8} = 3.3 \times 10^{-5}$$

$$\frac{u^2}{c^2} = 1.09 \times 10^{-9}$$

and

$$1 - \frac{u^2}{c^2} = 1 - 1.09 \times 10^{-9} \simeq 1.$$

The observer would measure a length of 30 m unless he could make a length measurement to 9 significant figures.

(b) to (e): The calculations are summarized in Table 5.1.

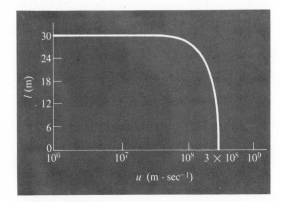

Fig. 5.8. The variation in the measured length of a rocket as a function of its speed relative to the frame of reference in which the measurement is made.

Table 5.1   Summary of calculations for parts (b) through (e) of the example

$u\ (m \cdot sec^{-1})$	$\dfrac{u}{c}$	$\dfrac{u^2}{c^2}$	$1 - \dfrac{u^2}{c^2}$	$\left(1 - \dfrac{u^2}{c^2}\right)^{1/2}$	$L(m)$
(b) 10^7	3.3×10^{-2}	1.1×10^{-3}	0.9989	0.9995	29.9(9)
(c) 10^8	0.33(3)	0.11(1)	0.88(9)	0.94(3)	28.0(3)
(d) 2×10^8	0.66(7)	0.44(4)	0.55(6)	0.74(5)	22.0(4)
(e) 2.9×10^8	0.96(7)	0.93(4)	0.06(6)	0.25(6)	7.6(8)

These results are shown in the Fig. 5.8, where the measured length is plotted against the logarithm of the speed. We see that the contraction of the length does not become perceptible until speeds greater than 10^7 m · sec^{-1} are attained. Contraction sets in rapidly as u approaches c, with the measured length becoming zero if $u = c$.

It should be stressed that length contraction is an observational effect only and that no physical changes in the structure of the rocket occur. An observer at rest relative to the rocket will still measure a length of 30 m whatever the speed of the rocket may be.

In the previous discussion we have developed the idea of contraction with regard to the length of a physical object. However, we are really considering only the distance between the two points represented by the ends of the rod; removing the rod would not affect the fact that the distance between the two points is contracted in the direction of motion. Therefore, we can immediately see that all lengths measured by a moving observer are contracted in the direction of his motion. His knowledge of distant objects is obtained by detection of light waves coming from them. These light rays are measured to travel toward him at a speed c regardless of his motion. It is no wonder then that his distance measurements will depend upon the speed at which he is traveling toward the point in question.

Example. A rocket ship sets out for the nearby star α-Centauri, which is measured to be 4 light years from the earth by an observer on the earth (light takes four years to travel to us from it). By the time the rocket ship leaves the solar system, it is traveling at a speed of 0.9c. What is the distance to α-Centauri that would be measured by an observer on the rocket ship?

Solution. The speed of the rocket ship is $u = 0.9c$. Therefore,

$$\frac{u}{c} = 0.9, \qquad \frac{u^2}{c^2} = 0.81,$$

and

$$1 - \frac{u^2}{c^2} = 0.19,$$

so that

$$\left(1 - \frac{u^2}{c^2}\right)^{1/2} = 0.44.$$

The observer on the rocket ship will measure the distance to be $0.44 \times 4 = 1.7(6)$ light years.

5.4 TIME DILATION

Now we turn to a comparison of measurements of time made by observers in the frames of reference R and R'. Let us suppose that an unstable particle, such as may be produced by the proton synchrotron mentioned in Section 4.5, is created at rest in R' (that is, R' moves at a speed u equal to the speed of the particle). The particle lives for a short time and then decays. An observer in R' would measure a time interval τ' for the lifetime of the particle. What time interval would an observer in the system R attached to the laboratory measure for the lifetime?

The observer in R' makes his measurements on a single standard clock located in R' at the position of the particle. The observer in R must make use of synchronized clocks such as those described in the previous section. Let us suppose that the clock in R situated at the point where the moving particle is observed to appear reads time t_1, and the clock located where the moving particle disappears reads t_2. Since the clocks are properly synchronized, observers in R conclude that the lifetime of the particle is

$$\tau = t_2 - t_1.$$

The times measured in R are related to the time measured in R' by the Lorentz transformation equations

$$t_2 = \gamma\left[t'_2 + \frac{ux'_2}{c^2}\right], \qquad t_1 = \gamma\left[t'_1 + \frac{ux'_1}{c^2}\right].$$

Therefore,

$$\tau = t_2 - t_1 = \gamma(t'_2 - t'_1) + \frac{\gamma u}{c^2}(x'_2 - x'_1).$$

However, the particle is at rest in R', so that

$$x'_2 = x'_1.$$

Since the lifetime of the particle measured in R' is

$$\tau' = t'_2 - t'_1,$$

$$\tau = \gamma\tau'.$$

The time interval measured by the observer in R is longer than that measured in R' by the factor γ. This effect is known as **time dilation**.

The shortest time of occurrence that can be measured for an event is that time interval recorded by a single clock moving with the event and is known as the **proper time**. An observer moving relative to the event records a longer time interval using clocks that are stationary with respect to himself. He concludes that the clock moving with the event must be running more slowly than the clocks at rest in his frame because it measures a shorter time interval than his clocks. In fact, all clocks moving with respect to the observer run more slowly than a clock at rest with respect to the observer. Moreover, since we are considering uniform relative motion, an observer in the frame R' would see the reference frame R moving away from him at speed u and would conclude by observation that clocks attached to frame R run more slowly than a clock held in his hand. Two observers moving with respect to each other both observe the other's clock to run more slowly than his own. Time dilation is due solely to relative motion; a time measurement is not absolute but is relative to the observer.

Convincing confirmation of time dilation comes from observations on cosmic-ray particles[2], which are believed to come from outer space and consist mainly of fast-moving hydrogen nuclei, or protons as they are commonly called. When these cosmic particles enter the earth's atmosphere and collide with nuclei of the constituent nitrogen and oxygen molecules, new particles called μ-mesons are produced with a mass of the order of 0.1 proton masses. The μ-meson is an unstable particle that decays into other particles after a short lifetime. Relatively slow-moving μ-mesons may be produced in the laboratory by means of the large accelerators of modern high-energy physics research. In a given sample of μ-mesons at rest or moving very slowly, half of them will decay to other particles in 1.5×10^{-6} sec. Half of the remainder will decay in the next 1.5×10^{-6} sec, and so on. The time for one-half of the μ-mesons to decay is called the **half-life**. The half-life of these slowly moving μ-mesons is essentially the proper half-life. Even if a μ-meson traveled at the speed of light, the average distance traveled during a half-life would be only

$$c\tau_0 = 3 \times 10^8 \times 1.5 \times 10^{-6} = 450\,\text{m}.$$

When we determine experimentally the relative abundances of μ-mesons at a height of 6×10^4 m and at sea level, we find that there are approximately eight times as many μ-mesons at 6×10^4 m as at sea level.

[2] K. W. Ford, *The World of Elementary Particles* (New York: Blaisdell Publishing Co., 1963), Chapter 1. See also B. Rossi, "Where Do Cosmic Rays Come From?" *Scientific American*, September, 1953. Available as *Scientific American Offprint 239* (San Francisco: W. H. Freeman and Co., Publishers).

However, we expect half of the μ-mesons to have disappeared after traveling a distance of 450 m. If we divide 6×10^4 by 450 we obtain 133, which is the number of half-lives in the travel time to sea level. After 133 half-lives there should be only the fraction

$$\tfrac{1}{2} \times \tfrac{1}{2} \times \cdots = (\tfrac{1}{2})^{133} \simeq 10^{-40}$$

remaining. The observer on earth measures the half-life $\tau = \gamma\tau_0$ rather than τ_0. Since there are approximately eight times as many μ-mesons at 6×10^4 m than at sea level, there are only about three half-lives in the travel time to sea level. If we take u to be the speed of the high-energy μ-mesons, then

$$3\tau = \frac{s}{u} = \frac{6 \times 10^4}{u}$$

$$\tau = \frac{2 \times 10^4}{u}$$

and

$$\tau = \gamma\tau_0 = \frac{1}{\left(1 - \dfrac{u^2}{c^2}\right)^{1/2}}(1.5 \times 10^{-6}) = \frac{6 \times 10^4}{u}$$

or

$$u = 4 \times 10^{10}\left(1 - \frac{u^2}{c^2}\right)^{1/2}.$$

This expression can be solved to give

$$u = c(1 - 2.8 \times 10^{-5})$$

for the speed of the μ-mesons. Since the cosmic-ray μ-mesons are traveling very close to the speed of light, γ is large, and τ, their half-life, is large.

Another way of looking at this problem is to note that, according to the moving meson's clock, the half-life is the proper time τ_0, but the distance from the point at which the particle is created to the surface of the earth is contracted. Since

$$u = c(1 - 2.8 \times 10^{-5}), \qquad \gamma = 45,$$

and the μ-meson has a distance of only 1330 m to travel in its frame of reference. This distance is much shorter than the distance an observer on the earth would measure[3].

---

[3] A similar experiment is presented in the film by D. H. Frisch and J. H. Smith, "Time Dilation: an Experiment with Mu-mesons." (ESI film.)

5.5 SIMULTANEITY

When we say that two events are simultaneous, we mean that they occur at the same time. Suppose that the two events in question occur at points x_1 and x_2 in system R. The simultaneity of the events is dependent upon the times t_1 and t_2 of the two events being equal; that is,

$$t_1 = t_2.$$

The situation is illustrated in Fig. 5.9. If observers stationary at x_1 and x_2 note the same time of occurrence of the events on the synchronized clocks located at those points, then the inhabitants of reference frame R are justified in denoting the two events to be simultaneous. However, the inhabitants of reference frame R' moving with speed u with respect to R may view the events differently. The times of occurrence for the events as measured in R' are

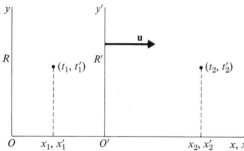

Fig. 5.9. Two events occurring at x_1 and x_2 in system R are simultaneous if $t_1 = t_2$. Can the same events be simultaneous in R'?

$$t'_1 = \left(t_1 - \frac{ux_1}{c^2}\right), \qquad t'_2 = \left(t_2 - \frac{ux_2}{c^2}\right).$$

Therefore,

$$t'_2 - t'_1 = \gamma(t_2 - t_1) - \frac{\gamma u}{c^2}(x_2 - x_1).$$

The two events will be simultaneous in R' only if

$$x_1 = x_2.$$

If

$$x_1 \neq x_2,$$

then the events are not simultaneous in R', and

$$t'_2 - t'_1 = -\frac{\gamma u}{c^2}(x_2 - x_1).$$

5.6 CAUSALLY CONNECTED EVENTS

Suppose two events occur at different times t_1 and t_2 as measured in reference frame R. The order of occurrence of these two events as viewed in R' may be reversed from that in R. This depends upon whether $(\gamma u/c^2)(x_2 - x_1)$ is less than or greater than $\gamma(t_2 - t_1)$. For the events to occur in the same order in R' as in R, we must have $t_2' > t_1'$ if $t_2 > t_1$, which requires that

$$t_2 - t_1 > \frac{u}{c^2}(x_2 - x_1).$$

Now, the largest possible value that u/c may have is unity. Therefore, the condition that the order of events will not be reversed in any inertial frame is

$$t_2 - t_1 > \frac{1}{c}(x_2 - x_1)$$

or

$$x_2 - x_1 < c(t_2 - t_1).$$

But $c(t_2 - t_1)$ is just the distance that a light signal will travel in time $t_2 - t_1$. The order of the events will not be reversed if the distance between their points of occurrence in R is less than the distance that a light signal will travel in the time interval between the events. Events are said to be **causally connected** if the occurrence of event 2 could have been triggered by the occurrence of event 1. A light signal could be sufficient to do this. Two causally connected events will not have their order reversed in any inertial reference frame. However, if two events are not causally connected, their order of occurrence may be reversed in different reference frames. For reversal of events we must have

$$t_2 - t_1 < \frac{u}{c^2}(x_2 - x_1).$$

Example. Io, a satellite of the planet Jupiter, is observed from the Earth to be eclipsed by Jupiter at 9:58 P.M., the time being measured by a clock in the observatory. If there were an observer on Jupiter with a clock synchronized to the earth clock, he would observe the Earth's moon to be eclipsed by the Earth at 9:27 P.M. of the same day. At this time the Earth and Jupiter are 8×10^{11} m apart.

(a) At what minimum speed would a spaceship have to be traveling between the Earth and Jupiter toward Jupiter so that astronomers in the spaceship would observe the eclipse of Io by Jupiter to occur first?

(b) What is the shortest possible time interval between the two eclipses that would allow them to be causally connected?

Solution.

(a) The motion of both the Earth and Jupiter in their orbits (3.0×10^4 m·sec^{-1} and 1.3×10^4 m·sec^{-1}, respectively) is much less than the speed of light (3×10^8 m·sec^{-1}), so that we may neglect their motion in this problem. For the events to be reversed for the spaceship observers, we must have

$$\frac{u}{c^2}(x_2 - x_1) > t_2 - t_1,$$

where

$$t_2 - t_1 = 31 \text{ min} = 1860 \text{ sec}$$

$$x_2 - x_1 = 8 \times 10^{11} \text{ m}$$

$$u = \text{required minimum spaceship speed.}$$

Therefore,

$$u > \frac{c^2(t_2 - t_1)}{x_2 - x_1} = \frac{(3 \times 10^8)(1860)}{(8 \times 10^{11})} c = 0.7c.$$

The minimum required speed is $0.7c$.

(b) For causal connection

$$t_2 - t_1 > \frac{x_2 - x_1}{c} = \frac{8 \times 10^{11}}{3 \times 10^8} = 2.67 \times 10^3 \text{ sec.}$$

Therefore, the eclipse of Io must occur a minimum of 2.67×10^3 sec or approximately 45 min after the eclipse of the moon.

5.7   THE CLOCK (OR TWIN) PARADOX

Consider the (as yet) hypothetical experiment in which two identical clocks are first synchronized and then separated, with one clock remaining on the earth while the second clock is put aboard a very fast spaceship, sent to a nearby star, and then brought back. The question concerns whether the clocks will record the same time when the spaceship returns from its journey. It is widely believed that the spaceship clock will record a shorter time interval than the clock that has remained on the earth. When stated in terms of identical twins, the result is even more surprising: the twins are no longer the same age!

The paradox involved is this. Although an observer on the earth realizes that he should observe the clock on the spaceship to run at a slower rate than the clock on the earth, he knows that, because the time dilation effect depends only on the relative velocity of the two clocks, an observer traveling on the spaceship should observe the clock on the earth to run at a slower rate than his clock. Since the problem at first sight appears symmetrical in relation to the two clocks, it may seem at first sight that no time difference between the clocks should occur. In fact, an asymmetry clearly exists in the problem, since the star is at rest relative to the earth but is not at rest relative to the spaceship.

Although it is quite straightforward to calculate the time difference using the special theory of relativity[4], it can be argued (and, indeed, very

---

[4] See, for example, E. F. Taylor and J. A. Wheeler, *Spacetime Physics* (San Francisco: W. H. Freeman & Co., Publishers, 1966), Chapter 1.

often is) that a proper solution to the problem lies outside of the special theory, since accelerations are involved in the problem and the special theory deals with uniform relative motion only. The reader is referred to articles by G. D. Scott and J. Bronowski[5] for further reading.

PROBLEMS

1. A K-meson is created by a high-energy proton of speed $0.6c$ striking a thin target. If the K-meson has a speed of $0.8c$ in a frame of reference moving with the incident protons, what is the speed of the K-mesons relative to a frame of reference fixed in the laboratory?

2. K-mesons of speed $0.9c$ relative to the laboratory frame of reference are produced by protons of speed $0.7c$ relative to the laboratory. What is the speed of the K-mesons relative to a frame of reference traveling with the protons?

3. A spaceship traveling at a speed $u = 0.8c$ is on its way to a star that is 20 light years from the earth. When the spaceship is 2 light years from the earth, it sends back a message at the speed of light to its base on earth. The message is answered within a few seconds. Will the reply, which also travels at the speed of light, reach the spaceship before it arrives at its destination? Perform the calculations both in a reference frame attached to the earth and in a reference frame attached to the spaceship.

*4. An electron with speed $0.8c$ is emitted from a source moving with speed $u = 0.6c$ with respect to the laboratory. What are the components of the electron's velocity as measured by an observer in the laboratory, both in the direction of the moving source and perpendicular to the direction of the moving source, if the electron is emitted
   (a) in the direction of motion of the source?
   (b) perpendicular to the direction of motion of the source?
   (c) at 45° with respect to the direction of the source as measured in a frame of reference moving with the source? What is the direction of emission of the electron as measured in the laboratory?

5. Repeat the calculations of Problem 4 using the Galilean transformation, and compare the results.

6. The length of a meter stick is measured to be 0.60 m. Calculate its minimum velocity.

7. A square of side 1 m in length is placed in a moving coordinate system, as shown in Fig. 5.10. What are the lengths of the sides as measured by a stationary

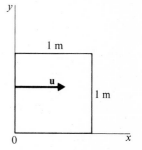

Fig. 5.10

[5] G. D. Scott, "On Solutions of the Clock Paradox," *The American Journal of Physics,* **27** (November 1959), 580. See also J. Bronowski, "The Clock Paradox," *Scientific American,* February, 1963, available as *Scientific American Offprint 291* (San Francisco: W. H. Freeman & Co., Publishers).

observer when the coordinate system is moving in the x direction at speed $u = 0.8c$, $u = 0.9c$, $u = 0.99c$, $u = 0.9999c$?

8. An ellipse with semi-major and semi-minor axes of 6 m and 2 m, respectively, is located in a moving coordinate system, as shown in Fig. 5.11. With what velocity must the coordinate system move past an observer in order that the ellipse is measured to be a circle by the observer?

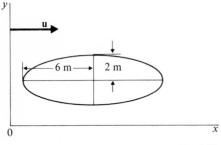

Fig. 5.11

9. A **linear accelerator** is a device that accelerates electrons to a very high energy in a long, straight, evacuated copper tube[6]. A linear accelerator two miles long (3200 m) has been constructed at Stanford University in California. Assuming that the mean speed of the electrons in the accelerator is $0.9999999c$, calculate the length of the accelerator in a frame of reference moving with electrons of this speed.

10. A meter stick is at rest in a coordinate system moving at speed $u = 0.6c$ with respect to an observer. If the meter stick is lying at an angle of 45° with respect to the direction of its motion, what length would the observer measure for the meter stick, and what angle with respect to its direction of motion would he measure?

11. A farmer wishes to stack a pile of lumber in his barn, but unfortunately the length of the lumber exceeds the length of the barn. The man's son, who is studying the special theory of relativity, reasons as follows. An object in motion experiences a contraction. If he runs fast enough with the lumber towards the barn, the length of the lumber as measured by his father can be made less than that of the barn. Once the lumber is inside, his father can quickly close the door. Comment.

12. There was a young lass called Fay
Who went traveling in a relative way
She left home one night
At nine-tenths the speed of light
Returning on the previous day.

Comment on the possibility or impossibility of the facts stated in this limerick.

13. A proton passes through a chamber that is 2 m in length. The speed of the proton relative to the chamber is $0.98c$. Calculate the time required for the proton to pass through the chamber
(a) according to the proton's clock.
(b) according to an observer in the laboratory.

---

[6] W. Panofsky, "The Linear Accelerator," *Scientific American*, October, 1954, available as *Scientific American Offprint 234* (San Francisco: W. H. Freeman & Co., Publishers), and E. L. Ginzton and W. Kirk, "The Two-mile Accelerator," *Scientific American*, November, 1961, available as *Scientific American Offprint 322* (San Francisco: W. H. Freeman & Co., Publishers).

14. A high-energy physicist measures the speed of μ-mesons by observing the time required for two bursts of μ-mesons from an accelerator to pass between two detectors C_1 and C_2, which are 2.7 m apart. If the time observed in the laboratory is 9.3×10^{-9} sec, what time interval would be measured by an observer traveling with the same speed as the μ-mesons? What distance would he measure between C_1 and C_2?

15. The μ-meson is an unstable particle, so that some of the particles passing through detector C_1 of Problem 14 do not reach detector C_2. By studying the rate at which μ-mesons are lost from the beam as a function of distance traveled, it was found that the μ-mesons of Problem 14 had a half-life of 6.0×10^{-6} sec. What half-life would be observed if the μ-mesons were at rest relative to the laboratory?

16. The Λ-baryon is an unstable particle with a proper half-life of 1.7×10^{-10} sec. What is the half-life measured by an observer moving with respect to the Λ-baryon at speeds of $u = 0.2c, 0.4c, 0.6c, 0.8c, 0.9c, 0.95c, 0.99c$? Plot a graph of the observed half-life as a function of speed.

17. What must be the relative velocity between two frames of reference in order that a time interval of t_0 occurring in one of them is observed to be equal to $2t_0$ in the second?

*18. An Albert-in-Wonderland Adventure[7]: one night I dreamed I was traveling along in a train in a world where the speed of light was only $5.00 \text{ m} \cdot \text{sec}^{-1}$. As the train approached a station platform I noticed the stationmaster awaiting the train's arrival and a suspicious looking individual some distance away sitting on a bench reading a newspaper. Suddenly the man jumped from the bench and fired a gun at the stationmaster, who to my amazement nimbly jumped to one side and watched the bullet harmlessly strike a nearby post. Given that the speed of the train relative to the platform was $4.50 \text{ m} \cdot \text{sec}^{-1}$, the speed of the bullet relative to the platform was $4.90 \text{ m} \cdot \text{sec}^{-1}$, the bullet and the train were essentially traveling along the same straight line toward one another, and the distance between the stationmaster and his assailant measured along the platform was 9.30 m, calculate the required time of flight of the bullet according to
(a) an observer on the platform awaiting the train.
(b) a demon observer traveling with the bullet.
(c) a passenger at rest relative to the train.

*19. The Andromeda Galaxy, which is the closest spiral galaxy to our own, is approximately 2,000,000 light years from the earth. If the occupants of a spacecraft can expect to live for another 50 years, at what minimum constant speed must the spacecraft travel towards the Andromeda galaxy from the earth in order that it arrive there during the astronauts' lifetimes?

20. The earth is at a mean distance of 1.497×10^{11} m from the sun. A magnetic storm occurs on the earth (a) 480.0 sec, (b) 499.0 sec, (c) 520.0 sec after an eruption takes place on the sun's surface. In which of (a), (b), and (c) are the events causally connected?

[7] A similar situation is discussed by G. Gamow in his entertaining book *Mr. Tompkins in Wonderland* (Cambridge: Cambridge University Press, 1962).

21. Two events are observed to occur 5 sec apart by an observer moving at a speed of 0.6c relative to the frame of reference R in which the events occur. What is the distance between the points at which the events occur, measured in R, if

   (a) the events are simultaneous in R?

   (b) the events are 4 sec apart in R?

   (c) the events are 8 sec apart in R?

A more general approach 6
to classical kinematics

6.1 INTRODUCTION

In this chapter we return to a more general discussion of single-particle kinematics from the Newtonian point of view. It is much simpler to consider the motion of a single particle rather than to try to follow individually all the atomic particles of which even the smallest familiar object is composed. In many cases, the relative motion of the constituent particles is unimportant, and it is only the motion of the object as a whole that concerns us. As we shall show in Chapters 13 and 14, a real object can be treated as if it is a point particle located at the center of mass of the object, insofar as the response of the object to external conditions is concerned. For this reason a wide variety of natural phenomena can be treated from the single-particle point of view; we shall restrict ourselves to a discussion of these phenomena for the next several chapters.

6.2 CURVILINEAR MOTION

Fig. 6.1. The displacement of a particle is a vector.

In Chapter 3 we investigated straight-line motion. The motion of a particle traveling along a straight line is known if we know the particle's displacement from any arbitrarily chosen origin along that line as a function of time. We did not use vector notation in Chapter 3, as the only direction involved was along the straight line. Figure 6.1 shows the displacement from the origin of a particle along a straight line as a vector.

The motion of a particle along a circular path is also familiar to us. A person seated on a Ferris wheel, for example, travels in a circular path, as shown in Fig. 6.2. The Ferris wheel rotates about an axis at the center of the circular rim of the wheel. The occupant of one of the seats remains at a constant distance from the axis of rotation and is said to move in a circular path.

In Fig. 6.3 we consider a particle moving in a circular path of radius r. We may specify the position of the particle at any time by giving its Cartesian coordinates (x, y) or its polar coordinates (r, θ) at that time. We should very likely prefer to use polar coordinates because r remains constant and only θ varies with time.

We note that, when we use polar coordinates, straight-line motion through the origin is described by a constant θ and a time-varying r, whereas for circular motion about the origin r is constant, and

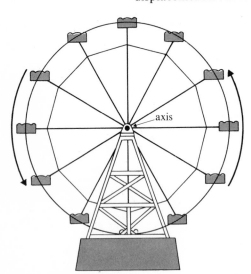

Fig. 6.2. A Ferris wheel rotates about an axis through its center.

θ varies with time. We see that straight-line motion and circular motion are just two special cases of a more general motion for a single particle, in which both r and θ vary with time. This general motion is known as **curvilinear motion**.

We shall restrict our discussion for the time being to two dimensions for simplicity. In Fig. 6.4 we consider the motion of a particle along a curved line. The position and velocity of the particle at two different times t_1 and t_2 are indicated. The distance traveled by the particle between times t_1 and t_2 must be measured along the path of its motion, the curve. This is difficult to handle mathematically, and, in fact, there is an alternative and much more useful method of describing the particle's behavior.

The coordinates of point (1) are r_1 and θ_1, and those of point (2) are r_2 and θ_2. If we knew how the r and θ coordinates varied with time, this would tell us where the particle was at all times. This is equivalent to knowing how the particle moves along its curved path. We can simplify our approach even further by writing the coordinate as a vector **r**. The direction of **r** gives the coordinate angle θ, and the magnitude of **r** is the position coordinate r. Therefore, knowing how the vector **r** varies with time will tell us how r and θ vary with time. Since

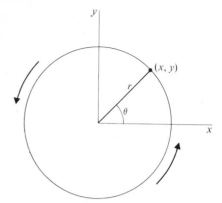

Fig. 6.3. A particle moving in a circular path whose center is the origin.

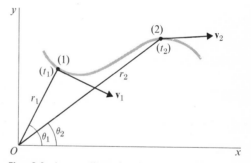

Fig. 6.4. A two-dimensional motion of a particle in the x, y plane.

r varies with time, it is often written as $\mathbf{r}(t)$, where the t in parentheses merely indicates that **r** is a function of time. The vector $\mathbf{r}(t)$ is known as the **position vector**. As the particle moves along the curved path, the position vector changes its magnitude and direction.

Let us consider two points along the path of motion separated by a small time interval Δt. The position vectors at the beginning and end of the time interval are $\mathbf{r}(t)$ and $\mathbf{r}(t + \Delta t)$, respectively, and are shown in Fig. 6.5. The quantities $\mathbf{r}(t)$ and $\mathbf{r}(t + \Delta t)$ are to be interpreted as the position vector at some time t and the position vector a short time Δt later, respectively. During the time Δt, the position vector of the particle has changed from $\mathbf{r}(t)$ to $\mathbf{r}(t + \Delta t)$, where $\mathbf{r}(t + \Delta t)$ denotes the position vector at time $t + \Delta t$. We have joined the tips of the position vectors by a vector $\Delta \mathbf{r}$ as shown,

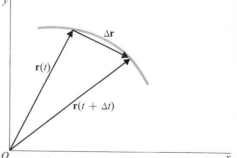

Fig. 6.5. The variation in the position vector of a particle as the particle moves along a curved path.

the direction being chosen so that

$$\mathbf{r}(t) + \Delta\mathbf{r} = \mathbf{r}(t + \Delta t)$$

or

$$\Delta\mathbf{r} = \mathbf{r}(t + \Delta t) - \mathbf{r}(t).$$

Therefore, we interpret $\Delta\mathbf{r}$ as the change in the position vector of the particle during time Δt. As in Section 3.3, we define the **average velocity** $\bar{\mathbf{v}}$ to be the change in position divided by the time interval, so that

$$\bar{\mathbf{v}} = \frac{\mathbf{r}(t + \Delta t) - \mathbf{r}(t)}{\Delta t} = \frac{\Delta\mathbf{r}}{\Delta t}.$$

The **instantaneous velocity** is found by letting $\Delta t \to 0$, that is, by taking the limit of an infinitesimally small time interval. Therefore, the instantaneous velocity is

$$\mathbf{v}(t) = \lim_{\Delta t \to 0} \frac{\mathbf{r}(t + \Delta t) - \mathbf{r}(t)}{\Delta t}$$

$$= \lim_{\Delta t \to 0} \frac{\Delta\mathbf{r}}{\Delta t}$$

or

$$\mathbf{v}(t) = \frac{d\mathbf{r}}{dt}.$$

This is the same form as we found for the instantaneous velocity in straight-line motion in Section 3.3.

As $\Delta t \to 0$, the vectors $\mathbf{r}(t)$ and $\mathbf{r}(t + \Delta t)$ approach each other, and $\Delta\mathbf{r}$ becomes more nearly identical to the segment of the curve between the ends of the position vectors. Therefore, in the limit, $d\mathbf{r}$ has the direction of the curve at a point or is tangent to the curve so that the instantaneous velocity at a point on the curve is tangent to the curve. This is illustrated in Fig. 6.6.

The velocity at time t is $\mathbf{v}(t)$, which is tangent to the curve at the point corresponding to position vector $\mathbf{r}(t)$. Similarly, at time $t + \Delta t$ the velocity is $\mathbf{v}(t + \Delta t)$, which is tangent to the curve at the point corresponding to position vector $\mathbf{r}(t + \Delta t)$. The vectors are shown in Fig. 6.7(a); in Fig. 6.7(b) the velocity vectors only are shown and are placed tail to tail. The change in velocity during time Δt is the vector $\Delta\mathbf{v}$, where

$$\Delta\mathbf{v} = \mathbf{v}(t + \Delta t) - \mathbf{v}(t).$$

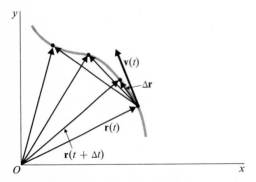

Fig. 6.6. The instantaneous velocity at a point on the curve is tangent to the curve.

We now define the **average acceleration** to be

$$\mathbf{a} = \frac{\mathbf{v}(t + \Delta t) - \mathbf{v}(t)}{\Delta t} = \frac{\Delta \mathbf{v}}{\Delta t},$$

and the **instantaneous acceleration** at a point is

$$\mathbf{a} = \lim_{\Delta t \to 0} \frac{\mathbf{v}(t + \Delta t) - \mathbf{v}(t)}{\Delta t}$$

$$= \lim_{\Delta t \to 0} \frac{\Delta \mathbf{v}}{\Delta t}$$

or

$$\mathbf{a} = \frac{d\mathbf{v}}{dt}.$$

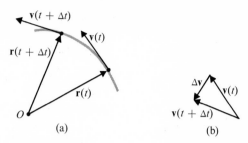

(a)   (b)

Fig. 6.7. (a) The position and velocity vectors of a particle at times t and $t + \Delta t$. (b) The vector $\Delta \mathbf{v} = \mathbf{v}(t + \Delta t) - \mathbf{v}(t)$.

In general, the acceleration vector is not in the direction of either the velocity or position vector of the particle at the point in question, and the velocity vector is not usually in the direction of the position vector. In order to solve problems involving curvilinear motion, it is necessary to know the variations in $\mathbf{r}$, $\mathbf{v}$, or $\mathbf{a}$ as a function of time. If $\mathbf{v}(t)$ is known, for example, $\mathbf{r}(t)$ may be found from

$$\int d\mathbf{r} = \int \mathbf{v}(t)\, dt$$

(that is, by integration), and $\mathbf{a}(t)$ may be found from

$$\mathbf{a}(t) = \frac{d\mathbf{v}(t)}{dt}$$

(that is, by differentiation).

The general treatment of curvilinear motion in two dimensions may be readily extended to three dimensions. A particle may be located in three-dimensional space by spherical coordinates (r, θ, ϕ), as shown in Fig. 6.8. The position vector $\mathbf{r}(t)$ again has the magnitude r and a direction given by the angles θ and ϕ. The instantaneous velocity and acceleration are then defined exactly as for two-dimensional motion.

6.3 RECTILINEAR MOTION

Straight-line or **rectilinear motion** is just curvilinear motion in one dimension. Vector notation is not necessary, and $\mathbf{r}(t)$ is replaced by $s(t)$, the displacement, or just by s. The velocity and

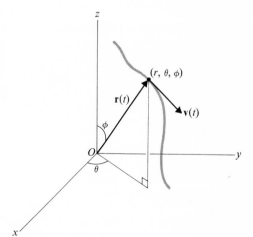

Fig. 6.8. Curvilinear motion in three dimensions.

acceleration are the familiar expressions

$$v = \frac{ds}{dt} \quad \text{and} \quad a = \frac{dv}{dt}.$$

Constant acceleration

We have already dealt with constant acceleration in Section 3.6. There we discovered the following relations involving $s, v, a,$ and t:

$$s = \left(\frac{v + v_0}{2}\right)t, \qquad v = v_0 + at,$$

$$v^2 = v_0^2 + 2as, \qquad s = v_0t + \tfrac{1}{2}at^2,$$

where s_0 and v_0 are the initial displacement and speed at time $t = 0$.

Using the formalism of the calculus,

$$\int_{v_0}^{v} dv = \int_{0}^{t} a\, dt,$$

where it is assumed that the speed changes from v_0 to v as the time changes from 0 to t. Therefore,

$$v \Big|_{v_0}^{v} = at \Big|_{0}^{t},$$

and

$$v = v_0 + at.$$

Similarly,

$$\int_{0}^{s} ds = \int_{0}^{t} v\, dt = \int_{0}^{t} (v_0 + at)\, dt,$$

where it is assumed that the displacement changes from 0 to s as the time changes from 0 to t. Therefore,

$$s \Big|_{0}^{s} = (v_0t + \tfrac{1}{2}at^2) \Big|_{0}^{t},$$

and

$$s = v_0t + \tfrac{1}{2}at^2.$$

This pair of equations for v and s is identical to two of the equations previously deduced in Section 3.6. The other equations from that section can be derived from these two.

Nonconstant acceleration

When the acceleration is a function of time, we must know the variation with time of one of the three basic quantities s, v, or a in order to determine the motion completely.

Example. A particle, initially at rest, is given an acceleration of $3t$ m · sec^{-2}, where t is the time. Determine the acceleration, speed, and displacement 5 sec after the particle starts to move.

Solution.

(a) When $t = 5$ sec, $\quad a = 3(5) = 15$ m · sec^{-2}.

(b) $\qquad v(t) = \int a \, dt.$

When $t = 5$ sec,

$$v = \int_0^5 3t \, dt = \left. \frac{3t^2}{2} \right|_0^5$$

$$= 37.5 \text{ m} \cdot \text{sec}^{-1}.$$

(c) $\qquad s(t) = \int_0^t v \, dt.$

When $t = 5$ sec

$$s = \int_0^5 \frac{3t^2}{2} \, dt = \left. \frac{t^3}{2} \right|_0^5$$

$$= 62.5 \text{ m}.$$

The acceleration, speed, and displacement after 5 sec are 15 m · sec^{-2}, 37.5 m · sec^{-1}, and 62.5 m, respectively.

Example. A particle's displacement from rest varies with time, as $s = 0.5t^4$ m. What is the displacement, speed, and acceleration after 5 sec?

Solution. Since

$$s = 0.5t^4,$$

$$v = \frac{ds}{dt} = 2t^3$$

and

$$a = \frac{dv}{dt} = 6t^2.$$

For $t = 5$ sec,

$$s = 312.5 \text{ m}$$

$$v = 250 \text{ m} \cdot \text{sec}^{-1}$$

$$a = 150 \text{ m} \cdot \text{sec}^{-2}.$$

The displacement, speed, and acceleration after 5 sec are 312.5 m, 250 m · sec^{-1}, and 150 m · sec^{-2}, respectively.

6.4   TWO-DIMENSIONAL MOTION—GENERAL SOLUTION

In general, two-dimensional curvilinear-motion problems may be solved by determining the components of the acceleration, velocity, and position vectors in appropriate coordinate systems.

Cartesian components

Figure 6.9 shows the Cartesian component vectors of the acceleration, velocity, and position vectors for a particle undergoing curvilinear motion.

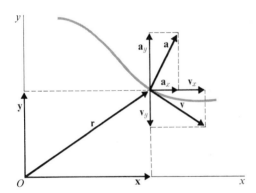

A given problem may be solved by determining the equations of motion in the x and y directions independently, using the methods outlined in Section 6.3 for rectilinear motion. The results may then be combined to determine the curvilinear motion.

Example. An object traveling with an initial speed of 500 m · sec^{-1} at time $t = 0$ is subject to an acceleration of 9.80 m · sec^{-2} in a direction at 135° to its initial direction of motion. Determine its velocity at $t = 50$ sec.

Fig. 6.9. The Cartesian component vectors of the acceleration, velocity, and position vectors in two dimensions.

Solution. The initial conditions are shown in Fig. 6.10(a). If Cartesian axes x and y are chosen as shown in Fig. 6.10(a), there will be no acceleration in the x direction and an acceleration of 9.8 m · sec^{-2} in the negative y direction. We shall solve the problem by working with the x and y components of the motion separately. In practice, problems involving curvilinear motion are solved by working with the component equations rather than with the formal vector equations.

The component vectors for the acceleration, velocity, and position vectors are shown in Fig. 6.10(b) for some instant of time $t > 0$. The object follows a curved path (it is actually a parabolic path, as we shall see in Section 7.3). The x component of the initial velocity is

$$v_x = v_0 \cos 45°,$$

where v_0 is the initial speed of 500 m · sec^{-1}. Since there is no acceleration in the x direction, v_x remains constant.

$$v_x = 500(0.707) = 353(.5) \, \text{m} \cdot \text{sec}^{-1}.$$

In the y direction the initial upwards speed is

$$v_y(0) = v_0 \cos 45° = 353(.5) \, \text{m} \cdot \text{sec}^{-1}.$$

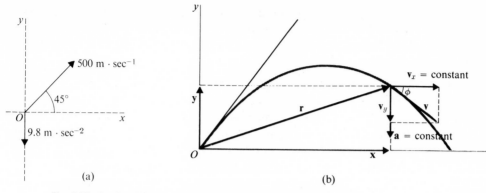

(a)   (b)

Fig. 6.10. (a) An object is subject to an acceleration of 9.8 m · sec^{-2} at 135° to its initial direction of motion. (b) The curved path of motion of a particle experiencing a constant acceleration in the negative y direction.

Since the acceleration is in the $-y$ direction, it is negative; for $t = 50$ sec,

$$v_y = v_y(0) - at$$

$$= 353(.5) - 9.80(50)$$

$$= 353(.5) - 490$$

$$= -137 \text{ m} \cdot \text{sec}^{-1}.$$

Therefore, the y component of the velocity is 137 m · sec^{-1} downward. The magnitude of the velocity vector of the object at $t = 50$ sec is

$$v = (v_x^2 + v_y^2)^{1/2}$$

$$= (12.5 + 10^4 + 1.9 \times 10^4)^{1/2}$$

$$= 379 \text{ m} \cdot \text{sec}^{-1}.$$

The angle that the vector **v** makes with the horizontal is given by [see Fig. 6.10(b)]

$$\tan \phi = \frac{v_y}{v_x} = \frac{-137}{353.5} = -0.387.$$

The angle ϕ is then

$$\phi = -21.2°.$$

Therefore, after 50 sec the object is traveling at a speed of 379 m · sec^{-1} and heading at an angle of 21.2° below the x axis or at an angle of $45 + 21.2 = 66.2°$ to its initial direction of motion. Note that the position vector of the object can be determined in a similar manner.

Polar coordinates

Figure 6.11 shows the component vectors of the acceleration, velocity, and position vectors in polar coordinates. The velocity and acceleration vectors are resolved into components lying along the line of **r** (**the radial component**) and perpendicular to **r** (**the transverse component**). In Chapter 13 we shall deal with the motion of two particles that attract each other with a force that varies inversely as the square of the distance between them. This is a very important example of a particular type of two-dimensional curvilinear motion that is most easily solved using polar coordinates.

Another very important example of two-dimensional curvilinear motion is circular motion; it will be dealt with in the next section.

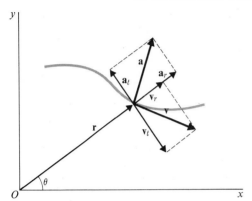

Fig. 6.11. The component vectors of the velocity and acceleration in polar coordinates are along the line of **r** and perpendicular (transverse) to **r**.

6.5   CIRCULAR MOTION

In circular motion, the position vector has constant magnitude, but its direction varies; the origin is chosen to be at the center of the circular path, as illustrated in Fig. 6.12. We still write

$$\mathbf{r}(t + \Delta t) - \mathbf{r}(t) = \Delta \mathbf{r}$$

and

$$\mathbf{v} = \frac{d\mathbf{r}}{dt}, \qquad \mathbf{a} = \frac{d\mathbf{v}}{dt}.$$

As $\Delta \mathbf{r} \to 0$, **v** becomes tangent to the path as expected. However, as the path is circular, the vector **v** is perpendicular to the vector **r**, since the tangent to a circle is perpendicular to the radius at that point.

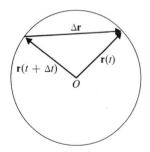

Fig. 6.12. In circular motion the position vector has constant magnitude and only its direction changes with time.

Motion at constant speed

Figure 6.13(a) shows the position and velocity vectors for three different points on a circular path for the case of constant speed (constant v). The velocity vectors are all equal in magnitude but are different in direction. In Fig. 6.13(b) the velocity vectors are brought together at a point. As the time interval $\Delta t \to 0$, $\Delta \mathbf{v}$ becomes perpendicular to $\mathbf{v}(t)$, since $\mathbf{v}(t)$ and $\mathbf{v}(t + \Delta t)$ have the same magnitude. Therefore, the vector

$$\mathbf{a} = \frac{d\mathbf{v}}{dt}$$

is perpendicular to **v**, which in turn is perpendicular to **r** such that **a** points in the direction of $-\mathbf{r}$, as is evident from Fig. 6.13. This is shown in

Fig. 6.14 for two positions of the particle. The vector **a** has been displaced to the side of the vector **r** for greater clarity, although both vectors are acting along the same line. Since **v** is constant in magnitude, **a** is constant in magnitude; it is known as the **centripetal acceleration**.

Note that there is an acceleration even though the speed (the magnitude of the velocity) is constant. This illustrates the vector nature of acceleration and velocity. Since acceleration is the rate of change of velocity and the velocity is a vector with both magnitude and direction, a change in either the magnitude or the direction of the velocity vector is sufficient to give rise to an acceleration. In rectilinear motion, acceleration is due to a change in speed while the direction of motion remains constant; in uniform circular

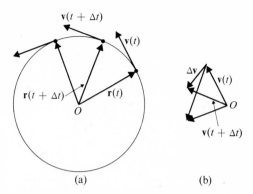

Fig. 6.13. (a) Position and velocity vectors at times t and $t + \Delta t$ for a particle in uniform circular motion. (b) The vector $\Delta \mathbf{v} = \mathbf{v}(t + \Delta t) - \mathbf{v}(t)$ becomes perpendicular to $\mathbf{v}(t)$ as $\Delta t \to 0$.

motion, acceleration is due to a change in direction of the velocity while the magnitude remains constant; in general curvilinear motion, acceleration is due to simultaneous change in both magnitude and direction of the velocity.

During uniform circular motion the radius vector of the particle changes its orientation with respect to a fixed axis periodically with time. The **angular frequency of rotation** ω is defined as the angle swept out per unit time by the radius vector and is measured in the units of radians per second (rad · sec^{-1});

$$\omega = \frac{d\theta}{dt}.$$

For uniform circular motion, ω is a constant, and θ varies uniformly with time. If the particle begins its motion at time $t = 0$ from the x axis, then, at any time t, θ is given by $\theta = \omega t$, as indicated in Fig. 6.15.

The **period** T of the motion is the time required for the particle to make one complete trip around the circumference of the circle. Since **r** turns through an angle of 2π radians in this time, we have

$$\omega = \frac{2\pi}{T}.$$

The particle has moved a distance of $2\pi r$ in the time T, so that the speed v is

$$v = \frac{2\pi r}{T} = \omega r.$$

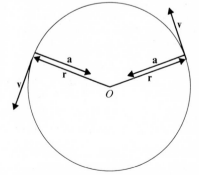

Fig. 6.14. The acceleration vector **a** acts in the direction of $-\mathbf{r}$. It is displaced to the side in the diagram for clarity.

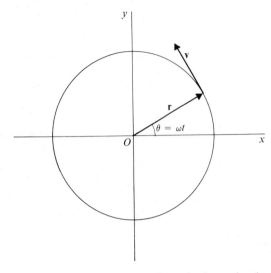

Fig. 6.15. During uniform circular motion the radius vector to the particle changes its orientation with respect to a fixed axis periodically with time.

The **frequency** of the motion is defined by the relation

$$v = \frac{1}{T}$$

and has the unit of rev · sec^{-1} for circular motion.

While the particle is making one complete revolution in its circular path, the velocity vector is changing its orientation in space uniformly with time, as seen in Fig. 6.16(a). In Fig. 6.16(b) the velocity vectors are brought together at one point in a "velocity space," where the coordinates of a point are given by the magnitude and direction of the velocity. It is clearly seen that, as the particle moves in its circular path at constant speed, the tip of the velocity vector in Fig. 6.16(b) moves uniformly about a circular path of radius v in "velocity space." The distance traveled per unit time gives the rate of change of the orientation of the velocity vector, which is the acceleration.

In time T, the tip of the velocity vector moves through a distance of $2\pi v$, and

$$a = \frac{2\pi v}{T} = \frac{2\pi}{T}\left(\frac{2\pi r}{T}\right) = \frac{4\pi^2 r}{T^2}$$

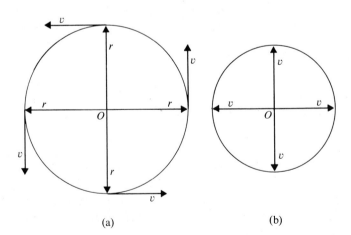

(a)   (b)

Fig. 6.16. (a) The velocity vector changes its orientation in space uniformly with time. (b) When the velocity vectors are brought together at one point in a "velocity space" it is easy to see that the tip of the velocity vector moves uniformly about a circular path of radius v.

or

$$a = \omega^2 r = \frac{v^2}{r} = \omega v,$$

where we have used

$$\omega = \frac{2\pi}{T}, \qquad v = \frac{2\pi r}{T}, \qquad \text{and} \quad v = \omega r.$$

Therefore, the centripetal acceleration is given by

$$a = \omega^2 r = \frac{v^2}{r} = \omega v = \frac{4\pi^2 r}{T^2}.$$

Example. An important step in the project to send men to other planets in the solar system is to provide a space station orbiting the earth as a stepping-off point. One form proposed for such a space platform is a large doughnut-shaped structure rotating in the plane of the doughnut about an axis through its center. What is the angular frequency of rotation, the centripetal acceleration, and the (linear) speed at a point 50 m from the center of rotation, if the period of rotation is 14.2 sec?

Solution. We have $r = 50$ m, $t = 14.2$ sec. Therefore,

$$v = \frac{2\pi r}{T} = \frac{2\pi(50)}{14.2} = 22.2 \text{ m} \cdot \text{sec}^{-1}.$$

The angular frequency is

$$\omega = \frac{2\pi}{T} = \frac{2\pi}{14.2} = 0.444 \text{ rad} \cdot \text{sec}^{-1}.$$

The centripetal acceleration is

$$a = \frac{v^2}{r} = \frac{(22.2)^2}{50} = 9.9 \text{ m} \cdot \text{sec}^{-2}$$

or

$$a = \omega v = (0.444)(22.2) = 9.9 \text{ m} \cdot \text{sec}^{-2}$$

or

$$a = \omega^2 r = (0.444)^2(50) = 9.9 \text{ m} \cdot \text{sec}^{-2}$$

or

$$a = \frac{4\pi^2 r}{T^2} = \frac{4\pi^2(50)}{(14.2)^2} = 9.9 \text{ m} \cdot \text{sec}^{-2}.$$

The velocity is 22.2 m · sec^{-1}, the angular frequency is 0.444 rad · sec^{-1}, and the centripetal acceleration is 9.9 m · sec^{-2}. This acceleration is approximately equal to the acceleration due to gravity on the earth's surface. Thus, this space station would have an artificial earth's gravity for its inhabitants (at $r = 50$ m).

Motion at variable speed

If the motion is not uniform, the particle in its circular orbit has non-constant speed, and the velocity varies both in magnitude and direction. The resultant acceleration vector must have a component $\mathbf{a}_t$ in the direction of the motion in order for the speed of the particle to change. This component is known as the **tangential component**, since it is tangent to the curve. (It is the transverse component of the general case outlined in the last section.) The component of the acceleration $\mathbf{a}_r$ perpendicular to the tangential component is the **radial component**; both components are illustrated in Fig. 6.17.

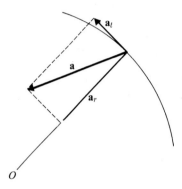

It is convenient to define an **angular acceleration** α such that α is the rate of charge of angular frequency ω with time; that is,

$$\alpha = \frac{d\omega}{dt}.$$

The angular frequency ω and the speed v of the particle are related by

$$v = \omega r.$$

Fig. 6.17. For nonuniform circular motion there is a tangential acceleration in the direction of motion as well as a radial acceleration toward the center of the circle.

Therefore,

$$\frac{d\omega}{dt} = \frac{1}{r}\frac{dv}{dt},$$

since r is constant. The quantity dv/dt is the rate of change of speed, or the tangential acceleration a_t, of the particle. Therefore,

$$\frac{d\omega}{dt} = \frac{a_t}{r}$$

and

$$\alpha = \frac{a_t}{r}$$

or

$$a_t = \alpha r.$$

The angular acceleration times the radius yields the tangential acceleration; the centripetal acceleration, which is the radial component of the acceleration vector, is not constant, since v is varying with time.

If α is a constant, then

$$\int_{\omega_0}^{\omega} d\omega = \int_{0}^{t} \alpha \, dt,$$

where it is assumed that the angular frequency changes from ω_0 to ω as the time changes from 0 to t. Therefore,

$$\omega = \omega_0 + \alpha t.$$

Similarly,

$$\int_0^\theta d\theta = \int_0^t \omega\, dt = \int_0^t (\omega_0 + \alpha t)\, dt,$$

where it is assumed that the angular displacement changes from 0 to θ as the time changes from 0 to t. Therefore,

$$\theta = \omega_0 t + \tfrac{1}{2}\alpha t^2.$$

Example. A carousel, or merry-go-round, is a popular midway ride at fairs and exhibitions. A carousel, starting from rest, attains a frequency of one revolution every 5 sec after 10 sec. What is the angular acceleration of the carousel, assuming that the angular frequency increases uniformly with time? What is the acceleration in the direction of motion of a person seated on the carousel at a point 5 m from the center of rotation? What is the speed and centripetal acceleration at this point at $t = 10$ sec?

Solution. The frequency after 10 sec is one revolution every 5 sec or $\frac{1}{5}$ revolutions $\cdot$ sec^{-1}. In one revolution an angle of 2π rad is swept out, so that $\omega = 2\pi(\frac{1}{5}) = 2\pi/5$ rad $\cdot$ sec^{-1}. Since

$$\omega = \omega_0 + \alpha t$$

with $\omega_0 = 0$,

$$\alpha = \frac{\omega}{t} = \left(\frac{2\pi}{5}\right)\left(\frac{1}{10}\right) = \frac{\pi}{25} = 0.13 \text{ rad} \cdot \text{sec}^{-2}.$$

The tangential acceleration at $r = 5$ m is

$$a_t = \alpha r = \left(\frac{\pi}{25}\right) 5 = \frac{\pi}{5} = 0.63 \text{ m} \cdot \text{sec}^{-2}.$$

The speed after 10 sec is

$$v = \frac{2\pi r}{T} = \frac{2\pi(5)}{5} = 2\pi = 6.28 \text{ m} \cdot \text{sec}^{-1},$$

and the centripetal acceleration is

$$a = \frac{v^2}{r} = \frac{4\pi^2}{5} = 7.9 \text{ m} \cdot \text{sec}^{-2}.$$

The centripetal acceleration is much greater than the tangential acceleration, as it is increasing proportional to v^2 while the tangential acceleration remains constant.

Example. A stagecoach is being attacked by Indians and accelerates uniformly from rest up to a speed of 8 m $\cdot$ sec^{-1} in 5 sec. An arrow strikes a spoke halfway between the axle and the rim of one of the wheels. The radius of the wheel is 1.0 m, and the arrow is initially directly above the axle. Calculate

(a) the angular acceleration of the wheel after one revolution.
(b) the angular frequency of the wheel after 4 sec.
(c) the velocity of the arrow after one revolution (magnitude and direction).
(d) the acceleration of the arrow after one revolution (magnitude and direction).

Solution. The motion of any point on the wheel relative to the ground is most readily determined by considering separately the motion of the point on the wheel relative to the axle and the motion of the axle relative to the ground. The former is simply circular motion; the latter is linear motion (see Fig. 6.18). The circular motion problem can be stated in terms of the angular quantities θ, ω, and α, or the corresponding orbital quantities s_0, v_0, and a_0, where

$$s_0 = r\theta, \qquad v_0 = r\omega, \qquad a_0 = r\alpha$$

if the point of interest is a distance r from the axle. The linear motion problem can be stated in terms of the linear quantities s, v, a. These also can be related to θ, ω, and α by imposing the condition that the wheel rolls along the ground. That is, every time the wheel rotates through 2π rad about the axle, the axle moves a distance $2\pi R$ along the ground, where R is the radius of the wheel. It follows readily that

$$s = R\theta, \qquad v = R\omega, \qquad a = R\alpha.$$

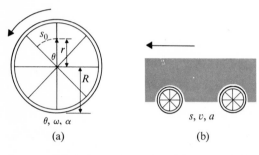

Fig. 6.18. (a) The motion of the arrow relative to the axle is circular motion. (b) The motion of the axle relative to the ground is linear motion.

(a) By the definition of linear acceleration,

$$a = \frac{\Delta v}{t} = \frac{8}{5} = 1.6 \text{ m} \cdot \text{sec}^{-2}.$$

Since $R = 1$ m,

$$\alpha = \frac{a}{R} = 1.6 \text{ rad} \cdot \text{sec}^{-2}.$$

(b) Since α is a constant and the stagecoach starts from rest,

$$\omega = \alpha t$$
$$= 1.6 \times 4$$
$$= 6.4 \text{ rad} \cdot \text{sec}^{-1}.$$

(c) The velocity of the arrow $\mathbf{v}_A$ is given by

$$\mathbf{v}_A = \mathbf{v} + \mathbf{v}_0,$$

where v and v_0 are the appropriate values after one revolution of the wheel. When $\theta = 2\pi$,

$$\omega^2 = 2\alpha\theta$$
$$= 2 \times 1.6 \times 2\pi$$

and

$$\omega = 4.5 \text{ rad} \cdot \text{sec}^{-1}.$$

Therefore,

$$v = 4.5 \text{ m} \cdot \text{sec}^{-1} \quad \text{and} \quad v_0 = \frac{4.5}{2} \text{ m} \cdot \text{sec}^{-1}.$$

After one revolution $\mathbf{v}$ and $\mathbf{v}_0$ are parallel, so that

$$v_A = \tfrac{3}{2} \times 4.5$$

$$= 6.7 \text{ m} \cdot \text{sec}^{-1}.$$

The direction of $\mathbf{v}_A$ is the direction of travel of the stagecoach.

(d) The acceleration of the arrow is given by

$$\mathbf{a}_A = \mathbf{a} + \mathbf{a}_0 + \mathbf{a}_c,$$

where $\mathbf{a}_c$ is the centripetal acceleration. After one revolution of the wheel, the vector $(\mathbf{a} + \mathbf{a}_0)$ has the magnitude

$$\tfrac{3}{2}\alpha = \tfrac{3}{2} \times 1.6 = 2.4 \text{ m} \cdot \text{sec}^{-2}$$

and is in the direction of motion of the stagecoach. The vector $\mathbf{a}_c$ has magnitude

$$\omega^2 r = \frac{2 \times 1.6 \times 2\pi}{2} = 3.2\pi \text{ m} \cdot \text{sec}^{-2}$$

and is directed toward the axle. The acceleration vector is illustrated in Fig. 6.19. The resultant acceleration $\mathbf{a}_A$ has magnitude

$$[(2.4)^2 + (3.2\pi)^2]^{1/2} = 10.3 \text{ m} \cdot \text{sec}^{-2}$$

and makes an angle χ with the horizontal, where

$$\chi = \tan^{-1}\left(\frac{3.2\pi}{2.4}\right) = \tan^{-1}\left(\frac{4\pi}{3}\right) = 76.6°.$$

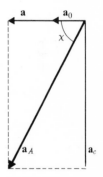

Fig. 6.19. The acceleration vector experienced by the arrow after one complete revolution of the wheel is the sum of the three component accelerations.

Vector representation of angular quantities

We have spoken two or three times about the axis of rotation. When a wheel spins in the manner shown in Fig. 6.20, any particular minute

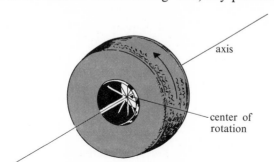

Fig. 6.20. A wheel rotates about an axis through the center of rotation.

particle of matter in the wheel moves in a circular path of a given radius, except for those particles lying along a line through the center of rotation and perpendicular to the plane of rotation. This line is known as the **axis of rotation**.

It is often convenient to consider the angular frequency and angular acceleration as vector quantities and to write them as ω and α, respectively. Their directions are taken to be along the axis of rotation, since this is a unique line in space defined by the plane of rotation; ω and α determine the rotational motion in this plane. A difficulty arises now because the axis of rotation defines only a line in space, and we are free to choose either sense of direction along this line as indicated in Fig. 6.21, where we show only one particle for simplicity. Physicists and mathematicians have settled on the convention shown in the (a) part of Fig. 6.21. The direction

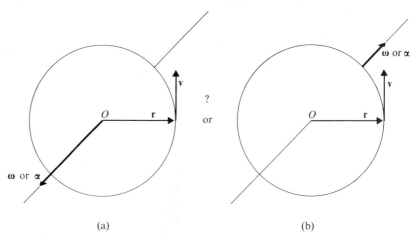

(a)   (b)

Fig. 6.21. The angular frequency and angular acceleration are taken to be vectors lying along the axis of rotation. The direction shown in (a) is chosen by convention.

of ω or α is then given by the **right-hand rule**, which operates as follows: point your right hand in the direction of **r** with the palm facing in the same direction as **v**; the thumb then points in the direction of ω or α. The rule is sometimes stated as: curl the fingers of your right hand in the direction of rotation; the thumb then points in the direction of ω or α. The right-hand rule is illustrated in Fig. 6.22.

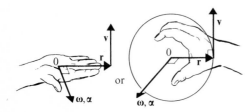

Fig. 6.22. The direction of ω and α is given by the right-hand rule.

In summation ω and α are vectors lying along the axis of rotation in a direction given by the right-hand rule. They have magnitudes given by $\omega = v/r$ and $\alpha = a_t/r$, respectively. We note also that ω is a vector perpendicular to both **r** and **v** and that $v = \omega r$. Referring to the definition of the vector product given in Section 2.5, we see that we can write

$$\mathbf{v} = \boldsymbol{\omega} \times \mathbf{r}.$$

Since $\mathbf{a}_t$ is in the direction of $\mathbf{v}$, $\boldsymbol{\alpha}$ is in the direction of $\boldsymbol{\omega}$, and $a_t = \alpha r$, we may also write

$$\mathbf{a}_t = \boldsymbol{\alpha} \times \mathbf{r}.$$

These are the vector equations for angular motion.

PROBLEMS

1. A particle starts to move in a straight line with zero initial velocity from $s = 0$ at $t = 0$, and its displacement varies with time according to the equation $s = 0.12t^3$, where s is in meters and t in seconds. Calculate the displacement, velocity, and acceleration of the particle at $t = 3$ sec.

2. A particle initially at rest is subjected to an acceleration that varies with time according to the equation $a = 500 - 800t$, where a is in $m \cdot sec^{-2}$ and t is in seconds. Determine the speed and displacement after 0.1 sec.

3. A particle travels along a straight-line path. Its displacement varies with time according to the equation

$$s = At[1 - \exp(-Bt)],$$

where A and B are positive constants.
(a) Derive expressions for the velocity and acceleration as functions of the time.
(b) What values do the velocity and acceleration approach for large values of t?

4. A particle with an initial displacement of 0.05 m is subjected to an acceleration $a = -0.20\pi^2 \cos 2\pi t$. What is the velocity and displacement at $t = 0.5$ sec, assuming the initial velocity is zero?

*5. The displacement of a particle varies according to the equation

$$s^3 = \tfrac{9}{2}GMt^2,$$

where G and M are constants. Show that the acceleration is given by

$$a = -\frac{GM}{s^2}.$$

*6. A particle travels along a path such that the x and y components of its displacement are given by

$$x = A \cos Ct, \qquad y = B \sin Ct,$$

where A, B, C are positive constants.
(a) Sketch the curve described by the particle.
(b) Show that the acceleration is directed towards the origin.

7. A stone is thrown horizontally at a speed of 15 m · sec^{-1} from a cliff 50 m above a quiet lake. What is the velocity of the stone at the instant it strikes the surface of the lake? Assume the acceleration due to gravity to be 9.8 m · sec^{-2}.

*8. Starting from

$$x = r \cos \theta, \qquad y = r \sin \theta,$$

calculate

$$\frac{d^2x}{dt^2}, \qquad \frac{d^2y}{dt^2}.$$

Then, by resolving the acceleration along the radius vector and perpendicular to the radius vector, show that the radial component of the acceleration is

$$\frac{d^2r}{dt^2} - r\left(\frac{d\theta}{dt}\right)^2$$

and that the transverse component of the acceleration is

$$\frac{1}{r}\frac{d}{dt}\left[r^2\frac{d\theta}{dt}\right].$$

9. A particle travels in a circular path starting from rest, as the result of an acceleration $3t$ rad · sec^{-2}, where t is the time. Determine the angular acceleration 4 sec after the motion starts.

10. A particle's angular displacement varies with time, as $\theta = 0.5t^4$ rad. What is the angular displacement, angular frequency, and angular acceleration after 6 sec?

11. The performance of airplane pilots under various magnitudes of acceleration is tested by placing the person under test in a small room situated at the end of a long boom that can rotate about one end. If the person is sitting 5 m from the center of rotation, at what speed must the device rotate in order that an acceleration of $6g$ towards the center is felt, where g is the acceleration due to gravity at the earth's surface.

12. It is often convenient to think of the hydrogen atom to consist of a single electron moving in a circular orbit about a stationary proton. If the period of the motion is 1.5×10^{-16} sec and the centripetal acceleration of the electron is 9.2×10^{22} m · sec^{-2}, determine the radius of the orbit and the speed of the electron in the orbit.

13. If a discus thrower releases his discus at a point 0.85 m from the thrower's center of rotation and at a speed of 42.5 m · sec^{-1}, what is the centripetal acceleration and angular frequency of the discus just prior to the moment of release?

14. If the discus thrower of Problem 13 reaches his maximum angular frequency 1.5 sec after beginning his rotational motion, determine the angular acceleration of the discus during this period, assuming uniform increase in speed of the discus. What other assumption must you make in order to solve this problem?

15. A sports car accelerates around a circular track, starting from rest at $t = 0$ with initial angular displacement $\theta = 0$. Its displacement varies with time according to the equation

$$\theta = C_1 t^2 - C_2 t^3,$$

where C_1 and C_2 are positive constants. The equation is valid from $t = 0$ until the car is again at rest. Draw graphs to scale to illustrate the angular displacement, angular frequency, and angular acceleration as functions of the time.

*16. A wheel of radius R moves along a flat horizontal plane at constant speed. Find an expression for the path through space of a point on the circumference of the wheel. (Hint: set up a two-dimensional rectangular coordinate system with the x axis in the direction of motion of the wheel and the y axis perpendicular to the plane. Take the origin to be at a position where the point on the circumference of the wheel is in contact with the plane.)

*17. A cart travels along a straight path with constant speed v. The wheels of the cart are of diameter D. One wheel has a red spot on its circumference. Write expressions to describe the acceleration, velocity, and displacement of the spot as functions of the time.

18. A car moves at a constant speed of $15 \text{ m} \cdot \text{sec}^{-1}$ around a curve having a radius of 100 m. What is the magnitude of the centripetal acceleration of the car?

19. A charged particle will travel in a circular path when acted on by a magnetic force acting perpendicular to the electron's direction of motion (see Section 11.6). An electron is forced to undergo such circular motion in a particle accelerator known as the betatron[1]. If the electron travels in a circular path of radius 22 cm at an average speed of $2.99 \times 10^8 \text{ m} \cdot \text{sec}^{-1}$ for 10^{-3} sec,
(a) how many revolutions will each electron make?
(b) what linear distance in miles does each electron travel?

20. A motorcycle is speeding around a circular race track of diameter 500 m. At one particular moment in its travel, the motorcycle is traveling at a speed of $35 \text{ m} \cdot \text{sec}^{-1}$ but is increasing its speed at the rate of $2 \text{ m} \cdot \text{sec}^{-2}$. What is the magnitude and direction of the acceleration of the motorcycle?

21. An airplane is parked with the engine off, and a fly lands on the tip of the propeller. The propeller is then accelerated with uniform angular acceleration to a speed of $200 \text{ rev} \cdot \text{min}^{-1}$ in 10 sec, at which time the fly comes off. If the propeller has a radius of 1 m, determine
(a) the angular acceleration of the propellor.
(b) the speed of the fly just before it comes off.
(c) the centripetal acceleration of the fly just before it comes off.
(d) the tangential acceleration of the fly just before it comes off.

[1] R. R. Wilson and R. Littauer, *Accelerators: Machines of Nuclear Physics* (Garden City, N.Y.: Doubleday & Company, Inc., 1960). See also R. Gouiran, *Particles and Accelerators* (New York: McGraw-Hill Book Company, 1967).

22. A car having tires of radius 30 cm is traveling at 70 km·hr^{-1}. Calculate the velocity and acceleration of a point on the circumference of the tire
(a) at the top of the tire.
(b) at the bottom of the tire.
(c) midway between top and bottom of the tire.

*23. A railroad car undergoes uniform acceleration from rest up to a speed of 20 m·sec^{-1} in 10 sec. The wheels are flanged, with the radius of the wheel being 50 cm and the flange extending 5.0 cm beyond the wheel. There is a paint spot on the outer edge of the flange, and the spot is initially at the bottom, as shown in Fig. 6.23.

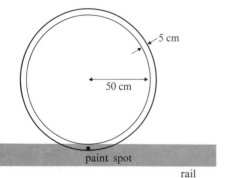

5 cm

50 cm

paint spot

rail

Fig. 6.23

(a) Calculate the angular acceleration after one revolution, the angular frequency of the wheel after 4 sec, the velocity in the forward direction of the paint spot after one revolution, and the angle that the acceleration vector of the paint spot makes with the horizontal after half a revolution.
(b) Deduce the angular displacement of the paint spot when its horizontal velocity becomes zero for the first time after the train begins to move.

24. A missile is launched from the earth's surface from a giant cannon and, after reaching its maximum height, falls back to the earth **exactly reversing its path in space**. If the missile were launched from a point of latitude 25° N, how far from its point of launch would it strike the earth's surface if it reaches a height of 50 km? Assume the acceleration due to gravity to be constant at 9.8 m·sec^{-2}, and give your answer both in kilometers and in degrees. What would the result have been if the missile had been launched vertically?

An introduction to forces   7

7.1 INTRODUCTION

Dynamics deals with the causes of motion or of the change of motion. You are familiar with the effects of forces in nature and undoubtedly have a very good intuitive idea of what constitutes a **force**. The force of the wind is capable of bending treetops, the force of falling water is able to generate electric power to light city streets and keep the machines of industry humming, the muscular force exerted by a baseball player on a baseball via his bat is capable of projecting the baseball out of the stadium, and the force exerted on a satellite by the earth is capable of keeping the satellite in orbit. Surprising as it may seem, there are only four fundamental forces known to man. All the forces mentioned above are simply manifestations of one of these fundamental forces.

7.2 MASS AND NEWTON'S SECOND LAW

We wish to study experimentally the effect of forces on a system constrained to move in one dimension. First of all we look for a system that can execute almost force-free motion. Such a system will move in nearly perfect response to experimental forces imposed upon it. The linear air track shown schematically in Fig. 7.1 provides just such an

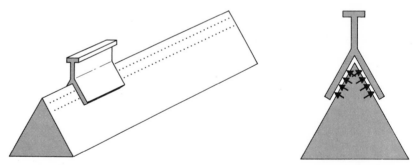

Fig. 7.1. A glider on a linear air track "floats" on a relatively frictionless pad of escaping air. The air cushion is actually only a few thousandths of an inch thick but is exaggerated for clarity.

experimental tool. A linear air track is a long straight tube with a supply of air escaping from rows of tiny perforations. "Gliders" of an appropriate mating shape "float" on the relatively frictionless pads of escaping air.

As a first experiment we observe the motion of a glider with no external forces applied to it. That is, we simply give the glider a shove to start it and then let it move freely. Such a glider is said to be **isolated**, for it does not interact with any other objects. The fact that the glider is constrained to move on a horizontal, (almost) frictionless track makes the gravitational effect on the motion of the glider negligible. Therefore, in effect, we have removed the interaction of the glider with the earth, and this interaction need no longer be considered.

Using the techniques of stroboscopic photography[1] we can photograph the position of the glider at successive equal intervals of time and obtain a record similar to that shown in Fig. 7.2. We take the initial position of the glider to be x_0 and the position at successive intervals of time to be x_1, x_2, and x_3. The distances traveled by the glider between flashes are given by

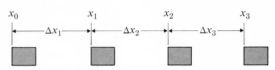

Fig. 7.2. The glider moves equal distances in equal intervals of time when no external force is applied to it.

$$\Delta x_1 = x_1 - x_0, \qquad \Delta x_2 = x_2 - x_1, \qquad \Delta x_3 = x_3 - x_2.$$

A measurement of Δx_1, Δx_2, and Δx_3 shows them to be equal within the limits of experimental accuracy. By performing experiments using gliders of different size and the same glider given different initial speeds, we are able to verify **Newton's first law** (sometimes known as **Galileo's principle**), which says that "in the absence of a net force, an object set in motion will continue in motion in a straight line with constant speed."

The property of the object responsible for its tendency to resist changes in its state of uniform motion is known as its **inertial mass**. We shall now turn to more experiments in order to illustrate the concept of inertial mass.

If we repeat the experiments described above with a fan blowing against the gliders so as to oppose their motion, then a light glider would slow up appreciably in its motion, whereas a more massive glider would be less affected.

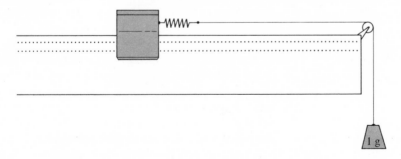

Fig. 7.3. A constant force is applied to a glider by means of a spring kept at constant extension.

[1] The Physical Science Study Committee, *College Physics*, 2nd Edition (Boston: D. C. Heath & Company, 1968), Chapter 2.

In the next experiment we apply a constant force to a glider and calculate its acceleration. To measure the force we attach a spring to the glider; a constant extension implies a constant force. The source of the force is unimportant; we shall use the force of gravity acting on a standard l-g mass (see Fig. 7.3). Note that the glider is no longer an isolated object. Using the techniques of stroboscopic photography, we can photograph the position of the glider at successive equal intervals of time and obtain a record similar to that shown in Fig. 7.4. The extensions of the spring are equal in each case to within the limits of experimental error, indicating a constant applied force.

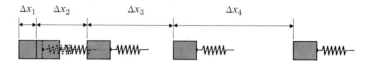

Fig. 7.4. The distance moved in equal time intervals by a glider acted upon by a constant force increases with time.

The distances traveled by the glider between successive light flashes from the stroboscope are measured and plotted as a function of the time in Fig. 7.5, and a linear relation is indicated. The average speeds during the intervals between stroboscopic flashes will be

$$\bar{v}_1 = \frac{\Delta x_1}{\Delta t}, \qquad \bar{v}_2 = \frac{\Delta x_2}{\Delta t}, \qquad \bar{v}_3 = \frac{\Delta x_3}{\Delta t}, \qquad \bar{v}_4 = \frac{\Delta x_4}{\Delta t},$$

respectively, where Δt is the (constant) time between light flashes. Since $\bar{v}$ is proportional to Δx, the result of the experiment indicates a linear increase in the speed of the glider with time and therefore a constant acceleration. The experiment may be repeated with various multiples of the accelerating force by attaching two or more springs to the glider, each spring being identical to the original spring and kept at the same constant extension as the original spring. The results of these measurements show that the resulting acceleration of the glider is proportional to the number of springs employed and therefore to the applied force.

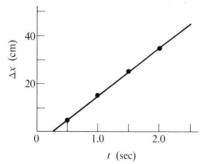

Fig. 7.5. The successive distances Δx increase linearly with time.

Next, we may repeat the experiment using two identical gliders coupled together. The acceleration of the coupled gliders is found to be one-half the acceleration of a single glider acted on by the same force. Similarly, the acceleration drops to one-third for three gliders coupled together, and so on. Since the gliders are identical, we assign the same inertial mass to each and conclude that the acceleration is inversely proportional to the inertial mass m_i of the object being accelerated.

In mathematical notation

$$a \propto \frac{F}{m_i}.$$

The units of force, mass, and acceleration are chosen so that the constant of proportionality can be taken as unity and

$$F = m_i a.$$

Note that it is the inertial mass that determines the acceleration in the dynamic measurement that we have just discussed. The mass of an object that is most familiar to us is its **gravitational mass**, as measured by an equal-arm balance, as shown in Fig. 7.6. If an object with gravitational mass m_g is placed on one pan of an equal-arm balance, m_g standard-unit masses must be placed on the other pan to balance the object.

The inertial mass of an object is a measure of its reluctance to change its state of motion; the gravitational mass is a measure of the force exerted on the object by the earth. Actually, in measuring gravitational mass, we are really comparing the gravitational pulls on objects when they are in the same place in relation to all other pieces of matter in the universe. The inertial mass of an object is a dynamic property, whereas the gravitational mass

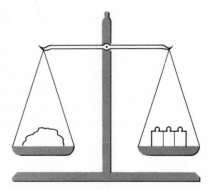

Fig. 7.6. This apparatus measures the gravitational mass of an object.

is determined when the object is at rest. Since the inertial and gravitational masses are measured in quite different ways that have no connection with each other, we might think that there is no connection between inertial mass and gravitational mass. In a brilliant series of experiments performed about 1900, Baron Roland von Eötvös showed that the gravitational and inertial masses of an object are proportional to each other to an accuracy of one part in 10^9. R. H. Dicke[2] has recently repeated the experiment and established the proportionality to one part in 10^{11}. Therefore, if we choose the same unit of mass for both the inertial and gravitational mass, we can make them numerically equal. This allows us to remove the subscript on the mass in the equation relating force and acceleration and write

$$F = ma,$$

where m is the mass as measured using an equal-arm balance. This equation, known as **Newton's second law of motion**, may then be taken as a

[2] R. H. Dicke, "The Eötvös Experiment," *Scientific American*, December, 1961.

definition of force. If the mass is measured in kilograms and the acceleration in meters per second per second, Newton's second law gives the
magnitude of the force in newtons (N).

If we define the **momentum** p of a particle as

$$p = mv,$$

we may write Newton's second law as

$$F = ma$$

$$= m\left(\frac{\Delta v}{\Delta t}\right)$$

$$= \frac{\Delta(mv)}{\Delta t}$$

$$= \frac{\Delta p}{\Delta t}.$$

The average time rate of change of a particle's momentum is equal to the
average applied force. If no force is applied, the momentum of the particle is a **constant of the motion** (that is, it remains constant when the
particle is in motion). When $\Delta t \to 0$, we replace average values by instantaneous values and have

$$F = m\frac{dv}{dt} = \frac{dp}{dt} = ma,$$

where F, v, a, and p are now instantaneous values. Note that the equality
of the two forms of Newton's second law implicitly assumes that the mass
of a particle is a constant of the motion. If the mass is not a constant of the
motion, the two forms are not identical, and we must turn to experiment to
choose between them. We might imagine an arrangement in which a
glider is fitted with a water reservoir that can be drained as the glider is
being accelerated. Once the rate at which the total mass (glider + water)
decreases is deduced, this arrangement can be used to choose between the
two forms of Newton's law. In this way we could verify that

$$F = \frac{dp}{dt}$$

is the correct form of Newton's second law.

The change in momentum (and the acceleration) are in the direction of
the applied force, so that Newton's second law may be written in vector
form as

$$\mathbf{F} = \frac{d\mathbf{p}}{dt}.$$

We have been studying the motion of one or more gliders coupled together under the influence of forces of various magnitudes as exerted by a spring or springs. The collection of gliders coupled together is an example of a **mechanical system**. We have been studying the motion of various systems of gliders under the influence of various **external forces** (that is, forces originating outside the system). Any collection of objects considered as a unit forms a system. Newton's second law tells us what acceleration to expect for a system subject to a given external force. Of course, objects within a system may exert forces on each other, called **internal forces**. Internal forces may give rise to internal motions (see Chapter 14, for example). However, for the moment we shall be concerned with motion due to external forces except when we are dealing with work (see Section 7.8).

Problems in Newtonian mechanics, in principle, consist of writing down Newton's second law for the system under consideration and solving the resultant **equation of motion**. Since, in many cases, the equation so obtained is difficult to handle mathematically, other techniques are often used to avoid mathematical difficulties, but Newton's second law remains the foundation of Newtonian mechanics.

Example. A 1000-kg car moves northward along a horizontal road at a constant speed of $12 \, \text{m} \cdot \text{sec}^{-1}$.

(a) What resultant force acts on the car?

(b) If the brakes are applied and the car is brought to rest in 6 sec with uniform deceleration, what constant resultant force acts on the car?

Solution.

(a) The force is zero, since the speed is constant.

(b) Let the initial speed at time $t = 0$ be $v_0 = 12 \, \text{m} \cdot \text{sec}^{-1}$. When $t = 6 \, \text{sec}$, $v = 0$. Since

$$v = v_0 + at,$$

where a is the acceleration, we have

$$0 = 12 + 6a$$

$$a = \frac{-12}{6}$$

$$= -2 \, \text{m} \cdot \text{sec}^{-2}.$$

The force F is given by

$$F = ma,$$

since m is constant. Therefore,

$$F = 1000(2)$$

$$= 2000 \, \text{N}.$$

The force causing the constant deceleration has a magnitude of 2000 N.

Example. In the "airplane ride" at a carnival, cars are suspended from cables 5 m long attached to arms at points 7 m from the vertical axis of rotation. At what angular frequency of rotation will the cars swing out so that the cables make an angle of 45° with the horizontal?

Solution. The airplane ride is pictured schematically in Fig. 7.7, along with the forces acting on one of the cars (assumed to be a particle of mass m). The forces acting on the car are the force of gravity $\mathbf{F}_g$, directed vertically downward, and the

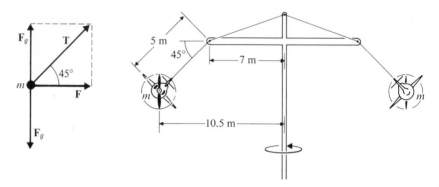

Fig. 7.7. The car in an airplane ride is acted upon by the force of gravity $\mathbf{F}_g$ and the tension in the cable $\mathbf{T}$ to give a resultant force $\mathbf{F}$.

tension in the cable, acting in the direction of the cable. When the ride is rotating uniformly with the cables at 45°, the cars have no vertical motion. Therefore, the vertical component of $\mathbf{T}$ must be equal and opposite to $\mathbf{F}_g$, so that the net vertical force on the cars is zero. The only acceleration experienced by the cars is the centripetal acceleration directed toward the axis of rotation. The centripetal force $\mathbf{F}$ that produces this acceleration is the horizontal component of the tension $\mathbf{T}$. The force $\mathbf{F}_g$ has magnitude

$$F_g = mg,$$

where $g = 9.80 \text{ m} \cdot \text{sec}^{-2}$ is the acceleration due to gravity. The magnitude of the vertical component of T is

$$T \cos 45°.$$

Therefore,

$$0.707T = 9.80m.$$

The magnitude of the centripetal force is also

$$F = T \cos 45° = 0.707T$$

$$= 9.80m.$$

From Newton's second law, the centripetal acceleration a is

$$a = \frac{F}{m} = \frac{9.80m}{m} = 9.80 \text{ m} \cdot \text{sec}^{-2}.$$

The centripetal acceleration may be written as

$$a = \omega^2 r,$$

where ω is the angular frequency of rotation and r is the radius of the circular motion. Here

$$r = 7 + 5 \cos 45° = 10.5m,$$

so that

$$a = \omega^2(10.5) = 9.80$$

and

$$\omega^2 = 0.933$$

$$\omega = 0.966 \text{ rad} \cdot \sec^{-1}.$$

The cables will make an angle of 45° to the horizontal when the angular frequency of rotation is 0.966 rad $\cdot$ sec^{-1}.

7.3 PROJECTILE MOTION

The problem is usually formulated in terms of an object, usually an artillery shell or missile of some sort, traveling over the surface of the earth and being subject only to the constant gravitational attraction of the earth. As we shall see in Section 7.6, the force of attraction between the projectile and the earth is not really constant, but is a function of the distance between them (or, more precisely, between their centers of mass as defined in Section 13.3). However, as the radius of the earth is rather large (6.4×10^6 m) compared to the change in separation of the earth and the projectile during the motion of the projectile (approximately 10^3 to 10^4 meters), it is a good approximation to consider the force and the acceleration to be constant. The problem could just as well be approached from the point of view of a charged particle experiencing a constant electric force. We shall speak of a projectile moving under a constant acceleration g in the development below.

We should note that while an artillery shell is in the gun barrel during firing or while a missile is burning its fuel supply, it is also being accelerated in the direction of its motion. In both cases, neither the force causing the initial motion nor the resulting acceleration is constant. A general treatment of motion under nonconstant forces is given in Section 7.4, and a brief treatment of the motion of rockets (or missiles) is given in Section 7.5. The initial acceleration in the direction of motion is present for only a very short time at the beginning of the flight of an artillery shell; this is often true also for missiles. In this section we shall neglect initial accelerations and consider only the motion of a projectile launched with some initial velocity and subject only to a constant downward acceleration.

We shall start by asking how far a shell fired at an angle of 45° to the horizontal with an initial speed of 500 m · sec⁻¹ will go before striking the ground. The problem is illustrated in Fig. 7.8. This is a particular example of the more general problem of the motion of a particle having initial speed v_0 and an initial direction of motion along a line making an angle of $(90 - \theta)$ degrees with the direction of a constant acceleration. This situation is illustrated in Fig. 7.9. We would like to find a mathematical expression involving v_0 and θ that will describe the path of the particle.

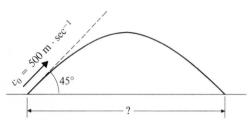

Fig. 7.8. How far will a shell fired at 45° to the horizontal with a speed of 500 m · sec⁻¹ travel before striking the ground?

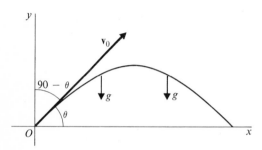

Fig. 7.9. The motion of a particle experiencing a constant acceleration in the $-y$ direction.

We shall put the origin of a Cartesian coordinate system at the initial position of the particle. The constant acceleration due to gravity g acts in the $-y$ direction, and there is no acceleration along the x axis. The initial speed of the particle in the y direction is $v_0 \cos (90 - \theta) = v_0 \sin \theta$ and in the x direction is $v_0 \cos \theta$. This latter speed remains constant, as there is no acceleration along the x axis. Therefore, the x coordinate at any time t will be given by

$$x = (v_0 \cos \theta)t,$$

and the y coordinate at any time t will be given by

$$y = (v_0 \sin \theta)t - \tfrac{1}{2}gt^2,$$

where we have used the equations given in Section 3.6.

The path of the particle in space (that is, y as a function of x) is found by eliminating the time t from the equations for x and y.

From

$$x = (v_0 \cos \theta)t$$

we have

$$t = \frac{x}{v_0 \cos \theta}.$$

Therefore,

$$y = (v_0 \sin \theta)\left(\frac{x}{v_0 \cos \theta}\right) - \frac{1}{2}g\left(\frac{x}{v_0 \cos \theta}\right)^2$$

$$= (\tan \theta)x - \left(\frac{g}{2v_0^2 \cos^2 \theta}\right)x^2.$$

This is the equation of a parabola

$$y = ax + bx^2,$$

where

$$a = \tan \theta$$

$$b = -\frac{g}{2v_0^2 \cos^2 \theta}.$$

The negative sign for the coefficient b indicates that the parabola opens out in the $-y$ direction.

The parabola cuts the x axis at two points. If we substitute the value $y = 0$ into the equation for the parabola, we have

$$(\tan \theta)x = \left(\frac{g}{2v_0^2 \cos^2 \theta}\right)x^2.$$

One solution is $x = 0$, which merely confirms that the path of the particle passes through the origin; the second solution is

$$x = \frac{2v_0^2}{g} \tan \theta \cos^2 \theta$$

$$= \frac{2v_0^2}{g} \sin \theta \cos \theta$$

$$= \frac{v_0^2}{g} \sin 2\theta.$$

If the particle were an artillery shell fired at an angle θ over a level region of countryside, then

$$x = \frac{v_0^2}{g} \sin 2\theta$$

is the horizontal distance that the shell would travel before striking the ground. For this reason the quantity $(v_0^2 \sin 2\theta)/g$ is called the **range**.

Example. What is the range of the shell in the question posed near the beginning of this section?

Solution. The shell was fired at an angle of 45° to the horizontal with a speed of 500 m · sec^{-1}. The range is

$$x = \frac{(500)^2}{9.8} \sin 90°$$

$$= \frac{25 \times 10^4}{9.8}$$

$$= 2.5(5) \times 10^4 \text{ m}.$$

The shell travels $2.5(5) \times 10^4$ m or $25(.5)$ km.

The maximum value of y, or the **maximum height** y_{max}, of the particle occurs when the speed in the y direction becomes equal to zero. Since the speed in the y direction is given by

$$v^2 = (v_0 \sin \theta)^2 - 2gy,$$

we have, for $v = 0$,

$$2gy_{max} = (v_0 \sin \theta)^2$$

$$y_{max} = \frac{v_0^2 \sin^2 \theta}{2g}.$$

As an alternate derivation for y_{max}, we notice in Fig. 7.10 that when y has its maximum value, the slope of the tangent to the parabolic curve is zero. The tangent to the curve $y = f(x)$ is just dy/dx, so that

$$\frac{dy}{dx} = 0$$

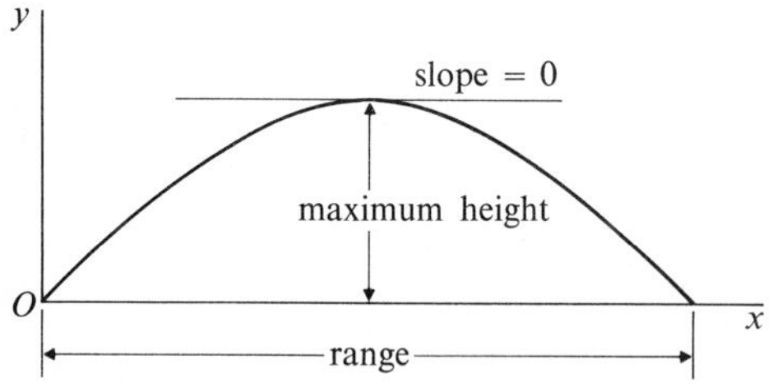

Fig. 7.10. The range and maximum height of a projectile.

for maximum y.

Therefore,

$$\frac{dy}{dx} = \frac{d}{dx}\left[(\tan \theta)x - \frac{gx^2}{2v_0^2 \cos^2 \theta} \right]$$

$$= \tan \theta - \frac{gx}{v_0^2 \cos^2 \theta}$$

$$= 0,$$

or

$$x = \frac{v_0^2 \cos^2 \theta \tan \theta}{g}$$

$$= \frac{v_0^2 \sin \theta \cos \theta}{g}$$

$$= \frac{v_0^2 \sin 2\theta}{2g}.$$

Substituting this value of x into the equation for the path of the particle, we obtain

$$y_{max} = (\tan \theta)\left(\frac{v_0^2 \sin 2\theta}{2g} \right) - \left(\frac{g}{2v_0^2 \cos^2 \theta} \right)\left(\frac{v_0^2 \sin 2\theta}{2g} \right)^2$$

$$= \frac{v_0^2 \sin^2 \theta}{2g}.$$

Example. A golfer hits a ball with a five iron so that the ball makes an angle of 25° with the horizontal as it leaves the ground. On level terrain the ball travels 150 m in the air. On a hole having a tee-to-green distance of 150 m, there is a tree

15 m high in the middle of the fairway 50 m from the tree. The golfer decides to use his five iron. Will the ball hit the tree or pass on safely to the green?

Solution. The range, which is 150 m, is given by

$$x = \frac{v_0^2 \sin 2\theta}{g}.$$

Therefore,

$$\frac{v_0^2}{g} = \frac{x}{\sin 2\theta} = \frac{150}{\sin 50°} = \frac{150}{0.766} = 196 \text{ m}.$$

The equation for the trajectory of the golf ball is

$$y = (\tan \theta)x - \frac{gx^2}{2v_0^2 \cos^2 \theta}$$

$$= (\tan 25°)x - \frac{x^2}{2(196)(\cos^2 25°)}$$

$$= (0.466)x - \frac{x^2}{(392)(0.906)^2}$$

$$= 0.466x - (3.11 \times 10^{-3})x^2.$$

We are interested in the value of y when $x = 50$ m. Substituting into the equation for the trajectory, we have

$$y = 0.466(50) - (3.11 \times 10^{-3})(2500)$$

$$= 23.3 - 7.8$$

$$= 15.5 \text{ m}.$$

Since the tree is only 15 m high, the ball will pass on safely to the green.

This discussion of particle motion has been an idealized one in that only one force is considered to act on the particle; for a projectile this is the force due to gravity. In reality, since a projectile passes through air, which tends to resist the motion of any object passing through it, there must be a component of force acting on the projectile due to the **air resistance**. This force acts in a direction opposite to the direction of motion of the projectile. The actual path will be similar to the dotted line in Fig. 7.11. The resistive force is approximately proportional to the speed of the particle, so that the departure from the ideal parabolic path will increase with the speed of the projectile.

A mathematical treatment of motion in a resisting medium is much more complex than we are able to deal with at this time. In addition, we should note that the motion of the air itself (that is, wind) will also affect the path of the projectile.

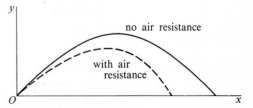

Fig. 7.11. The projectile's path is actually not parabolic due to air resistance.

This is almost impossible to treat mathematically except in very special cases, but it is usually such a small effect that it may be ignored.

7.4 IMPULSE

Let us consider two particles that interact with each other during some time interval Δt. We shall focus our attention upon one of the particles. During the time interval Δt a force has been applied to this particle. Newton's second law for the motion of the particle may be written in the form (omitting subscripts)

$$\bar{F} \, \Delta t = m \, \Delta v,$$

where $\bar{F}$ is the magnitude of the average applied force. The quantity $\bar{F} \, \Delta t$ is called the **impulse of force** during the time Δt; $m \, \Delta v$ is the change in momentum. If we consider smaller and smaller time intervals, $\Delta t \to dt$, $\Delta v \to dv$, and $\bar{F} \to F$, the instantaneous value of the force. We now write

$$F \, dt = m \, dv.$$

We are now prepared to consider the case of applying a time-varying force to a particle beginning at time $t = 0$ and ending at, say, time $t = t_0$. We may integrate the impulse equation to obtain

$$\int_0^t F \, dt = \int_{v_1}^{v_2} m \, dv,$$

where v_1 is the speed of the particle in the direction of the applied force at time $t = 0$, and v_2 is the speed in the direction of the applied force at time $t = t$.

Case 1. Suppose F is a constant, independent of time. Then

$$\int_0^t F \, dt = F \int_0^t dt = Ft = m(v_2 - v_1).$$

The result

$$Ft = m(v_2 - v_1)$$

is just another way of writing the equation

$$\bar{F} \, \Delta t = m \, \Delta v.$$

Case 2. F is a function of time. Here we have

$$\int_0^t F \, dt = m(v_2 - v_1),$$

where the integral may be evaluated if the explicit dependence of F upon time is known.

Example. A particle is subjected for 3 sec to a force that varies as shown in Fig. 7.12. What is the change in momentum of the particle?

Solution. We note that the force is 0 at $t = 0$ and that $F = 4N$ at $t = 3$ sec. Therefore, we may write

$$F = \tfrac{4}{3}t$$

to describe the dependence of the force upon time. Therefore,

$$\int_0^t F \, dt = \int_0^3 \tfrac{4}{3}t \, dt$$

$$= \frac{4}{3}\left(\frac{t^2}{2}\right)\Big|_0^3$$

$$= 6 \, \text{N} \cdot \text{sec.}$$

The change in momentum is 6 N · sec or 6 kg · m · sec^{-1}.

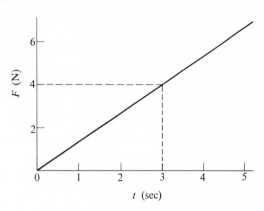

Fig. 7.12. The variation of the force with time.

Average values

An alternative approach when F is not constant is to replace the integral of F over the requisite time interval by an average force $\bar{F}$ acting during the time interval. Let us consider the case in which a constant force F_1 acts for time Δt_1, a constant force F_2 acts for time Δt_2, etc. The total impulse is

$$F_1 \, \Delta t_1 + F_2 \, \Delta t_2 + \cdots,$$

which spans a time period of

$$\Delta t = \Delta t_1 + \Delta t_2 + \cdots.$$

The average impulse $\bar{F} \, \Delta t$, is **defined** by the equation

$$\bar{F} \, \Delta t = F_1 \, \Delta t_1 + F_2 \, \Delta t_2 + \cdots,$$

where $\bar{F}$ is the average force acting during the total time interval Δt. If we let $\Delta t_1, \Delta t_2, \Delta t_3$, etc. $\to 0$, then

$$F_1 \, \Delta t_1 + F_2 \, \Delta t_2 + \cdots \to \int F \, dt$$

and

$$\Delta t_1 + \Delta t_2 + \cdots \to \int dt,$$

so that

$$\bar{F} \int dt = \int F \, dt$$

and

$$\bar{F} = \frac{\int F \, dt}{\int dt}$$

defines the average force. For a time interval from $t = 0$ to $t = t$, we have

$$\bar{F} = \frac{\int_0^t F \, dt}{\int_0^t dt} = \frac{\int_0^t F \, dt}{t}$$

or

$$\bar{F}t = \int_0^t F \, dt.$$

Example. Replace the time-varying force of the previous example by an average force, and find the change in momentum of the particle.

Solution. Since the force varies linearly from $F = 0$ to $F = 4N$ during the time interval, the average force $\bar{F}$ will be $2N$. Therefore,

$$m(v_2 - v_1) = \bar{F}t = 2 \times 3$$
$$= 6 \, \text{N} \cdot \text{sec}^{-1}$$

in agreement with the previous calculation using the integral. It is obvious that determining an average value is equivalent to evaluating the appropriate integral as defined above.

Impulsive forces are very common. As defined, an impulsive force is any force acting for a period of time. However, the term is usually used only for those forces acting for a very short time (for example, the force applied to a golf ball by a golf club).

Fig. 7.13. A rocket exhausts hot gas at a velocity **u** relative to the rocket which is traveling in the opposite direction with velocity **v**.

7.5 ROCKET MOTION

A rocket propels itself through space by burning fuel stored on board and then thrusting the heated fuel away from the rocket as a stream of exhaust gas (see Fig. 7.13). During the vaporization process some of the mass of the rocket is ejected through the engine nozzle with velocity **u relative to the rocket**, where **u** may or may not be independent of time.

Let us suppose that a mass dm is ejected in time dt, the rocket having a residual mass m and forward velocity $\mathbf{v}$ at the end of this time interval as measured by an observer in an inertial frame of reference. The speed of the rocket at the beginning of the time interval will be $v - dv$, where dv is the increase in speed during time dt. The total momentum before the mass dm is ejected is

$$(m + dm)(v - dv) = mv + v\,dm - m\,dv - dm\,dv.$$

The total momentum after mass dm is ejected is

$$mv + dm[(v - dv) - u] = mv + v\,dm - u\,dm - dm\,dv,$$

where $(v - dv) - u$ is the change in the magnitude of the velocity of the mass dm. The momentum of the system is a constant only if no external forces are acting on the system. In this case

$$m\,dv = u\,dm.$$

For a rocket ascending vertically upwards from the earth's surface, there is a force F acting in the direction opposite to the direction of motion due to the earth's gravity and resistance of the atmosphere to the passage of the rocket. This force acts in the direction of $\mathbf{u}$, so that the rocket equation for this particular case is

$$m\,dv = u\,dm + F\,dt$$

or

$$m\frac{dv}{dt} = u\frac{dm}{dt} + F.$$

F varies with height above the earth's surface, since the gravitational force decreases and the air resistance first increases as the rocket speed increases and later decreases as the atmosphere becomes thinner. The rocket equation is not easy to solve even for this fairly simple situation.

If we consider the rocket to be in interplanetary space and subject to negligible external force, we may use the relation

$$m\,dv = u\,dm$$

to describe the rocket's motion. It is a good approximation to assume that u, the speed of the rocket exhaust, is constant. Then

$$dv = u\frac{dm}{m}$$

may be integrated to give

$$\int_{v_0}^{v} dv = u\int_{m_0}^{m} \frac{dm}{m}$$

or

$$v - v_0 = u \ln\left(\frac{m}{m_0}\right).$$

We may rewrite the solution in the form

$$v - v_0 = -u \ln\left(\frac{m_0}{m}\right).$$

The quantities m_0 and v_0 are the initial mass and speed of the rocket; m and v are the final mass and speed after the rocket exhaust is turned off or after the fuel burns out, as the case may be. The minus sign indicates that $\mathbf{v}$ and $\mathbf{u}$ are in opposite directions.

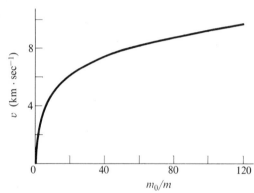

Fig. 7.14. The variation in final speed of a rocket starting from rest as a function of (m_0/m) for an exhaust speed of $2 \, \text{km} \cdot \text{sec}^{-1}$.

The speed v is plotted as a function of (m_0/m) in Fig. 7.14 for $v_0 = 0$ and $u = -2 \, \text{km} \cdot \text{sec}^{-1}$ (a typical exhaust speed). It is obvious that the increase in speed is very rapid during the early period of the rocket's firing, but as the mass of the rocket decreases, the increase in speed decreases. To attain a rocket speed of $9 \, \text{km} \cdot \text{sec}^{-1}$ requires that over 99% of the rocket mass be ejected in the exhaust. (Actually, the figure would be higher in any real situation, since we have neglected the retarding action of the external forces.)

Example. In order to launch a rocket to the moon, we must give the rocket an initial upward speed of $11 \, \text{km} \cdot \text{sec}^{-1}$ near the earth's surface so that the rocket will "escape" from the earth's gravitational pull. What mass must the rocket have initially so that a payload of $2000 \, \text{kg}$ reaches the required speed? (Neglect external forces.)

Solution. We have

$$-\frac{v}{u} = \ln\left(\frac{m_0}{m}\right).$$

Taking the exponential of each side of the equation gives

$$\exp\left(-\frac{v}{u}\right) = \frac{m_0}{m}$$

or

$$m_0 = m \exp\left(-\frac{v}{u}\right).$$

If we assume $u = -2\,\text{km}\cdot\text{sec}^{-1}$, $v/u = -\frac{11}{2} = -5.5$, so that

$$m_0 = 2000\exp(5.5)$$

$$= 2000(245)$$

$$= 4.9 \times 10^5\,\text{kg.}$$

Therefore, the initial mass of the rocket must be (at least) 4.9×10^5 kg (540 tons).

7.6   CLASSIFICATION OF FORCES

There are only four types of forces known to man. They are: gravitational, electromagnetic, nuclear, and the weak interaction. The gravitational force is a force of attraction and is associated with mass. The electromagnetic force, which may be either attractive or repulsive, is associated with charges. For convenience, one often speaks of an electric force associated with charges at rest relative to one another and a magnetic force associated with charges in relative motion; these two forces are merely different aspects of the electromagnetic force. The nuclear force[3] is responsible for the binding together of the protons and neutrons in the nucleus of an atom and acts only over distances comparable to the size of the nucleus ($\sim 10^{-15}$ m). The weak interaction force[4] is responsible for radioactive decay of atomic nuclei and is the least well understood force in nature.

Most of the forces common to us in our everyday experiences are gravitational or electromagnetic in origin. Muscular force, for example, is produced by cells in muscle fibers. The cells, on the other hand, are collections of molecules composed of atoms that owe their dimensions to the electrical attraction between negative electrons and a positive nucleus. Muscular exertion represents adjustments in the configurations of countless numbers of atoms (mostly carbon, hydrogen, oxygen, and nitrogen), so that the electric force is the basis of muscular force and, indeed, of pushing and pulling in general.

The strengths of the basic forces are very different in magnitude. They are listed in Table 7.1 in decreasing order of strength. The number after each force represents the relative strength of the force assuming that each force is acting between a pair of protons in a nucleus. You are probably surprised at the weakness of the gravitational force, since it is such a strong force in our ordinary experience and is responsible, for

[3] H. A. Bethe, "What Holds the Nucleus Together," *Scientific American*, September, 1953. Available as *Scientific American Offprint 201* (San Francisco: W. H. Freeman & Co., Publishers).

[4] S. B. Treiman, "The Weak Interactions," *Scientific American*, March, 1959. Available as *Scientific American Offprint 247* (San Francisco: W. H. Freeman & Co., Publishers).

example, for the moon revolving about the earth and for the structure of the solar system and indeed of the whole universe. It is only because these objects are so large that the gravitational attraction between their individual constituent particles adds up to produce an appreciable force. At the same time, since the other forces are so great, the constituent particles arrange themselves so as to tend to cancel the effects of the stronger forces between particles. For example, atoms are normally found in a state of neutral charge because of the strong electromagnetic force.

Table 7.1 The relative strengths of the fundamental forces acting between a pair of protons in a nucleus.

1. Nuclear	1
2. Electromagnetic	10^{-2}
3. Weak Interaction	10^{-13}
4. Gravity	10^{-38}

The nuclear force

As mentioned above, the nuclear force is of extremely short range, and its operation is not normally noticeable. It is responsible for the stability of the atomic nucleus. A mathematical description of the behavior of particles interacting under nuclear forces is well beyond the scope of this text. Indeed, an important part of present-day physics research is devoted to the study of the interactions between "elementary" particles, both inside and outside of the atomic nucleus.

The electromagnetic force

Electromagnetic phenomena are well understood. Over 100 years ago Clerk Maxwell wrote down the basic equations of electromagnetism. Since then his theory has been extended to include relativistic effects and quantum effects (see Chapter 17), and the whole scheme works very well. At this time we shall describe only the basic electric and magnetic interactions between particles.

The **electric force** between two charged particles is given by **Coulomb's law** (after Charles Augustin Coulomb),

$$F = \frac{kq_1q_2}{r^2},$$

where q_1 and q_2 are the electrical charges of the two particles, k is a constant, and r is the distance between the charged particles. Experiments have shown that this **inverse square law** behavior is true even down to separations of 10^{-15} m or less. For simplicity, however, we normally consider the charged particles to have separations much greater than their own dimensions, so that we are usually dealing with **point particles** and r is just the distance between the two points in space where the particles are situated. In the MKS system of units, charge is measured in coulombs (C), and the constant $k = 9 \times 10^9 \, \text{N} \cdot \text{m}^2 \cdot \text{C}^{-2}$.

As we shall see very shortly, the form of Coulomb's law is identical to that of Newton's law of gravitation. Since gravitational effects are much more familiar to the average person than electromagnetic phenomena, we are going to deal quite extensively with the motion of particles under mutual gravitational interaction. Therefore, we shall not treat the motion of charged particles explicitly at this time but shall refer to the electrical interaction from time to time during our discussion of motion under gravitational forces. We should note, however, that the electrical force between two charged particles may be attractive (if the particles are oppositely charged) or repulsive (if the particles carry the same kind of charge).

Moving charges can also interact with each other through the **magnetic force**. We are all familiar with so-called permanent magnets, which attract metallic objects containing iron but which do not affect most non-ferrous materials. The magnetic force that these magnets exert is a result of the motion of atomic electrons in the iron atoms in the magnet. In most elements the motions of the various electrons making up the atoms produce magnetic forces that tend to act in random directions and as a result give a net force of zero over the entire volume of the material. Iron, however, has atoms that join together in a solid in such a way (especially if some alloying material such as nickel is present) that the magnetic forces produced by the moving electrons in the various atoms are not randomly oriented and result in a net force in some direction. The strength of a magnet is a function of how great is the departure from random orientation.

A permanent magnet is normally composed of a very large number of atoms (10^{23} or more), each of which can be regarded as a permanent magnet, since an individual atom contains circulating electric charges, that give rise to a magnetic force. A useful model for an atomic magnet is shown in Fig. 7.15. The **current** I represents the circulating negatively charged electrons; this atomic magnet has a north pole N and south pole S directed as shown. If two of these atomic magnets are brought together so that their N poles point in the same direction, a force of repulsion between the magnets occurs. If they are brought together so that the N pole of one is in the same direction as the S pole of the other, a force of attraction occurs. This same behavior takes place when permanent magnets interact with each other. The N and S labels for magnets derive from the fact that a permanent magnet freely suspended in the earth's magnetic field will orient itself so that one end points toward the earth's north magnetic pole; this end of the magnet is designated the N pole of the magnet. The N poles of all other magnets can be determined in a similar way or by comparison with magnets whose N pole has already

Fig. 7.15. An atom may be considered as a small permanent magnet produced by the circulating electron current.

been determined. We want to keep in mind that individual atoms behave as if they were little magnets, for we shall have occasion to refer to this fact in later chapters.

Magnetic forces also arise between a flow of charges (a current), for example, in electric circuits or in evacuated tubes. In the very simple case of currents flowing in two parallel straight wires separated by a distance r, the magnitude of the force between the wires is given by

$$F = k' \frac{I_1 I_2}{r},$$

where I_1 and I_2 are the currents in the two wires and k' is a constant. In the MKS system of units, currents are measured in the units of amperes (A). A flow of 1 C of charge past a point in space each second constitutes a current of 1 A through that point in space. The constant k' has the value of 10^{-7} N $\cdot$ m $\cdot$ A^{-2}. The force is attractive for currents flowing in the same direction and is repulsive for currents flowing in opposite directions. The force between currents when the currents are not parallel to each other is not given by the above simple expression but may be calculated for each particular configuration according to the laws of magnetism.

The weak interaction

The weak interaction force is the one about which the least is known. That such a force must exist is indicated by the fact that in radioactive decay, the speed of the reaction is much slower than it would be if either nuclear or electromagnetic forces are involved. Characteristic times for nuclear reactions are about 10^{-23} sec, and the interaction involved in radioactive decay has a characteristic time of about 10^{-10} sec. To explain this very lethargic behavior it has been necessary to assume a force weaker than electromagnetism but stronger than the extremely weak force of gravity. The study of weak interactions is being pursued vigorously in contemporary research.

The gravitational force

The gravitational force of attraction between two point particles of masses m_1 and m_2, respectively, is given by **Newton's law of gravitation**,

$$F = \frac{G m_1 m_2}{r^2},$$

where r is the distance between the particles and G is a universal constant known as the **gravitational constant**. For objects with finite dimensions, the distance r is measured between the centers of mass of the objects, and the force law given above is a very good approximation only when r is much greater than the dimensions of the objects, so that any irregularities of shape have a negligible effect. This general situation is pictured in Fig. 7.16. When an object of finite dimensions moves under the influence of external forces, its path through space is just that path that a point particle of mass equal to the mass of the object and situated at some **average point** within the object would follow under the influence of the external forces. The average position at which the equivalent point particle is situated is called the **center of mass** of the object. A general technique for calculating its position is given in Sections 13.3 and 14.2.

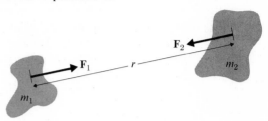

Fig. 7.16. Two objects separated by a distance $r \gg$ dimensions of the objects exert gravitational forces on each other proportional to the product of their masses and inversely proportional to the square of the distance between their centers of mass.

The law of gravitation was first enunciated by Isaac Newton, who was able to show that the motion of the moon about the earth and of the planets about the sun would be as observed only if the force between them is of the form stated above[5]. The observational laws concerning the planets were enunciated by Johannes Kepler and are discussed in Chapter 12. The force of attraction between small spheres of various substances was later measured in the laboratory by Robert Cavendish[6], who verified the relation stated by Newton and evaluated the constant G. The presently accepted value of G is $6.67 \times 10^{-11} \, \text{N} \cdot \text{m}^2 \cdot \text{kg}^{-2}$. Precise observations of the motion of the planets in the solar system have indicated that the exponent of r in the gravitational law is equal to 2 to about one part in one million.

The force of attraction acts along the line joining the two particles. The law of gravitation may be written in vector form by making use of the **separation vector r**, which is a vector originating at particle m_2 and terminating on particle m_1, as shown in Fig. 7.17. The force $\mathbf{F}_1$ on particle m_1 due to particle m_2 acts in the direction of $-\mathbf{r}$; the force $\mathbf{F}_2$ on particle m_2 due to particle m_1 acts in the direction of $\mathbf{r}$. We write

$$\mathbf{F}_1 = -\frac{Gm_1m_2}{r^2}\hat{\mathbf{r}}, \qquad \mathbf{F}_2 = \frac{Gm_1m_2}{r^2}\hat{\mathbf{r}}$$

for the two forces, where $\hat{\mathbf{r}}$ is a **unit vector** in the direction of the separation vector.

Fig. 7.17. The separation vector $\mathbf{r}$ begins on m_2 and terminates on m_1. $\hat{\mathbf{r}}$ is a unit vector in the direction of the separation vector.

[5] The Physical Science Study Committee, *College Physics*, 2nd Edition (Boston: D. C. Heath & Company, 1968), Chapter 21.

[6] M. H. Shamos, ed., *Great Experiments in Physics* (New York: Holt, Rinehart & Winston, Inc., 1960), Chapter 6.

We notice that

$$\mathbf{F}_1 = -\mathbf{F}_2,$$

which is consistent with the observation that two particles exert equal and opposite gravitational forces on each other. It is, in fact, always observed that, when two particles interact with each other, they exert equal and opposite forces on each other. This is often put in the form "for every action there is an equal and opposite reaction" and is known as **Newton's third law**.

Example. Deduce a value for the acceleration due to gravity at the Earth's surface.

Solution. Applying Newton's second law,

$$F = ma,$$

to an object of mass m on the Earth's surface, we can write

$$\frac{GmM_e}{R_e^2} = mg_e,$$

where $M_e = 5.98 \times 10^{24}$ kg is the mass of the earth, $R_e = 6.37 \times 10^6$ m is the radius of the earth, and g_e is the acceleration due to gravity at the earth's surface. Therefore

$$g_e = \frac{6.67 \times 10^{-11} \times 5.98 \times 10^{24}}{(6.37)^2 \times 10^{12}}$$

$$= 9.81 \ \text{m} \cdot \text{sec}^{-2}.$$

Example. Compare the gravitational force on the moon due to the earth with the gravitational force on the moon due to the sun.

Solution. Let us denote the masses of the sun, earth, and moon by m_s, m_e, and m_m, respectively, and the sun-moon and earth-moon separations by r_s and r_e, respectively. Using the law of gravitation, we write for the force of attraction of the earth for the moon

$$F_e = \frac{Gm_e m_m}{r_e^2}$$

and for the force of attraction of the sun for the moon

$$F_s = \frac{Gm_s m_m}{r_s^2}.$$

The ratio of F_e to F_s is

$$\frac{F_e}{F_s} = \left(\frac{Gm_e m_m}{r_e^2}\right)\left(\frac{r_s^2}{Gm_s m_m}\right) = \left(\frac{m_e}{m_s}\right)\left(\frac{r_s}{r_e}\right)^2.$$

Taking

$$m_e = 5.98 \times 10^{24} \text{ kg}$$
$$m_s = 1.98 \times 10^{30} \text{ kg}$$
$$r_s = 1.49 \times 10^{11} \text{ m}$$
$$r_e = 3.8 \times 10^8 \text{ m},$$

we obtain

$$\frac{F_e}{F_s} = \left(\frac{5.98 \times 10^{24}}{1.98 \times 10^{30}}\right)\left(\frac{1.49 \times 10^{11}}{3.8 \times 10^8}\right)^2$$
$$= 0.464.$$

The two forces of attraction are comparable; the nearness of the earth to the moon almost compensates for the effect of the much greater mass of the sun compared to that of the earth.

Weight

Weight is normally defined as the magnitude of the gravitational attraction of the earth for an object. The weight of an object will vary according to its location on the surface of the earth, due to the fact that the earth is not truly spherical and has irregular features such as mountains and deep oceans and because the earth rotates about its axis. A variation of 0.5% occurs from pole to equator. According to this definition an object is never **weightless**, since gravitational forces extend to large distances. An object will always be experiencing gravitational forces, however minute.

Why then is so much said about astronauts being weightless in space? When we are on the surface of the earth, we are normally in contact with some part of it (the ground, a chair, a floor, etc.). The earth exerts a gravitational force on us, and we exert a force on the ground, or chair, or floor, etc. The ground, chair, or floor, etc. in turn pushes back at us with the same force. An astronaut in a spaceship in orbit around the earth is still subject to a gravitational force. This gravitational force provides the centripetal acceleration that keeps the astronaut circling the earth. However, any surface that the astronaut comes in contact with is also revolving about the earth with the same centripetal acceleration as the astronaut. In a sense, they are both falling around the earth with the same acceleration. The astronaut, therefore, does not exert any force on his spaceship by virtue of the gravitational attraction that the earth has for him. The spaceship does not exert any force on the astronaut, and the absence of this normal reaction force is readily apparent to him. The best description of such a condition is given by the term **weightless**; the astronaut does not feel that he has weight under these circumstances. On the surface of the earth we experience the same state whenever we jump off the surface. However, we normally remain out of contact with the earth's surface for so little time that we do not notice the absence of reaction forces.

Weight is often specified in mass units rather than in force units. The weight of an object in kilograms is defined as

$$W\,(\text{kg}) = \frac{W\,(\text{newtons})}{g_e} = \frac{mg}{g_e},$$

where m is the mass in kilograms, g is the local acceleration due to gravity, and g_e is an average value of the acceleration due to gravity over the earth's surface.

7.7 WORK AND ENERGY

Work is a familiar concept to all of us. The furniture mover does a considerable amount of work in moving a piano from his van into a house, the farmer does work when he pumps water from his well, the student does work when he struggles to obtain adequate solutions to his assigned physics problems. Examples such as these provide us with an intuitive feeling for the concept of work. In physics, however, the term has a rather more restricted meaning. For instance, if a constant force $\bar{F}$ applied to an object results in a displacement s of the object in the direction of the force, then the work done on the system by the force is defined as

$$W = \bar{F}s.$$

The unit of work is the joule (J) for a force expressed in newtons and a displacement expressed in meters. That is, a force of 1 N acting on an object and causing a displacement of 1 m does 1 J of work.

Work is defined to be either positive or negative, as determined by the direction of the displacement relative to an internal force. If the displacement is in the direction of the internal force, the work is positive; if the displacement is in the direction opposite to the internal force, the work is negative. This can be illustrated by a simple example, that of picking up a book from a table. If we take the book and the earth as a system, the gravitational attraction of the earth on the book constitutes an internal force. The displacement of the book is in a direction opposite to the internal force and the work done is negative. If we take the book and the person lifting it as a system, the force causing the displacement is an internal force. The displacement of the book is now in the same direction as the internal force, and the work done is positive. In practice, one selects the system in which the calculations are the simplest.

Considering the earlier intuitive examples of work in terms of our physics definition, we see that only the moving of the piano and the pumping of water from the well require an appreciable amount of work. Problem solving may be a mentally exhausting task but does not require a great deal of work as defined above. We might designate work involving any sort of physical or mental exertion as **physiological work** and reserve the term **physical work** for work as defined above. However, the physical definition of work, although not obviously related at this moment to physiological work, is tied to it by energy concepts that are a consequence of the physical definition of work. The connection will become clear later as we develop our concepts of energy and heat.

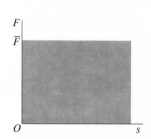

Fig. 7.18. The force vs displacement graph for a constant force.

By analogy with the kinematic considerations developed in Chapter 3, we can see immediately that in the plot of F versus s, shown in Fig. 7.18, the work done by the constant force $\bar{F}$ is represented by the area of the shaded rectangle. If a nonconstant force is responsible for the work being done, as shown in Fig. 7.19, then in analogy with our earlier considerations we can see that the work done is represented by the shaded area under the curve in Fig. 7.19 and can be represented symbolically by

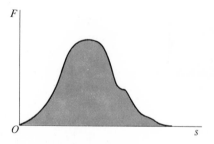

$$W = \int_0^s F \, ds.$$

Fig. 7.19. The force vs displacement graph for a nonconstant force.

If the displacement of an object resulting from the application of a force does not take place in the direction of the force, then only the component of the force in the direction of the displacement is effective in causing the displacement and should be used in calculating the work. We have already discussed this in Section 2.5, where we defined the dot or scalar product of two vectors. The general formula for the work done is

$$W = \int \mathbf{F} \cdot d\mathbf{s} = \int F \cos \theta \, ds,$$

where θ is the angle between $\mathbf{F}$ and $d\mathbf{s}$, and the integral is to be evaluated along the line of displacement of the object. This expression is valid for both positive and negative work, for if, as is shown in Fig. 7.20, $\theta > 90°$, the force opposes the displacement $d\mathbf{s}$. But $\cos \theta$ is negative for $\theta > 90°$, so that the work done is negative, indicating work done against a force.

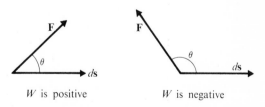

W is positive W is negative

Fig. 7.20. If the angle θ between an applied force $\mathbf{F}$ and a resulting displacement of $d\mathbf{s}$ is $<90°$ the work done is positive while if θ is $>90°$ the work done is negative.

Kinetic energy

Suppose that a constant force $\bar{F}$ acts on a particle initially at rest, causing it to suffer a displacement s in time t in the direction of the force. The work done on the particle is

$$W = \bar{F}s.$$

From Newton's second law we know that, for constant mass m,

$$\bar{F} = ma,$$

where a is the acceleration resulting from the applied force. Since $\bar{F}$ is constant, so is a, and we know from straight line kinematics that

$$v^2 = v_0^2 + 2as$$

$$= 2as$$

for $v_0 = 0$, where v is the speed of the particle after the displacement s. Substitution yields

$$W = mas = \frac{mv^2}{2}$$

for the work done on the particle.

The particle that was initially at rest now has a speed v, which results from an amount of work $mv^2/2$ having been done on it. The quantity $mv^2/2$ is called the **kinetic energy** of the particle moving with speed v. That is,

$$K = \tfrac{1}{2}mv^2,$$

where K is the kinetic energy of a particle of mass m moving at speed v. Kinetic energy will also have units of joules.

If the particle is moving with an initial speed v_0 and the force is applied **in the direction in which the particle is moving**, then

$$W = mas$$

$$= \frac{m}{2}(v^2 - v_0^2)$$

$$= \tfrac{1}{2}mv^2 - \tfrac{1}{2}mv_0^2.$$

The work done is just the difference between the final kinetic energy and the initial kinetic energy. If the force is not applied in the direction of motion, the work must be calculated from the general expression, and the final velocity determined by vector analysis. In any event, we can define the kinetic energy of a particle of mass m moving with speed v to be $mv^2/2$ regardless of how the particle attained the speed v.

Potential energy

Energy can also be associated with the position of a particle. This type of energy is known as potential energy. For example, let us consider a particle at rest on the earth's surface. If the particle is raised through a distance that is small compared to the earth's radius, the force of gravitational attraction is approximately a constant with magnitude

$$F = mg,$$

where g is the local acceleration due to gravity. The work done on the particle shown in Fig. 7.21 in raising it through a height h is, as computed in the earth-particle system,

$$W = - Fh = - mgh.$$

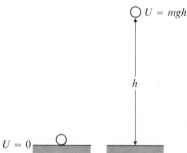

Fig. 7.21. Potential energy is energy due to position.

If the particle were then released, it would fall back to its initial position, the gravitational force of the earth doing work mgh on the particle in the process. The particle is said to have **potential energy** of mgh with respect to the point on the earth's surface from which it was raised. As the particle falls back it acquires kinetic energy at the expense of its initial potential energy.

There are many other familiar examples of potential energy. When an arrow is placed in a bow and the bow string is drawn back and held, the arrow acquires potential energy, for the release of the string will do work on the arrow and give it considerable kinetic energy. Model airplanes are often powered by an elastic band attached to the propeller. The elastic band is wound up tightly and held until we wish to let the airplane fly. During the flight the potential energy stored in the deformed elastic band is gradually transformed into kinetic energy in the form of motion of the airplane. A particle, or any object, has potential energy when it is in a condition such that some force would do work on it if the force were allowed to operate freely and were not counterbalanced by some opposing force. A ball held above a table has potential energy because it will fall under the earth's gravitational attraction if it is released from your hand. The elastic band in a model airplane has potential energy because twisting it out of its normal shape distorts the molecular structure, thereby upsetting the balance of interatomic forces that is normally present. When the propeller is released, the electrical forces in the atoms do work to restore the elastic band to its original undistorted shape. Potential energy may be regarded quite generally as energy due to position of a particle or collection of particles in a region of force in which the force is momentarily counterbalanced by some other external force. When the balancing force is removed, the particle or particles are free to move under the influence of the primary force.

7.8 GRAVITATIONAL POTENTIAL ENERGY

We have had an introduction to the concept of gravitational potential energy in the previous section, in which an object of mass m raised a distance h against the earth's gravitational attraction is shown to have acquired a potential energy mgh. The use of this expression is restricted, however, to small regions of space near the earth's surface, since the acceleration g is assumed to be constant. The force of gravity on an object can be considered constant only if the change in separation $\Delta r = h$ involved in changing the object's position is very small compared to the initial separation r. For objects near the earth's surface this is a good approximation, since $r = 6 \times 10^6$ m and Δr is normally of the order of 10^2 m or less. However, if we wish to discuss energy changes for objects moving through distances not negligible compared to the separation, we must take account of the variation of force with separation.

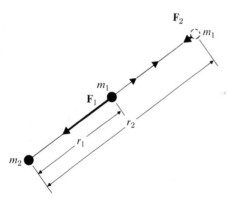

Let us consider a system of two particles of masses m_1 and m_2. Suppose that we wish to move the object of mass m_1 situated a distance r_1 from the object of mass m_2 to a position a distance r_2 from the object of mass m_2. As shown in Fig. 7.22, we are assuming that we move m_1 directly away from m_2, so that the force always acts directly opposite to the direction of motion. This will simplify our discussion, although it is not necessary to restrict the result to this special case, as we shall show in the next section. The force on m_1 when the separation from m_2 is r is

Fig. 7.22. A particle of mass m_1 is moved along a line directly away from a particle of mass m_2; the original separation is r_1 and the final separation is r_2.

$$F = \frac{Gm_1m_2}{r^2}.$$

The work that must be done in moving m_1 from r_1 to r_2 against the gravitational attraction of m_2 is

$$W = \int_{r_1}^{r_2} F \cos\theta \, dr = -\int_{r_1}^{r_2} F \, dr = -\int_{r_1}^{r_2} \frac{Gm_1m_2}{r^2} \, dr,$$

where θ, the angle between $\mathbf{F}$ and $d\mathbf{r}$, is 180°. Evaluating the integral in the expression for work, we have

$$W = -Gm_1m_2 \int_{r_1}^{r_2} \frac{dr}{r^2}$$

$$= -Gm_1m_2 \left[-\frac{1}{r_2} - \left(-\frac{1}{r_1} \right) \right]$$

$$= -Gm_1m_2 \left(\frac{1}{r_1} - \frac{1}{r_2} \right).$$

We have assumed in the development of this expression for work that $r_2 > r_1$. Therefore, the term in brackets is always positive, which means that the work is always negative in agreement with the convention that work done against an internal force is negative. If we let r_2 be less than r_1, the term in brackets will become negative, and the work will be positive. Thus the expression for the work is valid in general, regardless of whether m_1 is being moved away from m_2 or is moving closer to m_2.

If potential energy is defined as energy due to the position of a particle in the region of a source of force, the work done in moving m_1 is equal to the change in potential energy. We define

$$U_1 = -\frac{Gm_1m_2}{r_1}$$

as the potential energy at separation r_1 and

$$U_2 = -\frac{Gm_1m_2}{r_2}$$

as the potential energy at separation r_2. In general, the **potential energy** U **at separation** r is

$$U = -\frac{Gm_1m_2}{r}.$$

By this definition, the potential energy at infinite separation is zero, and the potential energy decreases as the separation decreases. The fact that the potential energy is always negative by this definition should not worry us, because actually only changes in potential energy can be observed. The change will be either positive or negative depending upon whether the separation increases or decreases.

Potential energy is energy that a particle may have in addition to kinetic energy (K); it follows that the total mechanical energy of the particle of mass m_1 is

$$E = K + U$$

$$= \tfrac{1}{2}m_1 v^2 - \frac{Gm_1m_2}{r},$$

where v is the speed of m_1 when the separation between m_1 and m_2 is r.

Example. Assuming that the moon moves about the earth in a circular orbit of radius 3.8×10^8 m and of period 27.3 days, calculate the total energy E of the moon in its orbit.

Solution. The kinetic energy is $K = \tfrac{1}{2}mv^2$, where $m = 7.3 \times 10^{22}$ kg and

$$v = \frac{2\pi r}{T} = \frac{2\pi(3.8 \times 10^8)}{27.3 \times 86,400}$$

$$= 1.02 \times 10^3 \text{ m} \cdot \text{sec}^{-1}.$$

Therefore,

$$K = \tfrac{1}{2}(7.3 \times 10^{22})(1.02 \times 10^3)^2$$

$$= 3.8 \times 10^{28} \text{ J}.$$

The potential energy is

$$U = -\frac{GmM}{r},$$

where $M =$ mass of the earth $= 6.0 \times 10^{24}$ kg. Therefore,

$$U = -\frac{(6.67 \times 10^{-11})(6.0 \times 10^{24})(7.3 \times 10^{22})}{3.8 \times 10^8}$$

$$= -7.7 \times 10^{28} \text{ J}.$$

Therefore, the total energy is

$$E = K + U = 3.8 \times 10^{28} - 7.7 \times 10^{28}$$

$$= -3.9 \times 10^{28} \text{ J}.$$

The total energy of the moon calculated in the above example is seen to be negative. A total energy of less than zero indicates that the moon is **bound** to the earth, for the following reason. If the moon were not bound to the earth, it could go off to infinity and have some velocity greater than or equal to zero. Since the potential energy at infinity is zero, the total energy $E = K + U$ would be equal to or greater than zero. But E is less than zero, so that the moon would require some additional energy to be able to move an infinite distance from the earth. For this reason the moon is said to be bound to the earth and has a **binding energy** of -3.9×10^{28} J.

A graph of potential energy versus separation for a particle acted on by a gravitational force is shown in Fig. 7.23. Notice that the potential energy approaches an infinite negative value as the particles move very close together. This is an indication of the fact that the gravitational force approaches infinity as the two particles approach each other. Thus, a graph of gravitational potential energy versus separation gives an indication of the strength of the gravitational attraction between the two particles. We note that the force F is given by

$$F = \frac{Gm_1 m_2}{r^2}$$

and that

$$\frac{dU}{dr} = \frac{d}{dr}\left(-\frac{Gm_1 m_2}{r}\right) = \frac{Gm_1 m_2}{r^2} = F.$$

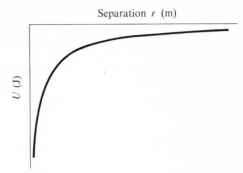

Fig. 7.23. The gravitational potential energy of a particle approaches $-\infty$ as the separation approaches zero and approaches zero at infinite separation.

Therefore, the potential energy and force are related by the fact that the force is equal to the rate of change of potential energy with separation. Graphs of potential energy versus separation are often used in discussing the interaction between particles instead of graphs of force versus separation.

We noted in Section 7.6 that the electric force between two charged particles is given by

$$F = \frac{kq_1 q_2}{r^2}.$$

By analogy with the gravitational case we can immediately write for the **electric potential energy**

$$U = -\frac{kq_1 q_2}{r}.$$

We shall have occasion to use this expression in Chapter 18.

7.9 CONSERVATIVE FORCES

Let us consider the general case of the motion of a particle of mass m_1 from separation r_1 with respect to a particle of mass m_2 (assumed stationary) to a separation r_2, where r_1 and r_2 are not in the same direction. This is outlined in Fig. 7.24.

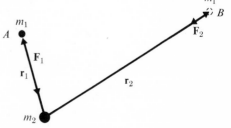

Fig. 7.24. The particle of mass m_1 is to be moved from A to B.

Case 1. Let us assume that the particle m_1 moves from position A to position B along a straight line, as in Fig. 7.25(a). Since the direction of the force is not along the direction of the motion, we must take the component of the force in the direction of the motion when calculating the work done. The work done in moving from A to B is

$$W = \int_{r_1}^{r_2} \mathbf{F} \cdot d\mathbf{s} = \int_{r_1}^{r_2} F \cos \theta \, ds,$$

where θ is the angle between the force vector $\mathbf{F}$ and the infinitesimal displacement ds along the direction of motion at an arbitrary point [labeled C in Fig. 7.25(a)]. When

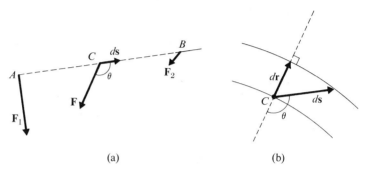

(a)   (b)

Fig. 7.25. (a) The particle is moved from A to B along the straight line ACB. (b) When the particle moves a distance ds along AB it moves a distance dr away from m_2.

the particle moves a distance ds along AB, it moves a distance dr away from m_2, where

$$\frac{dr}{ds} = \cos (180 - \theta) = -(\cos \theta),$$

from Fig. 7.25(b). Therefore,

$$ds = -\frac{dr}{\cos \theta}$$

and

$$W = \int_{r_1}^{r_2} \frac{Gm_1 m_2}{r^2} \cos \theta \left(\frac{-dr}{\cos \theta} \right)$$

$$= -Gm_1 m_2 \int_{r_1}^{r_2} \frac{dr}{r^2}$$

$$= -Gm_1 m_2 \left(\frac{1}{r_1} - \frac{1}{r_2} \right),$$

which is the same result as obtained in Section 7.8 for the motion of m_1 directly away from m_2.

Case 2. Let us assume that m_1 follows some arbitrary curved path as indicated in Fig. 7.26(a). The work done is again

$$W = \int_{r_1}^{r_2} \mathbf{F} \cdot d\mathbf{s} = \int_{r_1}^{r_2} F \cos \theta \, ds,$$

where ds is measured along the curved path AB. If we look at an infinitesimal displacement along AB at some point C, we obtain Fig. 7.26(b), since ds is so small that the curvature of the line disappears and ds appears to be a linear displacement.

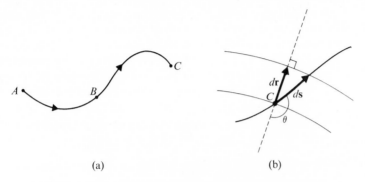

          (a)   (b)

Fig. 7.26. (a) The particle moves from A to B along some arbitrary curved path. (b) For an infinitesimal displacement ds along the curved path, the particle again moves a distance dr away from m_2.

Figure 7.26(b) is identical to Fig. 7.25(b), so that the work will again be

$$W = -Gm_1m_2\left(\frac{1}{r_1} - \frac{1}{r_2}\right).$$

We can now state the general result that the work done in moving a particle from position r_1 to position r_2 in the region of a gravitational force is independent of the path followed. A force is said to be a **conservative force** when the work done in moving a particle between two points in space when the force is acting is independent of the path followed in space between the points. From this definition we see that both the gravitational force and the electric force are conservative forces.

7.10 FRICTION

When a force is applied to a particle or to a collection of particles forming an object of finite dimensions, the particle or object does not in general respond perfectly to the applied force. The degree of departure of the actual response to the expected response varies considerably depending upon the exact situation of the particle or object in question. For example, when you attempt to pick up a heavy box from the floor, the box responds almost perfectly to the force that you apply. However, if you attempt to move the same box sideways over the surface of the floor by applying the same force used to lift the box, the box does not respond nearly so well to the applied force; in fact, often it may not respond but may refuse to move at all. This is one particular example of the fact that when one portion of matter attempts to move over another portion of matter, resistance to the motion occurs. We use the term **friction** when referring to this resistance to motion when solids are in contact. When we consider fluids (liquids or gases), we use the term **viscosity**.

Friction is due partly to the intermolecular forces between two surfaces and partly to the nature and condition of the surfaces. It has the nature of a force acting when surfaces are in contact and opposing their sliding or rolling on each other. The force of friction is found to be proportional to the force with which the two surfaces are pressed together but independent of the area of the surfaces in contact and independent of the speed of the motion **once motion has begun**. It is found that the force required to overcome friction in order to start a body in motion upon some surface is greater than the force required to keep it in motion once motion has commenced.

If a force **F** is required to overcome the frictional force between the two surfaces when the two surfaces are pressed together with a force **N**, the **coefficient of friction** μ for these two surfaces in contact is

$$\mu = \frac{F}{N},$$

where μ has two values: one associated with starting the motion (called **the static coefficient**) and one associated with the force necessary to maintain uniform motion (called **the kinetic coefficient**). The force of friction acts in a direction opposite to the force that produces the motion (or opposite to the component of the applied force in the direction of the motion if the applied force is not in the direction of motion). In the example of a box sitting on a floor mentioned above, the force pressing the surfaces together is just the force of gravity acting on the box, as shown in Fig. 7.27. Typical kinetic coefficients of friction for several pairs of surfaces are given in Table 7.2.

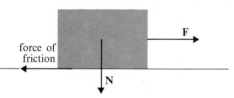

Fig. 7.27. The force of friction opposes the motion of one surface over a second surface.

7.11   THE NATURE OF FORCES

The earth and the moon exert gravitational forces on each other. It is this mutual gravitational attraction that keeps them bound together

Table 7.2 Coefficients of friction

Surfaces in contact	Kinetic coefficient of friction
Nickel and steel	0.66
Aluminum and steel	0.47
Carbon and glass	0.18
Diamond and steel	~ 0.05

as they move around the sun. On the other hand, we are told that the space between these objects is relatively empty of material objects. Yet the fact that there is a virtual vacuum between them does not affect in any manner the existence and magnitude of the force between them. This is an example of **action at a distance**. One possible explanation of the transfer of the effect of forces through a vacuum is to postulate the existence of some medium (the ether) that transmits the force (like a fluid in a hydraulic system). The ether would fill the whole of space and forces would be transmitted as mechanical stresses in this material ether. This theory was put forward in the nineteenth century. However, as the theory was developed, the properties of the ether became more and more extraordinary and self-contradictory. Einstein abandoned the concept of the ether in the formulation of the special theory of relativity and discovered that he could get along very well without it.

An alternate, more popular, view is to say that some kind of tension or stress can exist in empty space. This tension or stress is called a **field** and reveals itself by producing forces that act on material objects that lie in the space the field occupies. From this point of view, the four fundamental forces can be viewed as manifestations of four different fields in space, any or all of which can be present at any given time.

In addition, the mathematical theories that have been applied to describe the properties of fields predict that certain particles are associated with fields. In this picture, the **light quantum,** or **photon** (see Chapter 18), is the particle of the electromagnetic field, and various types of the **elementary particles** known as mesons are associated with the nuclear field. The corresponding particles for the gravitational field (the **graviton**) and the weak interaction field (the **vector meson**) have not yet been observed, although their properties have been predicted mathematically.

The photon travels at the speed of light, has zero rest mass (see Section 18.5), and has the theoretical capability of lasting for an indefinite time unless it interacts with some matter (light that has been traveling for billions of years from far regions of our universe is reaching us all the time). For this reason, electromagnetic forces have an infinite range, as is apparent, for example, from Coulomb's law, given in Section 7.6. The mesons, on the other hand, are fairly massive particles and live for very short times, which results in the nuclear and weak interaction forces extending over only very small regions of space of less than nuclear dimensions.

Some idea of the connection of particles with fields and the forces with which they are associated may be gained by considering the following analogy. Two boys are separate persons, being able to travel about independently of each other. Suppose, however, that they are both interested in baseball and enjoy throwing a baseball back and forth to each other. This is pictured in Fig. 7.28. As long as the boys are throwing the ball back and forth or, to put it a different way, are exchanging the ball, they are bound together into a single stable system of two boys and one ball. In a similar way we can think of an electromagnetic force existing between two charged particles due to an exchange of photons between them, and nuclear forces to exist between nuclear particles due to an exchange of mesons. The same idea can be extended to the weak interaction; however, no one has as yet detected or proved the existence of the particle of the weak interaction field, so that we shall say nothing further about it.

Fig. 7.28. Two boys throwing a ball back and forth to each other are "tied" together as long as the game is continued.

The graviton is predicted to have zero rest mass, to travel at the speed of light, and to be extremely difficult to detect. In fact, the properties of the gravitational field can be explained quite adequately without postulating the existence of a field particle at all. This was done by Einstein in 1916 in his general theory of relativity. In this theory, the existence of gravitational force is attributed to a curvature of the four-dimensional space-time produced by the presence of matter in that space-time. Gravitational forces between material objects result from the superposition of strains in the four-dimensional space-time. It can be noted here that the special theory of relativity is appropriate only for a "flat" four-dimensional space-time in which there are no gravitational forces and, hence, no gravitational accelerations present.

A simple two-dimensional analogy to the general theory is as follows. Consider a blanket pulled out tautly in a horizontal plane as shown in Fig. 7.29(a). Placing a baseball on this taut, flat, two-dimensional surface produces a depression as shown in Fig. 7.29(b); note that **the depression or curvature can be observed only by viewing the surface in a third dimension**. The existence of the curvature or depression can be sensed, however, by placing an object such as a Ping-Pong ball on the blanket and observing its motion. The Ping-Pong ball will produce a slight depression of its own and will commence to roll toward the baseball. (Actually, the two

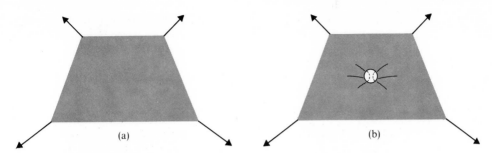

Fig. 7.29. (a) A blanket will form a flat two-dimensional surface when pulled out tautly at the corners. (b) A baseball placed on the blanket will cause a depression or curvature of the surface which can be observed directly only by a three-dimensional observer.

balls roll toward each other, but the motion of the baseball is small due to its much heavier mass. See Chapter 12 for general two-body motions.)

The general theory of relativity considers a four-dimensional space-time to be curved by the presence of matter, the curvature being directly observable only in some unimaginable five-dimensional space. The existence of the curvature can be sensed, however, by noting that material objects tend to move toward each other just as the Ping-Pong ball and baseball move toward each other on the blanket. Since the gravitational force is so weak, it is difficult to find phenomena that will help to differentiate between various theories of gravity. The general theory, however, has been reasonably successful in explaining the few observations that can be made to test it. (One example is the bending of light from a star from a straight-line path as it passes very close to the sun.)

Before leaving the general theory, we should note that the Newtonian description of gravity as embodied in the equation

$$F = \frac{Gm_1 m_2}{r^2}$$

is a special case of the general theory and is valid so long as one does not consider a region of space very close to a massive object like the sun.

PROBLEMS

1. An electron moving under the influence of a constant electric force travels 0.01 m starting from rest in 3.3×10^{-9} sec. Calculate the force experienced by the electron.

2. What force must be applied to a car traveling at $20 \text{ m} \cdot \text{sec}^{-1}$ to bring it to a halt in 50 m, assuming that it decelerates uniformly?

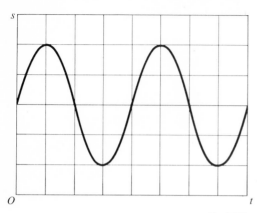

Fig. 7.30

3. The nuclear separation for a vibrating hydrogen molecule varies with time as shown in Fig. 7.30. Derive the force-time graph for the vibrating molecule.

4. A ball is dropped vertically from a point 2 m above the floor of a laboratory at the same time as a second ball is projected horizontally from the same point with a speed of 3 m · sec^{-1}. Which ball will strike the floor first? Explain your answer.

5. A discus thrower hurls a discus 180 m along fairly level ground. The discus is projected at an angle of 45° and strikes the ground at a point 1 m lower than the point at which it was released. Determine the initial speed of the discus.

6. A gun fires a projectile horizontally with a speed of 100 m · sec^{-1} from the edge of a 150 m cliff. How many seconds after firing does a person standing at the gun hear the splash of the projectile in the water below the cliff? The air temperature is 20° C, and the speed of sound in air at 20° C is 331 m · sec^{-1}.

7. The equation for the trajectory of a baseball is

$$y = x - (2.88 \times 10^{-3})x^2,$$

where y is the vertical height in feet and x is distance in feet traveled in a horizontal direction from the point of impact.
(a) Assuming that the ball started out at an angle of 45° to the ground, find the initial speed.
(b) Determine the maximum height the baseball reaches.
(c) If the baseball is struck at a point 4 ft above the ground level, where does it hit the ground, assuming the playing field to be flat?

8. Plot the trajectory of the baseball of the previous problem. Determine the direction of the velocity vector of the ball at the points $x = 170$ ft and $x = 300$ ft
(a) by drawing tangents to the curve at the points in question and determining the slopes.
(b) by differentiating the equation for the trajectory and determining the slopes at the two points.
Compare the results from parts (a) and (b).

9. A golfer hits a ball with a five iron so that it makes an angle of 25° with the horizontal as it leaves the ground. On level terrain the ball travels 150 m. On a hole with the tee 18 m higher than the green and a tee-to-green horizontal distance of 180 m, he decides to use his five iron. Will he hit the ball too far, within 5 m of the correct distance, or not far enough?

*10. A rocket, starting from rest, constantly accelerates at 100 m · sec^{-2} in a direction of 60° to the earth's surface. The fuel burns out after 10 sec. Find
(a) the maximum height the rocket rises.
(b) the flight time and range of the rocket.

11. For what value of θ is the range of a projectile at a maximum?

12. A quarterback in a football game throws a forward pass to his left end, the ball traveling a total of 50 yd horizontally. Assuming that the quarterback launches the ball at an angle of 20° to the horizontal, find the initial speed that was given to the ball by the quarterback. Is there any other assumption that must be made to solve this problem?

13. A diver takes a running dive from a tower 10 m above the surface of the water, launching himself horizontally with a speed of $2 \text{ m} \cdot \sec^{-1}$ in so doing. What is his speed when he hits the water?

14. Using the fact that the maximum height of a particle following a parabolic trajectory above a level surface occurs at half the value of the range, find the expression for the maximum height.

***15.** The left fielder in a baseball game fields a ball hit over the third baseman's head and throws it toward home plate, attempting to catch a runner who is heading toward home plate from third base at a speed of $25 \text{ ft} \cdot \sec^{-1}$ at the moment when the ball is thrown. Assuming that the runner continues toward home plate at constant speed and has 65 ft to go and that the left fielder is 200 ft from home plate and throws the ball with an initial speed of $95 \text{ ft} \cdot \sec^{-1}$ at an angle of 30° to the horizontal, will the runner be safe or out?

***16.** An artillery station situated on top of a cliff has a gun that can impart to a shell a maximum initial speed $v_0 = 100 \text{ m} \cdot \sec^{-1}$ (see Fig. 7.31). How close to the bottom of the cliff can a ship venture and still be in no danger of being hit by a shell?

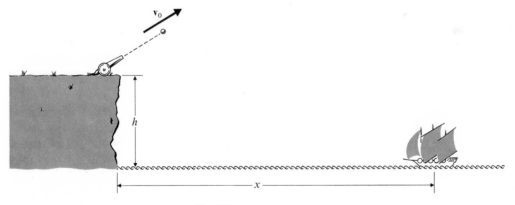

Fig. 7.31

***17.** In a demonstration experiment a physics instructor aims a compressed-air gun directly at a tin can suspended by an electromagnet at height h above the lecture table, as shown in Fig. 7.32. As the projectile emerges from the barrel of the gun, the current to the electromagnet is momentarily cut off, and the can begins to fall. Show that the projectile will always hit the can independent of its muzzle speed v_0, so long as that speed is greater than some minimum value.

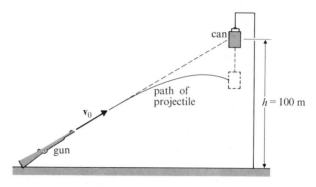

Fig. 7.32

18. High-speed photography shows that a golf club during an average swing is in contact with the golf ball for periods of the order of 10^{-4} sec. Suppose that a golf ball of mass 0.1 kg is in contact with a golf club for 2.0×10^{-4} sec and has a speed of 44 m · sec^{-1} upon leaving the club face.
(a) Calculate the impulse of the force required to give the ball its initial speed.
(b) Calculate the average force exerted on the ball during the period of impact.

19. A molecule of mass m and speed v strikes a wall and rebounds with the same speed. If the time of collision is Δt, what is the average force exerted by the molecule on the wall
(a) if it strikes the wall perpendicularly?
(b) if it strikes the wall at an angle θ to the normal to the wall?

20. A billiard ball, initially at rest, is struck by a cue and acquires a speed of 2.5 m · sec^{-1}. If the billiard ball has a mass of 0.20 kg and an average force of 50 N is exerted on the ball by the cue, over what interval of time is the blow struck?

*21. A bullet of mass 2×10^{-3} kg emerges from the muzzle of a gun with a speed of 300 m · sec^{-1}. The force on the bullet while it is in the gun barrel is given by

$$F = 400 - \frac{4 \times 10^5}{3} t,$$

where F is in newtons and t in seconds. Calculate the time required for the bullet to travel the length of the barrel.

22. For a certain nuclear-heated hydrogen rocket, the exhaust speed is 7000 m · sec^{-1}. Ignoring gravitational forces, calculate the increase in speed of the rocket if the initial mass is 3.2 times the final mass.

23. A rocket of mass 2×10^6 kg (including payload) when at rest on the surface of the earth is to be used to boost a payload of 2×10^4 kg to a speed of 1.1×10^4 m · sec^{-1}. What is the required exhaust speed? (Neglect external forces.)

24. For what values of m_0/m is the rocket speed
(a) equal to the exhaust speed?
(b) equal to twice the exhaust speed?
(c) equal to n times the exhaust speed?

25. A 10^4-kg rocket is to be launched vertically. Assuming an exhaust speed of 2 km · sec^{-1}, what mass of gas must be ejected per second
(a) to overcome the weight of the rocket?
(b) to give the rocket an initial upward acceleration of $3g$?

26. An electron moving in an electric field experiences a force of 1.2×10^{-15} N. If the electron starts from rest and travels a distance of 0.5 cm through the electric field, calculate
(a) the final velocity of the electron.
(b) the work done on the electron by the electric field.

27. The gravitational force between two particles varies with their separation as shown in Fig. 7.33. Determine the work required to increase their separation from 1 to 4 units.

28. The force between the two atoms of a molecule varies with their separation as shown in Fig. 7.34. Calculate the work required to increase the separation of the atoms from 3 units to 6 units.

29. Making use of the force-separation graph of the (vibrating) molecule given in Problem 28, derive a potential energy-separation graph.

30. Early in 1965 a Russian astronaut emerged from his space capsule while orbiting the earth at $\sim$7600 m · sec^{-1}. How would you answer the following questions asked by a layman?
(a) Why was the astronaut weightless?
(b) Why was the astronaut not seriously injured by stepping from such a fast-moving vehicle?
(c) Could the astronaut have "swum" back to his spacecraft?

31. Calculate the force of attraction between the earth ($M_e = 6.0 \times 10^{24}$ kg) and an object of mass (a) 1 kg, (b) 10^3 kg, (c) 10^6 kg situated at a distance of 6.4×10^6 m from the center of the earth. Calculate the acceleration of the object and of the earth in each case, and comment on the results.

32. Explain briefly how each of the following "common" forces is related to one (or more) of the fundamental forces:
(a) friction.
(b) wind.
(c) elastic.
(d) the force exerted on an object by a hydraulic press.

33. Calculate the binding energy of the earth in its orbit about the sun.

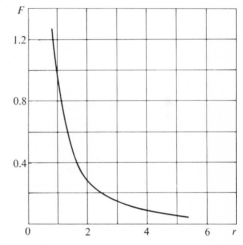

Fig. 7.33

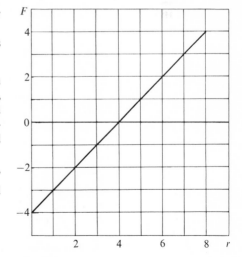

Fig. 7.34

34. Given that the electron in a hydrogen atom has a negative charge of 1.6×10^{-19} C and moves in an orbit of radius 5.3×10^{-11} m about a proton of positive charge 1.6×10^{-19} C, determine the binding energy of the electron to the proton in joules and in electron volts. An electron volt (eV) is the amount of energy gained by an electron accelerated through a distance of 1 m by an electric field exerting a force of 1 N on each coulomb of charge present in the field. Since the charge on an electron is 1.6×10^{-19} C, an electron will experience a force of 1.6×10^{-19} N in such a field, and the work done will be 1.6×10^{-19} N $\times$ 1 m $= 1.6 \times 10^{-19}$ J. Therefore,

$$1 \text{ eV} = 1.6 \times 10^{-19} \text{ J}.$$

Also,

$$1 \text{ MeV} = 1.6 \times 10^{-13} \text{ J}.$$

$$1 \text{ GeV} = 1.6 \times 10^{-10} \text{ J}.$$

35. Calculate the work done in raising a 100-kg body from the earth's surface to a height of 5000 km.

36. In Section 7.6 it is shown that the gravitational attraction between the earth and the moon is less than the gravitational attraction between the sun and the moon. However, the moon is observed to revolve about the earth. Comment on the relative motions of the earth, moon, and sun.

37. Show that the ratio of the gravitational force between two charged particles to the Coulomb force between the particles is independent of the separation of the particles. Calculate the ratio for two electrons.

38. Calculate the work done in moving a 1000-kg object from the surface of the earth to the surface of the moon.

39. The nearest star to the earth is α-Centauri, which is about 4.4 light years distant and has a mass 2.1 times the sun's mass. Calculate the gravitational force between the sun and α-Centauri, and compare it with the force between the earth and the sun.

*40. Draw a graph of gravitational potential energy versus separation for two particles of mass 2 kg and 5 kg. By determining the slope of the resulting curve at several points, construct a force-separation curve for the two particles. Compare the resulting force-separation curve with the theoretical force-separation curve.

41. Calculate the force between a charge of $+2 \mu$C (microcoulombs) and a charge of -5μC, separated by a distance of 1 m. How much work must be done to change the separation of the two charges to 5 m? Calculate the potential energy in each case.

*42. An astronaut travels to planet X, where he conducts the following experiments. First, he selects a flat terrain and sets up a gun 1 m above the surface to fire a bullet horizontally. The muzzle velocity of the gun is $200 \text{ m} \cdot \text{sec}^{-1}$. He measures the range of the bullet as 200 m. Second, he measures the distance of a small satellite of planet X above the planet's surface to be 1.0×10^6 m. Then he measures the period of the same satellite to be $(13.5)^{1/2}\pi \times 10^3$ sec. Using these data calculate

(a) the value of the acceleration due to gravity at the surface of planet X.

(b) the radius of planet X.

(c) the mass of planet X.

*43. A small pendulum of length l is used to control a clock. Assuming that the period of the pendulum is given by

$$T = C\left(\frac{l}{g}\right)^{1/2},$$

where C is a constant, calculate the change in the rate of the clock (in seconds gained or lost per day) when a lead sphere 4.0 m in diameter is brought as close as possible to the pendulum and is directly below it.

*44. A skier weighing 80 kg travels 35 m down a slope making an angle of 60° with the horizontal. His skis are coated with Teflon, thus reducing the kinetic coefficient of friction to 0.040. The slope terminates in a jump so constructed that the skier takes off in a horizontal direction. If the jump is 6 m higher than the slope directly below it and the slope below the jump falls away so as to make an angle of 30° with the horizontal, determine the length of the jump measured along the slope. Assume that he starts from rest and that the only forces acting on him are the forces of gravity and friction.

45. The speed of an object moving through a retarding medium (such as air or water) can sometimes be written as a function of time t as

$$v = \frac{A}{B} - \frac{A}{B}\exp(-Bt),$$

where A and B are constants. In such a case, show that the retarding force on the body will be $F = m(A - Bv)$, that is, that the force will be composed of a constant (frictional) force mA and a force increasing with speed.

*46. Three particles are located in the x, y plane: a particle of mass m at the point $(\frac{1}{2}, 0)$, a particle of mass m at the point $(-1, 0)$, and a particle of mass $3m$ at the point $(0, 1)$.

(a) Calculate directly the gravitational force on a particle of mass m_1 located at the origin.

(b) Calculate the gravitational potential energy of the particle at the origin.

(c) Calculate the gravitational potential energy of a particle of mass m_1 at a point (x, y) very close to the origin.

(d) By differentiation of the potential energy determine the force at the origin.

47. A particle of mass m is constrained to move along the right bisector of the line joining two fixed particles of mass M (see Fig. 7.35). Take the midpoint O of the line joining the particles of mass M as origin.

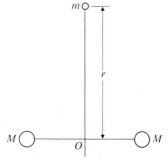

Fig. 7.35

(a) Derive a formula for the gravitational potential energy of the particle of mass M.

(b) Derive a formula for the gravitational force experienced by the particle of mass m.

(c) Locate the point at which the acceleration experienced by m reaches a maximum as m moves from infinity to 0.

Rotational dynamics

8

8.1   INTRODUCTION

Newton's second law can always be used to determine the resultant rate of change of momentum of a particle in terms of the forces applied to it. However, when we are dealing with rotational motion, it is very convenient to rewrite Newton's second law in a manner that is more compatible with the rotational quantities introduced in Chapter 6 (angular frequency, angular acceleration, etc.). In this chapter we shall deal primarily with the motion of a single particle about some center of rotation. The concepts developed here, however, can be applied to describe the rotational motion of macroscopic objects, for they are just collections of a large number of particles.

8.2   CIRCULAR MOTION

In Chapter 6 we discussed the kinematics of circular motion and introduced the angular frequency ω and the angular acceleration α. If α is zero ω is a constant, and there can be no force acting on the particle with a component in the direction of motion. The only force acting on a particle in uniform circular motion is that force necessary to provide the centripetal force required to keep the particle traveling in a circular path. That is, a force of constant magnitude always acting at right angles to the direction of motion of a particle causes the particle to move along a circular path with constant speed.

If a force $\mathbf{F}_t$ is applied in the tangential direction (that is, in the direction of motion of the particle), a tangential acceleration $\mathbf{a}_t$ occurs, the velocity $\mathbf{v}$ changes both in magnitude and direction, and an angular acceleration α results. This is pictured in Fig. 8.1 for circular motion. It is convenient to define two new quantities for the discussion of rotational motion: **torque** and **angular momentum**. We define the torque τ to have magnitude

$$\tau = rF_t,$$

where r is the magnitude of the position vector, and to point in the direction given by the right-hand rule [that is, in the direction of ω and α (see Fig. 8.1)]. Torque is the product of force times distance. We note that $\mathbf{F}_t$ is perpendicular to $\mathbf{r}$, so that τ, $\mathbf{r}$, and $\mathbf{F}_t$ are three mutually perpendicular vectors for circular motion. Therefore, we can write the relation

$$\tau = \mathbf{r} \times \mathbf{F}_t$$

for the torque.

The angular momentum $\mathbf{J}$ is defined to have magnitude

$$J = rp,$$

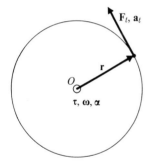

Fig. 8.1. A tangential force $\mathbf{F}_t$ produces a tangential acceleration $\mathbf{a}_t$ of a particle traveling in a circular path. The direction of the torque is perpendicular to the plane of the paper as given by the right-hand rule.

142

where p is the magnitude of the momentum of the particle, and to point in the same direction as τ, ω, and α. Angular momentum is a product of linear momentum and distance. Again, $\mathbf{J}$, $\mathbf{r}$ and $\mathbf{p}$ are three mutually perpendicular vectors, and we can write the relation

$$\mathbf{J} = \mathbf{r} \times \mathbf{p}.$$

The vectors $\mathbf{r}$, $\mathbf{p}$, and $\mathbf{J}$ are pictured in Fig. 8.2.

Differentiating the equation

$$J = rp$$

with respect to time yields

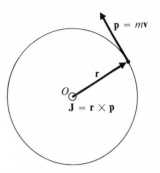

Fig. 8.2. The direction of the angular momentum $\mathbf{J}$ is given by the right-hand rule and $\mathbf{J} = \mathbf{r} \times \mathbf{p}$.

$$\frac{dJ}{dt} = \frac{d}{dt}(rp) = r\frac{dp}{dt} = rF_t,$$

where we have used Newton's second law. But rF_t is just the torque τ. Therefore,

$$\tau = \frac{dJ}{dt};$$

the rate of change of angular momentum is equal to the (applied) torque. This is just Newton's second law in a form appropriate for rotational motion. It may also be written in vector form as

$$\tau = \frac{d\mathbf{J}}{dt},$$

as we have defined τ and $\mathbf{J}$ to be vectors pointing in the same direction.

If there is no torque, $\tau = 0$, and

$$\frac{d\mathbf{J}}{dt} = 0,$$

which means that $\mathbf{J}$ is a constant independent of time. It is often referred to as a **constant of the motion**. If there is no torque, **angular momentum is conserved**. Since the angular momentum is a vector, both the direction and magnitude of the angular momentum are conserved. A further discussion of the conservation of angular momentum is given in Chapter 9.

Example. A 10-g sample is placed in a centrifuge at a point 20 cm from the center of rotation. The centrifuge is turned on and accelerates uniformly to 100 rev·sec^{-1} after 30 sec. What is the torque acting on the sample? What is the angular momentum at the end of the 30-sec interval?

Solution. The final angular frequency is $2\pi \times 100 = 200\pi$ rad·sec^{-1}. Since $v = \omega r$, the final speed of the sample is

$$v = 200\pi(0.20) = 40\pi \text{ m·sec}^{-1}.$$

The initial speed is zero, and the acceleration is uniform, so that the tangential acceleration is found from

$$a_t = \frac{v}{t} = \frac{40\pi}{30} = \frac{4\pi}{3}\,\text{m} \cdot \text{sec}^{-2}.$$

The tangential force is

$$F_t = ma_t = 0.01 \times \frac{4\pi}{3} = 4.2 \times 10^{-2}\,\text{N}.$$

The torque acting on the sample during this time interval is

$$\tau = rF_t = 0.20(4.2 \times 10^{-2}) = 8.4 \times 10^{-3}\,\text{N} \cdot \text{m}.$$

The angular momentum after 30 sec is

$$J = rp = rmv = 0.20(0.01)(40\pi) = 0.25\,\text{kg} \cdot \text{m}^2 \cdot \text{sec}^{-1}.$$

8.3 CURVILINEAR MOTION

For curvilinear motion, the force on a particle will not in general bear any special relationship to the position vector of the particle. Let us consider first the case of curvilinear motion in two dimensions in a plane. We now define the torque to be

$$\tau = r'F,$$

where r' is the perpendicular distance from the center of rotation to the line of action of the force **F**, as shown in Fig. 8.3. If θ is the angle between the directions of **r** and **F**, we see that

$$r' = r \sin(\pi - \theta) = r \sin \theta,$$

so that

$$\tau = rF \sin \theta.$$

However, $rF \sin \theta$ is the magnitude of the vector product $\mathbf{r} \times \mathbf{F}$, which is a vector perpendicular to the plane containing **r** and **F**. Therefore, we can write

$$\boldsymbol{\tau} = \mathbf{r} \times \mathbf{F}$$

as the general defining equation for the torque. A similar analysis shows us that the angular momentum **J** may be written as

$$\mathbf{J} = \mathbf{r} \times \mathbf{p},$$

where **p** is the momentum of the particle. For planar two-dimensional curvilinear motion, the vectors $\boldsymbol{\tau}$ and **J** are both perpendicular to the plane of motion.

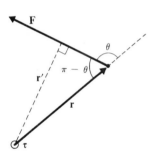

Fig. 8.3. For general curvilinear motion the torque is $\tau = r'F$ and the angular momentum is $J = r'mv$; again, $\tau = dJ/dt$.

Let us take the time derivative of $\mathbf{J}$:

$$\frac{d\mathbf{J}}{dt} = \frac{d}{dt}(\mathbf{r} \times \mathbf{p}) = \mathbf{r} \times \frac{d\mathbf{p}}{dt} + \frac{d\mathbf{r}}{dt} \times \mathbf{p}.$$

However,

$$\frac{d\mathbf{r}}{dt} = \mathbf{v} \quad \text{and} \quad \frac{d\mathbf{p}}{dt} = \mathbf{F},$$

so that

$$\frac{d\mathbf{J}}{dt} = \mathbf{r} \times \mathbf{F} + \mathbf{v} \times \mathbf{p}.$$

But

$$\mathbf{v} \times \mathbf{p} = \mathbf{v} \times m\mathbf{v} = m\mathbf{v} \times \mathbf{v} = 0$$

since the vectors $\mathbf{v}$ and $\mathbf{v}$ are the same vector, so that $\theta = 0$ and $\sin \theta = 0$. Therefore,

$$\frac{d\mathbf{J}}{dt} = \mathbf{r} \times \mathbf{F} = \boldsymbol{\tau}$$

or

$$\boldsymbol{\tau} = \frac{d\mathbf{J}}{dt}.$$

We see that the torque is again equal to the rate of change of angular momentum.

For three-dimensional curvilinear motion (and for two-dimensional nonplanar curvilinear motion), the description of rotational motion becomes a little more complicated. For two-dimensional planar motion, the vectors $\boldsymbol{\tau}$, $\mathbf{J}$, $\boldsymbol{\omega}$, and $\boldsymbol{\alpha}$ have all been perpendicular to the plane of motion. However, it is not true in general that these vectors all point in the same direction. As a simple example, we shall consider the motion of a particle at the end of a string of negligible mass.

Case 1. Let us suppose that the string is fixed at one end, the particle is suspended at the other end, and only the earth's gravitational field is acting on the particle. The particle would then come to equilibrium at a point directly below the point of suspension. Suppose that the particle is now moved from its equilibrium position, keeping the string taut, so that the string moves through some angle θ less than 90° as shown in Fig. 8.4(a). If the particle is then released, it moves back toward its equilibrium position, and an oscillation results. This system is known as a **simple pendulum**; a mathematical description of the oscillation is given in Section 10.3. However, for the time being we are interested only in the torque and angular momentum of the particle.

The forces acting on the particle are the **tension** in the string $\mathbf{T}$ and the force of gravity $\mathbf{F}_g$. These forces sum to give a resultant force $\mathbf{F}$, which acts to bring the

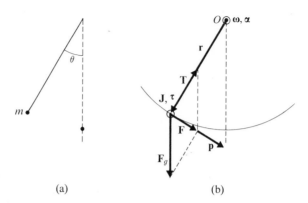

Fig. 8.4. (a) A simple pendulum displaced through an angle θ. (b) The torque and angular momentum vectors τ and $\mathbf{J}$ are both perpendicular to the plane of oscillation as are the vectors $\boldsymbol{\omega}$ and $\boldsymbol{\alpha}$.

particle back toward its equilibrium position. We see from Fig. 8.4(b) that the torque $\boldsymbol{\tau} = \mathbf{r} \times \mathbf{F}$ is perpendicular to the plane of oscillation, as is the angular momentum $\mathbf{J} = \mathbf{r} \times \mathbf{p}$. The vectors $\boldsymbol{\omega}$ and $\boldsymbol{\alpha}$ are also perpendicular to the plane of oscillation, and $\boldsymbol{\tau}, \mathbf{J}, \boldsymbol{\omega}$, and $\boldsymbol{\alpha}$ all point in the same direction. (Note that in one half of the oscillation the four vectors point in one direction in space, and in the other half of the oscillation they are all reversed in direction.) We also see that $\boldsymbol{\tau} = 0$ when the particle is passing through its equilibrium position, whereas $\mathbf{J} = 0$ at the end points of the oscillation. The torque has a maximum value at the end points of the oscillation; the angular momentum is maximum at the equilibrium position. The vector $\mathbf{J}$ always lies along a line perpendicular to the plane of oscillation, so that the vectors $\Delta\mathbf{J}$, which represent changes in $\mathbf{J}$, lie along this line also. The torque vector $\boldsymbol{\tau}$ lies along this same line, and

$$\boldsymbol{\tau} = \frac{d\mathbf{J}}{dt}.$$

Case 2. Let us suppose that the string is fixed at one end and the particle moves in a circular path at constant speed v in a horizontal plane, as shown in Fig. 8.5. This system is known as a **conical pendulum**. The forces acting on the particle are the force of gravity $\mathbf{F}_g$ and the tension in the string $\mathbf{T}$. The vector sum of $\mathbf{F}_g$ and $\mathbf{T}$ is a vector

$$\mathbf{F} = \mathbf{F}_g + \mathbf{T}$$

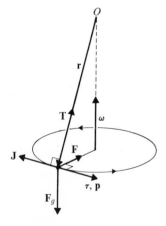

Fig. 8.5. A conical pendulum. The torque $\boldsymbol{\tau}$ is in the same direction as the momentum $\mathbf{p}$ but is not in the same direction as the angular momentum $\mathbf{J}$ or the angular frequency $\boldsymbol{\omega}$.

lying in the plane of rotation of the particle and providing the centripetal force necessary to keep the particle moving in a circular path. The torque $\boldsymbol{\tau}$ is in the direction of $\mathbf{r} \times \mathbf{F}$, is tangent to the path of the particle, and consequently points in the same direction as the momentum vector $\mathbf{p}$. The angular momentum $\mathbf{J}$ is in the direction of $\mathbf{r} \times \mathbf{p}$, so that $\boldsymbol{\tau}$ and **J are not in the same direction** for a conical pendulum. These vectors are pictured in Fig. 8.5. Also, the angular-frequency vector $\boldsymbol{\omega}$ lies along the axis of rotation, which is perpendicular to the plane of motion.

Since the vectors **r** and **p** have constant magnitude, **J** has a constant magnitude, but its direction in space varies periodically with time. This is pictured in Fig. 8.6(a). If at some time t the angular momentum is $\mathbf{J}(t)$ and at some time Δt later it is $\mathbf{J}(t + \Delta t)$, the vector $\Delta\mathbf{J}$ points in a horizontal direction. In the limit $\Delta t \to 0$, $\Delta\mathbf{J} \to d\mathbf{J}$, and $d\mathbf{J}$ is in the same direction as $\boldsymbol{\tau}$, since both $\boldsymbol{\tau}$ and $d\mathbf{J}$ are perpendicular to **J**, as can be seen from Figs. 8.5 and 8.6. The angular-momentum vector is said to **precess** about a vertical axis. Note that **J** precesses in the direction of the torque vector $\boldsymbol{\tau}$, that is, in a horizontal direction.

The torque and angular momentum are still related by the equation

$$\boldsymbol{\tau} = \frac{d\mathbf{J}}{dt},$$

since

$$\frac{d\mathbf{J}}{dt} = \mathbf{r} \times \frac{d\mathbf{p}}{dt} + \frac{d\mathbf{r}}{dt} \times \mathbf{p}$$
$$= \mathbf{r} \times \mathbf{F};$$

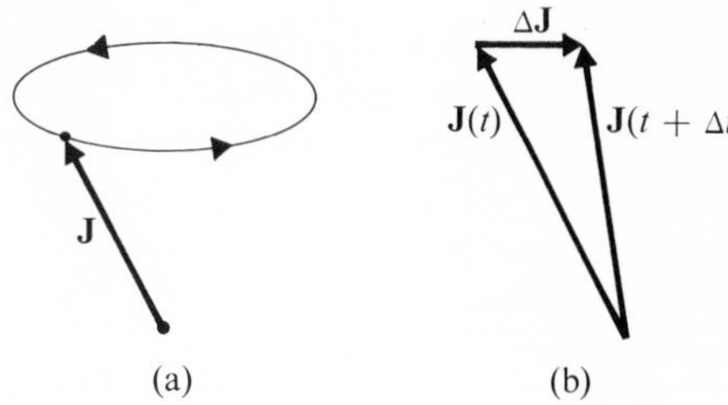

Fig. 8.6. (a) The angular momentum **J** changes direction periodically with time. (b) The change $\Delta\mathbf{J}$ in angular momentum during time Δt is in the horizontal direction.

the centripetal force **F** is responsible for the change in orientation of the vector **p**. Note that the centripetal force **F** is perpendicular to **p** at all times and changes only the direction of **p**, not its magnitude.

Example. A conical pendulum of length 1.00 m and mass 1.00 kg has a period of 1.00 sec, and the speed of the particle in its circular motion is $1.00 \text{ m} \cdot \text{sec}^{-1}$.
(a) Determine the radius of the circular path of the particle and the half-angle of the cone.
(b) Show numerically that the torque is equal to the rate of change of angular momentum.

Solution.
(a) The angular frequency is given by

$$\omega = \frac{2\pi}{T} = \frac{2\pi}{1.00} = 2.00\pi \text{ rad} \cdot \text{sec}^{-1}.$$

The radius of the circular path is b, where

$$v = \omega b$$

or

$$b = \frac{v}{\omega} = \frac{1.00}{2.00\pi} = 0.159 \text{ m}.$$

The half-angle θ of the cone (see Fig. 8.7) is given by the relation

$$\tan \theta = \frac{0.159}{1.00} = 0.159$$

$$\theta = 9.04°.$$

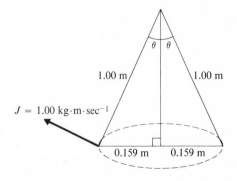

Fig. 8.7. The string of a conical pendulum sweeps out a cone of half-angle θ.

The half-angle of the cone is 9.04°.
(b) The angular momentum about the point of suspension is

$$J = rp = rmv = 1.00(1.00)(1.00) = 1.00 \, \text{kg} \cdot \text{m}^2 \cdot \text{sec}^{-1}.$$

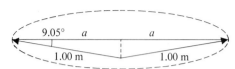

Fig. 8.8. The motion of the angular moment-um vector of the conical pendulum about the point of suspension.

The tip of the angular-momentum vector moves about a circular path of radius a (see Section 6.5 and Fig. 8.8), where a is given by

$$\cos 9.04 = \frac{a}{1.00}$$

$$a = 0.988 \, \text{m}$$

The period of rotation is 1.00 sec, so that the tip of the angular-momentum vector moves through a distance $2\pi a$ in 1.00 sec. Therefore,

$$\frac{|\Delta \mathbf{J}|}{\Delta t} = \frac{2\pi(0.988)}{1.00} = 6.20 \, \text{kg} \cdot \text{m}^2 \cdot \text{sec}^{-2}.$$

The torque about the point of suspension (see Fig. 8.9) is

$$\tau = rF \sin(90 + \theta) = \frac{rmv^2}{b} \cos \theta,$$

since the centripetal force is given by

$$F = \frac{mv^2}{b}.$$

Therefore,

$$\tau = \frac{(1.00)(1.00)(1.00)(0.988)}{0.159} = 6.20 \, \text{kg} \cdot \text{m}^2 \cdot \text{sec}^{-2}.$$

The torque is indeed equal to the rate of change of angular momentum.

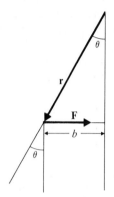

Fig. 8.9. The centripetal force acting on the bob of the conical pendulum is in a horizontal direction.

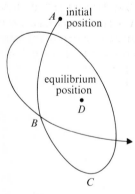

Fig. 8.10. A projection on a horizontal plane of the general motion of a particle on the end of a massless string.

Case 3. Let us suppose that the string is fixed at one end and the particle is moved from its equilibrium position, keeping the string taut. However, instead of letting the particle fall back towards its equilibrium position, as in simple pendulum motion, or giving it a horizontal circular motion, as in the conical pendulum, let us give it a push in a direction intermediate to these two special cases. The particle will then start to move in some curved path, but its motion will not be planar as it was for the simple and conical pendulums. Figure 8.10 shows a projection of its path on a horizontal plane. From points A to B the particle is falling below its initial position, from B to C it is rising, from C to D it is falling, and so forth.

A mathematical description of this general motion is beyond the scope of this text. We can note, however, that the resultant force **F** on the particle is no longer horizontal, in general, and has a component vector along the line of motion of the particle. This component vector changes with time and is responsible for the up-and-down portion of the particle's motion. The angular-momentum vector **J** again precesses but in a more complicated manner than for the conical pendulum, since both τ and **J** vary in both magnitude and direction due to the variation in **F**. This more complicated precession is evident from Fig. 8.10, where it is seen that the particle path is not a closed one.

8.4   MOMENT OF INERTIA

Let us consider again the motion of a conical pendulum. The particle at the end of the string executes circular motion at speed v, the circular path having some radius b depending upon the initial condition when motion began. The kinetic energy associated with the rotational motion is

$$E_r = \tfrac{1}{2}mv^2,$$

where m is the mass of the particle. Since $v = \omega b$, we can write E_r as

$$E_r = \tfrac{1}{2}mb^2\omega^2 = \tfrac{1}{2}I\omega^2,$$

where

$$I = mb^2$$

is the **moment of inertia** of the particle about an axis through the center of the circle and perpendicular to the plane of the motion. In general, the moment of inertia of a particle about any axis of rotation is defined as

$$I = mr^2,$$

where r is the perpendicular distance from the particle to the axis of rotation.

We should note here that the moment of inertia of a particle about an axis is not necessarily a constant of the motion. For example, in the case of a particle undergoing circular motion, if one tries to calculate the moment of inertia about an axis inclined to the plane of motion, it is immediately obvious that the perpendicular distance of the particle from the axis varies with time. However, the moment of inertia about the axis of rotation is a constant of the motion.

If we compare the expressions for translational kinetic energy K and rotational kinetic energy E_r,

$$K = \tfrac{1}{2}mv^2 \quad \text{and} \quad E_r = \tfrac{1}{2}I\omega^2,$$

we see that the moment of inertia plays a role in rotational motion similar to that of mass in translational motion if the moment of inertia is a constant of the motion, whereas angular frequency plays a role similar to that of speed in translational motion.

8.5   COLLECTIONS OF PARTICLES INTO RIGID BODIES

A rigid body is simply a collection of a very large number of particles into a system in which the separation between particles remains constant with time. The moment of inertia of a rigid body about a given axis of rotation can be calculated as the sum of the moments of the individual particles comprising the body. In general, an integration over three-dimensional space is required.

As we pointed out in Section 7.3, the mass of an object can be considered to be concentrated at one point in space known as the center of mass of the object. When an external force is applied to the object, the resulting motion may involve both translational motion of the object plus rotation about the center of mass. The mathematical description of this behavior is given in Chapters 13 and 14. It is also possible for a rigid body to rotate about some point other than its center of mass if a proper torque is applied.

Gravitational force normally acts on a rigid body to produce translational motion only. A baseball thrown into the air will have translational kinetic energy that will first decrease as the ball rises into the air and then increase as the ball falls to the ground, plus rotational kinetic energy about the center of mass imparted in the act of throwing the ball by a torque produced by the thrower's hand.

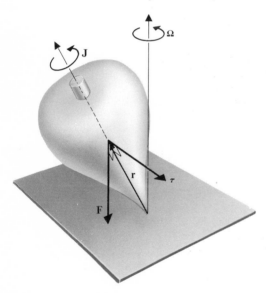

There is a particularly interesting case of rotational motion in which the force of gravity does produce a torque on a rotating body and, in so doing, alters its motion in a striking manner. When a top is set spinning in such a manner that the axis of symmetry about which it spins is not vertical, the force of gravity on the top produces a torque about its point of contact with the floor. This torque τ is in a horizontal direction as shown in Fig. 8.11. This horizontal torque causes the angular-momentum vector **J** of the spinning top to precess about a vertical axis, just as did the angular-momentum vector for the conical pendulum. The angular frequency of precession is Ω and is related to **J** and τ by the relation[1]

$$\tau = \Omega \times J.$$

Fig. 8.11. The torque on a spinning top causes the angular momentum vector **J** to precess about a vertical axis with angular frequency Ω.

The surprising behavior of the spinning top[1] is that, instead of falling over under the action of the torque due to gravity, it precesses about a vertical axis. Precession continues with a real top while the top gradually loses its rotational energy due to friction at its point of contact with the floor. This frictional effect produces a retarding torque that reduces the value of **J** gradually. As

[1] A. Lemonick, "Angular Momentum, a Vector Quantity," a film from Educational Services Incorporated, Watertown, Mass.

rotational energy is lost the inclination of the top to the vertical gradually increases, until the energy is spent and the top comes to rest on the floor.

Since in this text we are primarily concerned with the behavior of single particles and the way in which the behavior of systems of particles, both small and large, can be described in terms of the constituent particles, we shall not go into detailed discussion of the motion of rigid bodies at this point. The reader interested in more details can consult Feynman, et al.[2], or almost any textbook on mechanics.

PROBLEMS

1. (a) A 2000-kg satellite is in a circular orbit about the earth at an altitude of 500 km. What is its angular momentum if the period is 93 min?
   (b) A small rocket engine on the satellite can be fired to provide a thrust in a direction opposite to the direction of motion of the satellite. The engine is turned on for 10 sec and provides a constant force of 1.4×10^4 N on the satellite during this period. What is the torque on and the change in angular momentum of the satellite?

2. In its normal state, the electron in a hydrogen atom in the simple Bohr model moves in a circle of radius 0.528×10^{-10} m with a frequency of 6.58×10^{15} rev $\cdot$ sec^{-1}. The electron mass is 9.11×10^{-31} kg. What is the orbital angular momentum of the hydrogen atom in the normal state? Sketch the path of the electron, indicating the direction of its velocity and of the angular-momentum vector.

3. A man sitting on a swivel chair equipped with a frictionless pivot holds a rapidly spinning wheel with its axle horizontal. What happens when he turns the axis of the wheel into the vertical direction?

4. A cyclotron is a device for accelerating charged particles of interest to the nuclear physicist. In a cyclotron the particles travel in circular paths in a magnetic field. The magnetic force on a particular particle is 1.0×10^{-11} N, the mass of the particle is 1.7×10^{-27} kg, and the radius of the path it describes is 0.33 m.
   (a) What is the centripetal force experienced by the particle?
   (b) What is the speed of the particle?
   (c) What is the rotational kinetic energy of the particle?
   (d) How long does it take the particle to go once around its circular path?
   (e) What is the angular momentum of the particle?

5. A proton is accelerated by an electric field and acquires a kinetic energy of 1000 electron volts (eV). It then moves horizontally into a region of uniform magnetic field acting on the proton with a horizontal force of $(2.0 \times 10^{-20})v$ N (where v

[2] R. P. Feynman, R. B. Leighton, and M. Sands, *The Feynman Lectures on Physics* (Reading, Mass.: Addison-Wesley Publishing Company, Inc., 1963), Chapter 20.

is the speed of the proton) directed always at right angles to the motion of the proton. Determine the motion of the proton in the magnetic field. Does the force due to the magnetic field do work on the proton? What is the torque on the proton?

6. A constant torque of 10^{-2} N·m is applied in an ultracentrifuge for 45 sec to a sample initially at rest. If the sample is situated 25 cm from the axis of rotation, determine the angular frequency and angular momentum of the particle after 45 sec. What is the centripetal force on the sample?

7. An alpha particle (the nucleus of a helium atom) is a particle emitted in the decay of some heavy radioactive elements. Ernest Rutherford in the first few years of this century discovered the nucleus of the atom by bombarding atoms with alpha particles and studying the way in which their paths through space were affected by interactions with the atoms. In Fig. 8.12, the path of an alpha particle being deflected by the electric force between the alpha particle and the

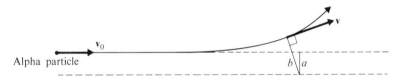

Fig. 8.12

nucleus of an atom is shown. If the alpha particle were not deflected from its path, it would pass the nucleus at a distance a known as the **impact parameter**. If an alpha particle with an energy of 8.8×10^{-13} J is incident on a nucleus with an impact parameter of 3.8×10^{-15} m, calculate the initial angular momentum of the alpha particle about the nucleus before any deflection takes place. At the closest approach b of the alpha particle to the nucleus during the scattering, the velocity vector $\mathbf{v}$ is perpendicular to the line joining the alpha particle to the nucleus. If $b = 4.7 \times 10^{-15}$ m for the alpha particle mentioned above, calculate the speed of the alpha particle at the closest approach to the nucleus.

*8. Show that

$$\tau = \Omega \times J$$

for the conical pendulum in the example in Section 8.3, where Ω is the angular frequency of precession of $\mathbf{J}$.

9. Show that the motion of a planet about the sun must be in a plane containing the planet and the sun, neglecting the presence of the other planets.

10. A simple pendulum of mass 1.0 kg and length 0.75 m is oscillating about its equilibrium position with a maximum angle of 10°. Plot a graph of torque versus angular displacement of the pendulum. Comment on the form of the graph.

11. A particle of mass 0.5 kg is suspended on the end of a 0.50 m long string of negligible mass.

(a) Determine the moment of inertia of the particle about a horizontal axis through the point of suspension and perpendicular to the plane of motion when the particle and string are moving as a simple pendulum.

(b) The particle is set into motion as a conical pendulum with a half-angle of 50°. Calculate the moment of inertia of the particle about the axis of rotation.

12. Calculate the rotational energy of the conical pendulum of Problem 11(b).

13. Two particles of mass 2.0 g and 5.0 g are joined together by a rigid rod of negligible mass of length 14 cm. Calculate the moment of inertia of the two-particle system

(a) about an axis perpendicular to the line joining the particles and midway between them.

(b) about an axis perpendicular to the line joining the particles and a distance 4 cm from the 5.0-g particle.

(c) about an axis through a point 4 cm from the 5.0-g particle along the line joining the particles and inclined at an angle of 60° to the line joining the particles.

14. A conical pendulum of length l is rotating in a circle of radius a. If the half-angle of the cone is θ, show that the angular momentum $\mathbf{J}_1$ about the point of suspension is related in magnitude to the angular momentum $\mathbf{J}_2$ about an axis through the point of suspension and perpendicular to the plane of rotation by the equation

$$J_2 = J_1 \sin \theta.$$

What is the angle between $\mathbf{J}_1$ and $\mathbf{J}_2$?

*15. A **gyroscope** is a small, well-balanced wheel mounted on bearings of very low friction, with the wheel and its axis in turn mounted on a low-friction pivot such that the axis of the gyroscope is free to move about in space. As a joke, a physicist once concealed a gyroscope, which was spinning at a very high rate, in a suitcase in such a manner that the axis of the gyroscope was pointing along the length of the suitcase. He then checked the bag with a baggage clerk. Describe what would happen when the clerk picked up the suitcase to send it on to its destination. Consider all the possible motions he might try to make with the suitcase.

16. A gyroscope such as the one described in Problem 15 is spinning at a high rate of speed and has its axis horizontal. In what direction must a force be applied to turn the axis into the vertical direction?

9 Conservation laws

9.1 INTRODUCTION

A number of the properties of particles that we have introduced in the first few chapters of this text are found to be constant independent of particle motion under suitable conditions. Momentum, angular momentum, and energy are examples of quantities that may be conserved. The conservation laws embodying the conditions under which various quantities remain constant have been established experimentally. However, they can be shown to be consequences of certain symmetries that exist in nature.

9.2 CONSERVATION OF MOMENTUM

Let us discuss a linear-air-track experiment to obtain further insight into the concept of momentum. We make use of two gliders with magnets attached to them so that they tend to repel one another. Initially the gliders are a large distance apart in the sense that the force on each one due to the magnetic interaction is negligible. Glider (1) of mass $2M$ is moving with speed v relative to the track and is approaching glider (2) of mass M from the right in Fig. 9.1. Glider (2) is at rest relative to the track. During the ensuing collision energy is exchanged between the gliders via their magnetic interaction.

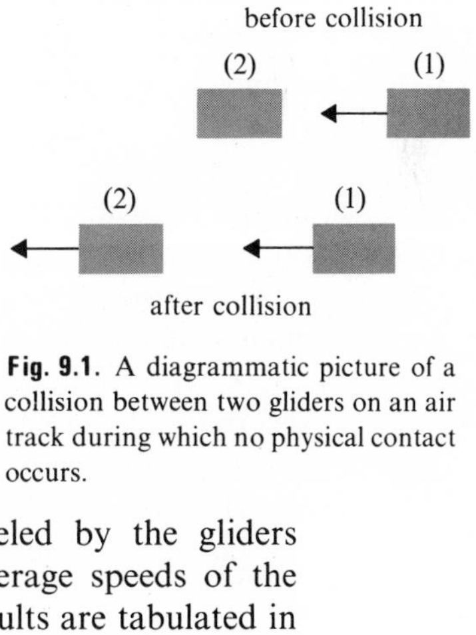

Following the collision, both gliders are observed to be moving toward the left. We consider this example both to emphasize that a collision does not necessarily involve physical contact and also to emphasize that interactions may extend over large enough distances for measurements to be made.

The positions of both gliders were recorded at equal intervals of time by using the techniques of stroboscopic photography. The distances traveled by the gliders between strobe flashes are proportional to the average speeds of the gliders in the time intervals between flashes. The results are tabulated in Table 9.1.

The results are seen to fall into three regions. In region I the speed of glider (1) is essentially constant and directed to the left and glider (2) is at rest. Region II is the region of interaction; both gliders are changing their speeds continuously. Finally, in region III the speeds of both gliders are again essentially constant, and the gliders are moving in the same direction.

before collision

(2)   (1)

after collision

Fig. 9.1. A diagrammatic picture of a collision between two gliders on an air track during which no physical contact occurs.

155

Table 9.1   Momentum and energy changes in a collision involving no physical contact

Region	Time interval	Δl_1 (cm)	Δl_2 (cm)	Δp_1	Δp_2	K_1	K_2	K
	1	20.5	—	$0.0M$	—	$420M$	—	$420M$
I	2	20.0	—	$-1.0M$	—	$400M$	—	$400M$
	3	20.0	—	$0.0M$	—	$400M$	—	$400M$
	4	19.0	1.5	$-2.0M$	$1.5M$	$380M$	M	$381M$
	5	16.0	7.0	$-6.0M$	$5.5M$	$255M$	$25M$	$280M$
II	6	12.5	14.5	$-7.0M$	$7.5M$	$155M$	$105M$	$260M$
	7	9.0	21.5	$-7.0M$	$7.0M$	$80M$	$230M$	$310M$
	8	7.0	25.0	$-4.0M$	$3.5M$	$50M$	$315M$	$365M$
	9	6.5	26.5	$-1.0M$	$1.5M$	$40M$	$350M$	$390M$
III	10	6.5	27.0	$0.0M$	$0.5M$	$40M$	$365M$	$405M$
	11	7.0	27.0	$1.0M$	$0.0M$	$50M$	$365M$	$415M$

Δl_1 and Δl_2 are the distances traveled by gliders (1) and (2), respectively. Δp_1 and Δp_2 are momentum changes of gliders (1) and (2), respectively, in relative units. K_1 and K_2 are kinetic energies of gliders (1) and (2), respectively, in relative units. K is the total kinetic energy ($K = K_1 + K_2$).

From Table 9.1 we see that the changes in momentum Δp_1 and Δp_2 of the two gliders are, for every interval of time Δt, equal in magnitude and opposite in sign within the experimental uncertainty of $\pm 0.5M$; that is,

$$\Delta p_1 = -\Delta p_2$$

for all intervals Δt. The general statement of this result is that the linear momentum of an isolated system (the two gliders in our case) is a constant of the motion. This result is known as the **law of conservation of linear momentum**. It holds for all systems that experience no forces other than those forces acting between the particles in the system; such systems are referred to as isolated systems.

According to Newton's second law of motion, the average force $\bar{F}_{21}$ exerted on glider (2) by glider (1) during the time interval Δt is equal to the change in momentum Δp_2 of glider (2)/Δt; that is,

$$\bar{F}_{21} = \frac{\Delta p_2}{\Delta t}.$$

Similarly, $\bar{F}_{12}$, the average force exerted on glider (1) by glider (2), is equal to the change in momentum Δp_1 of glider (1)/Δt; that is,

$$\bar{F}_{12} = \frac{\Delta p_1}{\Delta t}.$$

Since $\Delta p_1 = -\Delta p_2$, we may also write

$$\frac{\Delta p_1}{\Delta t} = -\frac{\Delta p_2}{\Delta t}.$$

Therefore, from the experimental data we can conclude that

$$\bar{F}_{12} = -\bar{F}_{21}$$

for all intervals Δt. It is reasonable to extrapolate this result to arbitrarily small time intervals between strobe flashes, so that the average forces become instantaneous forces and

$$F_{12} = -F_{21}$$

for all times. What we have just verified is the general property of forces known as **Newton's third law**, namely, that forces occur in pairs, equal in magnitude but oppositely directed. Note that the pairs of forces *never* act on the same body. We may consider this glider experiment to have verified Newton's third law for the particular system of two gliders interacting via magnetic forces.

We shall now consider two particles of masses m_1 and m_2 and initial velocities v_1 and v_2, respectively. Suppose they interact through some mutual force of interaction and after a time Δt their velocities are v_3 and v_4, respectively. This situation is pictured in Fig. 9.2. The change in velocity of m_1 is

$$\Delta v_1 = v_3 - v_1,$$

and the change in velocity of m_2 is

$$\Delta v_2 = v_4 - v_2.$$

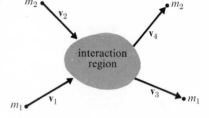

Fig. 9.2. Particles of masses m_1 and m_2 and velocities v_1 and v_2 undergo a mutual interaction which changes the velocity of m_1 to v_3 and the velocity of m_2 to v_4.

From Newton's second law the average forces felt by the particles during the interaction are given by the equations

$$F_{12} = m_1 \frac{\Delta v_1}{\Delta t} \quad \text{and} \quad F_{21} = m_2 \frac{\Delta v_2}{\Delta t},$$

respectively. If the two-particle system is isolated, these two forces are the only forces acting on the particles. We now invoke Newton's third law and put

$$F_{12} = -F_{21}$$

or

$$F_{12} + F_{21} = 0.$$

Therefore,

$$m_1 \frac{\Delta v_1}{\Delta t} = - m_2 \frac{\Delta v_2}{\Delta t}$$

or

$$m_1(\mathbf{v}_3 - \mathbf{v}_1) = - m_2(\mathbf{v}_4 - \mathbf{v}_2)$$

and

$$m_1\mathbf{v}_1 + m_2\mathbf{v}_2 = m_1\mathbf{v}_3 + m_2\mathbf{v}_4.$$

The left-hand side of this equation gives the total momentum before the interaction, and the right-hand side is the total momentum after the interaction; therefore, the total momentum is conserved. Because we have not considered the time explicitly, we must have the total momentum conserved at any instant of time. We would have obtained the same result for isolated systems of more than two particles interacting, since the sum of all the forces of interaction between the particles is always zero.

Example. A billiard ball of mass m is propelled at a speed of $5\,\mathrm{m \cdot sec^{-1}}$ against an identical second stationary billiard ball, as shown in Fig. 9.3. The second ball moves off at an angle of 60° to the initial direction of motion, whereas the first ball is deflected through an angle of 30°. What is the speed of each ball immediately after the collision?

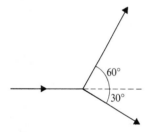

Fig. 9.3. After a collision between two billiard balls, one moves off at 60° and one at 30° to the direction of the incident ball.

Solution. Let v_1 and v_2 be the speeds of the first and second balls, respectively, immediately after the collision. The momentum of the first ball before the collision is $5m\,\mathrm{kg \cdot m \cdot sec^{-1}}$. The momentum of the two balls after the collision may be written in terms of components in the same direction as, and perpendicular to, the direction of motion of the incident ball. The components in the forward direction are $mv_1 \cos 30°$ and $mv_2 \cos 60°$ and in the perpendicular direction are $mv_1 \sin 30°$ and $mv_2 \sin 60°$. From conservation of momentum

$$5m = mv_1 \cos 30° + mv_2 \cos 60°$$

$$mv_1 \sin 30° = mv_2 \sin 60°.$$

Therefore,

$$0.866v_1 + 0.500v_2 = 5$$

$$0.500v_1 = 0.866v_2,$$

which yields

$$v_1 = 4.33\,\mathrm{m \cdot sec^{-1}}, \qquad v_2 = 2.50\,\mathrm{m \cdot sec^{-1}}$$

for the speeds of the two balls immediately after the collision.

9.3   CONSERVATION OF ENERGY

The kinetic energies of the gliders involved in the collision described in Section 9.2 are also tabulated in Table 9.1. In region II it is apparent that glider (1) is losing kinetic energy whereas glider (2) is gaining kinetic energy, but the sum of the kinetic energies is not constant. However, after the interaction is over, the total kinetic energy is again equal to the initial kinetic energy, or at least almost so, within the experimental uncertainties involved in making the measurements. During the interaction kinetic energy is lost temporarily but is then regained.

Before pursuing the question of the variable kinetic energy further, let us first consider what would happen if the magnets were removed from the gliders and the gliders were provided with sticky couplers that lock together upon collision. The momentum of glider (1) before the collision is $2Mv$, and the momentum of the coupled system of mass $2M + M = 3M$ after the collision is $3Mv'$, where v is the speed of glider (1) before the collision and v' is the speed of the coupled system. Invoking conservation of momentum we have

$$3Mv' = 2Mv$$

$$v' = \tfrac{2}{3}v.$$

The total kinetic energy before the collision is

$$\tfrac{1}{2}(2M)v^2 = Mv^2,$$

whereas the total kinetic energy after the collision is

$$\frac{1}{2}(3M)\left(\frac{2v}{3}\right)^2 = \frac{2}{3}Mv^2.$$

Therefore, kinetic energy **is not conserved in this collision**.

We have information now about two different types of collision: one in which kinetic energy appears to be conserved (within experimental uncertainties) when the collision is over, but in which kinetic energy is variable during the collision; and one in which kinetic energy is not conserved. Let us consider the magnetic-glider collision first. We can explain the temporary loss of kinetic energy by saying that some of the kinetic energy was changed into potential energy. We see from the results of Section 7.7 that this is a reasonable explanation, since during the collision the magnetic force between the two gliders is not negligible and work must be done by the gliders as they approach each other. This work is provided by the loss in kinetic energy but is stored in the system as potential energy. When the two gliders separate, the force does work on the gliders, and potential energy is changed back into kinetic energy. The energy changes are illustrated in Fig. 9.4.

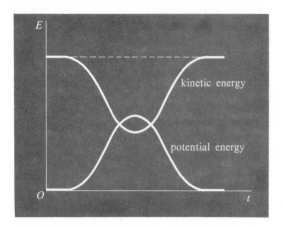

Fig. 9.4. During a collision between two gliders on an air track the kinetic energy of the gliders is reduced momentarily; this energy is stored in the system as potential energy.

If the measurement of the kinetic energy before and after the collision of the gliders is carried out more precisely, it becomes apparent that the total kinetic energy after the collision is actually a little bit less than the total kinetic energy before the collision. Collisions for which kinetic energy is very nearly conserved after the collision are said to be **elastic**. A **perfectly elastic** collision is one in which no kinetic energy is lost. The closest examples to such collisions are those that take place between atoms in a gas, or between neutrons and protons in a reactor, or generally many of the collisions involving nuclear or atomic particles. Those collisions in which more than a very small fraction of the kinetic energy is lost are termed **inelastic**.

What happens to the kinetic energy that is lost in most collisions? A close investigation shows that objects that collide tend to be hotter and to vibrate as a result of the collision. As we shall see later (beginning in Chapter 17), the heat content of an object is associated with the fact that atoms making up the object are jiggling in a random manner about their normal equilibrium positions. The kinetic energy associated with these random atomic motions is known as **heat energy**. During a collision some of the kinetic energy associated with the motion of an object as a whole is transformed into kinetic energy associated with the random motions of the atoms in the object. Thus, the heat energy rises, and the object becomes warmer. Also, some objects will tend to vibrate as a whole as a result of a collision, so that some kinetic energy can go into vibrational energy. Objects (such as those made of steel) in which very little heat and vibration is generated as a result of collisions are said to be elastic.

There are many other forms of energy, such as electrical energy, radiant energy (e.g., that of the sun), chemical energy, nuclear energy, mass energy (see Chapter 12), etc. If any number of objects are brought together and allowed to interact with each other through any one or any combination of the fundamental forces discussed in Section 7.6, and if we measure the initial amount of energy available to the system so assembled, taking into account all the forms in which energy may appear, and if we then measure the amount of energy that the system contains at some later time, we find that the total energy of the system is a constant independent of the time (provided that we take account of the fact that some energy may cross the boundary between the system and its surroundings, for example, in the form of heat energy or radiant energy). It is a well-established experimental fact that energy is conserved; it may appear in any one of several forms, and the amount appearing in any one form may change with time. Thus, it is observed that energy is always conserved, at least in that region of the universe that is accessible to us. The **law of the conservation of energy** is a formal recognition of this fundamental behavior of nature. The conservation of the mechanical energy of a system is a consequence of Newton's laws and the concept of a conservative force. For a conservative force, the work done in moving an object from one point to another in the presence of the force is independent of the path followed between the points. If this were not so, the potential energy could not be defined uniquely, and the total energy would in general depend upon the path followed. When we deal with nonconservative forces we must bring in other forms of energy, such as heat, in order to conserve energy. Friction, for example, is a nonconservative force. If we move a box from point A to point B by sliding it over the floor, the amount of work we have to do obviously depends upon the path, and the total mechanical energy of the system consisting of us and the box is not conserved. However, if we take account of the heat generated due to friction, the total energy is found to be conserved.

The conservation laws are particularly useful in situations in which we either cannot or do not wish to follow the time development of a system in detail. In this sense they go beyond Newton's laws. In the glider interaction via magnetic forces, for example, conservation of energy tells us how the potential energy must vary during the collision, and there is no need to calculate the potential energy as a function of time, which requires knowing the magnitude and direction of the magnetic force as a function of separation. However, we should emphasize that the use of forces and Newton's laws allows us to solve any problem (in principle) and to obtain the details of the motion among observable quantities, whereas conservation laws merely give general relations. In this sense Newton's laws are more useful than conservation laws.

Example. Suppose two gliders of masses M and $2M$ are moving along an air track with velocities $10 \, \text{m} \cdot \text{sec}^{-1}$ and $-5 \, \text{m} \cdot \text{sec}^{-1}$, respectively. Find the final velocities v_1 and v_2 of the two gliders, assuming an elastic collision and using the two conservation laws that we have just discussed.

Solution. From the law of conservation of linear momentum

$$10M - 5(2M) = M(v_1 + v_2),$$

and from the law of conservation of energy

$$\tfrac{1}{2}M(100) + \tfrac{1}{2}2M(25) = \tfrac{1}{2}M(v_1^2 + 2v_2^2).$$

From these equations

$$v_1 + 2v_2 = 0, \qquad v_1^2 + 2v_2^2 = 150.$$

Solving yields

$$v_1 = -10\ \text{m} \cdot \text{sec}^{-1} \quad \text{and} \quad v_2 = 5\ \text{m} \cdot \text{sec}^{-1}$$

for the final velocities of the gliders. This prediction, when checked using a linear-air-track apparatus, is found to be correct.

Example.   An alpha particle (the nucleus of a helium atom) of mass 6.69×10^{-27} kg having a speed of 2.19×10^7 m · sec^{-1} is incident upon the stationary nucleus of a ^{12}C atom, the nucleus having a mass of 2.01×10^{-26} kg. The alpha particle is scattered through an angle of 60°, and the ^{12}C nucleus is left with a nuclear excitation energy of 4.43 MeV plus the kinetic energy of recoil from the collision. Calculate the speeds of the alpha particle and the ^{12}C nucleus immediately after the collision and determine the direction in which the ^{12}C nucleus is traveling.

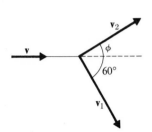

Fig. 9.5. An alpha particle colliding with a ^{12}C nucleus scatters through an angle of 60° while the ^{12}C nucleus recoils at an angle ϕ.

Solution. The details of the collision are pictured in Fig. 9.5. Let m_1 = mass of alpha particle, m_2 = mass of ^{12}C nucleus, v = initial speed of alpha particle, v_1 = speed of alpha particle immediately after the collision, v_2 = speed of ^{12}C nucleus immediately after the collision, ϕ = direction that velocity vector of ^{12}C nucleus makes with the initial direction of the alpha particle, E = nuclear excitation energy. From conservation of momentum, we have two equations,

$$m_1 v = m_1 v_1 \cos 60° + m_2 v_2 \cos \phi \tag{1}$$

$$m_1 v_1 \sin 60° = m_2 v_2 \sin \phi, \tag{2}$$

representing conservation of momentum in the initial direction of the alpha-particle motion and in the direction perpendicular to the initial direction of alpha-particle motion, respectively. From conservation of energy, we have

$$\tfrac{1}{2} m_1 v^2 = \tfrac{1}{2} m_1 v_1^2 + \tfrac{1}{2} m_2 v_2^2 + E. \tag{3}$$

Substituting known quantities into equation (3) yields

$$(6.69 \times 10^{-27})(2.19 \times 10^7)^2 = (6.69 \times 10^{-27}) v_1^2 + (2.01 \times 10^{-26}) v_2^2$$
$$+ 2(4.43)(1.6 \times 10^{-13})$$

$$3.21 \times 10^{-12} = 6.69 \times 10^{-27} v_1^2 + 2.01 \times 10^{-26} v_2^2$$
$$+ 1.42 \times 10^{-12}$$

$$6.69 \times 10^{-27} v_1^2 + 2.01 \times 10^{-26} v_2^2 = 1.79 \times 10^{-12}. \tag{4}$$

Squaring equations (1) and (2), we have

$$(m_2 v_2)^2 \cos^2 \phi = (m_1 v - m_1 v_1 \cos 60°)^2$$

$$(m_2 v_2)^2 \sin^2 \phi = (m_1 v_1 \sin 60°)^2.$$

Adding gives

$$(m_2 v_2)^2 = (m_1 v - m_1 v_1 \cos 60°)^2 + (m_1 v_1 \sin 60°)^2. \tag{5}$$

Substituting known quantities into equation (5) yields

$$(2.01 \times 10^{-26})^2 v_2^2 = (6.69 \times 10^{-27})^2 [(2.19 \times 10^7) - (0.500) v_1]^2$$
$$+ (6.69 \times 10^{-27})^2 (0.866 v_1)^2$$

$$v_2^2 = 0.111[2.19 \times 10^7 - 0.500 v_1]^2 + 0.0832 v_1^2. \tag{6}$$

Substituting from (6) into (4) for v_2^2 and solving gives

$$v_1 = 1.21 \times 10^7 \text{ m} \cdot \text{sec}^{-1}$$
$$v_2 = 6.36 \times 10^6 \text{ m} \cdot \text{sec}^{-1}.$$

Substituting for v_1 and v_2 in equation (2) yields

$$\sin \phi = \frac{m_1 v_1}{m_2 v_2} \sin 60°$$

$$= \frac{(6.69 \times 10^{-27})(1.21 \times 10^7)(0.866)}{(2.01 \times 10^{-26})(6.36 \times 10^6)}$$

$$= 0.549;$$

$$\phi = 33.3°.$$

The alpha particle has a final speed of $1.21 \times 10^7 \text{ m} \cdot \text{sec}^{-1}$, and the ^{12}C nucleus recoils with a speed of $6.36 \times 10^6 \text{ m} \cdot \text{sec}^{-1}$ at an angle of $33.3°$.

Example. Determine the height attained by a projectile launched with an initial velocity $\mathbf{v}_0$ at an angle θ to the horizontal.

Solution. The component of the velocity $\mathbf{v}_0$ in the vertical direction (see Fig. 9.6) is

$$v_y = v_0 \sin \theta.$$

The initial kinetic energy $K_y(0)$ in the vertical direction is

$$K_y(0) = \tfrac{1}{2} m v_0^2 \sin^2 \theta,$$

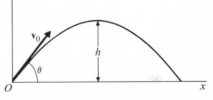

where m is the mass of the projectile. When the projectile reaches its maximum height h, the vertical component of the velocity is zero. Therefore, $K_y(h) = 0$, and all the energy has become potential energy, so that

$$mgh = \tfrac{1}{2} m v_0^2 \sin^2 \theta$$

or

$$h = \frac{v_0^2 \sin^2 \theta}{2g}$$

Fig. 9.6. A projectile fired with initial velocity **v** at an angle θ to the horizontal attains a maximum height h.

in agreement with the result of Section 7.3. The initial kinetic energy in the horizontal direction is

$$K_x(0) = \tfrac{1}{2} m v_0^2 \cos^2 \theta,$$

which remains constant since the horizontal component of the velocity is constant.

9.4   CONSERVATION OF ANGULAR MOMENTUM

We showed in Sections 8.2 and 8.3 that the torque acting on and the angular momentum of a particle are related by the expression

$$\tau = \frac{d\mathbf{J}}{dt},$$

where τ is the torque and $\mathbf{J}$ is the angular momentum. When $\tau = 0$, $\mathbf{J}$ is constant, so that angular momentum is conserved in an isolated system. Therefore, the law of conservation of angular momentum replaces the law of conservation of linear momentum when rotational motion is under discussion.

When two particles interact under mutual gravitational or electrical forces only, there is no torque on the two-particle system, since the forces act along the line joining the two particles. Therefore, the angular momentum of such a system is a constant of the motion. Since angular momentum is a vector, the direction and magnitude are both constant, so that the two-particle motion takes place in a plane of constant orientation. We shall make use of this fact in Chapter 13 when we discuss planetary motion.

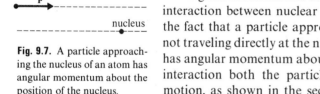

Fig. 9.7. A particle approaching the nucleus of an atom has angular momentum about the position of the nucleus.

Angular momentum is an important concept in the description of the interaction between nuclear particles. It is very often convenient to use the fact that a particle approaching the nucleus of an atom is normally not traveling directly at the nucleus, as illustrated in Fig. 9.7, and therefore has angular momentum about the position of the nucleus. In the ensuing interaction both the particle and the nucleus change their states of motion, as shown in the second example in Section 9.3. However, the angular momentum about the original position of the nucleus is conserved (see Problem 7 in Chapter 8). This fact is of great importance in the detailed description of such reactions according to the wave-mechanics description of nature (see Section 18.6 for a brief description of wave mechanics).

9.5 CONSERVATION LAWS AND INVARIANCE PRINCIPLES

A conservation law is a statement of constancy in nature. The most significant aspect of the conservation laws that we have discussed in this chapter is that they are laws of constancy during change; that is, during a process of change something is remaining constant. The classical laws of physics are expressed primarily as laws of change rather than as laws of constancy. Newton's second law, for example, describes how the motion of an object responds to the force that acts upon it, and Maxwell's equations of electromagnetism connect the rate of change of electric and magnetic fields in space and time.

Conservation laws have gradually risen to the top of the hierarchy of natural laws. They are very simple in form and thus appeal to the scientist, who is continually striving to describe nature in the most simple and most general manner. Much more important than this, however, is the connection between conservation laws and principles of invariance and symmetry in nature. The connection between conservation laws and invariance principles can be illustrated by the simple example of a collision of billiard balls worked out in Section 9.2. The paths of the two billiard balls are reproduced in Fig. 9.8. Also shown are three coordinate systems with origins O, O', O'' located at different points in the plane of the collision. Now it is an obvious fact that the details of such a collision are independent of the position of the coordinate axes used to describe the collision. Regardless of our choice of coordinate axes, the relation between the velocities of the balls before and after the collision is unchanged. The interaction is said to be **invariant** under the translation of the coordinate system. When the laws of motion are formulated mathematically, it can be proved explicitly that the invariance of a physical system under translation in any direction gives rise to the conservation of the component of linear momentum in that direction. Unfortunately, an adequate discussion of this principle requires the use of mathematics beyond the scope of this text.

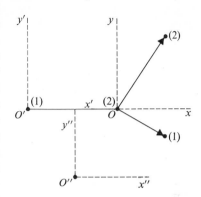

Fig. 9.8. A collision between two billiard balls is independent of translation of the coordinate system in which measurements are made.

Invariance with respect to translations in space is a consequence of the **homogeneity** of space. No matter where we are in space (at least in the regions to which we have so far had access) the properties of space seem to be the same. Experiments performed in one location give the same results (within experimental error) as identical experiments performed in other locations. The conservation of linear momentum is thus a consequence of the homogeneity of space.

Let us refer again to Fig. 9.8 and consider the angular momentum of the two billiard balls with respect to coordinate systems with origins O' and O''. For O' the initial angular momentum of each ball is zero; after the collision the angular moment of the two balls are equal and opposite. For O'' the total angular momentum after the collision is found to be equal to the initial angular momentum of the incident ball. The collision is shown again in Fig. 9.9, along with two sets of coordinate axes with the same origin, but with one set rotated with respect to the first. Now, it is an obvious fact that the details of the collision are independent of the orientation of the coordinate axes used to describe the collision. The interaction is invariant under rotation of the coordinate system. It can be shown that invariance under rotation requires that angular momentum be conserved.

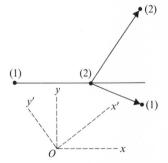

Fig. 9.9. A collision between two billiard balls is independent of the rotation of the coordinate axes.

Invariance with respect to rotations in space is a consequence of the **isotropy** of space. We do not expect, and indeed we do not find, that the results of experiments carried out in an isolated system depend upon the orientation of the system in space (that is, space has the same properties in all directions). The law of conservation of angular momentum is thus a consequence of the isotropy of space.

We have previously discussed the invariance of the form of the laws of nature for systems in uniform motion (see Chapter 4). This invariance is connected with the isotropy of four-dimensional space-time. What appears to us as uniform linear motion would appear to an observer able to perceive four-dimensional space-time to be a rotation in space-time. Moving uniformly in a smooth-riding vehicle, for example, is from the general point of view merely a turning from space partly toward the time direction. The laws of nature, according to relativity, should be no more changed by "turning" toward the time direction than by turning apparatus in a laboratory.

Turning back to the billiard ball collision, we note that if we were to carry out the experiment at some later time doing everything the same way, we would obtain the same result. In fact, identical experiments carried out at any time always give identical results. The details of the collision are found, then, to be independent of displacement or translation in time. This invariance with respect to displacement in time is found to give rise to the conservation of the energy of a system.

In summation, we see that the conservation laws of linear momentum, angular momentum, and energy are a consequence of the uniformity of space and time. The laws are independent of, or go beyond, the detailed mechanical description of real processes such as collisions, etc., which can only be described fully by reference to the forces that are acting and their variation with position and time. The conservation laws are thus more general or more fundamental than Newton's laws. They are always true even when applied to atomic systems, where the quantum nature of matter becomes important[1].

PROBLEMS

1. A hydrogen molecule of mass 3.34×10^{-27} kg traveling normal to a wall with a speed of $1500 \, \text{m} \cdot \text{sec}^{-1}$ collides elastically with the wall.
   (a) Calculate the change in momentum of the molecule upon collision.
   (b) If the interaction between the wall and the molecule lasts for 10^{-6} sec, calculate the average force exerted by the wall on the molecule during the interaction.

2. A 10.0-kg block is initially sliding on a frictionless surface at $50 \, \text{m} \cdot \text{sec}^{-1}$. It explodes into two parts, one of 10-kg mass and the other of 9.0-kg mass. The

[1] Readers interested in a further discussion of this topic should refer to G. Feinberg and M. Goldhaber, "The Conservation Laws of Physics," *Scientific American*, October, 1963; K. W. Ford, *The World of Elementary Particles* (New York: Blaisdell Publishing Co., 1963), Chapter 4; and D. H. Frisch and A. M. Thorndike, *Elementary Particles* (Princeton, N.J.: D. Van Nostrand Co., Inc., 1964), Chapter 9.

1.0-kg part is at rest on the surface after the explosion. What is the speed of the 9.0-kg part?

3. What force is exerted on a stationary flat plate held normal to a jet of water 5 cm in diameter and having a horizontal velocity of $20 \text{ m} \cdot \text{sec}^{-1}$? After striking, the water moves parallel to the plate.

4. (a) A ball is released from a height of 4 m above hard, level ground and bounces to a height of 3 m. Determine the velocities just before and after impact.

(b) The ratio of the relative speed after impact to the relative speed before impact is called the **coefficient of restitution**, *e*. This coefficient is generally supposed to be the same for all initial relative velocities for a pair of bodies, such as the ball and ground in this problem. Note that $e < 1$. Calculate the height to which the same ball will bounce from the same area of ground if released from 8*m*.

*5. An astronaut sitting on a raft in interplanetary space fires 1001 pellets from a gun that gives the pellets a speed of $200 \text{ m} \cdot \text{sec}^{-1}$. The initial total mass of the system is 500 kg, and the mass per pellet is 0.1 kg.

(a) What is the change in speed of the astronaut and his raft as a result of firing the first pellet?

(b) What is the change in speed of the astronaut and his raft as a result of firing 1001 pellets?

*6. Two identical space ships of mass *M* are at rest with respect to each other in free space. An astronaut of mass *m* leaves one space ship for the other, which he boards, and then returns to the first. If the astronaut moves with a velocity *v* with respect to the spaceship he leaves, determine the velocities of the space-ships after his return trip.

7. Two gliders of equal mass are used in a linear-air-track experiment to study straight-line dynamics. The gliders are fitted with spring bumpers so that col-lisions between them are elastic. If one glider is given a speed of $15 \text{ m} \cdot \text{sec}^{-1}$ toward the other glider, which is left stationary, calculate the speeds of the two gliders following the interaction.

8. The same two gliders of Problem 7 are fitted with magnetic couplers that lock upon collision. If the one glider is given a speed of $9 \text{ m} \cdot \text{sec}^{-1}$ toward the other glider, calculate the speed of each of the gliders following the interaction. Cal-culate the loss of kinetic energy.

9. A glider of mass 0.2 kg is moving along a linear air track at a speed of $1 \text{ m} \cdot \text{sec}^{-1}$, when an additional mass of 0.3 kg is dropped on it with no horizontal component of velocity.

(a) What is the final speed of the glider?

(b) What is the kinetic energy before the collision?

(c) What is the kinetic energy after the collision?

Comment on your answers.

10. A particle having a speed of $10 \text{ m} \cdot \text{sec}^{-1}$ collides with a stationary particle of equal mass. After the collision the particles move at 25° and 65° to the incident direction. Find the two final velocities, and calculate the loss in kinetic energy for this collision.

11. The speed-time graph for two colliding particles is shown in Fig. 9.10.

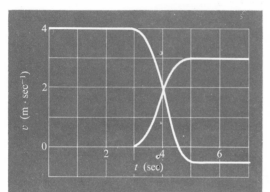

(a) What are the relative masses of the two particles?

(b) Calculate the loss of kinetic energy as a result of the interaction.

(c) Draw an acceleration-time graph for each of the particles.

*12. A particle of mass m_1 moves along the x axis with speed v_1, strikes a stationary particle of mass m_2, and continues with a speed v_1' along the x axis. The particle of mass m_1 has a speed v_2' after the collision. If $v_2' - v_1' = v_1$ (relative speeds are the same before and after the collision), show that the total kinetic energy does not change upon impact.

*13. In Problem 12, show that the assumption $v_2' - v_1' > v_1$ violates energy conservation.

Fig. 9.10

*14. Assuming that the interaction between a golf ball and a golf club is elastic and that the mass of the golf ball is much less than the mass of the golf club, show that the speed of the golf ball after the impact is approximately twice the speed of the club head just before impact.

15. Referring to the example of the scattering of an alpha particle by a ^{12}C nucleus given in Section 9.3, solve the problem for the case of elastic scattering of the same alpha particle. Calculate the kinetic energy of the alpha particle both before and after the collision and the kinetic energy of the ^{12}C nucleus after the collision.

*16. The force between two hydrogen atoms varies with their separation as shown in Fig. 9.11. If the two atoms are initially far apart and approaching one another with relative speed v, how could you calculate their distance of closest approach?

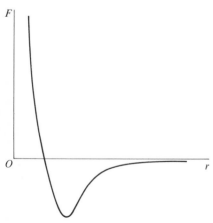

*17. A particle collides with a stationary particle of equal mass. If after the collision the particles move off at angles θ_1 and θ_2 to the direction of the incident particle, show that the loss in kinetic energy is zero as long as $\theta_1 + \theta_2 = 90°$.

*18. (a) A particle of mass m_1 and velocity $\mathbf{v}_1$ is incident upon a stationary particle of mass m_2. Discuss the variation in maximum scattering angle of the particle of mass m_1 as the mass m_2 changes from $m_2 \ll m_1$ to $m_2 = m_1$ assuming a perfectly elastic collision.

Fig. 9.11

(b) What is the maximum kinetic-energy transfer that can occur in a perfectly elastic collision?

19. A figure skater is able to spin at high angular frequency. Describe the procedure that a skater must follow in order to attain a condition of rapid spin.

20. A particle of mass M, attached to a cord of length R, rotates with speed v_0 (see Fig. 9.12). Calculate the work done in shortening the cord to a length r.

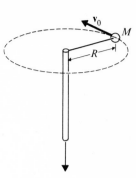

Fig. 9.12

10 Motion in accelerated (noninertial) reference frames

10.1 INTRODUCTION

Newton's second law of motion states the relation between the force acting on a particle and the resultant change in momentum (or change in acceleration if the mass of the particle is constant). The demonstration of the validity of Newton's second law in an inertial frame of reference is straightforward. In a noninertial frame, however, complications arise due to the presence of an acceleration that is experienced by all objects in that frame of reference. An observer in a noninertial frame of reference finds it desirable to attribute this acceleration to a force acting in the noninertial frame; an observer in an inertial frame of reference has no need for such forces (often called 'fictitious' forces or pseudoforces). In this chapter we shall look at several physical phenomena, both from the point of view of an observer in an inertial frame of reference and from the point of view of an observer in a noninertial frame of reference.

10.2 A LINEARLY ACCELERATING REFERENCE FRAME

Newton's laws of motion hold in frames of reference in constant relative motion[1]. We can demonstrate this by means of a simple experiment. Suppose we arrange for a trolley to travel with a constant speed relative to the laboratory by attaching a string to the trolley and around a record-player turntable. A support is attached to the trolley, and a metal sphere is suspended above the trolley by means of an electromagnet atop the support, as shown in Fig. 10.1. When the trolley is moving with a constant speed v, the sphere is released; it subsequently strikes the trolley at the base of the support. The motion of the sphere is recorded by means of a slow-motion moving-picture camera (a) at rest relative to the laboratory and then (b) at rest relative to the trolley. The results are shown in Fig. 10.2. In the first case the path of the sphere is a parabola, and in the second a straight line. In either case the motion is consistent with Newton's second law of motion, which assumes that the only force acting on the sphere is the gravitational force. In the laboratory frame of reference, the sphere has a constant horizontal component of velocity equal to that of the trolley to which it was initially attached and a linearly increasing vertical component of velocity resulting from the gravitational force acting on it. In the trolley frame of reference, the sphere has zero velocity in the horizontal direction, because the camera is moving horizontally with the sphere. The vertical component of velocity is the same in both cases.

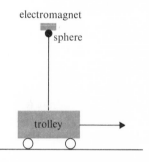

Fig. 10.1. A trolley with a steel ball suspended above it by means of an electromagnet.

[1] J. N. P. Hume and D. Ivey, "Frames of Reference" (The Physical Science Study Committee, Film Number MLA 0307). This is an excellent film, whose treatment of frames of reference parallels closely the material in this section.

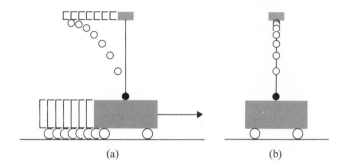

Fig. 10.2. (a) When the trolley is moving at constant speed an observer in the laboratory sees the dropping ball move in a parabolic path. (b) An observer on the trolley sees the ball dropping in a straight line with respect to the trolley.

In a linearly accelerating reference frame, Newton's second law of motion is not valid. To demonstrate this, we give the trolley a uniform acceleration relative to the laboratory in the manner illustrated in Fig. 10.3. A string attached to the trolley passes over a pulley to a weight, which experiences a constant acceleration because of the force of gravity.

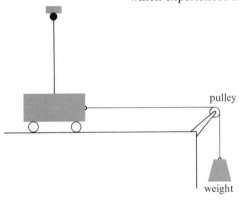

pulley

weight

Fig. 10.3. The trolley is given a constant acceleration by a weight attached to the trolley by a string which passes over a pulley.

Now when the sphere is released it strikes the trolley at a point to the left of the base of the support. The motion of the sphere is again recorded by the camera when at rest relative to the laboratory and when at rest relative to the trolley. In the laboratory reference frame, the path of the sphere is again a parabola. The motion is according to Newton's second law, the only force on the sphere being the gravitational force. Now, however, the sphere does not strike the trolley at the base of the support, because after the sphere is released the horizontal velocity of the trolley continues to increase, whereas that of the sphere remains at the value it had just when the sphere was released. The trolley therefore moves out from under the sphere.

In the trolley frame of reference, the sphere's path is as shown in Fig. 10.4. Just when the sphere is released it is at rest relative to the trolley. Subsequently, it acquires both horizontal and vertical components of velocity. According to Newton's second law, a force with both a horizontal and a vertical component must be acting on the sphere. But gravity is the sole force present, and it acts in the vertical direction. Newton's second law does not hold for the frame of reference attached to the accelerating trolley. The motion can be made consistent with Newton's second law if we invoke a hypothetical force acting horizontally toward the left and giving the sphere a horizontal acceleration equal in magnitude to the acceleration of the trolley but opposite in direction. Such a hypothetical or **fictitious force** is introduced so that motion in a linearly accelerating reference frame may be described by Newton's second law.

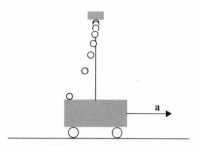

Fig. 10.4. When the trolley is uniformly accelerated the dropping ball follows a curved path in the trolley's frame of reference.

Consider an elevator descending with constant acceleration a. In a reference frame attached to the elevator, any particle of mass m not attached to the elevator moves as if it were acted on by an upward fictitious force ma as well as the downward force mg. When the constituent particles making up the human body are subjected to such fictitious forces, the reactions among the particles are altered, and we are conscious of an acceleration even though we cannot see outside of the elevator.

10.3   A ROTATING REFERENCE FRAME

For the discussion of rotating reference frames we introduce a rotating turntable on which certain experiments can be performed with **dry-ice pucks**. A dry-ice puck consists of a heavy brass disc on top of which a hollow spherical container is mounted. A small hole is drilled through the disc to the interior of the sphere. Dry ice is introduced into the sphere through an opening in the top, which is then corked. As the dry ice vaporizes, pressure is built up in the sphere. The gas then escapes through the hole in the bottom, and the puck floats on the layer of escaping carbon dioxide as indicated in Fig. 10.5.

If an observer situated on the turntable places such a dry-ice puck on the turntable, he finds that the dry-ice puck does not remain at rest when he lets go of it. He finds that in order to keep the puck at rest, he must exert a force on it toward the axis of rotation of the turntable (see Fig. 10.6). Therefore, Newton's second law does not apply in a frame of reference attached to the turntable. To an observer in the laboratory reference frame, the puck is not at rest but is traveling in a circular path.

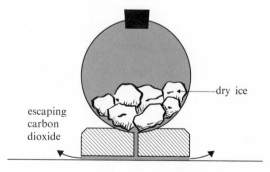

escaping carbon dioxide

dry ice

Fig. 10.5. A dry-ice puck consists of a hollow spherical container mounted on a brass disc. As dry ice evaporates in the container, gas is forced through a small hole in the disc and the puck rests on the thin layer of escaping gas.

In Section 6.5 we discussed the kinematics of circular motion and introduced the angular frequency ω and the angular acceleration α. If a particle is moving in a circular path with constant speed, α is zero, and ω is a constant. There can be no external force acting on the particle with a component in the direction of the particle's motion, since the speed is constant. The only acceleration experienced by the particle is toward the center of rotation and is known as the centripetal acceleration. Therefore, the only force acting on the particle is a force directed toward the center of rotation; it is called the **centripetal force** (see Fig. 10.7) and has magnitude

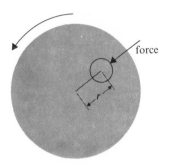

Fig. 10.6. To keep the puck stationary on a rotating turntable a force must be exerted on it toward the center of rotation.

$$F = \frac{mv^2}{r} = m\omega^2 r = m\omega v.$$

The centripetal force is the force required to keep the particle moving in its circular path. It is constant in magnitude if v is constant in magnitude.

The observer on the turntable exerts the required centripetal force on the puck to cause it to follow a circular path about the axis of rotation of the turntable.

The fact that the puck is at rest in the rotating reference frame can be made consistent with Newton's second law if we invoke a hypothetical force equal in magnitude to the centripetal force but directed outward from the axis of rotation. This force we call the **centrifugal force**.

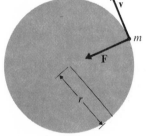

Fig. 10.7. The centripetal force on a particle moving in a circular path acts toward the center of rotation.

Now suppose that an observer at the center of the turntable releases a puck with speed v aimed toward a point A on the circumference. The path of the puck is shown in Fig. 10.8. Once again Newton's second law of motion does not hold; if it did the puck would have traveled in a straight line to A. In the laboratory reference frame the motion of the puck is in accord with Newton's second law; that is, the puck moves from the center of the turntable in a straight line to the circumference, as shown in Fig. 10.9. The time required is R/v sec, where R is the radius of the turntable. During this interval of time the point A moves through an arc length $\omega R t$ relative to the laboratory.

The motion of the puck in the rotating reference frame can be made consistent with Newton's second law if we invoke a hypothetical force to account for the deviation of the puck's motion from a linear path. A constant hypothetical force, called the **Coriolis force**[2], which acts perpendicular to the direction of the puck's motion relative to the turntable, can account for the observation. The magnitude of the Coriolis accelera-

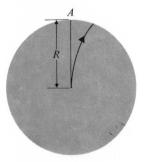

Fig. 10.8. A puck traveling toward A follows a curved path as observed by a person on the turntable.

[2] J. E. MacDonald, "The Coriolis Effect," *Scientific American*, May, 1952. Available as *Scientific American Offprint 839* (San Francisco: W. H. Freeman and Co., Publishers).

tion can be obtained from the equation

$$\omega Rt = \tfrac{1}{2}at^2$$

or

$$a = \frac{2\omega R}{t}.$$

The magnitude of the Coriolis force is given by

$$\frac{2m\omega R}{t} = 2m\omega v.$$

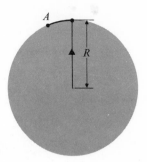

Fig. 10.9. An observer in the laboratory sees the puck travel in a straight line.

Motion as observed from a rotating reference frame can be made consistent with Newton's second law if a fictitious centrifugal force and fictitious Coriolis force are added to any real forces present.

The centrifugal and Coriolis force are called **fictitious forces** because they are needed only by an observer in a rotating reference frame. An observer in an inertial frame of reference who is observing the rotational motion has no need to invoke the centrifugal and Coriolis forces. However, to an observer in a rotating frame of reference these forces are indeed real. For example, every time you go around a corner in a car you are in a rotating frame of reference (the car is moving about some instantaneous center of rotation). The door or seat of the car exerts a force on you that is just the centripetal force required to deflect you from a straight-line motion into rotational motion. You push back at the door or seat of the car with the force you call the centrifugal force. It is very real to you indeed.

We may write vector equations for the centrifugal and Coriolis forces. If a particle of mass m has a position vector **r**, the centrifugal force is given by

$$\mathbf{F}_{\text{centrifugal}} = m\omega^2 r\hat{\mathbf{r}},$$

where $\hat{\mathbf{r}}$ is a unit vector in the direction of the position vector. The Coriolis force on a particle of mass m moving perpendicular to the axis of rotation as shown in Fig. 10.10 is given by

$$\mathbf{F}_{\text{Coriolis}} = -2m(\boldsymbol{\omega} \times \mathbf{v}).$$

This expression may be used to describe the Coriolis force on a particle in any direction relative to the axis of rotation by resolving the velocity of the particle into components parallel and perpendicular to the axis of rotation and using the perpendicular component to determine the Coriolis force.

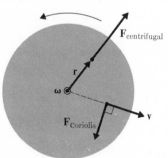

Fig. 10.10. The centrifugal force on a particle acts in the direction of the position vector **r** while the Coriolis force on a particle moving perpendicular to the axis of rotation is perpendicular to both the velocity **v** and the angular frequency **ω**.

10.4   EFFECTS OF THE EARTH'S ROTATION

The results of the previous section can be generalized to measurements made on the surface of a rotating sphere, such as the earth. Consider a particle of mass m on the earth's surface at a latitude λ. The particle experiences a centrifugal force of magnitude $m\omega^2 r$, where r is the perpendicular distance from the particle to the earth's axis of rotation (see Fig. 10.11). The centrifugal force acts outward from the axis and varies with latitude according to

$$F_{\text{centrifugal}} = m\omega^2 R_e \cos \lambda,$$

where R_e is the earth's radius. The effect of the centrifugal force is to slightly reduce the apparent gravitational acceleration. It may be shown that a good approximation to the magnitude of g_{app} is given by

$$g_{\text{app}} \simeq g - \omega^2 R_e \cos^2 \lambda$$

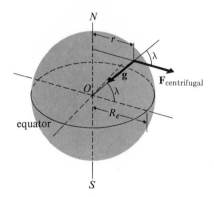

Fig. 10.11. A particle on the earth's surface at latitude λ experiences a fictitious centrifugal force of magnitude $m\omega^2 R_e \cos \lambda$.

(see Problem 10). The maximum value of the centrifugal acceleration is $3.4 \times 10^{-2}\,\text{m} \cdot \sec^{-2}$ and occurs at the equator. The value of the acceleration due to gravity at the equator is therefore reduced by 0.35% as a result of the earth's rotation. Although this is a small effect, it is not always a negligible one.

If a particle of mass m moves relative to the earth's surface with a speed v, then it will also experience a Coriolis force of magnitude $2m\omega v_\perp$, where $v_\perp$ is the component of v in the plane perpendicular to the earth's axis of rotation. The direction of the Coriolis force is easily obtained from a diagram, with a moment's thought (see Fig. 10.12). For a particle moving over the earth's surface with speed v in the northern hemisphere,

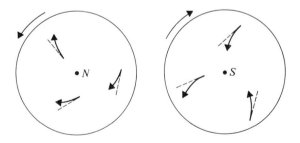

Fig. 10.12. The Coriolis force due to the earth's rotation is to the right in the northern hemisphere and to the left in the southern hemisphere.

the deflection is to the right; in the southern hemisphere, the deflection is to the left. We neglect any motion away from the earth's surface because this is not usually important. For an object traveling at $1\,\text{km}\cdot\text{sec}^{-1}$ perpendicular to the earth's axis, the Coriolis acceleration is 1.5×10^{-4} $\text{m}\cdot\text{sec}^{-2}$.

Coriolis effects are important only when high-speed objects are observed or when motion is observed over large distances or for long periods of time. For example, the Coriolis force is important in meteorological problems concerned with wind circulation. Wind is an air mass in motion and, for a nonrotating earth, would be directed from a high-pressure to a low-pressure region. In the northern hemisphere the Coriolis force deflects the winds to the right of this direction. This results in an equilibrium condition in which the wind patterns are stationary. The air attempting to move from a high-pressure region to a low-pressure region is deflected from its path by the Coriolis force. In the equilibrium state the winds travel parallel to regions of constant pressure (represented by isobars on a weather map), as shown in Fig. 10.13. Cyclones occur in regions of low pressure in which the lines of constant pressure are more or less concentric. The winds circulate in the counterclockwise direction in the northern hemisphere (see Fig. 10.14). In reality, because of the existence of frictional drag between the moving air mass and the earth's surface, the wind directions do not coincide exactly with the directions of constant pressure.

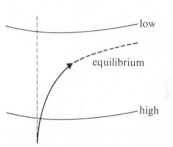

Fig. 10.13. Air moving from a high-pressure region to a low-pressure region is deflected to the right in the northern hemisphere.

The Foucault pendulum

Suppose that we mount a pendulum so that it can oscillate above a rotating disc, as illustrated in Fig. 10.15. Suppose further that the system is observed by means of a motion-picture camera mounted above the disc and at rest relative to the laboratory. One observes that the plane of oscillation is fixed but that the disc rotates beneath it. The arrow indicates the sense of rotation of the disc. Next, suppose that the system is observed with the camera mounted above the disc and at rest relative to the disc. The disc now appears to be stationary, but the plane of oscillation of the pendulum seems to rotate in the opposite sense to that of the disc relative to the laboratory. The plane of oscillation is shown at three consecutive times in Fig. 10.16.

This experiment shows that a pendulum can be used to demonstrate the rotation of a reference frame. The same sort of result is true for a

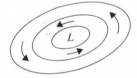

Fig. 10.14. Winds circulate in a counterclockwise direction about a low-pressure area in the northern hemisphere.

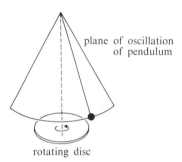

Fig. 10.15. A pendulum oscillating over a rotating disc oscillates in a plane fixed in space.

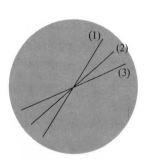

Fig. 10.16. An observer on the disc observes the plane of oscillation of the pendulum to rotate; the rotation is in a direction opposite to the direction of rotation of the disc.

rotating sphere, such as the earth. We see immediately that at the north pole of the earth the plane of motion of a pendulum would remain fixed in an inertial frame of reference while the earth rotated under the pendulum once every 24 hr. To an observer on the earth's surface, the plane of oscillation of the pendulum would rotate in the opposite sense once every 24 hr. The situation appears different and is more difficult to analyze for a pendulum located at another point on the earth's surface (other than the south pole, that is). The time for a complete rotation of the plane of oscillation is now longer than 24 hr. In this manner the rotation of the earth was first demonstrated publicly by Leon Foucault in 1851, under the great dome of the Panthéon in Paris. A 28-kg mass on a wire suspension nearly 70 m in length was used. Today, any pendulum used to demonstrate the earth's rotation is called a **Foucault pendulum**.

PROBLEMS

1. Discuss the possibility of playing Ping-Pong on the deck of an ocean liner.

2. A train moves over the surface of the earth with constant acceleration a. An object on the train is subjected to an acceleration

$$a'(t) = At$$

with respect to the train, where A is a constant. Write equations to describe the displacement of the particle with time with respect to a frame of reference
(a) attached to the train.
(b) attached to the earth.

3. A particle is moving in a straight-line path in an inertial frame such that its coordinates at any time are given by the equations

$$x = v_0 t, \qquad y = 0, \qquad z = 0,$$

where v_0 is a constant. What is the path of the particle in a reference frame rotating counterclockwise about the z axis with constant angular frequency ω?

4. A pistol, mounted horizontally at the center of a large turntable, is aimed directly at a target point at the edge of the turntable. Neglect the length of the gun compared to the radius of the turntable. The bullet, traveling at 300 m·sec^{-1}, is observed to fall 1.0×10^{-4} m to the left of and 1.0×10^{-4} m below the target point. Calculate the angular frequency of the turntable, its sense of rotation, and

its radius. Describe the trajectory of the bullet as seen by an observer on the turntable and an observer on the earth. Neglect the rotation of the earth.

5. Four guns are mounted at the corners of a square framework, and they are all aimed at the center of the square, as shown in Fig. 10.17. When the guns are fired simultaneously, all four bullets collide at the center. Describe what happens when the guns are fired simultaneously if
   (a) the framework moves with a constant speed in the x direction relative to the earth's surface.
   (b) the framework moves with a constant acceleration in the x direction relative to the earth's surface.
   (c) the framework rotates with a constant angular frequency in the x, y plane about its center.

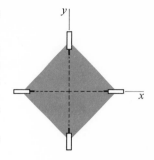

Fig. 10.17

*6. A bug is observed to run with constant relative speed v around the rim of a wheel of radius R rolling over level terrain with velocity V. Find the magnitude and direction of
   (a) the bug's acceleration relative to the wheel.
   (b) the bug's acceleration relative to the ground.

7. A 1-kg dry-ice puck is to be moved at a constant speed of $1 \text{ m} \cdot \text{sec}^{-1}$ along a straight-line path from the center to the edge of a horizontal, circular platform of radius 5 m, which is rotating with an angular frequency of $1 \text{ rad} \cdot \text{sec}^{-1}$. Find expressions, as a function of distance from the center of the platform, for the magnitude and for the direction of the force required to produce this motion. What is the magnitude and direction of the force 3 m from the center of the platform? *Hint*: A centripetal force is required to keep the puck rotating with the platform. A second force is required to balance the Coriolis force.

8. Calculate the value of g_{app} at various latitudes between 0° and 90°, and plot a graph of the results.

9. Calculate the magnitude of the centrifugal acceleration due to the earth's rotation at your latitude. What is the percentage reduction in your weight due to the earth's rotation?

*10. Show that the effect of the earth's rotation is to alter both the magnitude and direction of the apparent gravitational acceleration at the earth's surface. Derive approximate expressions for the magnitude and direction of $\mathbf{g}_{app}$ in terms of the corresponding quantities for a nonrotating earth.

11. An iceberg of mass 5×10^8 kg is near the north pole of the earth and is moving west at the rate of 5 km per day. Find the magnitude and direction of the Coriolis force, neglecting the curvature of the earth.

12. A Foucault pendulum is set into oscillation at a point directly over the earth's equator. What behavior would you expect to observe for its plane of oscillation?

13. Discuss the effect of the earth's rotation on a train traveling due south.

14. Write an essay on the effect of the earth's rotation on the motion of air masses over the earth's surface. List the sources of your information.

15. The direction of circulation of the water in whirlpools tends to be clockwise in the northern hemisphere and counterclockwise in the southern hemisphere. Discuss.

*16. Show that the period of rotation of the plane of oscillation of a Foucault pendulum at latitude λ is $(8.64 \times 10^4/\sin \lambda)$ sec.

*17. Calculate the horizontal deflection suffered as a result of the Coriolis force by an object falling through a vertical distance of 30 m at the equator. Assume a constant value of g. *Hint*: Taking the y, z plane as the north-south vertical plane with the z axis along the direction of the vertical, write the equation of motion in the x direction. Solve this equation by integrating twice.

*18. Calculate the horizontal deflection suffered by an object falling through a vertical distance of 100 m at a latitude of 45°. Assume that the object falls freely in an inertial frame and that the displacement is due to the rotation of the earth during the time of the object's fall.

Simple harmonic motion 11

11.1 INTRODUCTION

Any motion that is repetitive in time can be called a **periodic motion**. A child on a swing, the pendulum of a grandfather's clock, the motion of a satellite around the earth, the rising and falling of the tides, and the molecular motions that give rise to the sensation of sound are just a few examples of repetitive or periodic motions.

In this chapter we shall study in detail a very simple form of periodic motion known as **simple harmonic motion**. Any system undergoing this form of motion will be referred to as a **simple harmonic oscillator**. We shall introduce simple harmonic motion experimentally by considering the up-and-down motion that can be given to an object attached to a hanging spring.

The concept of a simple harmonic oscillator is an important one. In later chapters we shall use it to try to account for such phenomena as the radiation emitted by a heated solid and the specific heats of gases and solids. We shall also see that, although many physical systems are not strictly simple harmonic systems, they are conveniently described in terms of their departure from such motion.

11.2 HOOKE'S LAW

We begin by fastening one end of a spring to a support and adding, one by one, a series of standard masses to the other end. The free end of the unloaded spring is taken as the origin, and the displacements x of the end of the spring are measured as each mass is added. This is illustrated in Fig. 11.1.

A plot of the mass added to the spring as a function of the resulting displacements is given in Fig. 11.2; a line is drawn through the experimental points. For each measurement the system is in equilibrium. The downward force of gravity mg on the added mass m is balanced by an upward elastic force F_{el} developed in the springs. That is,

$$F_{el} = - mg,$$

and the resultant force on m is zero. From the graphs we see that for moderate extensions, the extension is linearly proportional to the added mass and hence to the applied force. For this linear region we can define a force constant k, where k is the force required to produce unit extension of the

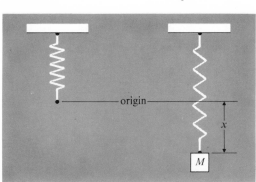

Fig. 11.1. An object of mass M attached to a hanging spring causes the spring to stretch a distance x.

spring. Therefore, we can write

$$mg = kx$$

or

$$F_{el} = -kx.$$

Example. Deduce the force constant of the spring used to obtain Fig. 11.2.

Solution. The force constant k is g times the slope of the linear portion of the graph. Inspection of the graph shows that a displacement of 0.5 m results from the addition of a mass of 1.5 kg. Therefore,

$$k = \frac{1.5 \times 9.8}{0.5} = 29.(4) \text{ N} \cdot \text{m}^{-1}.$$

For extension less than about 0.7 m,

$$F_{el} = -29.(4)x.$$

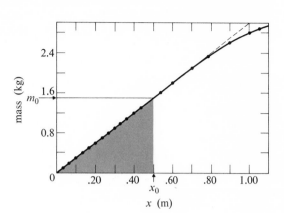

Fig. 11.2. The displacement of the end of a spring is proportional to the mass added to the spring for moderate extensions.

When the end of the spring undergoes a displacement x_0, the work done by the gravitational force to stretch the spring is stored in the spring as **elastic potential energy**. That is, the elastic potential energy U_{el} of a spring extended by an amount x_0 is

$$U_{el} = \int_{x=0}^{x=x_0} dW,$$

which is just the shaded area in Fig. 11.2. Therefore,

$$U_{el} = -\int_0^{x_0} F_{el} \, dx$$

$$= \int_0^{x_0} kx \, dx$$

$$= \tfrac{1}{2}kx_0^2.$$

The potential energy of the spring when it is extended by 0.5 m is

$$U_{el} = \tfrac{1}{2} \times 29.(4) \times (0.5)^2 = 3.7 \text{ J}.$$

11.3   THE EQUATION FOR SIMPLE HARMONIC MOTION

If the spring considered in the previous section is loaded with a 1.50-kg mass M and allowed to come to equilibrium, the spring will be extended by an amount

$$x_0 = \frac{1.50 \times 9.8}{29.(4)}$$

$$= 0.50 \text{ m.}$$

Now suppose that the spring is extended further by an amount x. If the mass M is then released, it will be subject to an unbalanced resultant force

$$F = Mg - k(x_0 + x)$$

$$= -kx.$$

This situation is illustrated in Fig. 11.3. When M is released, this force causes it to accelerate back toward its equilibrium position x_0. Since the force is proportional to displacement from equilibrium, the force decreases as M moves back toward x_0. However, when M reaches x_0 it is moving quite quickly and overshoots the mark, even though the net force on M is zero at x_0. As the extension of the spring becomes less than x_0, the resultant force on M is reversed in direction and serves to decelerate M and finally bring it to a stop. If no energy is lost from the system, we would expect M to stop at position $x_0 - A$ if it is released from position $x_0 + A$. M will now move back again toward x_0 and indeed will oscillate about x_0, the displacement x varying between $+A$ and $-A$ as measured from x_0. Since this force tends to return the mass M to the equilibrium position, it is called a **restoring force**. The maximum displacement A of the mass M from its equilibrium position is called the **amplitude** of the oscillation.

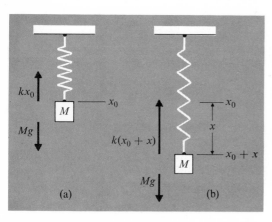

Fig. 11.3. (a) When the mass M suspended from a spring is in a position of equilibrium the force of gravity Mg is balanced by the elastic force kx in the spring. (b) When the mass M is further extended by an amount x and then released, the elastic force $k(x_0 + x)$ exceeds the force of gravity Mg by an amount kx.

The resultant motion of M is governed by Newton's second law of motion,

$$F = Ma.$$

Therefore, we may write

$$-kx = M\frac{d^2x}{dt^2}$$

or

$$\frac{d^2x}{dt^2} + \frac{k}{M}x = 0.$$

Motion governed by this form of equation is called **simple harmonic motion**. To determine whether a given motion is simple harmonic is to determine if its equation of motion can be put into this form.

It should be noted that the role of gravity in the above discussion is solely to fix the equilibrium extension. Gravity plays no role in the motion. To see this, one might consider a mass M attached by a spring to a wall and resting on a frictionless table. If the spring is the same as the one considered above, the motion of mass M will be described by the identical equation of motion but will take place about a different equilibrium position.

The simple pendulum

A simple pendulum consists of a point particle of mass m suspended from a massless string of length l. The pendulum is shown in Fig. 11.4 at an instant when the string makes an angle θ with the vertical. To what extent is the subsequent motion of the bob of the pendulum described by the equation for simple harmonic motion?

First we consider the forces acting on m. The force of gravity $m\mathbf{g}$ acts vertically downward; it may be resolved into a component $mg \cos \theta$ acting along the string and away from the point of support and a component $mg \sin \theta$ tangent to the arc along which the bob moves. The force of tension $\mathbf{T}$ in the string acts along the string toward the point of support. The component of gravity $mg \sin \theta$ is an unbalanced restoring force tending to return the bob to its equilibrium position vertically below the point of support; it is responsible for the motion of the bob along the circular arc. The displacement of the bob from equilibrium is measured along the arc of the pendulum's swing.

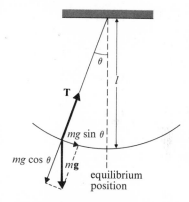

Fig. 11.4. The restoring force on the bob of a simple pendulum is the component of the gravitational attraction perpendicular to the string.

Using Newton's second law, we may write for the equation of motion of the bob

$$-mg \sin \theta = m\frac{d^2(l\theta)}{dt^2}.$$

Since l is a constant,

$$l\frac{d^2\theta}{dt^2} + g \sin \theta = 0.$$

If θ is a small angle, then, for θ measured in radians,

$$\sin \theta \simeq \theta,$$

as shown in Table 11.1. Therefore,

$$\frac{d^2\theta}{dt^2} + \frac{g}{l}\theta = 0,$$

which is of the same form as the equation for simple harmonic motion. That is, the motion of the bob of a simple pendulum is simple harmonic to the extent to which $\sin \theta$ can be approximated by θ. The net force acting towards the point of suspension, having the magnitude

$$T - mg\cos\theta,$$

is the centripetal force required to move the bob along the circular arc.

Table 11.1   A comparison of $\sin \theta$ and θ

θ (radians)	θ (degrees)	$\sin \theta$
0.0	0.0	0.0
0.0200	1.15	0.0200
0.0400	2.29	0.0400
0.0600	3.44	0.0600
0.0800	4.58	0.0799
0.1000	5.73	0.0998
0.1200	6.88	0.1197
0.2000	11.46	0.1987
0.2400	13.75	0.2377
1.57	90.0	1.000

11.4   THE SOLUTION OF THE EQUATION FOR SIMPLE HARMONIC MOTION

The equation of simple harmonic motion that we have deduced, namely

$$\frac{d^2x}{dt^2} + \frac{k}{M}x = 0,$$

is a second-order differential equation. Rather than solve this equation directly, we shall consider the useful relation between uniform circular motion and simple harmonic motion. Figure 11.5 shows a particle of mass M moving in a path of radius R. The component of the motion in the x direction is

$$x = R\cos(\omega t + \phi).$$

The angle ϕ gives the angular displacement of the particle at time $t = 0$. If the motion is uniform, the particle will experience only a centripetal acceleration

$$\mathbf{a} = -\omega^2\mathbf{R}.$$

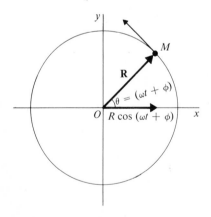

Fig. 11.5. The motion of a particle undergoing uniform circular motion may be projected on an axis through the center of motion and in the plane of motion. This projected motion is simple harmonic motion.

The component of this acceleration in the x direction is

$$a_x = -\omega^2 R \cos(\omega t + \phi)$$

$$= -\omega^2 x.$$

That is, the projection of the uniform circular motion on the x axis as described by Newton's second law

$$F = Ma$$

becomes

$$-M\omega^2 x = M\frac{d^2 x}{dt^2}$$

or

$$\frac{d^2 x}{dt^2} + \omega^2 x = 0.$$

This motion is simple harmonic motion of amplitude R and with ω^2 replacing k/m. Since A is normally used to designate the amplitude of a simple harmonic oscillation, we shall set

$$R = A.$$

In the process of obtaining this equation we have also obtained its solution, namely

$$x = A \cos(\omega t + \phi).$$

Referring back to the equation of motion for a mass M oscillating on the end of a spring, we see that the displacement of the spring is given by

$$x = A \cos\left[\left(\frac{k}{m}\right)^{1/2} t + \phi\right].$$

We define the **period** T as the time required for the particle to travel once around the circular path, that is,

$$T = \frac{2\pi A}{v}$$

$$= \frac{2\pi}{\omega}$$

and

$$T = 2\pi\left(\frac{M}{k}\right)^{1/2}$$

for the mass on the spring.

The **frequency** ν is the number of complete oscillations per second, so that

$$\nu = \frac{1}{T} = \frac{1}{2\pi}\left(\frac{k}{M}\right)^{1/2}.$$

The unit for frequency is the hertz (Hz), which is equivalent to one cycle per second (cycle $\cdot$ sec^{-1}).

The angle ϕ, called the **initial phase angle**, simply locates the oscillation with respect to the time axis. If $\phi = 0$, the oscillation is represented by a cosine wave; if $\phi = \pi/2$, it is represented by a sine wave. For arbitrary ϕ, the oscillation is represented by a linear combination of a sine wave and a cosine wave. That is,

$$A \cos\left[\left(\frac{k}{M}\right)^{1/2} t + \phi\right] = A \cos\phi \cos\left(\frac{k}{M}\right)^{1/2} t - A\sin\phi \sin\left(\frac{k}{M}\right)^{1/2} t$$

$$= A' \cos\left(\frac{k}{M}\right)^{1/2} t + A'' \sin\left(\frac{k}{M}\right)^{1/2} t,$$

where

$$A' = A\cos\phi \quad \text{and} \quad A'' = -A\sin\phi.$$

Figure 11.6 illustrates the solution of the equation of simple harmonic motion.

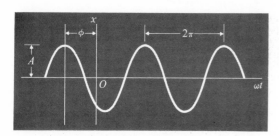

Fig. 11.6. The amplitude and initial phase angle for simple harmonic motion.

Example. A 1.175-kg mass oscillating on the end of a spring of force constant 29.4 N $\cdot$ m^{-1} is observed to have a displacement from equilibrium of 0.3 m at $t = 1$ sec and to be momentarily at rest. Write the equation of motion of the oscillating mass.

Solution. The equation of motion is

$$x = A \cos\left[\left(\frac{k}{M}\right)^{1/2} t + \phi\right].$$

When $x = A$, the particle is momentarily at rest, so that $A = 0.3$ m. Therefore, at $t = 1$ sec

$$0.3 = 0.3 \cos\left[\left(\frac{29.4}{1.175}\right)^{1/2} + \phi\right],$$

from which

$$\cos(5.0 + \phi) = 1.$$

Therefore,

$$5.0 + \phi = 0$$

and

$$\phi = -5.0 \text{ rad.}$$

The equation of motion of the oscillating mass is

$$x = 0.3 \cos(5.0t - 5.0) \text{ m.}$$

Example. What length of simple pendulum has a period of 1 sec for small-amplitude oscillations?

Solution. We saw in Section 11.3 that the equation of motion for the bob of a simple pendulum executing small-amplitude oscillations is

$$\frac{d^2\theta}{dt^2} + \frac{g}{l}\theta = 0.$$

The period of the oscillation is therefore

$$T = 2\pi \left(\frac{l}{g}\right)^{1/2}.$$

That is,

$$l = \frac{gT^2}{4\pi^2} = \frac{9.8 \times 1}{4\pi^2} = 0.24(8) \text{ m.}$$

A simple pendulum 0.24(8) m long has a period of 1 sec.

11.5 THROUGH THE EARTH IN FORTY MINUTES

It has been suggested[1] that rapid intercontinental travel could be achieved by crisscrossing the earth with frictionless subterranean passages. As a model of the earth, we consider a sphere of uniform density with radius $R = 6370$ km and surface gravitational acceleration $g = 9.80$ m·sec^{-2}. The gravitational force on a mass m at a distance r from the earth's center is[2]

$$F_g = -\frac{mgr}{R}$$

and acts towards the earth's center. The mode of propulsion is then merely to drop the mass m into a tunnel, exchanging gravitational potential energy for kinetic energy during the first half of the trip and then reversing the exchange during the second half.

[1] P. W. Cooper, "Through the Earth in Forty Minutes," *The American Journal of Physics*, **34** (January 1966), 68.

[2] See, for example, U. Ingard and W. L. Kraushaar, *Introduction to Mechanics, Matter, and Waves* (Reading, Mass.: Addison-Wesley Publishing Co., Inc., 1960), Chapter 10.

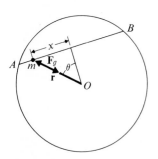

Fig. 11.7. An object traveling in a straight line through the earth from A to B undergoes simple harmonic motion with a period 84.5 minutes.

Consider a particular tunnel connecting cities A and B as shown in Fig. 11.7. The component of the gravitational force parallel to the tunnel and directed toward its midpoint is

$$-mgr\left(\frac{\cos(90-\theta)}{R}\right) = -mgr\left(\frac{\sin\theta}{R}\right) = -\frac{mgx}{R},$$

where x is the distance measured from the midpoint of the tunnel to mass m. The equation of motion of the particle is

$$m\frac{d^2x}{dt^2} = -\left(\frac{mg}{R}\right)x$$

or

$$\frac{d^2x}{dt^2} + \left(\frac{g}{R}\right)x = 0,$$

which is just the equation for simple harmonic motion.

The particle will oscillate in the tunnel with a periodic time

$$T = 2\pi\left(\frac{R}{g}\right)^{1/2}$$

$$= 5060 \text{ sec}$$

$$= 84.5 \text{ min},$$

independent of the choice of cities A and B.

One can envisage a most convenient worldwide transportation system in which cities are linked by tunnels and the departure time is universally on the hour and the arrival time forty-two and one-quarter minutes past the hour. However, the possibility of producing such a system, even if the enormous financial costs could be met, is reduced considerably by the fact that the temperature of the earth increases as the center of the earth is approached, there being very strong evidence that there is a sizeable molten core in the earth.

11.6   THE ENERGY ASSOCIATED WITH AN OSCILLATING OBJECT

If we take the free end of the unloaded spring of Section 11.2 as the origin, the potential energy U associated with stretching the spring is

$$U = U_g + U_{el}$$

$$= mgx + \tfrac{1}{2}kx^2,$$

the sum of gravitational and elastic contributions. (Note that x is always a negative number.) U is plotted as a function of x in Fig. 11.8. The resulting curve is a parabola. It is convenient to make a transformation of the origin of our coordinate system describing the oscillation to the position corresponding to the vertex of the parabola in Fig. 11.7. This is effected by setting

$$x' = x + \frac{Mg}{k}$$

and

$$U' = U + \frac{(Mg)^2}{2k},$$

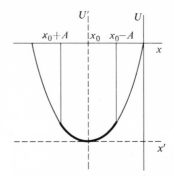

Fig. 11.8. The potential energy of an object oscillating on the end of a spring. The motion is treated most simply, mathematically, by using the U' and x' coordinate axes.

where x' and U' are the position and potential energy of the oscillating mass referred to the new origin. Substituting

$$U = Mgx + \tfrac{1}{2}kx^2$$

$$= Mg\left(x' - \frac{Mg}{k}\right) + \frac{1}{2}k\left(x' - \frac{Mg}{k}\right)^2$$

into

$$U' = U + \frac{(Mg)^2}{2k}$$

yields

$$U' = \tfrac{1}{2}kx'^2.$$

The total energy E of the oscillating mass is

$$E = \tfrac{1}{2}Mv^2 + \tfrac{1}{2}kx'^2$$

$$= \frac{1}{2}M\left(\frac{dx'}{dt}\right)^2 + \frac{1}{2}kx'^2.$$

Substituting

$$x' = A \cos\left[\left(\frac{k}{M}\right)^{1/2} t + \phi\right]$$

and

$$\frac{dx'}{dt} = -A\left(\frac{k}{M}\right)^{1/2} \sin\left[\left(\frac{k}{M}\right)^{1/2} t + \phi\right]$$

gives

$$E = \frac{1}{2}MA^2\frac{k}{M}\sin^2\left[\left(\frac{k}{M}\right)^{1/2}t + \phi\right] + \frac{1}{2}kA^2\cos^2\left[\left(\frac{k}{M}\right)^{1/2}t + \phi\right]$$

$$= \tfrac{1}{2}kA^2,$$

which is constant. As the mass oscillates, energy is changed from elastic potential to kinetic and back again but in such a way that the total energy remains constant. The kinetic energy and potential energy as a function of position are plotted in Fig. 11.9. Both the kinetic-energy and potential-energy curves are parabolic in shape but sum to give a constant energy independent of the position of the oscillating mass. In reality, the losses in the system, which we have neglected, gradually result in a dissipation of the energy, and the mass eventually comes to rest at the equilibrium position.

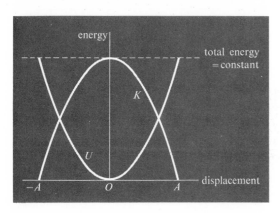

Fig. 11.9. The potential energy and kinetic energy of a mass oscillating in simple harmonic motion vary with time but sum to give a constant energy independent of position.

11.7 ROTATIONAL SIMPLE HARMONIC MOTION

Two equal masses M are fixed at the ends of a rigid but essentially weightless rod of length $2r$ to form a dumbbell. The dumbbell is suspended from above by a wire fastened to its midpoint. The situation is shown in Fig. 11.10. We shall assume that, if the dumbbell is rotated through an angle θ_0, the wire will exert a restoring torque τ on it proportional to θ_0. We can write

$$\tau = -C\theta_0,$$

where C is the **torsion constant of the wire** and is measured in the units $N \cdot m \cdot rad^{-1}$. When the dumbbell is released the restoring torque produces an angular acceleration

$$\alpha = \frac{d^2\theta}{dt^2}.$$

The restoring torque and angular acceleration are related by Newton's second law, written in the form

$$\tau = \frac{dJ}{dt},$$

where J is the angular momentum. The angular momentum of the

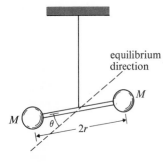

Fig. 11.10. A dumbbell suspended from above by a wire attached to its midpoint.

dumbbell about an axis of rotation lying along the string is

$$J = Mrv + Mrv$$

$$= 2Mr^2\omega$$

$$= 2Mr^2\frac{d\theta}{dt},$$

where $\omega = d\theta/dt$ is the angular frequency of rotation. Therefore, the torque at an angular displacement θ is given by

$$-C\theta = \frac{d}{dt}\left(2Mr^2\frac{d\theta}{dt}\right)$$

$$= I\frac{d^2\theta}{dt^2}$$

or

$$\frac{d^2\theta}{dt^2} + \frac{C}{I}\theta = 0,$$

where

$$I = 2Mr^2$$

is the moment of inertia of the dumbbell about an axis through its center and perpendicular to it. The motion of the dumbbell is therefore simple harmonic.

The differential equation that we have obtained is entirely general. That is, it is valid for an object of any configuration suspended from the wire, by the substitution of an appropriate value of I. The solution of the equation is

$$\theta = \theta_0 \cos(2\pi vt + \phi),$$

with

$$4\pi v^2 = \frac{C}{I}.$$

The total energy of the oscillating system is

$$E = \tfrac{1}{2}I\omega^2 + \tfrac{1}{2}C\theta^2$$

$$= \tfrac{1}{2}C\theta_0^2,$$

where θ_0 is the amplitude of the oscillations.

Example. A disc of mass 2.0 kg and radius 0.10 m suspended from its center by a wire, is observed to oscillate at a frequency of 5/4 Hz. What is the torsion constant of the wire? If the amplitude of oscillation is 30°, what is the maximum angular frequency of the disc?

Solution. The torsion constant of the wire is given by

$$C = 4\pi^2 v^2 I,$$

where for a disc of radius r suspended from its center[3]

$$I = \frac{Mr^2}{2}.$$

Therefore,

$$C = 4(3.14)^2 \left(\frac{5}{4}\right)^2 \left(\frac{2.0 \times 0.01}{2}\right)$$

$$= 0.62 \text{ N} \cdot \text{m} \cdot \text{rad}^{-1}$$

and

$$E = \tfrac{1}{2}c\theta_0^2 = \tfrac{1}{2}I\omega_{max}^2.$$

Therefore,

$$\omega_{max} = \left(\frac{C}{I}\right)^{1/2} \theta_0 = 2\pi v \theta_0.$$

Since $30° = \pi/6$ rad,

$$\omega_{max} = 2(3.14)\left(\frac{5}{4}\right)\left(\frac{3.14}{6}\right)$$

$$= 4.1 \text{ rad} \cdot \text{sec}^{-1}.$$

11.8   DAMPED OSCILLATIONS

If a mass hung from the end of a spring is set in oscillation, the amplitude of the oscillations will gradually decrease until the mass finally comes to rest. This gradual loss of energy to the surroundings is known as **damping**.

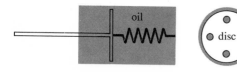

To study the effect of damping, we shall consider a spring connected to a perforated disc and mounted within a cylinder filled with oil. The arrangement is illustrated in Fig. 11.11. Suppose we displace the disc to the right by an amount A, thereby compressing the spring, which in turn will exert a force on the disc

Fig. 11.11. Spring connected to a perforated disc and mounted within a cylinder filled with oil.

$$F_{el} = -kA.$$

[3] J. B. Clark, *Physical and Mathematical Tables* (Edinburgh: Oliver and Boyd, Ltd., 1956).

This is shown in Fig. 11.12. How will the disc move if it is released? We assume for the moment that the mass of the disc may be neglected. Then the disc will move with a velocity limited only by the **internal friction** or **viscosity** of the oil, which must flow through the perforations in order for the disc to move at all. For relatively low speeds, it is found experimentally that the viscous force F_v will be proportional to the speed so that

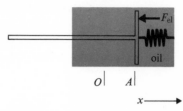

$$v = \frac{dx}{dt} = -\frac{F_v}{\gamma},$$

Fig. 11.12. The disc is displaced to the right by an amount A.

where γ is a constant and the minus sign indicates that the velocity is in the $-x$ direction while F_v is acting in the x direction. But if the disc has no mass, then according to Newton's second law the net force on the disc will be zero even though the acceleration of the disc is not zero. Therefore, the magnitude of the viscous force must just be equal to the elastic restoring force, so that

$$\frac{dx}{dt} = -\frac{k}{\gamma}x.$$

Therefore,

$$\int_A^x \frac{dx}{x} = -\frac{k}{\gamma}\int_0^t dt$$

and

$$\ln\left(\frac{x}{A}\right) = -\frac{k}{\gamma}t$$

or

$$x = A \exp\left(-\frac{k}{\gamma}t\right).$$

That is, the disc **relaxes** exponentially toward its equilibrium position with a **characteristic time** or **relaxation time** k/γ. The stiffer the spring (the larger k) or the less viscous the oil (the smaller γ), the shorter the relaxation time.

Now let us include the finite mass M of the disc. Because of the disc's mass we might expect its inertia to carry it beyond the equilibrium position and an oscillatory motion to occur. The sum of the restoring force of the spring and the viscous damping force

$$F_{el} + F_v = -kx - \gamma\frac{dx}{dt}$$

will then accelerate the disc so that, according to Newton's second law $(F = Ma)$,

$$-\left(kx + \gamma \frac{dx}{dt}\right) = M \frac{d^2x}{dt^2}$$

or

$$M \frac{d^2x}{dt^2} + \gamma \frac{dx}{dt} + kx = 0.$$

If we neglect the viscous damping (set $\gamma = 0$), this equation reduces to

$$M \frac{d^2x}{dt^2} + kx = 0,$$

which is just the equation for simple harmonic motion. Therefore, the solution for $\gamma = 0$ is

$$x = A \cos \left[\left(\frac{k}{M}\right)^{1/2} t + \phi\right].$$

For $\gamma > 0$, we seek a solution of the form

$$x = A(t) \cos \left[\left(\frac{k}{M}\right)^{1/2} t + \phi\right],$$

where $A(t)$ is a decreasing function of the time.

In the last section we saw that for simple harmonic motion the total energy E was a constant. If we include damping, the effect of the viscous force is to drain energy out of the oscillating system into the oil, causing a rise in the temperature of the oil. The energy drained by the viscous force as the disc moves through a distance dx at a speed v is given by

$$dE = F_v \, dx$$

$$= -\gamma v \, dx.$$

But

$$dx = v \, dt,$$

so that

$$dE = -\gamma v^2 \, dt.$$

Therefore, the rate at which energy is dissipated by the viscous force is given by

$$\frac{dE}{dt} = -\gamma \bar{v}^2.$$

If the viscous force is sufficiently weak, the damping will be slow, and we can substitute for v^2 the average value $\bar{v}^2$ over a cycle of the motion to give

$$\frac{dE}{dt} = -\gamma \bar{v}^2.$$

To obtain a value for $\bar{v}^2$, we note that the average kinetic energy over a cycle is just one-half the total energy E. Therefore,

$$\tfrac{1}{2} M\bar{v}^2 = \tfrac{1}{2} E$$

and

$$\frac{dE}{dt} = -\frac{\gamma}{M} E.$$

But

$$E = \tfrac{1}{2} k A^2(t),$$

so that

$$\frac{dE}{dt} = \frac{dA^2(t)}{dt}$$

$$= k A(t) \frac{dA(t)}{dt}$$

$$= -\frac{\gamma}{M} E$$

$$= -\frac{\gamma}{M} \frac{k}{2} A^2(t)$$

or

$$\frac{dA(t)}{dt} = -\frac{\gamma}{2M} A(t).$$

Therefore,

$$\int_{A(0)}^{A(t)} \frac{dA(t)}{A(t)} = -\frac{\gamma}{2M} \int_0^t dt$$

and

$$\ln \frac{A(t)}{A(0)} = -\frac{\gamma}{2M} t$$

or

$$A(t) = A(0) \exp\left(-\frac{\gamma}{2M} t\right).$$

The final solution for the **lightly damped oscillator** is

$$x = A(t) \cos \left[\left(\frac{k}{M} \right)^{1/2} t + \phi \right]$$

$$= A(0) \exp \left(-\frac{\gamma}{2M} t \right) \cos \left[\left(\frac{k}{M} \right)^{1/2} t + \phi \right].$$

Figure 11.13 illustrates the solution. The envelope of the oscillations dies away with a characteristic time $2M/\gamma$, and the angular frequency of the

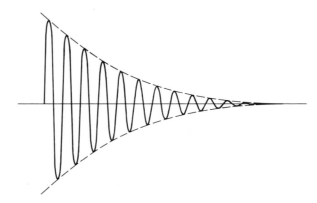

Fig. 11.13. The solution of the equation describing a lightly damped oscillator is a sinusoidal oscillation whose amplitude decreases exponentially with time.

oscillation is $(k/M)^{1/2}$. It should be emphasized that this solution is valid only if

$$\frac{\gamma}{2M} \ll \left(\frac{k}{M} \right)^{1/2}$$

or

$$\gamma \ll 2(kM)^{1/2}.$$

That is, the characteristic time must be much longer than the period of the oscillation.

PROBLEMS

1. A mass on the end of a spring vibrates in simple harmonic motion with amplitude 0.20 m and frequency 5 Hz. Calculate
   (a) the maximum values of the acceleration and speed.
   (b) the time required for the mass to move from its equilibrium position to a point 0.15 m from it.

2. A particle of mass 0.150 kg hangs from the end of a spiral spring. When pulled 0.10 m below its equilibrium position and released, it oscillates with a period of 2.5 sec.
   (a) Calculate the speed of the particle as it passes through the equilibrium position.
   (b) Calculate the acceleration of the particle when it is 0.050 m above the equilibrium position.
   (c) Calculate the time required for the particle moving downward to move from 0.050 m above its equilibrium position to 0.025 m below it.
   (d) Deduce the amount that the spring will shorten if the particle is removed.

3. A particle of mass 0.50 kg is suspended from a spring of force constant 38 N · m^{-1}. If the particle is displaced 0.040 m below its equilibrium position and given a downward velocity of 0.30 m · sec^{-1}, deduce the amplitude of the ensuing motion.

4. The exact equation for the period of a simple pendulum when the maximum angular displacement is θ is given by an infinite series, of which the first three terms are

$$T = 2\pi\left(\frac{l}{g}\right)^{1/2}\left[1 + \frac{1}{4}\sin^2\frac{\theta}{2} + \frac{9}{64}\sin^4\frac{\theta}{2} + \cdots\right]$$

   Deduce the relative importance of each term for $\theta = 15°$.

5. Except in the idealized case of a spring of zero mass, allowance should be made for the fact that when a particle at the end of the spring oscillates, the spring also oscillates. It may be shown that the frequency of oscillation of a mass m suspended from a spring of mass M with force constant k is given by

$$2\pi v = \left(\frac{k}{m + \dfrac{M}{3}}\right)^{1/2}.$$

   (a) Why is it reasonable that the denominator is not simply $(m + M)^{1/2}$?
   (b) Suppose a particle of mass 1.0 kg is suspended from a spring of mass 0.08 kg and force constant 65 N · m^{-1}. What error in the frequency would result from neglecting the mass of the spring?

6. A **ballistic pendulum** is used to measure the speed of a rifle bullet. It consists of a block of mass M suspended from above by a fine wire. A bullet fired horizontally into the block causes the block to swing about its point of suspension. By measuring the vertical height through which the block rises, the initial speed of the bullet may be determined. If a bullet of mass 0.02 kg fired into a ballistic pendulum of mass 5 kg causes the pendulum to rise through 0.10 m, calculate the initial speed of the bullet. Assuming that the pendulum undergoes simple harmonic motion and is of length 5 m, find the period of the motion and the displacement of the pendulum from rest as a function of time.

7. A mass of 1.0 kg is attached to a spring that is suspended in a vertical position and observed to increase its original length by 0.1 m. A 3.0-kg mass is substituted for the 1.0-kg mass, and the system is set into oscillation. What period is

expected for the motion? Ten measurements of the period of oscillation yield the following results: 1.178 sec, 1.187 sec, 1.203 sec, 1.184 sec, 1.201 sec, 1.199 sec, 1.192 sec, 1.193 sec, 1.181 sec, and 1.197 sec. Calculate the experimental period, assign an error to the measurement, and compare the result with the expected period. Your comparison should contain a full explanation for any discrepancy that may exist.

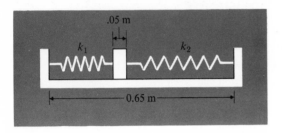

Fig. 11.14

*8. A cube 0.05 m to a side is in equilibrium under the action of two springs fixed to points 0.65 m apart, as shown in Fig. 11.14. The springs have force constants $k_1 = 1.0\,\mathrm{N \cdot m^{-1}}$ and $k_2 = 3.0\,\mathrm{N \cdot m^{-1}}$, and each has an unstretched length of 0.20 m. If the cube of mass 0.10 kg moves on a frictionless surface,
(a) find the lengths of the springs when the block is in its equilibrium position;
(b) find the period of vibration of the cube if it is displaced from equilibrium and then released.

*9. Suppose that the cube mentioned in Problem 8 is oscillating with an amplitude of 0.06 m. As it passes through its equilibrium position, a bullet of mass 0.010 kg is fired into it from directly above.
(a) Find the new period of oscillation.
(b) Find the new amplitude of oscillation.
(c) Account for any change in energy that occurs.

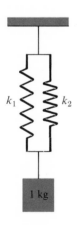

Fig. 11.15

*10. You are given two springs of different force constants k_1 and k_2. You are asked to suspend a 1-kg mass from the springs in the two configurations shown in Fig. 11.15 and to measure the period of the resulting oscillation in each case. What would be the ratio of the periods that you would measure if $3k_1 = 2k_2$?

11. Two identical masses are resting on a frictionless table and are connected by a spring. One of the masses is held fixed and the other vibrates at a certain frequency. When the fixed mass is released, the frequency changes. What is the ratio of the new period to the old?

*12. It is proposed that a straight tunnel be dug through the earth to link Toronto and New York in order to provide a rapid-transit system. How deep must the tunnel be at its deepest point?

13. We have seen that the time to travel between any two points on the earth's surface via a tunnel, utilizing only the force of gravity, is 42.3 min. How long would such trips take
(a) on the planet Mars?
(b) on the planet Jupiter?

14. Consider the model of a linear triatomic molecule shown in Fig. 11.16. In the equilibrium position the two atoms of mass m are symmetrically located on each side of the atom of mass M, and the equilibrium separation is X_e. The inter-atomic forces are represented by two identical springs of force constant k joining the three atoms. Write an expression for the potential energy of the molecule when the coordinates of the atoms are X_1, X_2, and X_3.

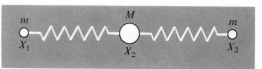

Fig. 11.16

15. A particle of mass m vibrates with amplitude A at the end of a spring of force constant k. Starting from the expression for the energy of the oscillating system, show that

$$\frac{dx}{(A^2 - x^2)^{1/2}} = \left(\frac{k}{m}\right)^{1/2} dt$$

and then by integration that the frequency of oscillation v is given by

$$v = \frac{1}{2\pi}\left(\frac{k}{m}\right)^{1/2}.$$

16. A disc of mass 0.75 kg and radius 1.12 m, suspended from a point on its circumference by a wire, is observed to oscillate at a frequency of 1.2 Hz and with an amplitude of 60°.
(a) Deduce the torsion constant of the wire.
(b) Calculate the angular frequency of the disc when $\theta = 15°$, 30°, 45°, and 60°.
(c) Calculate the angular acceleration of the disc when $\theta = 15°$, 30°, 45°, and 60°.

17. A small object of mass m revolving in a horizontal circle with speed v of constant magnitude at the end of a string of length l is known as a conical pendulum (see Section 8.3). This is shown in Fig. 11.17.
(a) Deduce the period of the conical pendulum.
(b) Show that for small θ the period of the conical pendulum is the same as that of a simple pendulum of the same length.

18. Compare the periods of oscillation of a disc suspended by the same wire from its center and then from a point on its circumference.

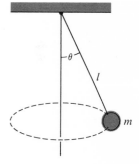

Fig. 11.17

*19. (a) A Cavendish experiment for the measurement of G (see Section 7.3) is illustrated in Fig. 11.18. Two small spheres are mounted on the ends of a beam. The beam is suspended by a wire from above. Two large spheres, placed in one of two positions near the small spheres, cause the beam to twist until the gravitational force is balanced by the elastic restoring force of the wire. The equilibrium positions are indicated by a spot of light on a scale a distance L from the wire. When the large spheres are switched from position (a) to position (b), the beam performs a damped oscillation with period T. Given the data[4] on p. 202, deduce a value of G.

---

[4] Film Loop 80–212, "The Measurement of G—the Cavendish Experiment" (Cambridge, Mass.: The Ealing Corporation).

$$b = 0.0476 \pm 0.0002 \text{ m}$$

$$d = 0.0494 \pm 0.0002 \text{ m}$$

$$M = 1.501 \quad \pm 0.001 \text{ kg}$$

$$T = 686 \quad \pm 3.0 \text{ sec}$$

$$S = 6.15 \quad \pm 0.10 \text{ cm}$$

$$L = 154.0 \quad \pm 0.03 \text{ cm}$$

(b) The value of G obtained in part (a) differs from the accepted value by much more than the calculated experimental error. Suggest a possible source or sources of systematic error that could account for the discrepancy.

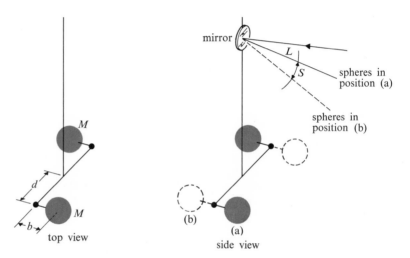

Fig. 11.18

*20. Electrical circuits in general contain three types of passive elements, known as capacitors (C), inductors (L), and resistors (R). A capacitor stores electrical energy, an inductor stores magnetic energy, and a resistor dissipates electromagnetic energy. Consider a circuit consisting only of a capacitor C, an inductor L, and a switch. Initially the switch is open, and the capacitor contains Q_0 C of electrical charge. At $t = 0$, the switch is closed, and a current

$$I = \frac{dQ}{dt}$$

begins to flow in the circuit as governed by the equation

$$L\frac{dI}{dt} + \frac{Q}{C} = 0.$$

(a) Discuss the nature of the flow of charge in this circuit after the switch is closed.

(b) For $L = 25 \times 10^{-3}$ H and $C = 1.0 \times 10^{-9}$ F, calculate the frequency of the oscillations in the circuit.

(c) If $Q_0 = 1.0 \times 10^{-6}$ C, calculate the maximum electrical (magnetic) energy in the system.

21. If in the previous problem we include the finite resistance R of a physical inductor, the flow of current in the circuit described is governed by the equation

$$L\frac{dI}{dt} + RI + \frac{Q}{C} = 0.$$

If $R = 3.5$ ohms, calculate the characteristic time for the decay of the current in the circuit. Justify the validity of any assumption that is made.

12 Relativistic dynamics

12.1 INTRODUCTION

In Section 4.7 the Lorentz transformation was derived using only the second postulate (concerning the constancy of the speed of light) of the special theory of relativity. Reference to the first postulate (concerning the identity of the laws of nature in all reference frames in uniform translatory motion) is not necessary for a discussion of kinematics. All the kinematical effects discussed in Chapter 5 (length contraction, time dilation, etc.) are consequences of the Lorentz transformation and therefore depend upon the assumption of constancy of the speed of light. Newton's laws of motion are the foundation of Newtonian dynamics. The Lorentz transformation and the Galilean transformation both connect observations made in reference frames in uniform relative motion. But only the Lorentz transformation satisfies the second postulate. We shall examine Newton's second law in this chapter in the light of the Lorentz transformation to see if it is invariant in form under the transformation.

12.2 THE TRANSFORMATION OF NEWTON'S SECOND LAW

We have already shown in Section 4.4 that acceleration is an invariant under a Galilean transformation. For example,

$$a'_x = a_x,$$

where a_x and a'_x are the x components of the acceleration of a particle as observed in two different frames of reference R and R' in uniform relative motion in the x, x' direction. An observer in R applying Newton's second law would write

$$F_x = ma_x$$

to describe the motion of the particle, whereas an observer in R' would write

$$F'_x = m'a'_x.$$

If we **assume** that the mass of the particle is invariant under a Galilean transformation, we have

$$m'a'_x = ma_x,$$

which implies that

$$F'_x = F_x$$

and that Newton's second law is invariant in form under a Galilean transformation.

We do know, however, that the speed of light is not invariant under a Galilean transformation but is invariant under a Lorentz transformation. We shall now test the invariance of Newton's second law under a Lorentz transformation. The observer in R again writes

$$F_x = ma_x = m\frac{dv_x}{dt},$$

whereas the observer in R' writes

$$F'_x = m'a'_x = m'\frac{dv'_x}{dt'}.$$

In Section 5.2 we developed a relation between v'_x and v_x; the relation is

$$v'_x = \frac{v_x - u}{1 - \dfrac{uv_x}{c^2}}.$$

We want the quantity dv'_x/dt', which can be written, by using the chain rule of calculus,

$$\frac{dv'_x}{dt'} = \frac{dv'_x}{dt}\frac{dt}{dt'}.$$

Now

$$t' = \gamma\left(t - \frac{ux}{c^2}\right),$$

where

$$\gamma = \left(1 - \frac{u^2}{c^2}\right)^{-1/2},$$

so that

$$\frac{dt'}{dt} = \gamma\left(1 - \frac{uv_x}{c^2}\right),$$

$$\frac{dt}{dt'} = \gamma^{-1}\left(1 - \frac{uv_x}{c^2}\right)^{-1}$$

and

$$\frac{dv'_x}{dt} = \frac{dv_x}{dt}\frac{1}{\left(1 - \dfrac{uv_x}{c^2}\right)} + (v_x - u)\left(-\frac{u}{c^2}\frac{dv_x}{dt}\right)\frac{(-1)}{\left(1 - \dfrac{uv_x}{c^2}\right)^2}.$$

Therefore,

$$
\begin{aligned}
\frac{dv'_x}{dt'} &= \gamma^{-1}\frac{dv_x}{dt}\left[\left(1-\frac{uv_x}{c^2}\right)^{-1}+\frac{u}{c^2}(v_x-u)\left(1-\frac{uv_x}{c^2}\right)^{-2}\right]\left(1-\frac{uv_x}{c^2}\right)^{-1} \\
&= \frac{1}{\gamma}\frac{dv_x}{dt}\left[\left(1-\frac{uv_x}{c^2}\right)^{-2}+\frac{(uv_x-u^2)}{c^2}\left(1-\frac{uv_x}{c^2}\right)^{-3}\right] \\
&= \frac{1}{\gamma}\frac{dv_x}{dt}\left[\left(1-\frac{uv_x}{c^2}\right)+\left(\frac{uv_x-u^2}{c^2}\right)\right]\left(1-\frac{uv_x}{c^2}\right)^{-3} \\
&= \frac{1}{\gamma}\frac{dv_x}{dt}\left[\frac{c^2-uv_x+uv_x-u^2}{\dfrac{(c^2-uv_x)^3}{c^4}}\right] \\
&= \frac{1}{\gamma}\frac{dv_x}{dt}\left[\frac{c^4(c^2-u^2)}{(c^2-uv_x)^3}\right],
\end{aligned}
$$

and the acceleration is *not* invariant under the Lorentz transformation.

The difficulty may arise from the fact that we have written Newton's second law in the form

$$
F = ma = m\frac{dv}{dt}
$$

rather than in the form

$$
F = \frac{dp}{dt}.
$$

In so doing we have assumed that the mass of the particle is independent of the time. Perhaps we should be paying attention to the momentum changes as seen from R and R' rather than the accelerations. Let us try this approach.

12.3   THE VARIATION OF MASS WITH VELOCITY

In Section 9.2 we considered the collision of two gliders on an air track and found that momentum was conserved. In particular, if the gliders have the same mass and are traveling in opposite directions with the same speed, as shown in Fig. 12.1, they are exactly reversed in direction after a collision and move off with the same speed, provided the collision is elastic (no energy lost to deformation of the bumper springs). Let us

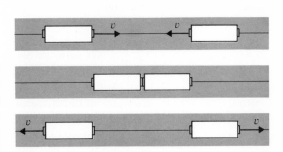

Fig. 12.1. An elastic collision between two identical gliders moving with equal speeds in opposite directions on an air track.

suppose that the air track is cut in half and that the two gliders are launched with equal but opposite velocities, so that they collide at the point where the track is cut (see Fig. 12.2). As long as the two cut ends of the track remain together, we may perform this experiment as often as we like with the same result as described above.

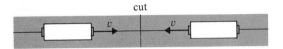

Fig. 12.2. The air track is cut at the point where collision occurs.

Now let us consider a situation in which the two halves of the track are separated, with one half placed on a stationary table and the second half placed on a moving table in such a manner that at some instant, as the table moves, the two halves of the air track will be opposite each other. This opposition would only be momentary, for the moving table would not stop. The layout is pictured in Fig. 12.3. If the two gliders are launched with equal but opposite velocities on the two halves of the track at the correct time, we see from Fig. 12.4 that we can arrange for them to collide at the instant that the two halves of the track line up. This is position (2) in Fig. 12.4. This situation is the same as that shown in Fig. 12.1, and we should expect the gliders to be exactly reversed on their tracks as before. We now want to discuss this experiment from a relativistic point of view.

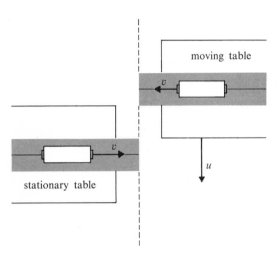

Fig. 12.3. One half of the air track is on stationary table while the second half is on a table which is moving such that the cut ends of the air track are briefly together at the moment when the gliders collide.

Let us attach frames of reference R and R' to the stationary and moving tables, respectively. We further assume that the x and x' axes point in the direction of motion of the moving table and that the y and y' axes are along the plane of the table, as shown in Fig. 12.5. Let us denote the glider in system R by A and the glider in R' by B. Before the collision glider A has velocity v_y as measured in R, and glider B has velocity $-v'_y$ as measured in R'.

We saw in Section 9.5 that the conservation laws of energy and momentum are connected with the uniform nature of space and time. We should expect that these conservation laws would be independent of any uniform translational motion of an observer. Therefore, believing that these conservation laws are fundamental, we should feel no hesitancy toward invoking these conservation laws in both frames of reference R and R'. If the collision is elastic in R, it is elastic in R', and the total momentum must be conserved in both R and R'.

We may use the velocity transformation equations (see Section 5.2) to write down the y component of velocity of glider B as seen by an observer

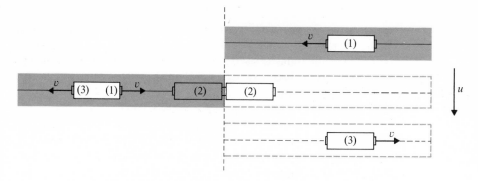

Fig. 12.4. Three positions of the moving table of Fig. 12.3 showing the gliders before, at the moment of, and after the collision.

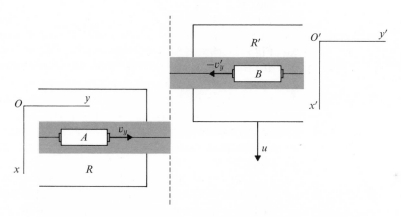

Fig. 12.5. Frames of reference R and R' are attached to the stationary and moving tables, respectively. The coordinate axes are so oriented that the gliders move along the y and y' axes and the table moves in the x' direction.

in R. Before the collision this velocity is

$$v_y = \frac{-v'_y}{\gamma\left(1 + \dfrac{uv'_x}{c^2}\right)}.$$

But $v'_x = 0$ (glider B is moving along the y' axis). Therefore,

$$v_y = \frac{-v'_y}{\gamma}$$

before the collision. After the collision, glider B has the velocity

$$v_y = \frac{v'_y}{\gamma}$$

as seen in R. In R', glider B has velocity $-v'_y$ before the collision and velocity v'_y after the collision.

We shall assume that the mass of the gliders may not be independent of their motion. We can take this into account by writing the momentum as

$$\mathbf{p} = m(u)\mathbf{u},$$

where $m(u)$ is the mass when the velocity is $\mathbf{u}$. Glider B has a speed v'_y in R' both before and after the collision (only the direction of motion changes). Therefore, the mass of glider B in R' is

$$m(v'_y).$$

As seen from R, the velocity of glider B has a magnitude v', where

$$v'^2 = v'^2_y + u^2;$$

this is shown in Fig. 12.6. The magnitude of v' is the same before and after collision, as $-v'_y$ merely changes to v'_y. Therefore, the mass of glider B in R is

$$m(v').$$

The velocity and momentum components in the y, y' direction as measured in both R and R' are tabulated in Table 12.1, along with the change in momentum.

Suppose we were to view this collision from R' rather than from R. Since the two coordinate systems are in uniform relative motion, an observer in R' could consider his table stationary and the table holding glider A to be moving past him at speed u. The collision as observed from both R and R' is outlined diagrammatically in Fig. 12.7. Since the collision is elastic, it is apparent that an observer in R' will measure an equal but

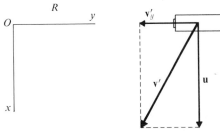

Fig. 12.6. Glider B has a velocity v' as observed in R, where $v'^2 = v'^2_y + u^2$.

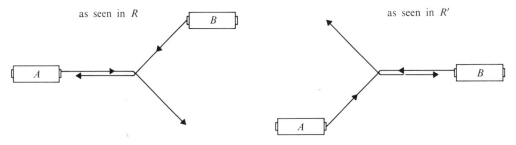

Fig. 12.7. The collision as viewed from R' and from R. The return paths of glider A in R and glider B in R' have been displaced to the side for clarity.

Table 12.1 The components of velocity and momentum of glider B in the y, y' direction as measured in the frames of reference R and R'.

		Frame R	Frame R'
Before collision	Velocity	$\dfrac{-v'_y}{\gamma}$	$-v'_y$
	Momentum	$\dfrac{-m(v')v'_y}{\gamma}$	$-m(v'_y)v'_y$
After collision	Velocity	$\dfrac{v'_y}{\gamma}$	v'_y
	Momentum	$\dfrac{m(v')v'_y}{\gamma}$	$m(v'_y)v'_y$
Change in momentum		$\dfrac{2m(v')v'_y}{\gamma}$	$2m(v'_y)v'_y$

opposite change in momentum for glider B compared to what an observer in R measures for glider A and will measure an equal but opposite change in momentum for glider A compared to what an observer in R measures for glider B. In addition, we see that the change in momentum of glider B as measured in R' is equal but opposite to the change in momentum of glider A as measured in R, since each observer sees a glider traveling down his track and back at the same speed.

Since we assume that momentum is conserved in both R and R', the change in momentum of glider B as measured in R must be equal and opposite to the change in momentum of glider A as measured in R and therefore must be equal to the change in momentum of glider B as measured in R'. Referring to Table 12.1, we see that

$$\frac{2m(v')v'_y}{\gamma} = 2m(v'_y)v'_y$$

$$m(v') = \gamma m(v'_y),$$

where

$$\gamma = \left(1 - \frac{u^2}{c^2}\right)^{-1/2}.$$

Note that in every case v'_y has canceled. Hence, we can now take the limit

$v'_y \to 0$, keeping u greater than zero. In this case

$$v'_y \to 0,$$

$$v' \to u,$$

and

$$m(u) = \gamma m(0)$$

or

$$m = \gamma m_0,$$

where we have set $m(0) = m_0$, the **rest mass of the glider**, and $m(u) = m$, the **effective mass of the glider when its speed is** u (often called the **relativistic mass**).

Although we have derived the relation giving the mass as a function of speed for the case of an elastic collision between gliders of equal mass, the relation is a general one and can be shown to hold for any particle of any mass whatsoever. This extension of the theory is a little too complicated to take up in this text, however.

The momentum of a particle having a speed v (reverting to our normal symbol for the speed of a particle) is given by

$$\mathbf{p} = m\mathbf{v}$$

$$= \gamma m_0 \mathbf{v}$$

$$= \frac{m_0 \mathbf{v}}{\left(1 - \dfrac{v^2}{c^2}\right)^{1/2}}.$$

The relation $m = \gamma m_0$ predicts an increase of mass with increasing relative speed of object and observer, with the observed mass approaching infinity as $v \to c$. The variation of m with v is shown in Fig. 12.8.

If a mass approaches infinity as the relative speed of the mass approaches the value c, then an infinite force is required to accelerate the mass to the speed c. The speed c therefore provides a limiting value that can never be achieved[1].

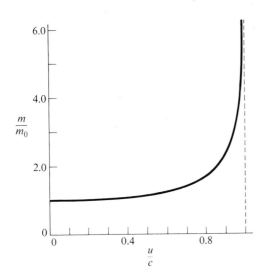

Fig. 12.8. The variation of the mass of an object as a function of its speed.

---

[1] For a treatment on film of this subject, see W. Bertozzi, "The Ultimate Speed, an Exploration with High-energy Electrons," an ESI film.

12.4   FORCE AND ENERGY

The results of the previous section suggest that we should define force for relativistic dynamics as

$$\mathbf{F} = \frac{d\mathbf{p}}{dt}$$

$$= \frac{d(\gamma m_0 \mathbf{v})}{dt}$$

$$= \frac{m_0 \, d(\gamma \mathbf{v})}{dt}.$$

It can be shown in a general way that Newton's second law, written in this manner, is invariant in form under a Lorentz transformation. However, before we can say very much more about the transformation of force, we must look at the relation between mass and energy when the mass is a function of the speed.

We recall from Section 7.7 that the work done on a particle traveling along some path between points A and B is given by the integral

$$W_{A \to B} = \int_A^B \mathbf{F} \cdot d\mathbf{s} = \int_A^B F \cos \theta \, ds,$$

where the integral is evaluated along the path that the particle follows (see Fig. 12.9). If the force is a conservative force, the work done is independent of the path and is equal to the change in potential energy $U_B - U_A$.

The work done is also equal to the change in kinetic energy in Newtonian dynamics, as we may easily show by using Newton's second law:

$$W_{A \to B} = \int_A^B \mathbf{F} \cdot d\mathbf{s},$$

$$= m \int_A^B \frac{d\mathbf{v} \cdot d\mathbf{s}}{dt}$$

$$= m \int_A^B \mathbf{v} \cdot d\mathbf{v}$$

$$= m \int_A^B v \, dv$$

$$= \frac{m}{2} \int_A^B d(v^2)$$

$$= \tfrac{1}{2} m v_B^2 - \tfrac{1}{2} m v_A^2.$$

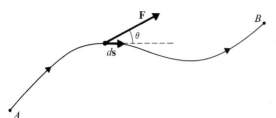

Fig. 12.9. The work done on a particle traveling between points A and B is equal to $\int_A^B \mathbf{F} \cdot d\mathbf{s}$.

Let us now carry out the calculation using the relativistic form of Newton's second law. The details of the calculation are given in the box below, where it is shown that the work done $W_{A \to B}$ is given by

$$W_{A \to B} = mc^2 \big|_A^B.$$

When the particle is initially at rest at A and has speed v at B, then, by analogy with Newtonian dynamics, we say that the particle has acquired a kinetic energy

$$K = mc^2 \bigg|_{v_A = 0}^{v_B = v}$$

$$= \frac{m_0 c^2}{\left(1 - \dfrac{v^2}{c^2}\right)^{1/2}} - m_0 c^2$$

or

$$K = mc^2 - m_0 c^2.$$

Derivation of the relativistic expression for kinetic energy

The work done on a particle traveling from A to B in a conservative field of force is independent of the path and is given by

$$W_{A \to B} = \int_A^B \mathbf{F} \cdot d\mathbf{s}$$

$$= \int_A^B \frac{d}{dt}(m_0 \gamma \mathbf{v}) \cdot d\mathbf{s}$$

$$= \int_A^B d(m_0 \gamma \mathbf{v}) \cdot \mathbf{v}.$$

Now

$$d(xy) = x\, dy + y\, dx$$

or

$$xy \bigg|_A^B = \int_A^B x\, dy + \int_A^B y\, dx$$

and

$$\int_A^B x\, dy = xy \bigg|_A^B - \int_A^B y\, dx.$$

Therefore,

$$\int_A^B \mathbf{v} \cdot d(m_0\gamma \, d\mathbf{v}) = \mathbf{v} \cdot (m_0\gamma\mathbf{v})\Big|_A^B - \int_A^B m_0\gamma\mathbf{v} \cdot d\mathbf{v}$$

$$= m_0\gamma v^2 \Big|_A^B - \int_A^B \frac{m_0 v \, dv}{\left(1 - \dfrac{v^2}{c^2}\right)^{1/2}}$$

$$= m_0\gamma v^2 \Big|_A^B - m_0 c^2\left(1 - \frac{v^2}{c^2}\right)^{1/2}\Big|_A^B$$

$$= \left[\frac{m_0 v^2}{\left(1 - \dfrac{v^2}{c^2}\right)^{1/2}} - m_0 c^2\left(1 - \frac{v^2}{c^2}\right)^{1/2}\right]\Bigg|_A^B$$

$$= \frac{m_0 v^2 - m_0 c^2\left(1 - \dfrac{v^2}{c^2}\right)}{\left(1 - \dfrac{v^2}{c^2}\right)^{1/2}}\Bigg|_A^B$$

$$= \frac{m_0 c^2}{\left(1 - \dfrac{v^2}{c^2}\right)^{1/2}}\Bigg|_A^B$$

$$= mc^2 \Big|_A^B.$$

Let us take the limit $v \ll c$ to see if this expression for kinetic energy reduces to the Newtonian expression as it should. Now

$$K = m_0 c^2\left[\left(1 - \frac{v^2}{c^2}\right)^{-1/2} - 1\right]$$

$$= m_0 c^2\left[\left(1 + \frac{1}{2}\frac{v^2}{c^2} + \frac{3}{8}\frac{v^4}{c^4} + \cdots\right) - 1\right]$$

$$= \tfrac{1}{2}m_0 v^2 + \tfrac{3}{8}m_0\frac{v^4}{c^2} + \cdots$$

$$= \tfrac{1}{2}m_0 v^2, \qquad v \ll c,$$

which is the Newtonian kinetic energy.

Evidently, the quantities mc^2 and $m_0 c^2$ must be energies also. Albert Einstein suggested that the term $m_0 c^2$ represents an intrinsic energy associated with the particle due to its rest mass; he called it the **rest energy** of the particle. The term mc^2 then represents the total energy of the particle, the sum of its rest energy and its kinetic energy. Denoting the total energy by E, we have

$$E = mc^2.$$

This formula suggests an equivalence between mass and energy and the possibility of conversion of energy to mass or mass to energy.

We note that

$$K = (m - m_0)c^2 > \tfrac{1}{2}m_0 v^2,$$

since all the terms in the expansion for K are positive. Since the relativistic mass increases with the speed of the particle, the force required to maintain a constant acceleration must also increase. Since the increase in kinetic energy is proportional to the applied force, it follows that the relativistic kinetic energy must exceed the classical value.

Example. Calculate the rest energy of the electron, proton, and neutron. This can also be stated as "calculate the **rest mass** of the electron, proton, and neutron in MeV."

Solution. For the electron:

$$m_0 = 9.109 \times 10^{-31} \, \text{kg}$$

$$m_0 c^2 = 9.109 \times 10^{-31}(2.998 \times 10^8)^2$$

$$= 8.189 \times 10^{-14} \, \text{J}$$

$$= \frac{8.189 \times 10^{-14}}{1.602 \times 10^{-13}}$$

$$= 0.5110 \, \text{MeV}.$$

For the proton:

$$m_0 = 1.673 \times 10^{-27} \, \text{kg}$$

$$m_0 c^2 = 1.673 \times 10^{-27} \times (2.998 \times 10^8)^2$$

$$= 1.504 \times 10^{-10} \, \text{J}$$

$$= 938.9 \, \text{MeV}.$$

For the neutron:

$$m_0 = 1.675 \times 10^{-27} \, \text{kg}$$

$$m_0 c^2 = 1.675 \times 10^{-27} \times (2.998 \times 10^8)^2$$

$$= 1.505 \times 10^{-10} \, \text{J}$$

$$= 939.3 \, \text{MeV}.$$

The rest mass of the electron, proton, and neutron are 0.5110 MeV, 938.9 MeV, and 939.3 MeV, respectively. (Note that a five-figure calculation gives 938.3 and 939.3 MeV for the proton and neutron, respectively.)

Example. An electron is traveling at a speed of $10^8 \, \text{m} \cdot \text{sec}^{-1}$. How big an error is made in calculating its kinetic energy from the Newtonian formula?

Solution. By the Newtonian formula:

$$K = \tfrac{1}{2}m_0 v^2$$
$$= \tfrac{1}{2}(9.11 \times 10^{-31})10^{16}$$
$$= 4.56 \times 10^{-15} \text{ J}$$
$$= \frac{4.56 \times 10^{-15}}{1.60 \times 10^{-16}} \text{ keV}$$
$$= 28.5 \text{ keV}.$$

By the relativistic formula:

$$K = (m - m_0)c^2$$
$$= (\gamma - 1)m_0 c^2,$$

where

$$\gamma = \left(1 - \frac{v^2}{c^2}\right)^{-1/2}$$
$$= \left(1 - \frac{1}{9.00}\right)^{-1/2} = \left(\frac{8}{9}\right)^{-1/2} = 1.06.$$

Therefore,

$$K = (1.06 - 1)(9.11 \times 10^{-31})(9.00 \times 10^{16})$$
$$= 4.(9) \times 10^{-15} \text{ J}$$
$$= 3(1) \text{ keV}.$$

Obviously we must calculate γ more carefully as three-figure accuracy is not sufficient. Let us use

$$c = 2.9979 \times 10^8 \text{ m} \cdot \text{sec}^{-1}.$$

Then

$$\gamma = \left(1 - \frac{1}{8.9871}\right)^{-1/2}$$
$$= (1 - 0.11127)^{-1/2}$$
$$= (0.88873)^{-1/2}$$
$$= 1.0608,$$

and

$$K = (1.0608 - 1)(9.11 \times 10^{-31})(9.00 \times 10^{16})$$
$$= 4.99 \times 10^{-15} \text{ J}$$
$$= 31.2 \text{ keV}.$$

The error made in using the Newtonian kinetic energy is $31.2 - 28.5 = 2.7$ keV or

$$\frac{\Delta K}{K} = \left(\frac{2.7}{31.2}\right)100 = 8.7\%.$$

An alternative method is to let K_N be the Newtonian kinetic energy and K_R be the relativistic kinetic energy. Now,

$$K_N = \tfrac{1}{2}m_0v^2$$

$$K_R = \frac{1}{2}m_0v^2 + \frac{3}{8}m_0\frac{v^4}{c^2} + \frac{5}{16}m_0\frac{v^6}{c^4} + \cdots.$$

The error may be expressed as

$$\text{Error} = \frac{K_R - K_N}{K_R}.$$

Considering terms in K_R up to v^4 only gives

$$\text{Error} = \frac{\dfrac{3}{8}m_0\dfrac{v^4}{c^2}}{\dfrac{1}{2}m_0v^2 + \dfrac{3}{8}m_0\dfrac{v^4}{c^2}}$$

$$= \frac{1}{1 + \dfrac{4}{3}\dfrac{c^2}{v^2}}$$

$$= \frac{1}{1 + \tfrac{4}{3}(9)}$$

$$= \tfrac{1}{13}$$

$$= 0.077$$

$$\equiv 7.7\%.$$

A similar calculation including terms in K_R up to v^6 yields 8.3% for the error. As more terms are included in the expansion for K_R, the error approaches the more exact value of 8.7%.

Note: Calculations of relativistic kinetic energies must be carried out very carefully for speeds for which the kinetic energy is appreciably less than the rest energy, since mc^2 is not much greater than m_0c^2 under these conditions.

12.5   TRANSFORMATION OF MOMENTUM, ENERGY, AND FORCE

The momentum and energy of a particle are given in terms of the rest mass by the expressions

$$\mathbf{p} = \gamma m_0\mathbf{v} \quad \text{and} \quad E = \gamma m_0c^2,$$

respectively, as measured by an observer in frame of reference R. We now inquire as to the relation between $\mathbf{p}$ and E as measured in R and $\mathbf{p}'$ and

E' measured in a frame of reference R' moving with velocity u relative to R in the x, x' direction.

The components of **p** in R are $p_x, p_y,$ and p_z, and the components of **p**$'$ in R' are $p'_x, p'_y,$ and p'_z. Now

$$p'_x = \gamma' m_0 v'_x$$

$$p'_y = \gamma' m_0 v'_y$$

$$p'_z = \gamma' m_0 v'_z$$

$$E' = \gamma' m_0 c^2,$$

where

$$\gamma' = \left(1 - \frac{v'^2}{c^2}\right)^{-1/2}$$

by definition. Since $v'_x, v'_y,$ and v'_z are kinematic quantities, we know the relation between v'_x and v_x, v'_y and v_y, v'_z and v_z, and v' and v from the velocity transformation equations derived in Section 5.2. Using this information we can derive the relation between p' and p and between E' and E. The calculations are carried out in the screened area below. The result is

$$p'_x = \gamma\left(p_x - \frac{uE}{c^2}\right)$$

$$p'_y = p_y$$

$$p'_z = p_z$$

$$\frac{E'}{c^2} = \gamma\left(\frac{E}{c^2} - \frac{up_x}{c^2}\right)$$

where

$$\gamma = \left(1 - \frac{u^2}{c^2}\right)^{-1/2}.$$

Transformation of momentum and energy

Let us consider p'_x first.

$$p'_x = m_0\left(1 - \frac{v'^2}{c^2}\right)^{-1/2} \frac{v_x - u}{1 - \frac{uv_x}{c^2}}.$$

But

$$\left(1 - \frac{v'^2}{c^2}\right) = \left[1 - \frac{(v_x'^2 + v_y'^2 + v_z'^2)}{c^2}\right]$$

$$= 1 - \frac{1}{c^2}\left[\left(\frac{v_x - u}{1 - \frac{uv_x}{c^2}}\right)^2 + \frac{v_y^2}{\gamma^2\left(1 - \frac{uv_x}{c^2}\right)^2} + \frac{v_z^2}{\gamma^2\left(1 - \frac{uv_x}{c^2}\right)^2}\right],$$

where

$$\gamma = \left(1 - \frac{u^2}{c^2}\right)^{-1/2}$$

Rearranging gives

$$\left(1 - \frac{v'^2}{c^2}\right) = 1 - \left(1 - \frac{uv_x}{c^2}\right)^{-2}\left[\frac{(v_x - u)^2}{c^2} + \left(1 - \frac{u^2}{c^2}\right)\left(\frac{v_y^2 + v_z^2}{c^2}\right)\right]$$

$$= \left(1 - \frac{uv_x}{c^2}\right)^{-2}\left[\left(1 - \frac{uv_x}{c^2}\right)^2 - \frac{(v_x - u)^2}{c^2} - \left(1 - \frac{u^2}{c^2}\right)\left(\frac{v_y^2 + v_z^2}{c^2}\right)\right]$$

$$= \left(1 - \frac{uv_x}{c^2}\right)^{-2}\left[\left(1 - \frac{v^2}{c^2}\right)\left(1 - \frac{u^2}{c^2}\right)\right].$$

Therefore,

$$p_x' = m_0\left(1 - \frac{uv_x}{c^2}\right)\left(1 - \frac{v^2}{c^2}\right)^{-1/2}\left(1 - \frac{u^2}{c^2}\right)^{-1/2}\frac{v_x - u}{1 - \frac{uv_x}{c^2}}$$

$$= \gamma m_0\left(1 - \frac{v^2}{c^2}\right)^{-1/2}(v_x - u)$$

$$= \gamma\left[p_x - m_0 u\left(1 - \frac{v^2}{c^2}\right)^{-1/2}\right]$$

$$= \gamma\left(p_x - \frac{uE}{c^2}\right),$$

since

$$p_x = m_0\left(1 - \frac{v^2}{c^2}\right)^{-1/2}v_x, \qquad E = \left(1 - \frac{v^2}{c^2}\right)^{-1/2}m_0 c^2.$$

A similar calculation shows that

$$p_y' = p_y, \qquad p_z' = p_z;$$

we leave this as an exercise.

We now consider the energy.

$$E' = \left(1 - \frac{v'^2}{c^2}\right)^{-1/2} m_0 c^2$$

$$= m_0 c^2 \left(1 - \frac{uv_x}{c^2}\right)\left(1 - \frac{v^2}{c^2}\right)^{-1/2}\left(1 - \frac{u^2}{c^2}\right)^{-1/2}$$

$$\frac{E'}{c^2} = \gamma m_0 \left(1 - \frac{v^2}{c^2}\right)^{-1/2}\left(1 - \frac{uv_x}{c^2}\right)$$

$$= \gamma\left(\frac{E}{c^2} - \frac{up_x}{c^2}\right).$$

The complete set of transformation equations for momentum and energy is

$$p'_x = \gamma\left(p_x - \frac{uE}{c^2}\right)$$

$$p'_y = p_y$$

$$p'_z = p_z$$

$$\frac{E'}{c^2} = \gamma\left(\frac{E}{c^2} - \frac{up_x}{c^2}\right).$$

Let us write down the Lorentz transformation equations. They are

$$x' = \gamma(x - ut)$$

$$y' = y$$

$$z' = z$$

$$t' = \gamma\left(t - \frac{ux}{c^2}\right).$$

The quantities p_x, p_y, p_z, and E^2/c transform exactly like the four space-time coordinates x, y, z, and t. It is common to think of x, y, z, and t as being the components of a vector in a four-dimensional space-time, where the distance between the points $(0, 0, 0, 0)$ and (x, y, z, t) is defined as

$$s^2 = -x^2 - y^2 - z^2 + c^2 t^2.$$

In this formulation of relativity, p_x, p_y, p_z, and E/c^2 are the components of a four-dimensional momentum vector. The force can also be expressed as a four-dimensional vector. In order to show, in general, that the form of Newton's second law is invariant under a Lorentz transformation, it is necessary to use the four-dimensional vector formulation. This is somewhat beyond our capabilities at this time, so that we shall restrict our discussion of the transformation to the following special case.

Let us consider the particular case of a particle instantaneously at rest in R' and acted on by a force F'_x acting in the x, x' direction only. We can show that the force F_x as observed in R, where the particle has a velocity u in the x direction, is equal to F'_x. The calculations are carried out in the box. In making the calculations, use is made of the following relation involving energy, momentum, and mass:

$$E^2 = p^2c^2 + m_0^2c^4.$$

The proof of this very useful relation is left as a problem (see Problem 12 in this chapter).

The relation between F'_x and F_x

The force F'_x is given by

$$F'_x = \frac{dp'_x}{dt'} = \frac{dp'_x}{dt}\frac{dt}{dt'}.$$

Now

$$\frac{dp'_x}{dt} = \frac{d}{dt}\left[\gamma\left(p_x - \frac{uE}{c^2}\right)\right]$$

$$= \gamma\frac{dp_x}{dt} - \frac{\gamma u}{c^2}\frac{dE}{dt},$$

since

$$\gamma = \left(1 - \frac{u^2}{c^2}\right)^{-1/2} = \text{constant}.$$

But, in analogy with the Lorentz transformation equations,

$$\frac{E}{c^2} = \gamma\left(\frac{E'}{c^2} + \frac{up'_x}{c^2}\right)$$

$$\frac{1}{c^2}\frac{dE}{dt} = \frac{\gamma}{c^2}\left(\frac{dE'}{dt} + u\frac{dp'_x}{dt}\right)$$

$$= \frac{\gamma}{c^2}\left(\frac{dE'}{dp'_x}\frac{dp'_x}{dt} + u\frac{dp'_x}{dt}\right)$$

$$= \frac{\gamma}{c^2}\frac{dp'_x}{dt}\left(\frac{dE'}{dp'_x} + u\right).$$

From the relation

$$E'^2 = p_x'^2c^2 + m_0^2c^4,$$

---

we have

$$2E'\frac{dE'}{dp'_x} = 2p'_x c^2$$

$$\frac{dE'}{dp'_x} = \frac{p'_x c^2}{E'}$$

$$= \frac{\gamma' m_0 v'_x c^2}{\gamma' m_0 c^2}$$

$$= v'_x.$$

In this particular example we take

$$v'_x = 0.$$

Therefore,

$$\frac{dE'}{dp'_x} = 0$$

and

$$\frac{1}{c^2}\frac{dE}{dt} = \frac{\gamma u}{c^2}\frac{dp'_x}{dt}$$

and

$$\frac{dp'_x}{dt} = \gamma \frac{dp_x}{dt} - \frac{\gamma^2 u^2}{c^2}\frac{dp'_x}{dt}$$

$$= \gamma \frac{dp_x}{dt}\left(1 + \frac{\gamma^2 u^2}{c^2}\right)^{-1}$$

$$= \frac{1}{\gamma}\frac{dp_x}{dt}.$$

Since

$$t = \gamma\left(t' + \frac{ux'}{c^2}\right),$$

$$\frac{dt}{dt'} = \gamma\left(1 + \frac{uv'_x}{c^2}\right) = \gamma.$$

Therefore,

$$F'_x = \frac{dp'_x}{dt}\frac{dt}{dt'}$$

$$= \frac{1}{\gamma}\frac{dp_x}{dt}\gamma$$

$$= \frac{dp_x}{dt}$$

and

$$F'_x = F_x.$$

Example. A particle of rest mass m_0 and speed 2.40×10^8 m $\cdot$ sec^{-1} strikes a particle of rest mass $2m_0$ that is initially at rest with respect to the laboratory. Calculate the momentum of the particles in the laboratory frame of reference and in a frame of reference moving in the same direction as the particle of rest mass m_0 at a speed of 1.09×10^8 m $\cdot$ sec^{-1}.

Solution. In the laboratory frame of reference, for particle m_0,

$$p(1) = \gamma(v)m_0 v,$$

where

$$\gamma(v) = \left(1 - \frac{v^2}{c^2}\right)^{-1/2}$$

$$= \left[1 - \left(\frac{2.40}{3.00}\right)^2\right]^{-1/2}$$

$$= (0.36)^{-1/2}$$

$$= 1.67.$$

Therefore,

$$p(1) = 1.67 \times 2.40 \times 10^8 m_0$$

$$= 4.01 \times 10^8 m_0 \text{ kg} \cdot \text{m} \cdot \text{sec}^{-1}.$$

For particle $2m_0$,

$$p(2) = 0.$$

In the moving frame of reference, for particle m_0,

$$p'(1) = \gamma(u)\left(p - \frac{uE}{c^2}\right),$$

where

$$\gamma(u) = \left(1 - \frac{u^2}{c^2}\right)^{-1/2}$$

$$u = 1.09 \times 10^8 \text{ m} \cdot \text{sec}^{-1}$$

$$E = \gamma(v)m_0 c^2.$$

Therefore,

$$p'(1) = \gamma(u)\left[\gamma(v)m_0 v - \frac{u\gamma(v)m_0 c^2}{c^2}\right]$$

$$= \gamma(u)\gamma(v)m_0(v - u).$$

Now,

$$\gamma(u) = \left(1 - \frac{u^2}{c^2}\right)^{-1/2}$$

$$= \left[1 - \left(\frac{1.09}{3.00}\right)^2\right]^{-1/2}$$

$$= (0.868)^{-1/2}$$

$$= 1.07.$$

Therefore,

$$p'(1) = (1.07)(1.67)m_0(2.40 - 1.09) \times 10^8$$

$$= 2.34 \times 10^8 m_0 \text{ kg} \cdot \text{m} \cdot \text{sec}^{-1}.$$

For particle $2m_0$,

$$p'(2) = \gamma(u)\left(0 - \frac{u\gamma(0)2m_0c^2}{c^2}\right)$$

$$= -\gamma(u)\gamma(0)2um_0$$

$$= -1.07(1.00)(2 \times 1.09 \times 10^8)m_0$$

$$= -2.33 \times 10^8 m_0 \text{ kg} \cdot \text{m} \cdot \text{sec}^{-1}.$$

The total momentum in the laboratory frame of reference is $4.01 \times 10^8 m_0 \text{ kg} \cdot \text{m} \cdot \text{sec}^{-1}$, whereas the total momentum in the moving frame of reference is $(2.34 - 2.33) \times 10^8 = 0.01 \times 10^8 \simeq 0 \text{ kg} \cdot \text{m} \cdot \text{sec}^{-1}$.

The velocity of the moving frame of reference was chosen so as to make the total momentum of the two-particle system equal to zero. (The fact that the momenta do not quite cancel is due to the fact that the third significant figure has a small "rounding" error in it, since all calculated numbers are "rounded off" to three figures.) The total momentum of the particles with respect to any point in space moving in the direction of the particle $2m_0$ from the particle m_0 with a speed of $1.09 \times 10^8 \text{ m} \cdot \text{sec}^{-1}$ is zero. In any isolated system of moving particles we can establish a frame of reference in which the total (linear) momentum is zero. We have already introduced the concept of the **center of mass** of a system of particles briefly in Chapters 6 and 7. If one calculates the center of mass of the two-particle system of this problem by the technique to be given in Section 12.4, it is found that the center of mass of the two-particle system is moving in the same direction as particle m_0 with a speed of $1.09 \times 10^8 \text{ m} \cdot \text{sec}^{-1}$ (making certain that the relativistic mass is used). The result of this example illustrates the general fact that the total (linear) momentum of any isolated system of particles measured with respect to the center of mass of the system is zero.

12.6   EXPERIMENTAL VERIFICATION

Convincing confirmation of the relativistic mass formula has been obtained by passing high-speed electrons through magnetic fields. If a moving electron enters a region of uniform magnetic field in which the direction of the field is perpendicular to the direction of motion of the particle, the electron proceeds along a circular path of radius R in a plane perpendicular to the magnetic field, as outlined in Fig. 12.10,

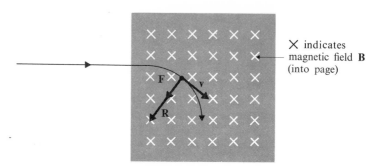

X indicates magnetic field **B** (into page)

Fig. 12.10. The path of an electron in a magnetic field which is perpendicular to the direction of its motion.

without any change in speed. The radius R of the circular path depends upon the particle's speed v, its mass m, and the strength of the magnetic field **B**. It is well established, experimentally, that the force on a charged particle moving with velocity **v** in a magnetic field **B** is given by

$$\mathbf{F} = q\mathbf{v} \times \mathbf{B},$$

where q is the charge on the particle. The force **F** is in the direction $\mathbf{v} \times \mathbf{B}$ if q is positive and in the direction $-\mathbf{v} \times \mathbf{B}$ if q is negative. In either case, **F** is always perpendicular to both **v** and **B** and has the magnitude qvB.

An electron of charge $q = -e$ (where $e = 1.602 \times 10^{-19}$ C) will experience a force evB in the direction of $-\mathbf{v} \times \mathbf{B}$. Since **F** is always perpendicular to **v**, there is no component of **F** in the direction of the particle's motion, and the force does no work on the particle. Since

$$\mathbf{F} = \frac{d\mathbf{p}}{dt},$$

we see that the momentum $\mathbf{p} = m\mathbf{v}$ is changing with time, whereas **v**, and therefore **p**, are constant in magnitude, so that only the direction of **p** (and **v**) is a function of time. We have already studied (in Section 7.7) the case of the motion of a particle under a force of constant magnitude directed at right angles to its direction of motion. The resulting motion is circular, and the force is the centripetal force. It follows then that the electron will travel in a circular path.

The centripetal force required to keep a particle of mass m and speed v traveling in a circular path of radius R is

$$F = \frac{mv^2}{R},$$

where $m = m_0$ in Newtonian dynamics and $m = \gamma m_0$ in relativistic dynamics. The electron will travel in a circular orbit whose radius R is such that the magnetic force equals the centripetal force required for circular motion with that radius. Therefore,

$$\frac{mv^2}{R} = evB$$

or

$$R = \frac{m_0 v}{eB} \qquad\qquad \text{(Newtonian)}$$

$$R = \frac{\gamma m_0 v}{eB}. \qquad\qquad \text{(relativistic)}$$

Experiments can be carried out wherein B is kept constant while v is varied or wherein R is kept constant while v is varied, and the variation in R or B required to keep B or R constant, respectively, is determined and compared to the predictions of the Newtonian and relativistic expressions for the radius R, which differ by the factor γ.

Electrons can be accelerated to high speeds in a particle accelerator known as the **electron synchrotron**. This accelerator is very similar to the proton synchrotron mentioned in Section 4.5 but is designed to accelerate electrons rather than protons, which are much heavier particles. Measurements of the circular paths of very high-energy electrons from an electron synchrotron have indicated that the factor γ must be included in the expression for the radius of the path, which verifies the relation

$$m = \gamma m_0$$

for the mass of the electrons. Values of γ up to of the order of 2000 were reached in these experiments. Since

$$\gamma = \left[1 - \frac{v^2}{c^2} \right]^{-1/2} \simeq 2000,$$

this means that the speed of the highest energy electrons was

$$v \simeq 0.999999875c.$$

The mass-energy relation is tested in every nuclear reaction. The nuclei of atoms are themselves composed of more fundamental building blocks called **nucleons**; protons and neutrons are both nucleons. The nuclei are bound together very tightly due to the strong attractive nuclear forces. The protons, in addition, repel each other due to the fact that each proton carries one unit of positive charge. This reduces the binding only slightly, since the electric force is considerably weaker than the nuclear force. The mass of a given nucleus is measured to be somewhat less than the sum of the masses of the constituent nucleons.

We know from our discussion of gravitational force that energy must be supplied to a two-particle system to overcome mutual gravitational attraction and separate the particles (that is, to give them kinetic energy at "very large" separation). In a similar manner, energy must be supplied to a nucleus in order to separate it into its constituent nucleons (that is, to give them kinetic energy at very large separation). The energy that must be added to a nucleus in order to decompose it into free nucleons is called the **binding energy** of the nucleus.

Let us consider a nucleus of mass M, having Z protons and N neutrons in its nucleus. The total mass of Z free protons and N free neutrons (assuming their speeds are much less than c) is

$$Zm_p + Nm_n,$$

where m_p and m_n are the rest masses of the proton and neutron, respectively. The mass difference Δm between the free nucleons and the nucleus is given by

$$\Delta m = Zm_p + Nm_n - M.$$

According to the prediction of the special theory of relativity, we should expect that the mass Δm lost in combining the $Z + N = A$ free nucleons into a nucleus with A nucleons is related to the binding energy E_B of the nucleus.

The binding energy of nuclei has been measured; the experimental results are shown in Fig. 12.11, where the **binding energy per nucleon** is plotted as a function of the total number of nucleons A. The binding energy per nucleon is a maximum in the region of $A = 60$. To test the equivalence of mass loss and binding energy, let us consider the nucleus ^{40}Ca (calcium 40), which has $A = 40$, $Z = 20$, and $N = 20$. From Fig. 12.11 we see that the binding energy per nucleon is about 8.4 MeV at $A = 40$. Therefore, for ^{40}Ca,

$$E_B \simeq 40(8.4) = 330 \text{ MeV}.$$

The mass of a nucleus, or more precisely of a nucleus plus atomic electrons, can be measured very accurately by measuring the curvature of

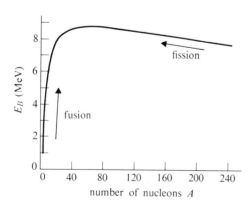

Fig. 12.11. The variation of the binding energy per nucleon with nucleon mass A.

the path of ionized atoms in a magnetic field. If the state of ionization of the atom is known, then the charge is known. The mass of the ionized atom is given by

$$M' = \frac{qBR}{v},$$

where q, B, R, and v can be measured very precisely in instruments known as **mass spectrometers**. The mass M of the nucleus is determined by subtracting the mass of the atomic electrons. In this way, the nucleus of ^{40}Ca has been measured to have a mass

$$M = 6.6353 \times 10^{-26} \text{ kg.}$$

The mass of 20 free protons and 20 free neutrons is

$$Zm_p + Nm_n = 20(1.6725 \times 10^{-27}) + 20(1.6748 \times 10^{-27})$$

$$= 3.3450 \times 10^{-26} + 3.3496 \times 10^{-26}$$

$$= 6.6946 \times 10^{-26} \text{ kg.}$$

Therefore,

$$\Delta m = Zm_p + Nm_n - M$$

$$= (6.6946 - 6.6353) \times 10^{-26} \text{ kg}$$

$$= 5.93 \times 10^{-28} \text{ kg,}$$

and

$$(\Delta m)c^2 = (5.93 \times 10^{-28})(3.00 \times 10^8)^2$$

$$= 5.34 \times 10^{-11} \text{ J}$$

$$= \frac{5.34 \times 10^{-11}}{1.60 \times 10^{-13}} \text{ MeV}$$

$$= 334 \text{ MeV.}$$

The close agreement between the values of E_B and $(\Delta m)c^2$ confirms the relation

$$E_B = (\Delta m)c^2$$

as predicted by the special theory of relativity. Similar calculations carried out with more precise binding energy values show that the result is true for all nuclei.

The curve of binding energy per nucleon as a function of A shows that nuclei of small or large nuclear mass have less binding energy per nucleon than nuclei of intermediate mass. Therefore, if a heavy atom such as uranium is split into two atoms of approximately equal mass, the total binding energy of the two is greater than that of the original uranium atom, so that a release of energy accompanies the breakup. This is known as a **fission process**. Power plants for nuclear submarines and nuclear generating stations rely on fission processes.

In the region of $A = 240$ the binding energy per nucleon is approximately 7.6 MeV, and in the region of $A = 120$ it is approximately 8.5 MeV. The total energy released in a fission process should be approximately

$$240(8.5 - 7.6) = 21(6) \text{ MeV}.$$

Actually about 200 MeV or less of energy is released in the fission of a heavy nucleus, since two or three neutrons are emitted in a typical fission process and some of the energy is used up in freeing these neutrons.

If a number of light nuclei combine to form a single heavier nucleus, the binding energy of the resultant nucleus is greater than that of the constituent nuclei, provided that the resultant nucleus still has a small atomic mass. This combination or **fusion process** also releases energy.[2] Fusion processes in the interior of stars are believed to provide the enormous energy outputs of the stars.

Example. One possible source of stellar energy concerns the fusion of four hydrogen nuclei (one proton each) into a helium nucleus (two protons plus two neutrons) through a series of nuclear reactions involving the nucleus of a carbon atom. The proton has a rest mass of 1.6725×10^{-27} kg, and the helium nucleus has a rest mass of 6.6459×10^{-27} kg. How much energy is released as a result of this fusion process?

Solution. The four protons have a total mass of 6.6900×10^{-27} kg, which is $(6.6900 - 6.6459) \times 10^{-27} = 0.0441 \times 10^{-27}$ kg greater than the mass of the helium nucleus. Since

$$E = (\Delta m)c^2,$$

the energy release is

$$E = 0.0441 \times (3.00 \times 10^8)^2$$

$$= 3.97 \times 10^{-10} \text{ J}$$

$$= 24.8 \text{ MeV}.$$

Therefore, 24.8 MeV of energy is released during this fusion process.

PROBLEMS

1. Calculate the increase in mass of an electron traveling at a speed of $0.6c$, $0.8c$, $0.9c$, $0.99c$.

[2] R. F. Post, "Fusion Power," *Scientific American*, December, 1957. Available as *Scientific American Offprint 236* (San Francisco: W. H. Freeman and Co., Publishers).

2. Determine the speed at which a particle is traveling if its mass is ten times its rest mass.

3. An electron travels at $\frac{7}{8}$ the speed of light. Calculate its kinetic energy.

4. At what speed must a particle move in order that its rest energy is only 1% of its total energy?

5. Under certain conditions, a γ-ray may produce a pair of particles, an electron and a positron, the γ-ray disappearing in the process. This is known as **pair production**. If the positron is identical to the electron except that it has the opposite charge, what is the minimum energy γ-ray that will produce a positron-electron pair?

6. An electron with a total energy of 20 MeV strikes a thin piece of metal and is quickly brought to rest. Some of its energy appears as X-rays emitted while the electron is decelerating to a stop in the target. The X-rays produced by this process are often referred to as **bremsstrahlung ("braking radiation")**. What is the maximum possible X-ray energy produced by a 20-MeV electron? (Neglect the motion of the atoms in the metal.)

7. When at rest a charged particle called the π-meson (or pion for short) decays into a μ-meson (muon) in a mean time of 2.5×10^{-8} sec. If a pion was traveling relative to you at a speed such that its kinetic energy was twice its rest energy, how far would you observe it to travel before decaying?

8. A **positron** (identical to the electron except that it has a positive charge) and an electron, each traveling at a speed of $0.88c$ as measured in the laboratory, approach one another. Calculate the speed of the positron measured by an observer moving with the electron. What mass would he measure for the positron? Upon collision, the two particles are annihilated to form two γ-rays of electromagnetic radiation. Calculate the total energy of the γ-rays in the units of MeV and of joules. Assume that the rest masses of both the positron and electron are 9.11×10^{-31} kg.

9. The **solar constant** (1.35×10^3 J · m^{-2} · sec^{-1}) gives the energy of solar radiation received per second at the earth by one square meter of surface normal to the line joining the sun and the earth. The sun loses mass as a result of radiation. How large is this mass loss per year if the earth-sun distance is 1.49×10^{11} m? State clearly any assumption you make.

10. If the speed of light were only 30 m · sec^{-1} rather than 3.0×10^8 m · sec^{-1}, relativistic effects would be commonplace in everyday life. Discuss some of the strange sights that you might expect to see. Assume that the sizes of atoms are unaffected by this large change in the speed of light, so that the sizes of physical objects would remain unchanged.

*11. Consider two beams of protons of energies 1.00 ± 0.10 GeV and 3.00 ± 0.05 GeV respectively approaching each other. What speed of approach would an observer traveling with the 1-GeV proton measure for the 3-GeV protons?

*12. A relativistic particle has a rest mass m_0, momentum p, and total energy E. Show that

$$E^2 = p^2c^2 + m_0^2c^4.$$

*13. Show that the y components of the momentum of a particle as viewed in two frames of reference R and R', where R' is moving in the x, x' direction with velocity u with respect to R, are related by

$$p'_y = p_y,$$

where

$$p'_y = m_0\left(1 - \frac{v'^2}{c^2}\right)^{-1/2} v'_y$$

and

$$p_y = m_0\left(1 - \frac{v^2}{c^2}\right)^{-1/2} v_y.$$

*14. A particle of rest mass m_0, initially at rest, is subjected to a constant force F for a period of t sec. Find an expression for the speed v as a function of time. An electron, initially at rest, is subjected to a force of 1.6×10^{-12} N for a period of 3×10^{-8} sec; the rest mass of the electron is 9.11×10^{-31} kg. What is the final speed of the electron? What is its kinetic energy in MeV?

15. Calculate the binding energy per nucleon of a helium nucleus. Assume that the nucleus consists of two protons of mass 1.007825 u (**unified mass units**) each and two neutrons of mass 1.008665 u each and that the mass of the helium nucleus is 4.002603 u (1 u $= 1.6604 \times 10^{-27}$ kg $= 1/N_0$, where $N_0 = $ Avogadro's number $= 6.0225 \times 10^{26}$ [kg $\cdot$ mole]$^{-1}$).

16. The energy liberation in the explosion of hydrogen bombs is measured in units of "megatons." A one megaton bomb releases the same amount of energy as does the detonation of one-million tons of TNT, that is, 5×10^{15} J. How much matter is converted into energy in the explosion of a 3-megaton bomb?

17. The first nuclear reaction to be studied by bombarding a nucleus with particles from an accelerator involved the capture of a proton by a nucleus of lithium, which subsequently broke up into two helium nuclei (^{4}He). The reaction can be represented by the equation

$$p + {}^7\text{Li} \rightarrow {}^4\text{He} + {}^4\text{He} + Q,$$

where Q represents the energy released in the reaction. Taking the mass of the proton to be 1.6725×10^{-27} kg, the mass of the ^{7}Li nucleus to be 1.1647×10^{-26} kg, and the mass of the ^{4}He nucleus to be 6.6459×10^{-27} kg, calculate the value of Q in MeV.

18. In Section 12.6 we showed that a charged particle entering a region of magnetic field with its velocity at right angles to the magnetic field travels in a circular path. What path would the particle follow if its velocity were not at right angles to the magnetic field? Draw a diagram of the path of the particle.

19. Calculate the radius of curvature of the path of an electron of kinetic energy 1.60×10^{-9} J in a magnetic field of 1.00 Wb $\cdot$ m^{-2} (Wb $\cdot$ m^{-2} $\equiv$ Weber $\cdot$ m^{-2} is the unit of magnetic field strength in the MKS system of units).

*20. What is the kinetic energy of a proton that travels along a path with radius of curvature of 2.0 m in a magnetic field of 1.0 Wb $\cdot$ m^{-2}?

*21. The nucleus of ^{235}U will undergo fission upon capture of a very slow neutron (having usually less than 1 eV of energy). The nucleus that fissions is actually ^{236}U. The fission process can be represented by the equation

$$^{235}\text{U} + \text{n} \rightarrow \, ^{236}\text{U} \rightarrow \text{X} + \text{Y} + v\text{n} + \text{energy},$$

where $v = 0, 1, 2, 3, \cdots$ is the number of neutrons emitted and X and Y represent the **daughter** nuclei formed as a result of the fission. (The average value of v is 2.47 for this reaction.) Given the following masses, calculate the energy released in the case in which X and Y represent the nuclei ^{139}Xe (xenon 139) and ^{94}Sr (strontium 94).

^{235}U	235.04392 u
^{139}Xe	138.91784 u
^{94}Sr	93.91538 u
n	1.00867 u

$$(1 \text{ u} = 931.5 \text{ MeV} = 1.6604 \times 10^{-27} \text{ kg.})$$

22. Calculate the increase in mass of the earth-moon system that would occur if the system became unbound (both earth and moon at rest at infinite separation). Express your answer as a percentage of the total mass of the earth-moon system in its present bound state (see the example in Section 7.5).

13 Planetary motions:

a two-particle problem

13.1 INTRODUCTION

During the latter half of the 16th century the Danish astronomer Tycho Brahe compiled a comprehensive body of observational data on the positions of the planets. In contrast to the haphazard methods characteristic of scientists of his time, Tycho Brahe made regular and systematic observations. His measurements of the angular positions of the planets in the sky over a period of 20 years contain no error larger than $\frac{1}{15}$ of a degree.

Early in the 17th century Johannes Kepler announced three fundamental laws of planetary motion satisfying Tycho Brahe's data.
1. All planets move in elliptical paths with the sun at one focus.
2. A line joining the sun to any of the planets (a radius vector) sweeps out equal areas of the orbit in equal times.
3. The squares of the periods of revolution of the several planets about the sun are proportional to the cubes of the semimajor axes of the ellipses.

These laws were presented without theoretical foundation, being empirically derived from Brahe's observations.

The importance of Kepler's laws should not be overlooked. The idea that the earth is the center of the universe with all the "heavenly bodies" rotating about the earth had prevailed for almost 2000 years. In 1543 Nicolaus Copernicus had published a theory in which he postulated that all the planets, including the earth, revolved about the sun. His theory was generally disregarded and indeed ridiculed because it removed the earth from the center of creation. Kepler's publication of "a new astronomy" in 1609 was, in one sense, a confirmation of the earlier work of Copernicus. More important, however, was the fact that Kepler's laws were the first "natural laws" in the modern sense—precise, verifiable statements about universal relations governing particular phenomena and expressed in mathematical terms[1].

In 1665 and 1666 Isaac Newton used Kepler's third law to deduce the fact that the gravitational force exerted by the sun on a planet must vary inversely as the square of the distance between the sun and the various planets. He generalized this statement to be a universal law giving the force between any two objects of nonzero mass. He also showed that Kepler's first two laws were a consequence of the inverse-square law of gravitational force. Ultimately, it was shown[2] that Newton's law of

---

[1] Louise B. Young [ed.], *Exploring the Universe* (New York: McGraw-Hill Book Company, 1963), Part 5. This reference contains a lengthy discussion of the work of Copernicus, Brahe, and Kepler.

[2] A. Einstein, "On the Generalized Theory of Gravitation," *Scientific American*, April, 1950. Also available as *Scientific American Offprint 209* (San Francisco: W. H. Freeman and Co., Publishers).

gravitation needed to be somewhat modified. This later modification, however, does not diminish the importance of the work of Kepler and Newton, which revolutionized the development of science.

In this chapter, we shall consider the theoretical basis of Kepler's laws as an example of a problem involving the mutual gravitational attraction between two particles.

13.2 THE TWO-PARTICLE PROBLEM

The sun contains 99.88 % of the entire mass of the solar system, and the remaining 0.12 % is distributed among the planets (the masses of the various asteroids, comets, etc., are insignificant). Therefore, it is a very good approximation to consider each of the planets to interact only with the sun and to neglect the interaction of the planets with each other when determining the paths of the planets through space (that is, their orbits). The planet-planet gravitational interactions lead to small perturbations in the orbits calculated by neglecting the existence of other planets and must be included if a very precise determination of the motion of a given planet is required. (The best example of the effect of planet-planet inter-actions is the following. Minute irregularities observed in the orbit of the planet Uranus led astronomers John Adams and Urbain Leverrier, working independently, to postulate an unknown planet and predict a position for it in the sky at a specified time. On September 23, 1846, the planet Neptune was found at the specified position.)

Furthermore, the relative motion of sun and planet is, to a high degree of approximation, independent of any internal motions (such as rotations) of either of the celestial bodies themselves. Therefore, the sun and planet may be considered as two point "particles" whose masses are equal to those of the actual sun and planet.

A mathematical description of the motion of a system of two (or more) particles is possible in which the translational motion of the system as a whole may be separated from the individual motions of the constituent particles relative to a well-defined point in the system known as the **center of mass**.

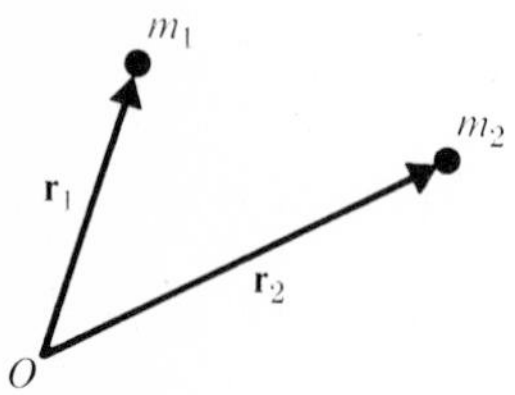

13.3 THE CENTER OF MASS

Fig. 13.1. A system of two particles having masses m_1 and m_2 and position vectors $\mathbf{r}_1$ and $\mathbf{r}_2$.

Consider the system of two particles shown in Fig. 13.1. We write for the force $\mathbf{F}_1$ acting on particle 1

$$\mathbf{F}_1 = \mathbf{F}_{ext}(1) + \mathbf{F}_{12},$$

where $\mathbf{F}_{ext}(1)$ is the force external to the system exerted on particle 1 and $\mathbf{F}_{12}$ is the force exerted on particle 1 by particle 2. Similarly, the force $\mathbf{F}_2$ acting on particle 2 can be written

$$\mathbf{F}_2 = \mathbf{F}_{ext}(2) + \mathbf{F}_{21},$$

where $\mathbf{F}_{ext}(2)$ is the force external to the system exerted on particle 2 and $\mathbf{F}_{21}$ is the force exerted on particle 2 by particle 1. From Newton's second law

$$\mathbf{F}_1 = m_1 \frac{d^2\mathbf{r}_1}{dt^2} \quad \text{and} \quad \mathbf{F}_2 = m_2 \frac{d^2\mathbf{r}_2}{dt^2}.$$

Therefore,

$$\mathbf{F}_1 + \mathbf{F}_2 = \mathbf{F}_{ext}(1) + \mathbf{F}_{ext}(2) + \mathbf{F}_{12} + \mathbf{F}_{21}$$

$$= m_1 \frac{d^2r_1}{dt^2} + m_2 \frac{d^2r_2}{dt^2}.$$

But, from Newton's third law,

$$\mathbf{F}_{12} = -\mathbf{F}_{21},$$

so that

$$\mathbf{F}_{ext} = \mathbf{F}_{ext}(1) + \mathbf{F}_{ext}(2)$$

$$= m_1 \frac{d^2r_1}{dt^2} + m_2 \frac{d^2r_2}{dt^2},$$

where $\mathbf{F}_{ext}$ is the total external force acting on the system of particles. This equation may be written in the form

$$\mathbf{F}_{ext} = M \frac{d^2\mathbf{R}}{dt^2},$$

where $M = m_1 + m_2$ is the total mass and $\mathbf{R}$ is the position vector of a suitable "average" position of the particles, called the center of mass (see Fig. 13.2). By comparing the two equations for $\mathbf{F}_{ext}$, we see that

$$\mathbf{R} = \frac{m_1\mathbf{r}_1 + m_2\mathbf{r}_2}{m_1 + m_2}.$$

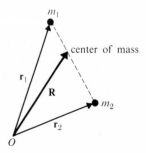

Fig. 13.2. The vector **R** is the position vector for the center of mass of the two-particle system.

The motion of the center of mass of the system through space is identical to that of a single particle having a mass equal to the total mass of the system, located at the center of mass, and acted on by a force equivalent to the total external force acting on the system. If there is no external force, then

$$\frac{d\mathbf{R}}{dt} = \text{constant},$$

and the center of mass moves through space with a constant velocity. Therefore, the translational motion of the system, which is given by the translational motion of the center of mass, may be calculated independently of the motions of the constituent particles relative to the center of mass.

Example

(a) Where is the center of mass of a dumbbell consisting of point particles of mass M and $3M$ separated by a rigid, weightless rod of length $4R$, as shown in Fig. 13.3?

(b) If a boy throws a dumbbell, what will be its path through space? Take $M = 1$ kg, $4R = 0.5$ m.

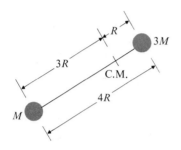

Solution

(a) We take the origin of the coordinate system to be at the center of mass (C.M.). Therefore, $\mathbf{r}_1$ and $\mathbf{r}_2$ point in opposite directions, and

$$|\mathbf{r}_1| + |\mathbf{r}_2| = r_1 + r_2 = 4R.$$

Also, $\mathbf{R} = 0$, so that

$$M\mathbf{r}_1 + 3M\mathbf{r}_2 = 0$$

or

$$M r_1 = 3M r_2.$$

Fig. 13.3. A dumbbell formed of particles of mass M and $3M$ separated by a distance $4R$.

Solving for r_1 and r_2 gives $r_1 = 3R$ and $r_2 = R$. The center of mass of the dumbbell lies on the axis of the dumbbell, as shown in Fig. 13.3.

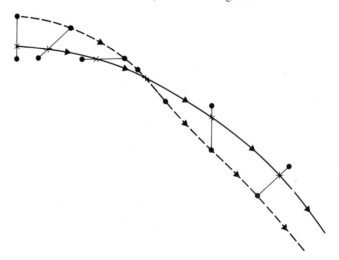

Fig. 13.4. The path through space of the particle of mass M in the dumbbell is a combination of the parabolic path of the center of mass and the circular path of the particle of mass M relative to the center of mass.

(b) Neglecting air resistance, gravity will be the only external force acting on the dumbbell. Therefore, the path of its center of mass through space will be parabolic. If the dumbbell is spinning, the point particles will not trace such simple paths through space. As an example, if the dumbbell is spinning in the direction of its motion through space with constant angular frequency, the path of the particle of mass M will be as shown in Fig. 13.4. This path is just the combination of the parabolic path of the center of mass relative to the earth and the circular path of the particle of mass M relative to the center of mass.

13.4   THE EQUATION OF MOTION OF TWO PARTICLES INTERACTING THROUGH MUTUAL GRAVITATIONAL ATTRACTION

Let us consider a point particle (1) of mass m, specified by a position vector $\mathbf{r}_1$, and a point particle (2) of mass M, specified by a position vector $\mathbf{r}_2$. The **separation vector** for the two particles (see Fig. 13.5),

$$\mathbf{r} = \mathbf{r}_2 - \mathbf{r}_1,$$

is directed from particle (1) to particle (2). The gravitational force $\mathbf{F}_{12}$ experienced by particle (1) as a result of the presence of particle (2) is

$$\mathbf{F}_{12} = \frac{GmM}{r^2}\hat{\mathbf{r}}.$$

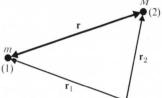

Fig. 13.5. The separation vector for two particles.

The gravitational force $\mathbf{F}_{21}$ experienced by particle (2) as a result of the presence of particle (1) is

$$\mathbf{F}_{21} = -\mathbf{F}_{12} = -\frac{GmM}{r^2}\hat{\mathbf{r}}.$$

That is, since $\mathbf{r} = r\hat{\mathbf{r}}$ is measured from particle (1) to particle (2), $\mathbf{F}_{12}$ is in the direction of $\mathbf{r}$, and $\mathbf{F}_{21}$ is in the direction of $-\mathbf{r}$. These are the only forces acting on the isolated two-particle system under consideration. The equation of motion of particle (1) is

$$m\frac{d^2\mathbf{r}_1}{dt^2} = \mathbf{F}_{12} = \frac{GmM}{r^2}\hat{\mathbf{r}};$$

the equation of motion of particle (2) is

$$M\frac{d^2\mathbf{r}_2}{dt^2} = \mathbf{F}_{21} = -\frac{GmM}{r^2}\hat{\mathbf{r}}.$$

In order to solve these coupled equations of motion, it is convenient to combine them in the following manner. Subtracting the first equation from the second gives

$$\frac{d^2\mathbf{r}_2}{dt^2} - \frac{d^2\mathbf{r}_1}{dt^2} = -\left(\frac{GmM}{Mr^2} + \frac{GmM}{mr^2}\right)\hat{\mathbf{r}}$$

or

$$\frac{d^2\mathbf{r}}{dt^2} = -\left(\frac{1}{M} + \frac{1}{m}\right)\frac{GmM}{r^2}\hat{\mathbf{r}}.$$

If we define the **reduced mass** μ of the system as

$$\frac{1}{\mu} = \frac{1}{M} + \frac{1}{m},$$

this equation becomes

$$\mu\frac{d^2\mathbf{r}}{dt^2} = -\frac{GmM}{r^2}\hat{\mathbf{r}}.$$

Fig. 13.6. The two-particle system may be replaced by a single particle with reduced mass μ and position vector **r**.

This equation of motion is identical to that of a single particle of mass μ and position vector **r** (illustrated in Fig. 13.6), which is acted on by a force

$$-\frac{GmM}{r^2}\hat{\mathbf{r}}.$$

That is, the two equations of motion for the actual particles m and M may be replaced by the single equation of motion for the fictitious reduced particle μ experiencing the gravitational force

$$-\frac{GmM}{r^2}\hat{\mathbf{r}}.$$

This latter equation may be solved using methods developed for a single-particle system. The solution will provide the **orbit** of the reduced particle, that is, its position as a function of time. Then, since the mass and position of the reduced particle are defined in terms of the masses and positions of the actual particles, the motions of the original particles may be deduced.

13.5 THE CONSERVATION LAWS AND THE TWO-BODY PROBLEM

For an isolated system the total angular momentum **J** and the total energy E are conserved. Since the angular momentum is a vector, both the direction and the magnitude of **J** must be constants of the motion.

A constant direction for **J** implies that the particles must move in a plane of constant orientation. It is convenient to describe the positions of the particles with respect to the center of mass by using polar coordinates, as shown in Fig. 13.7. The coordinates of particle (1) are r_1 and $(\pi + \theta)$, and the coordinates of particle (2) are r_2 and θ.

The magnitude of **J** is given by

$$|\mathbf{J}| = mr_1^2\frac{d\theta}{dt} + Mr_2^2\frac{d\theta}{dt}.$$

Since

$$mr_1 = Mr_2 \quad \text{and} \quad r_1 + r_2 = r,$$

it follows that

$$r_1 = \frac{M}{m + M}r \quad \text{and} \quad r_2 = \frac{m}{m + M}r.$$

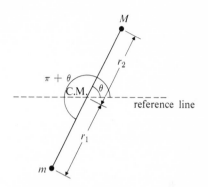

Fig. 13.7. Polar coordinates for the two-particle system.

Therefore,

$$|\mathbf{J}| = m\left(\frac{M}{m + M}\right)^2 r^2\frac{d\theta}{dt} + M\left(\frac{m}{m + M}\right)^2 r^2\frac{d\theta}{dt}$$

$$= \frac{mM}{(m + M)^2}(m + M)r^2\frac{d\theta}{dt}$$

$$= \mu r^2\frac{d\theta}{dt}$$

is also a constant of the motion. That is,

$$|\mathbf{J}| = \mu r^2\frac{d\theta}{dt} = \text{constant}.$$

The total energy E of the system is

$$E = \frac{1}{2}mv_1^2 + \frac{1}{2}Mv_2^2 - \frac{GmM}{r},$$

where the first two terms represent the kinetic energies of the particles (1) and (2), respectively, and the third term represents the potential energy of the interacting particles.

Since

$$r_1 = \frac{M}{m + M}r \quad \text{and} \quad r_2 = \frac{m}{m + M}r,$$

$$v_1 = \frac{dr_1}{dt} = \frac{M}{m + M}\frac{dr}{dt} = \frac{M}{m + M}v.$$

and

$$v_2 = \frac{dr_2}{dt} = \frac{m}{m+M}\frac{dr}{dt} = \frac{m}{m+M}v,$$

where v is the magnitude of the velocity of the reduced particle μ. Therefore,

$$E = \frac{1}{2}m\left(\frac{M}{m+M}\right)^2 v^2 + \frac{1}{2}M\left(\frac{m}{m+M}\right)^2 v^2 - \frac{GmM}{r}$$

$$= \frac{1}{2}\mu v^2 - \frac{GmM}{r}.$$

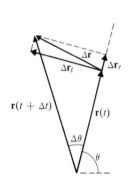

Fig. 13.8. The change in position vector **r** has a component $\Delta\mathbf{r}_r$ in the direction of **r** and a component $\Delta\mathbf{r}_t$ in the direction transverse to **r**.

In order to express the magnitude v of the velocity **v** of the reduced particle in terms of the polar coordinates r and θ, let us consider Fig. 13.8, which shows a position vector **r** at two instants of time t and $t + \Delta t$. The vector $\Delta\mathbf{r}$, which represents the change in the position vector, has components $\Delta\mathbf{r}_r$ in the direction of **r** and $\Delta\mathbf{r}_t$ in a direction perpendicular, or transverse, to **r**. $\Delta\theta$ is the angle swept out by the position vector in the time Δt.

For small $\Delta\theta$,

$$|\Delta\mathbf{r}_t| \simeq |\mathbf{r}(t)|\Delta\theta,$$

since $|\Delta\mathbf{r}_t|$ differs by only a small amount from the arc of a segment of a circle of radius $|\mathbf{r}(t)|$ and angle $\Delta\theta$. In the limit $\Delta\theta \to 0$,

$$|\Delta\mathbf{r}_t| \to r\,d\theta,$$

where $\Delta\mathbf{r}_t$ is a vector perpendicular to $\mathbf{r}(t)$, and

$$|\Delta\mathbf{r}_r| \to dr_r,$$

where $\Delta\mathbf{r}_r$ is a vector parallel to $\mathbf{r}(t)$. The quantity dr_r represents the change in the magnitude of $\mathbf{r}(t)$ in a time dt; from this point on we shall drop the subscript on dr_r and write simply dr. The quantity $r\,d\theta$ refers to the change in orientation of $\mathbf{r}(t)$. We can therefore define the **transverse and radial velocities** as having magnitudes

$$v_t = r\frac{d\theta}{dt} \quad \text{and} \quad v_r = \frac{dr}{dt},$$

respectively. Therefore,

$$v^2 = v_t^2 + v_r^2$$

$$= r^2\left(\frac{d\theta}{dt}\right)^2 + \left(\frac{dr}{dt}\right)^2.$$

As the particle moves through space, both the orientation and magnitude of its position vector will change with time. Therefore, the magnitude of $\mathbf{r}(t)$ is a function of θ. Using the chain rule of calculus we have

$$\frac{dr}{dt} = \frac{dr}{d\theta}\frac{d\theta}{dt}.$$

We then can write

$$v^2 = \left[\left(\frac{dr}{d\theta}\right)^2 + r^2\right]\left(\frac{d\theta}{dt}\right)^2$$

$$= \left[\left(\frac{dr}{d\theta}\right)^2 + r^2\right]\frac{J^2}{\mu^2 r^4},$$

where J is the constant magnitude of the angular momentum. Substituting this value of v^2 into the energy expression gives

$$E = \frac{\mu}{2}\left[\left(\frac{dr}{d\theta}\right)^2 + r^2\right]\frac{J^2}{\mu^2 r^4} - \frac{GmM}{r}$$

or, rearranging terms,

$$\left(\frac{dr}{d\theta}\right)^2 = \frac{2\mu r^4}{J^2}\left(E + \frac{GmM}{r}\right) - r^2.$$

Either the solution of this energy equation for the reduced particle or the solution of the equation of motion for the reduced particle derived in Section 13.4 will yield the orbit of the reduced particle. The solution of the energy equation carried out in the screened area is considerably easier than the solution of the equation of motion.

Determination of the orbit of the reduced particle

We have

$$\left(\frac{dr}{d\theta}\right)^2 = \frac{2\mu r^4}{J^2}\left(E + \frac{GmM}{r}\right) - r^2$$

$$= \frac{r^4}{J^2}\left[2\mu\left(E + \frac{GmM}{r}\right) - \frac{J^2}{r^2}\right].$$

Therefore,

$$\frac{dr}{d\theta} = \pm\frac{r^2}{J}\left[2\mu\left(E + \frac{GmM}{r}\right) - \frac{J^2}{r^2}\right]^{1/2}.$$

We use the substitution

$$u = \frac{1}{r}$$

$$du = -\frac{1}{r^2}\,dr$$

and obtain

$$\frac{du}{d\theta} = \pm\frac{1}{J}[2\mu(E + GmMu) - J^2u^2]^{1/2}$$

or

$$d\theta = \pm\frac{J\,du}{[2\mu(E + GmMu) - J^2u^2]^{1/2}}.$$

If we define the quantity e as

$$e = \left(1 + \frac{2EJ^2}{\mu G^2m^2M^2}\right)^{1/2},$$

we may rewrite the equation for $d\theta$ as (check by substitution)

$$d\theta = \frac{\pm\,du}{\left[\left(\frac{\mu GmMe}{J^2}\right)^2 - \left(u - \frac{\mu GmM}{J^2}\right)^2\right]^{1/2}}.$$

The equation is of the form

$$dx = \frac{\pm\,dy}{[a^2 - (y - b)^2]^{1/2}}$$

and has the solution (consult any table of integrals)

$$y = b - a\cos x.$$

Therefore, the solution for our equation is, by comparison,

$$u = \frac{1}{r}$$

$$= \frac{\mu GmM}{J^2} - \frac{\mu GmMe}{J^2}\cos\theta$$

or

$$r = \frac{l}{1 - e\cos\theta},$$

where

$$l = \frac{J^2}{\mu GmM}.$$

13.6 POSSIBLE PLANETARY ORBITS

The orbit of the reduced particle is given in polar coordinates by

$$r = \frac{l}{1 - e \cos \theta},$$

where

$$l = \frac{J^2}{\mu GmM}$$

and

$$e = \left[1 + \frac{2EJ^2}{\mu G^2 m^2 M^2} \right]^{1/2}$$

are constants of the orbit. These equations for l and e may be solved for the constants of the motion E and J^2 to yield

$$E = -\frac{(1 - e^2)}{2l} GmM$$

and

$$J^2 = \mu l GmM.$$

Let us investigate the nature of this solution.

Case 1 : $e = 0$.

When $e = 0$,

$$r = l,$$

and the reduced particle moves so that its radius vector is of a constant length; that is, the particle travels in a **circular orbit** of radius l, as shown in Fig. 13.9. The motions of the actual particles m and M relative to their center of mass may now be deduced. Their respective position vectors $\mathbf{r}_1$ and $\mathbf{r}_2$ satisfy

$$\mathbf{r} = \mathbf{r}_2 - \mathbf{r}_1.$$

The position of the center of mass is obtained from the equation

$$\mathbf{R} = \frac{m\mathbf{r}_1 + M\mathbf{r}_2}{m + M}.$$

We take the origin of the coordinate system at the center of mass (Fig. 13.10), so that

$$\mathbf{R} = 0$$

and

$$m\mathbf{r}_1 + M\mathbf{r}_2 = 0.$$

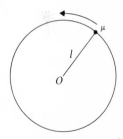

Fig. 13.9. When $e = 0$, the reduced particle travels in a circular orbit of radius l.

Fig. 13.10. For a two-particle system we take the origin at the center of mass, and $m\mathbf{r}_1 + M\mathbf{r}_2 = 0$.

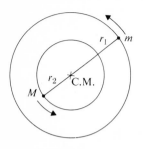

Fig. 13.11. When $e = 0$, the particles of mass m and M move in circular orbits of radius r_1 and r_2, respectively.

Therefore, $\mathbf{r}_1$ and $\mathbf{r}_2$ must be in opposite directions and have magnitudes that satisfy the equations

$$mr_1 = Mr_2$$

and

$$r_1 + r_2 = l.$$

These two equations can be solved for r_1 and r_2, showing that the two particles move in circular orbits of radii r_1 and r_2, respectively (see Fig. 13.11). If M is greater than m, then the particle of mass M will revolve in an orbit of smaller radius than that of the particle of mass m.

Case 2 : $0 < e < 1$.

The orbit in this case is an **ellipse**, as we may readily show. As an example we shall assume $e = 0.500$ and calculate r for a few values of θ. The calculations are shown in Table 13.1 and are plotted in Fig. 13.12. The constant e is the **eccentricity** and is defined geometrically from the relation

$$b = a(1 - e^2)^{1/2},$$

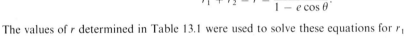

where a and b are the **semimajor and semiminor axes** of the ellipse (see Fig. 13.13) and are related to l and e by

$$l = a(1 - e^2) = b(1 - e^2)^{1/2}.$$

Fig. 13.12. When $0 < e < 1$, the reduced particle travels in an elliptical orbit with one focus of the ellipse at the center of mass of the two-particle system.

The orbits of the individual particles m and M are again found by noting that

$$mr_1 = Mr_2$$

$$r_1 + r_2 = r = \frac{l}{1 - e \cos \theta}.$$

The values of r determined in Table 13.1 were used to solve these equations for r_1

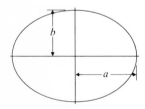

Fig. 13.13. The semimajor and semiminor axes of an ellipse are a and b, respectively.

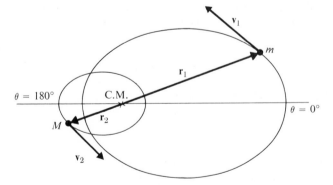

Fig. 13.14. When $0 < e < 1$, the particles of mass m and M travel in elliptical orbits with one focus of each ellipse at the center of mass. The particles move so as to satisfy the relation $mr_1 + Mr_2 = 0$ at all times.

Table 13.1 The orbit for eccentricity $e = 0.500$

θ	$\cos \theta$	$e \cos \theta$	$1 - e \cos \theta$	$r = \dfrac{l}{(1 - e \cos \theta)}$
0°	1.000	0.500	0.500	2.00l
30°	0.866	0.433	0.567	1.76l
45°	0.707	0.354	0.646	1.55l
60°	0.500	0.250	0.750	1.33l
90°	0.000	0.000	1.000	1.00l
120°	−0.500	−0.250	1.250	0.80l
150°	−0.866	−0.433	1.433	0.70l
180°	1.000	−0.500	1.500	0.67l
270°	0.000	0.000	1.000	1.00l
360°	1.000	0.500	0.500	2.00l

and r_2 as a function of θ for the case $M = 2m$. The individual orbits are plotted in Fig. 13.14 using the same scale as in Fig. 13.12.

Case 3: $e = 1$.
When $e = 1$,

$$r = \frac{l}{1 - \cos \theta}.$$

The reduced particle μ travels in a **parabolic** orbit, as shown in Fig. 13.15. The orbits of m and M may be deduced as described in Case 2. Their orbits are shown in Fig. 13.16 for $M = 2m$.

Case 4: $e > 1$.
The orbit now is a hyperbola and may be determined for a specific case in the manner used above. This is left as an exercise for the reader.

Example. A satellite in an elliptical orbit about the earth is 200 miles above the earth's surface at the lowest point in its orbit and 1000 miles above the earth's surface at the highest point. Use the theory of the two-particle system to determine the eccentricity of the satellite's orbit.

Solution. The orbit about the earth is illustrated in Fig. 13.17. Let us assume that the center of mass of the earth-satellite system is at the center of the earth. This is a good assumption, since the mass of the earth is 6.0×10^{24} kg, whereas the mass of the satellite is only of the order of 10^3 to 10^4 kg. Let the magnitude of the

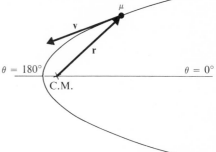

Fig. 13.15. When $e = 1$, the reduced particle travels in a parabolic orbit with the focus of the parabola at the center of mass of the two-particle system.

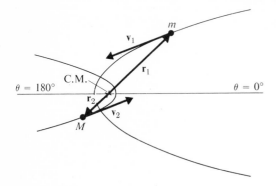

Fig. 13.16. When $e = 1$, the two particles of mass m and M move in parabolic orbits with the focus of each orbit at the center of mass of the system.

position vector at the lowest point in the orbit be denoted by r_{min} and the magnitude at the highest point in the orbit by r_{max}. When $r = r_{max}$, $\theta = 0°$ (refer to Fig. 13.18) and

$$r_{max} = \frac{l}{1 - e}.$$

When $r = r_{min}$, $\theta = 180°$, and

$$r_{min} = \frac{l}{1 + e}.$$

Eliminating l between these two equations gives

$$1 + e = \frac{l}{r_{min}}$$
$$= \frac{r_{max}(1 - e)}{r_{min}},$$

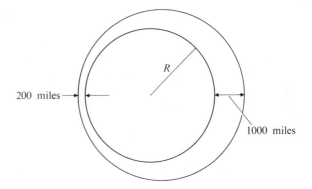

Fig. 13.17. The orbit of a typical earth satellite.

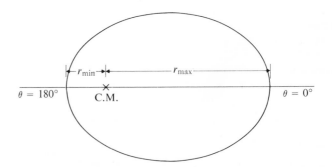

Fig. 13.18. The maximum and minimum values of r occur at $\theta = 0°$ and $\theta = 180°$, respectively.

which can be solved for e to give

$$e = \frac{r_{max} - r_{min}}{r_{max} + r_{min}}.$$

Since the radius of the earth is 3960 miles, r_{max} is 4960 miles and r_{min} is 4160 miles. Therefore, the eccentricity of the satellite's orbit is

$$e = \frac{4960 - 4160}{4960 + 4160}$$

$$e = 0.088.$$

13.7 KEPLER'S LAWS

We now wish to show that the mathematical solution of the two-particle problem outlined above contains Kepler's laws of planetary motion.

We note that even for Jupiter, the most massive of the planets, the distance of the center of mass of the sun-Jupiter system from the center of the sun is only 0.1 % of the sun-Jupiter separation. (The masses of the sun and of Jupiter are 1.98×10^{30} kg and 1.90×10^{27} kg, respectively.) In a scale drawing, if the sun-Jupiter separation is taken as 1 m, the sun's diameter will be 1 mm, and the center of mass of the system will lie 1 mm from the center of the sun. To within 0.1 %, the orbit of the reduced particle and the orbit of the planet Jupiter about the sun will be identical. Thus, we shall consider all the planets to move about a stationary sun.

Since the planets are observed to move in closed orbits about the sun, the only possible planetary orbits are the ellipse and the circle, which is, of course, just a special case of the ellipse. Therefore, the model invoked predicts Kepler's first law (see Section 13.1).

Let us consider a typical elliptic orbit as shown in Fig. 13.19. In a time Δt the line from the sun to the planet sweeps out an area ΔA, where

$$\Delta A \simeq \tfrac{1}{2}r(\Delta\theta)r.$$

Therefore,

$$\frac{\Delta A}{\Delta t} \simeq \frac{r^2}{2}\frac{\Delta\theta}{\Delta t}$$

and in the limit $\Delta\theta \to 0$,

$$\frac{dA}{dt} = \frac{r^2}{2}\frac{d\theta}{dt}.$$

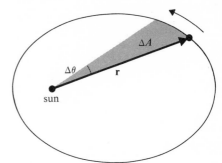

Fig. 13.19. The radius vector from the sun to a planet moves through an angle $\Delta\theta$ and sweeps out an area ΔA in time Δt.

But

$$J = \mu r^2 \frac{d\theta}{dt}$$

is a constant of the motion and therefore so is dA/dt, the rate at which the radius vector sweeps out area. We have

$$\frac{dA}{dt} = \frac{J}{2\mu}$$

$$= \text{constant},$$

which is Kepler's second law.

We may demonstrate Kepler's third law by integrating the second law

$$\frac{dA}{dt} = \frac{J}{2\mu}$$

over one period of the motion to obtain

$$A = \frac{J}{2\mu} T$$

or

$$T = \frac{2\mu A}{J},$$

where T is the period and $A = \pi ab$ is the area of the ellipse. Therefore,

$$T = \frac{2\mu \pi ab}{J}.$$

Now

$$a = \frac{r_{\text{max}} + r_{\text{min}}}{2}$$

$$= \frac{1}{2}\left(\frac{l}{1-e} + \frac{l}{1+e} \right)$$

$$= \frac{l}{1-e^2},$$

so that

$$b = a(1 - e^2)^{1/2}$$

$$= a\left(\frac{l}{a} \right)^{1/2}$$

and

$$T = \frac{2\pi\mu l^{1/2}}{J} a^{3/2}$$

or

$$T^2 = \left(\frac{2\pi\mu}{J}\right)^2 l a^3$$

$$= (\text{constant})a^3,$$

since l and J are constants of the motion. Therefore, the square of the period is proportional to the cube of the semimajor axis.

13.8   THE ESCAPE VELOCITY

The **escape velocity** may be defined as the velocity with which an object located on the surface of a celestial body must be projected in order that it escape the gravitational pull of the celestial body and move an infinite distance from it. From conservation of energy, the energy of the object on the surface of the celestial body must equal the energy at infinity, which, in this limiting case, is zero. Therefore, the escape velocity v_e is given by

$$\frac{1}{2}\mu v_e^2 - \frac{GmM}{r} = 0,$$

where r is the radius of the celestial body. That is,

$$v_e = \left(\frac{2GmM}{\mu r}\right)^{1/2}.$$

If the mass m of the object is very much less than that of the celestial body, then

$$\mu \simeq m$$

and

$$v_e \simeq \left(\frac{2GM}{r}\right)^{1/2}.$$

A more general derivation of the escape velocity proceeds from consideration of the planetary orbits. For $e < 1$, the particles m and M move in closed or **bound orbits**. For $e > 1$, the particles move in open or

unbound orbits. The condition $e < 1$ for a bound orbit means that

$$\left(1 + \frac{2EJ^2}{\mu G^2 m^2 M^2}\right)^{1/2} < 1.$$

Therefore,

$$1 + \frac{2EJ^2}{\mu G^2 m^2 M^2} < 1$$

or

$$\frac{2EJ^2}{\mu G^2 m^2 M^2} < 0.$$

This condition implies that $E < 0$, since all the other quantities are intrinsically positive. The condition for a bound orbit may be written

$$E = \frac{1}{2}\mu v^2 - \frac{GmM}{r} < 0,$$

so that

$$v < \left(\frac{2GmM}{\mu r}\right)^{1/2}.$$

The minimum speed v_e for which an orbit is unbound is

$$v_e = \left(\frac{2GmM}{\mu r}\right)^{1/2},$$

which is the **escape velocity** (or, more correctly, the **escape speed**).

Example. Calculate the escape velocity for a rocket of mass 10^3 kg at the earth's surface. The earth's mass M_E is 6.0×10^{24} kg; its radius r_E is 6.4×10^6 m.

Solution. The escape velocity at the earth's surface is

$$v_e = \left(\frac{2GM_E}{r_E}\right)^{1/2}$$

$$= \left(\frac{2(6.7 \times 10^{-11})(6.0 \times 10^{24})}{6.4 \times 10^6}\right)^{1/2}$$

$$= 1.1 \times 10^4 \text{ m} \cdot \text{sec}^{-1}$$

independent of the mass of the rocket. Note that at the earth's surface the acceleration g_E due to gravity is

$$g_E = \frac{GM_E}{r_E^2},$$

so that

$$v_E = (2g_E r_E)^{1/2}$$
$$= (2 \times 9.8 \times 6.4 \times 10^6)^{1/2}$$
$$= 1.1 \times 10^4 \text{ m} \cdot \text{sec}^{-1}.$$

PROBLEMS

1. At what point along the line from the earth to the moon is the gravitational pull of the moon equal to that of the earth? Express your result as a percentage of the distance from the center of the earth to the center of the moon.

2. A space traveler arrives on the moon and wishes to determine its mass relative to that of the earth. Before leaving the earth he had photographed the moon and thereby determined that its radius was $\frac{1}{4}$ that of the earth's radius. With him on the moon he has a rifle whose muzzle velocity he knows to be 300 m · sec^{-1}. He selects a flat, horizontal region and fires his rifle horizontally from a point 1 m above the ground. He measures the distance to the place where the bullet lands to be 308 m. Using this information, determine the mass of the moon relative to the mass of the earth.

3. The force of attraction between two charged particles is given by

$$\mathbf{F}_{12} = -\frac{Cq_1 q_2}{r^3}\mathbf{r},$$

where q_1 and q_2 are the magnitudes of the charges in coulombs, $\mathbf{r}$ is the separation vector between the charges as shown in Fig. 13.20, and C is a numerical constant of magnitude 9×10^9 (MKS units). Note that q_1 and q_2 can be either positive or negative. If q_1 and q_2 have the same sign, the force is one of repulsion; if q_1 and q_2 have opposite signs, the force is one of attraction. Deduce an expression for the orbit of the reduced charge q.

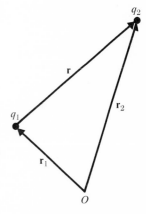

Fig. 13.20

4. In the next few years scientists will attempt to place a space station in orbit about the earth. One suggestion is that such a space station consist of a hollow doughnut-shaped satellite.

(a) How would you explain to a skeptical layman why it is possible for a satellite to orbit the earth?

(b) If the mass of the satellite is 1.20×10^6 kg and it is placed in a circular orbit 3.00×10^6 m above the earth's surface, what would be its period of revolution?

(c) In order for men to work within the station, an "artificial gravity" could be created by having the space station spin about its axis. The men could walk around inside with their heads pointing toward the axis of rotation. If the outside radius of the doughnut is 20 m, at what rate must the station spin about its axis to make this "gravity" equal to the earth's gravity at the earth's surface?

*5. The relative motions of a two-particle system correspond to an eccentricity $e = 1.5$. Sketch the orbits of the two particles, assuming that their masses are in the ratio of 3 to 1.

6. The earth's orbit around the sun has an eccentricity of 0.017. Draw an ellipse of this eccentricity, and superimpose on it a circle having radius equal to the average value of the major and minor axes of the ellipse.

*7. Using data given in Table 13.2 for the inner planets,
   (a) write equations for their orbits.
   (b) sketch their orbits on a single diagram to scale.

Table 13.2   Parameters for the orbits of the inner planets

Planet	Mean distance to sun (in millions of miles)	e
Mercury	36.0	0.206
Venus	67.2	0.007
Earth	92.9	0.017
Mars	141.6	0.093

8. (a) One of the planet Jupiter's moons is observed to move about the planet in a circular orbit of radius R with a period T. What is the mass of Jupiter in terms of R, T, and G?
   (b) Two of Jupiter's moons are observed to have the same angular momentum. Show that

   $$v_1 m_2 = v_2 m_1,$$

   where m_1 and v_1 are the mass and speed of one moon and m_2 and v_2 are the mass and speed of the other moon. (Neglect any rotation they may have about their own axes.)

9. A 1000-kg satellite is launched into an orbit of eccentricity 0.2 about the earth. The equation of the orbit is

   $$r = \left(\frac{7.8 \times 10^6}{1 - 0.2 \cos \theta} \right) \text{ m.}$$

   (a) What are the minimum and maximum values of r? What is the minimum and maximum distance of the satellite from the surface of the earth? (The minimum and maximum values of r are called the points of **perigee** and **apogee**, respectively.)
   (b) Determine the angular momentum and the total energy of the system.
   (c) What are the kinetic energy and the potential energy at minimum and maximum r?

*10. (a) Show that an earth satellite will take the same amount of time to move from its perigee to its apogee as it will to continue from its apogee back to its perigee.

(b) Two earth satellites of equal mass are in orbits having the same perigee. Show that the orbit of lower total energy is also that of lower angular momentum. If one of the satellites is in a circular orbit and one is in an elliptical orbit, which one will have the shorter period?

11. Halley's comet is the only conspicuous one having a period of less than 100 years. At perigee its distance from the sun is about one-half that of the earth; at apogee its distance from the sun is approximately equal to the mean of the distances of Neptune and Pluto from the sun. Estimate the eccentricity of the orbit of Halley's comet.

12. (a) The earth is 1.52×10^{11} m from the sun at apogee and 1.47×10^{11} m from the sun at perigee in its elliptical orbit. What is the difference in potential energy at apogee and at perigee?

(b) Calculate an average speed for the earth in its orbit, assuming a circular orbit of radius 1.495×10^{11} m, and calculate the average kinetic energy of the earth. Compare the magnitude of the kinetic energy with the magnitude of the potential energy of part (a).

13. **Binary stars** consist of two stars held close together by mutual gravitational attraction. The bright star Sirius and its faint companion have masses of 6.4×10^{30} kg and 1.6×10^{30} kg, respectively.

(a) What is the reduced mass of the binary star?

(b) The two component stars follow elliptical orbits of eccentricity $e = 0.59$. Sketch the orbits of the two stars, assuming that their maximum separation is 2.3×10^{12} m. Use a scale of 1 cm = 2×10^{11} m. Indicate the position of the center of mass on the diagram. From your diagram estimate the minimum separation of the two stars.

14. A binary star system consists of components of mass $m_1 = 1.1 \times 10^{30}$ kg and $m_2 = 1.9 \times 10^{30}$ kg. The total energy of the system is 3.2×10^{37} J, and the total angular momentum is 8.9×10^{45} kg $\cdot$ m$^2 \cdot$ sec^{-1}.

(a) Deduce the parameters of the orbit of the equivalent one-particle system.

(b) Sketch the orbits of the component stars.

15. A satellite of mass 5.0×10^3 kg travels in an elliptical orbit about the earth. At the lowest point in its orbit it is 320 km above the earth's surface and at its highest point 1600 km. Determine the period of the satellite.

*16. A satellite, mass 100 kg, is in a circular orbit of 7000-km radius. How much energy does it give up (frictional losses due to the atmosphere) in the process of re-entry? Assume that it reaches sea level with a small velocity. *Hint*: it is necessary to work out the kinetic energy of the satellite in orbit, as well as the potential energy lost in falling.

17. Show that all elliptical orbits of the same major axis have the same total energy independent of the eccentricity.

18. Two identical satellites orbiting the earth have their perigees 150 km above sea level. Their apogees are at 7000 and 200 km above sea level. If the Earth's diameter is 12,735 km, what is the ratio of total energies for the two orbits, and what is the ratio of the angular momenta?

19. (a) Find the speeds of the two satellites of Problem 18 at perigee.

(b) It is desired to accelerate the satellite in the more circular orbit so that it gets into the more eccentric orbit of the other satellite. This is done by firing a rocket engine at perigee. If the mass of the satellite and (unfueled) rocket motor is 400 kg and the fuel exhaust speed is 4000 m · sec^{-1}, how much fuel must be carried?

20. Two stars forming a binary system are observed to rotate about their center of mass in circular orbits of period 3.0 years. The center of mass lies $\frac{3}{5}$ of the way between them and their separation is 2.0 A.U. (1 A.U. = distance from earth to sun, assuming the earth's orbit is circular). What is the ratio of the mass of the lighter star to the mass of the heavier star? What is the ratio of the mass of the lighter star to the mass of the sun?

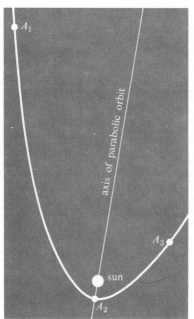

Fig. **13.21**

*21. Most comets travel around the sun in orbits that are nearly parabolas. Would Kepler's second law apply to these orbits? Comet A has angular momentum 2×10^{18} J · sec at point A_1 in Fig. 13.21. Determine graphically its speeds at A_2 and A_3 if its speed at A_1 is 20 km · sec^{-1}. Would the law of conservation of energy apply to this orbit? Check this numerically from measurements in the figure.

22. Use Kepler's second law to calculate the orbital angular momentum associated with the earth's motion around the sun.

*23. The planets have the orbital constants given in Table 13.3. Use this information to make an accurate graphical verification of Kepler's third law. Compute the mass of the sun.

24. Compare the escape velocity for a rocket at the earth's surface with that

(a) at the surface of Mars.

(b) at the surface of Jupiter.

25. The planet Saturn has a mass 95 times that of the earth and a diameter 9.0 times that of the earth. Calculate

(a) the acceleration due to gravity at the surface of Saturn compared with that at the surface of the earth (that is, g_s/g_e).

(b) the relative densities of Saturn and the earth.

(c) the escape velocity from Saturn compared with that from the earth's surface. Saturn is encircled by a system of thin rings lying in the plane of its equator. It has been postulated that the rings consist of tiny particles, each revolving about Saturn in a separate orbit. Measurements show that the outer parts of the ring revolve more slowly than the inner parts. By means of an equation show that this fact is consistent with the above postulate.

Table 13.3   Parameters describing the orbits of the planets

Planet	Semimajor axis (km)	Period	Eccentricity
Mercury	57.85×10^6	87.97 days	.206
Venus	108.11	224.70 days	.007
Earth	149.46	365.26 days	.017
Mars	277.4	686.98 days	.093
Jupiter	777.6	11 years 314 days	.048
Saturn	1426	29 years 167 days	.056
Uranus	2868.3	84 years 5 days	.047
Neptune	4494.3	164 years 288 days	.009
Pluto	5900	247 years 255 days	?

14 Molecular motions:
a many-particle problem

14.1 INTRODUCTION

In this chapter we shall consider simple molecules as an example of the many-particle problem. Molecules are systems of particles that, although they contain relatively few particles, cannot be approximated by two-particle systems. Molecules provide a good example for discussion, since an understanding of molecular motions[1] is essential to scientists working in chemistry and the life sciences.

As a typical molecule we might consider the carbon-monoxide (CO) molecule. To understand the structure and internal motions of the molecule, it is necessary to consider the mutual interactions between the carbon nucleus, the oxygen nucleus, and the twenty-eight extranuclear electrons. Since these thirty particles are electrically charged, the dominant interaction forces between them are electromagnetic in nature. The forces between all pairs of particles are comparable, in contrast to the situation in the planetary-motion problem. Therefore, in order to deduce the motion of any one of the particles, the simultaneous solution of 30 coupled equations is required. This problem has never been solved exactly; even approximate solutions are difficult to obtain.

Faced with such an intractable problem, it is convenient to invoke a **model** that in some sense represents the molecule. Two criteria that must be satisfied by an acceptable model are the following:

1) the model must be amenable to mathematical analysis, and
2) the predicted behavior of the model must approximate some aspects of the actual behavior of molecules.

Usually one starts with very simple models in an attempt to understand the general behavior of the system and then proceeds to consider progressively more sophisticated models to explain the detailed behavior.

14.2 THE CENTER OF MASS OF AN N-PARTICLE SYSTEM

We saw in Section 13.3 that the concept of the center of mass of a two-particle system allowed us to separate the translational motion of the system as a whole from the internal motions of the system with respect to the center of mass. The development of Section 13.3 may readily be generalized to a system of N particles, such as the one shown in Fig. 14.1. For such a system

$$\mathbf{F}_{ext} = m_1 \frac{d^2\mathbf{r}_1}{dt^2} + m_2 \frac{d^2\mathbf{r}_2}{dt^2} + \cdots + m_N \frac{d^2\mathbf{r}_N}{dt^2},$$

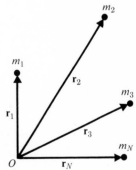

Fig. 14.1. A many-particle system.

[1] B. J. Adler and T. E. Wainwright, "Molecular Motions," *Scientific American*, October, 1959. Available as *Scientific American Offprint 265* (San Francisco: W. H. Freeman and Co., Publishers).

where

$$\mathbf{F}_{ext} = \mathbf{F}_{ext}(1) + \mathbf{F}_{ext}(2) + \cdots + \mathbf{F}_{ext}(N).$$

This equation may be written in the form

$$\mathbf{F}_{ext} = M\frac{d^2\mathbf{R}}{dt^2},$$

where

$$M = m_1 + m_2 + \cdots + m_N$$

is the total mass of the system of particles and $\mathbf{R}$ is the position vector of the center of mass. From a comparison of these two equations of motion, we see that

$$\mathbf{R} = \frac{m_1\mathbf{r}_1 + m_2\mathbf{r}_2 + \cdots + m_N\mathbf{r}_N}{m_1 + m_2 + \cdots + m_N}$$

or

$$\mathbf{R} = \frac{\displaystyle\sum_{i=1}^{N} m_i\mathbf{r}_i}{\displaystyle\sum_{i=1}^{N} m_i}.$$

Example. Determine the position of the center of mass of a system of three particles situated at the vertices of the isosceles triangle shown in Fig. 14.2. Particles (1) and (2) are of mass m, and particle (3) is of mass αm. The sides of the triangle have lengths a and b.

Solution. Let us choose a coordinate system such that the triangle lies in the x, y plane with the x axis along the base of the triangle and the origin coincident with particle (1). In this reference frame particles (1), (2), and (3) have coordinates $(0, 0, 0)$, $(a, 0, 0)$, and $(a/2, [b^2 - a^2/4]^{1/2}, 0)$, respectively. The coordinates of the center of mass (X, Y, Z) are given by

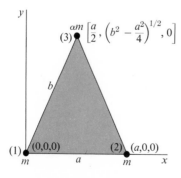

$$X = \frac{\displaystyle\sum_{i=1}^{3} m_i x_i}{\displaystyle\sum_{i=1}^{3} m_i}$$

$$= \frac{0 + ma + \dfrac{\alpha m a}{2}}{m + m + \alpha m}$$

$$= \frac{\left(\dfrac{a}{2}\right)(2 + \alpha)m}{(2 + \alpha)m}$$

$$= \frac{a}{2},$$

Fig. 14.2. A system of three particles forming an isosceles triangle.

$$Y = \frac{\sum\limits_{i=1}^{3} m_i y_i}{\sum\limits_{i=1}^{3} m_i}$$

$$= \frac{0 + 0 + \alpha m \left(b^2 - \frac{a^2}{4} \right)^{1/2}}{m + m + \alpha m}$$

$$= \left(b^2 - \frac{a^2}{4} \right)^{1/2} \left(\frac{\alpha}{2 + \alpha} \right),$$

$$Z = \frac{\sum\limits_{i=1}^{3} m_i z_i}{\sum\limits_{i=1}^{3} m_i}$$

$$= \frac{0 + 0 + 0}{m + m + \alpha m}$$

$$= 0.$$

The center of mass lies on the line bisecting the angle between the sides of equal length. If $\alpha \gg 1$, then the center of mass is essentially coincident with particle (3); if $\alpha \ll 1$, the center of mass is essentially on the x axis midway between particles (1) and (2).

We emphasize that the motion of the center of mass is independent of the internal forces between the particles in the system, since they cancel in pairs and therefore do not contribute a net force to the system as a whole.

14.3   THE RIGID-ROTATOR MODEL OF A DIATOMIC MOLECULE

Let us take as a model for a diatomic molecule two point particles of masses m_1 and m_2 fixed at the ends of a weightless rigid rod of length r (see Fig. 14.3). The masses of the particles are chosen to be equal to the nuclear masses of the constituent atoms and the length of the rigid rod to be equal to the internuclear separation in the molecule. The electrons are not explicitly accounted for in this model; however, they appear implicitly, since they are essential to the determination of the internuclear separation. This model is proposed as a possible one for predicting the rotational motion of the molecule as a whole relative to the center of mass. Therefore, it is often referred to as the **rigid-rotator model**.

Since the mass of the electrons is very small compared to the nuclear mass, the contribution of the electrons to the energy of rotation is neglig-

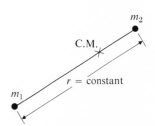

Fig. 14.3. The rigid rotator model for a diatomic molecule.

ible to a first approximation. The kinetic energy associated with rotation about some axis through the center of mass is given by

$$\begin{aligned} E_{\text{rot}} &= \tfrac{1}{2}m_1v_1^2 + \tfrac{1}{2}m_2v_2^2 \\ &= \tfrac{1}{2}m_1r_1^2\omega_1^2 + \tfrac{1}{2}m_2r_2^2\omega_2^2, \end{aligned}$$

where we have used $v = \omega r$. Here r_1 and r_2 are the perpendicular distances of m_1 and m_2 from the axis of rotation (Fig. 14.4). However,

Fig. 14.4. A rotation of a rigid rotator about an axis through the center of mass; r_1 and r_2 are the perpendicular distances of m_1 and m_2, respectively, from the axis of rotation.

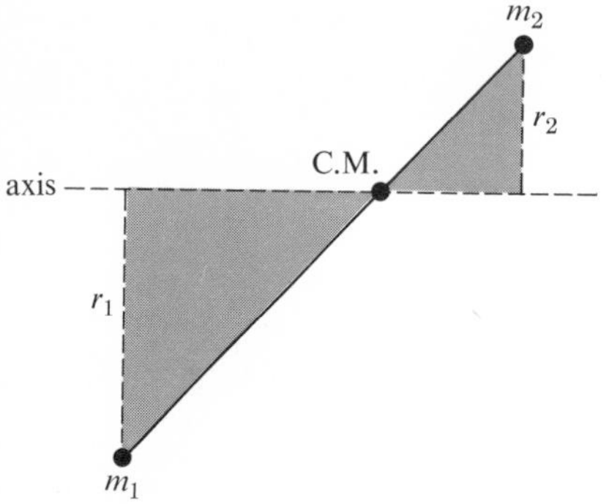

$$\omega_1 = \omega_2 = \omega,$$

since we are dealing with a rigid rotator, so that

$$\begin{aligned} E_{\text{rot}} &= \tfrac{1}{2}m_1r_1^2\omega^2 + \tfrac{1}{2}m_2r_2^2\omega^2 \\ &= \frac{1}{2}\left(\sum_{i=1}^{2} m_ir_i^2\right)\omega^2 \\ &= \tfrac{1}{2}I\omega^2, \end{aligned}$$

where

$$I = \sum_{i=1}^{2} m_ir_i^2$$

is the **moment of inertia** of the molecule about the axis of rotation. We may also write

$$\begin{aligned} I &= m_1r_1^2 + m_2r_2^2 \\ &= I_1 + I_2, \end{aligned}$$

where I_1 and I_2 are the moments of inertia of the individual particles about the axis of rotation.

Let us consider the special case of the axis of rotation perpendicular to the rotator, as shown in Fig. 14.5. From the definition of the center of mass,

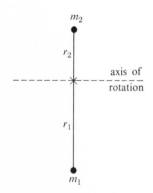

Fig. 14.5. The special case of rotation about an axis through the center of mass and perpendicular to the rigid rotator.

$$m_1r_1 = m_2r_2,$$

as well as

$$r_1 + r_2 = r;$$

it follows that

$$r_1 = \frac{m_2}{m_1 + m_2}r \quad \text{and} \quad r_2 = \frac{m_1}{m_1 + m_2}r.$$

Therefore,

$$I = \left(\frac{m_1 m_2^2}{(m_1 + m_2)^2} + \frac{m_2 m_1^2}{(m_1 + m_2)^2} \right) r^2$$

$$= \frac{m_1 m_2}{m_1 + m_2} r^2$$

$$= \mu r^2,$$

where μ is the reduced mass of the molecule. That is, the moment of inertia is the same as that of a point particle of reduced mass

$$\mu = \frac{m_1 m_2}{m_1 + m_2}$$

at a distance r from the axis. Therefore, the rotation of the rigid rotator is mathematically equivalent to the rotation of a single particle μ at a fixed distance r from the axis of rotation when the axis of rotation is perpendicular to the rotator. This is also true when the axis of rotation is not perpendicular to the rotator; the proof of this is left as an exercise for the reader.

We have for the energy of rotation

$$E_{\text{rot}} = \tfrac{1}{2} I \omega^2$$

$$= \frac{1}{2} \frac{(I\omega)^2}{I}$$

$$= \frac{1}{2} \frac{(\mu r^2 \omega)^2}{I}.$$

But the magnitude J of the angular momentum of the rotating molecule is

$$J = \mu r^2 \omega,$$

so that

$$E_{\text{rot}} = \frac{J^2}{2I}.$$

Classical mechanics shows that, in general, the kinetic energy of rotation of a three-dimensional body can be expressed as

$$E_{\text{rot}} = \frac{1}{2} \left(\frac{J_a^2}{I_a} + \frac{J_b^2}{I_b} + \frac{J_c^2}{I_c} \right),$$

where J_a, J_b, and J_c are the components of the total angular momentum $\mathbf{J}$ and I_a, I_b, and I_c are the moments of inertia relative to three mutually perpendicular axes. These axes, known as the **principal axes of inertia**,

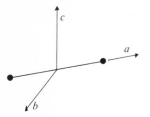

Fig. 14.6. Principal axes system for a diatomic molecule.

are chosen such that the isolated body would rotate about them with constant angular momentum[2]; when a body is rotating about a principal axis, **J** is parallel to **ω**. For a diatomic molecule the internuclear axis and any two other mutually perpendicular axes constitute a principal axes system (Fig. 14.6). Therefore, we may take

$$I_a = 0, \qquad I_b = I_c = I = \mu r^2,$$

and, since

$$J_a = 0,$$

$$E_{\text{rot}} = \frac{J_b^2 + J_c^2}{2I}.$$

But

$$J_a^2 + J_b^2 + J_c^2 = J^2,$$

so that

$$E_{\text{rot}} = \frac{J^2 - J_a^2}{2I}$$

$$= \frac{J^2}{2I}.$$

However,

$$E_{\text{rot}} = \tfrac{1}{2}I\omega^2,$$

so that the rotational frequency ω in radians per second is given by

$$\omega = \frac{J}{I}.$$

Hence the rotational frequency ν_{rot} in Hz is given by

$$\nu_{\text{rot}} = \frac{\omega}{2\pi}$$

$$= \frac{J}{2\pi I}.$$

This is a prediction of the rigid-rotator model. Since classical mechanics imposes no restrictions on the allowed values of the magnitude of the angular momentum, any observed value of the rotational frequency is in agreement with the theory. Experimentally, molecules are observed to rotate only with certain discrete frequencies, a result that does not agree

[2] C. Kittel, W. D. Knight, and M. A. Ruderman, *Mechanics* (New York: McGraw-Hill Book Company, 1965), Chapter 8.

with the classical prediction. A quantum-mechanics treatment of the rigid rotator (see Section 18.8) shows that a measurement of the rotational angular momentum can yield only certain discrete values. It therefore follows that only discrete values of the rotational frequency are allowed, in agreement with the experimental observations. For a particular value of the angular momentum we see that the rotational frequency is predicted to be inversely proportional to the molecular moment of inertia. This prediction is in substantial agreement with experimental findings.

It is generally found that motions that take place on an atomic scale of dimensions require a quantum-mechanical explanation (see Chapter 18).

Example. Compare the rotational frequencies of hydrogen chloride (HCl) molecules and deuterium chloride (DCl) molecules for the same value of the rotational angular momentum.

Solution

$$\frac{\nu_{HCl}}{\nu_{DCl}} = \frac{I_{DCl}}{I_{HCl}}$$

$$= \frac{\mu_{DCl} r_{DCl}^2}{\mu_{HCl} r_{HCl}^2}.$$

A hydrogen atom differs from a deuterium atom only in having one-half the nuclear mass; the electromagnetic interactions are essentially the same for an HCl molecule and a DCl molecule. Since the internuclear separation is determined by the electromagnetic interactions,

$$r_{HCl} \simeq r_{DCl}.$$

Therefore,

$$\nu_{HCl} \simeq \left(\frac{\mu_{DCl}}{\mu_{HCl}}\right) \nu_{DCl}$$

$$= \left(\frac{m_D m_{Cl}}{m_D + m_{Cl}}\right) \left(\frac{m_H + m_{Cl}}{m_H m_{Cl}}\right) \nu_{DCl}$$

$$\simeq 2\nu_{DCl}.$$

The rotational frequency of HCl is predicted to be twice that of DCl.

14.4   THE HARMONIC-OSCILLATOR MODEL OF A DIATOMIC MOLECULE

Another simple model for a diatomic molecule is one in which each nucleus is represented by a point particle and moves toward or away from the other along the internuclear axis. The restoring force experienced by each of the two atoms when they are displaced from their equilibrium positions is assumed to be proportional to the change in internuclear separation. The electrons again appear implicitly, since they are essential to the determination of the equilibrium internuclear separation and to the force constant characterizing the restoring force. No rotation of the molecule as a whole is considered. The nucleus of each atom may vibrate about its equilibrium position, as shown in Fig. 14.7. This model is proposed as a possible one for predicting the vibrational motion of the atoms in the molecule relative to the center of mass and is often called the **harmonic-oscillator model**.

The equations of motion of the two point particles for this model are

$$m_1 \frac{d^2 r_1}{dt^2} = - k(r - r_e)$$

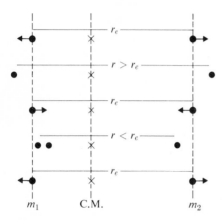

Fig. 14.7. The harmonic oscillator model of a diatomic molecule; the nucleus of each atom vibrates about its equilibrium position.

$$m_2 \frac{d^2 r_2}{dt^2} = - k(r - r_e),$$

where r_1 and r_2 are the distances of the two particles from their center of mass, r is the instantaneous internuclear separation, r_e is the equilibrium internuclear separation, and k is the molecular force constant. Substituting

$$r_1 = \frac{m_2}{m_1 + m_2} r \quad \text{and} \quad r_2 = \frac{m_1}{m_1 + m_2} r,$$

we obtain from both equations

$$\frac{m_1 m_2}{m_1 + m_2} \frac{d^2 r}{dt^2} = -k(r - r_e).$$

We may rewrite this equation as

$$\mu \frac{d^2 (r - r_e)}{dt^2} = -k(r - r_e),$$

where we have subtracted the constant r_e from r in the differential term. This introduces no change, since the constant disappears upon differentiation. Note that only $r - r_e$ appears in the equation of motion, irrespective of whether the amplitude of vibration is the same or not for m_1 and m_2.

This equation is identical to the general equation of a simple harmonic oscillator, except that x is replaced by $r - r_e$, the change of internuclear distance from its equilibrium value. We have therefore reduced the mutual vibrations of the two particles to the vibration of a single particle of mass μ (Fig. 14.8), whose amplitude equals the amplitude of the change of internuclear distance in the molecule. We can immediately say that the vibration frequency of the molecule is

$$\nu_{\text{vib}} = \frac{1}{2\pi}\left(\frac{k}{\mu}\right)^{1/2}$$

and that the energy associated with the vibration is

$$E_{\text{vib}} = \frac{1}{2}\mu\left[\frac{d(r-r_e)}{dt}\right]^2 + \tfrac{1}{2}k(r-r_e)^2.$$

That is, the classical-mechanics treatment of the harmonic-oscillator model predicts a single vibrational frequency; the same prediction is made by a quantum-mechanics treatment (see Section 18.8). Experiments confirm that this model gives a reasonable first approximation to the actual vibrational motion of diatomic molecules.

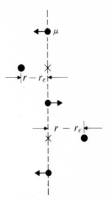

Fig. 14.8. The reduced particle vibrates about its equilibrium position.

Example. Estimate the ratio of the vibrational frequencies of the HF and DF molecules.

Solution. The ratio of the vibrational frequencies is

$$\frac{\nu_{\text{HF}}}{\nu_{\text{DF}}} = \left(\frac{k_{\text{HF}}}{k_{\text{DF}}}\right)^{1/2}\left(\frac{\mu_{\text{DF}}}{\mu_{\text{HF}}}\right)^{1/2}.$$

An H atom differs from a D atom only in having a different mass; the electromagnetic interactions are essentially the same for both molecules. Since the force constants k_{HF} and k_{DF} are determined by the electromagnetic interactions, it is reasonable to assume that

$$k_{\text{HF}} \simeq k_{\text{DF}}$$

and

$$\frac{\nu_{\text{HF}}}{\nu_{\text{DF}}} \simeq \left(\frac{\mu_{\text{DF}}}{\mu_{\text{HF}}}\right)^{1/2}.$$

Now

$$\mu_{\text{HF}} = \frac{m_{\text{H}}m_{\text{F}}}{m_{\text{H}} + m_{\text{F}}} = \frac{(1)(19)}{1 + 19} = \frac{19}{20}$$

$$\mu_{\text{DF}} = \frac{m_{\text{D}}m_{\text{F}}}{m_{\text{D}} + m_{\text{F}}} = \frac{(2)(19)}{2 + 19} = \frac{38}{21},$$

where the masses are given in atomic mass units, so that

$$\frac{\nu_{\text{HF}}}{\nu_{\text{DF}}} = \left(\frac{38}{21} \cdot \frac{20}{19}\right)^{1/2} = 1.38.$$

This agrees with the experimental ratio

$$\frac{v_{HF}}{v_{DF}} = \frac{12.42 \times 10^{13}}{8.99 \times 10^{13}} = 1.38.$$

Example. The vibrational frequency of a carbon-monoxide molecule is measured to be 6.53×10^{13} Hz. Calculate the force constant of the molecule if the mass of the carbon atom is 2.00×10^{-26} kg and the mass of the oxygen atom is 2.66×10^{-26} kg.

Solution. The reduced mass of the molecule is

$$\mu = \left(\frac{2.00 \times 2.66}{2.00 + 2.66}\right) \times 10^{-26}$$

$$= 1.14 \times 10^{-26}\,\text{kg}.$$

The force constant is

$$k = 4\pi^2 \mu v_{vib}^2$$

$$= 4 \times (3.14)^2 \times (1.14 \times 10^{-26}) \times (6.53 \times 10^{13})^2$$

$$= 1.92 \times 10^3\,\text{N}\cdot\text{m}^{-1}.$$

14.5 THE NONRIGID-ROTATOR MODEL OF A DIATOMIC MOLECULE

So far we have considered the rigid-rotator and harmonic-oscillator models of the diatomic molecule independently. It is quite obvious that a molecule cannot strictly speaking be a rigid rotator if it can vibrate in the direction of the internuclear axis. A better model for representing molecular rotations might be that of a nonrigid rotator, that is, a rotating system consisting of two point particles connected by a massless spring rather than a massless rigid bar (Fig. 14.9).

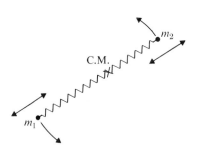

Fig. 14.9. The nonrigid rotator model of a diatomic molecule.

If the molecule is rotating, the internuclear separation r_c has a value such that the centripetal force necessary for the rotation is provided by the restoring force arising from the small displacement $r_c - r_e$ of the particles from their equilibrium separation r_e. The centripetal force F_c is given by

$$F_c = \mu\omega^2 r_c = \frac{J^2}{\mu r_c^3},$$

where ω is the angular frequency and $J = \mu r_c^2 \omega$ is the angular momentum.

Equating this to the restoring force yields for the change of internuclear separation

$$r_c - r_e = \frac{J^2}{\mu r_c^3 k}.$$

For actual molecules $r_c - r_e$ is very much smaller than r_e, so that

$$r_c \simeq r_e$$

and

$$r_c - r_e \simeq \frac{J^2}{\mu r_e^3 k}.$$

For a nonrigid rotator the energy of rotation is the sum of kinetic and potential (due to the stretching of the spring) contributions, so that

$$E = \frac{J^2}{2\mu r_c^2} + \tfrac{1}{2}k(r_c - r_e)^2$$

$$\simeq \frac{J^2}{2\mu r_e^2}\left(1 + \frac{J^2}{\mu r_e^4 k}\right).$$

Example. Calculate the change in internuclear separation $r_c - r_e$ for a CO molecule when it is rotating with an angular momentum of 1.46×10^{-34} kg·m·sec^{-1} if the separation for the nonrotating molecule is $r_e = 1.12 \times 10^{-10}$ m.

Solution. From the last example $\mu = 1.14 \times 10^{-26}$ kg, and $k = 1.92 \times 10^3$ N·m^{-1}. Therefore,

$$r_c - r_e = \frac{(1.46 \times 10^{-34})^2}{(1.14 \times 10^{-26})(1.12 \times 10^{-10})^3(1.92 \times 10^3)}$$

$$= 6.9 \times 10^{-16} \text{ m}.$$

That is,

$$\frac{r_c - r_e}{r_e} \simeq 6 \times 10^{-6}.$$

The change in internuclear separation is only six parts in one million.

14.6   THE ROTATIONAL MOTION OF POLYATOMIC MOLECULES

The rotational energy of a **polyatomic molecule**, in the rigid-rotator approximation, is given by the general expression

$$E_{\text{rot}} = \frac{1}{2}\left(\frac{J_a^2}{I_a} + \frac{J_b^2}{I_b} + \frac{J_c^2}{I_c}\right)$$

(see Section 14.3).

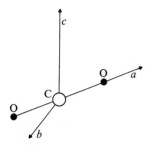

Fig. 14.10. Principal axes system for a linear polyatomic molecule.

For **linear polyatomic molecules**, such as CO_2, we may take (see Fig. 14.10)

$$I_a = 0, \qquad I_b = I_c = I = \mu r^2,$$

and since

$$J_a = 0 \quad \text{and} \quad J_a^2 + J_b^2 + J_c^2 = J^2,$$

it follows that

$$E_{\text{rot}} = \frac{J^2}{2I},$$

as for the diatomic molecule.

For molecules that have the configuration shown in Fig. 14.11,

$$I_a = I_b = I_c = I$$

and

$$E_{\text{rot}} = \frac{J^2}{2I},$$

as for the simple rotator (linear molecule). Molecules having the structure shown in Fig. 14.11 are called **spherical-top molecules**; methane (CH_4) is an example.

For molecules that have the configuration shown in Fig. 14.12,

$$I_a \neq 0, \qquad I_b = I_c,$$

$$E_{\text{rot}} = \frac{1}{2}\left(\frac{J_a^2}{I_a} + \frac{J_b^2 + J_c^2}{I_b}\right).$$

But

$$J_a^2 + J_b^2 + J_c^2 = J^2,$$

so that

$$E_{\text{rot}} = \frac{1}{2}\left(\frac{J_a^2}{I_a} + \frac{J^2 - J_a^2}{I_b}\right)$$

$$= \frac{1}{2}\left(\frac{J^2}{I_b} + \frac{J_a^2}{I_a} - \frac{J_a^2}{I_b}\right).$$

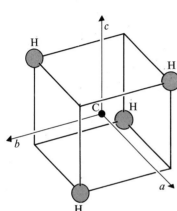

Fig. 14.11. Principal axes system for a spherical-top molecule.

and

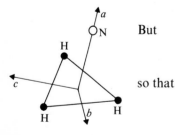

Fig. 14.12. Principal axes system for a symmetric-top molecule.

The energy of rotation now depends upon the total angular momentum of the molecule and the component of the angular momentum in the direction of the symmetry axis. Molecules having the structure shown in Fig. 14.12 are called **symmetric-top molecules**; ammonia (NH_3) is an example.

For **asymmetric-top molecules**, such as H_2O,

$$I_a \neq I_b \neq I_c \neq 0,$$

and the general expression for the rotational energy obtains.

Example. Derive a general expression for the rotational energy of the hypothetical molecule shown in Fig. 14.13. The molecule consists of three identical atoms of mass m situated at the vertices of an equilateral triangle. Each atom is at a distance r from the center of mass, as is apparent from the symmetry of the molecule. The principal axis a is perpendicular to the plane of the molecule. The principal axes b and c are as shown in Fig. 14.13.

Solution. The moments of inertia about the principal axes are

$$I_a = mr^2 + mr^2 + mr^2$$
$$= 3mr^2,$$
$$I_b = I_c = m(r \sin 15°)^2 + m(r \sin 75°)^2 + m(r \sin 45°)^2$$
$$= 0.067mr^2 + 0.933mr^2 + 0.500mr^2$$
$$= 1.50mr^2.$$

Therefore,

$$E_{rot} = \frac{1}{2mr^2}\left(\frac{J^2}{1.50} + \frac{J_a^2}{3.0} - \frac{J_a^2}{1.50}\right)$$
$$= \frac{0.333J^2 - 0.166J_a^2}{mr^2}.$$

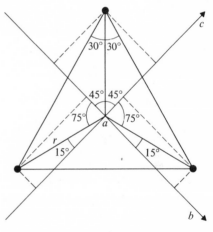

Fig. 14.13. A hypothetical molecule consisting of three atoms of equal mass forming an equilateral triangle.

14.7 THE VIBRATIONAL MOTION OF POLYATOMIC MOLECULES

A consideration of the vibrational motions of polyatomic molecules in the harmonic approximation is considerably more complex than that for diatomic molecules. To describe the motion of the N nuclei in a polyatomic molecule requires $3N$ coordinates, such as, for example, the $3N$ Cartesian coordinates x_i, y_i, and z_i, for $i = 1, \cdots, N$. We say there are $3N$ **degrees of freedom**. That is, $3N$ numbers are required to completely specify the positions of the N nuclei; this is illustrated in Fig. 14.14 for

$N = 3$. If we wish to study the relative vibrational motions of the system, we are not interested in either the translational motion of the system as a whole or the rotation of the system as a whole.

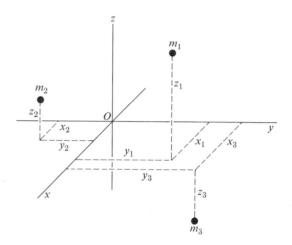

Fig. 14.14. A three-particle system requires nine coordinates to specify the positions of the three particles.

To eliminate the translational motion, we select the origin of coordinates at the center of mass; the coordinates of the particles then satisfy the conditions

$$\sum m_i x_i = \sum m_i y_i = \sum m_i z_i = 0.$$

To eliminate the rotational motion we require that the angular momentum of the system be zero about each coordinate axis, as given by the conditions

$$\sum m_i \left(y_i \frac{dz_i}{dt} - z_i \frac{dy_i}{dt} \right) = \sum m_i \left(z_i \frac{dx_i}{dt} - x_i \frac{dz_i}{dt} \right)$$

$$= \sum m_i \left(x_i \frac{dy_i}{dt} - y_i \frac{dx_i}{dt} \right) = 0.$$

Therefore, $3N - 6$ additional conditions are required to completely specify the positions of the N nuclei. These conditions specify the vibrational motion of the molecule; that is, there are $3N - 6$ vibrational degrees of freedom. For linear molecules, the angular momentum about the internuclear axis is identically zero, so that only two conditions are required to ensure that the angular momentum of a linear molecule is zero. Therefore, for linear molecules there are $3N - 5$ vibrational degrees of freedom.

The vibrational motion of a polyatomic molecule is described in terms of **normal modes**. In order to illustrate the concept of a normal mode, let us consider the system shown in Fig. 14.15. An object of mass m moves under the influence of four springs, two identical ones lying in the x direction and two other identical ones lying in the y direction. If the object is displaced slightly from its equilibrium position in the x direction and then left to itself, it will carry out simple harmonic motion with frequency

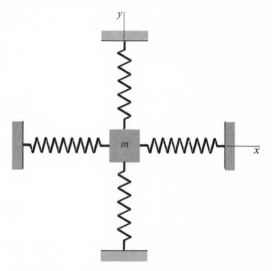

$$v_x = \frac{1}{2\pi}\left(\frac{2k_x}{m}\right)^{1/2},$$

where k_x is the force constant for the springs in the x direction. If the object is slightly displaced in the y direction and then released, it will carry out simple harmonic motion with frequency

$$v_y = \frac{1}{2\pi}\left(\frac{2k_y}{m}\right)^{1/2},$$

Fig. 14.15. An object of mass m moving under the influence of four springs.

where k_y is the force constant for the springs in the y direction. Unless the two springs lying in the x direction are identical to the ones in the y direction,

$$v_x \neq v_y.$$

Suppose the object is displaced in an arbitrary direction to point P, as shown in Fig. 14.16. The restoring force $\mathbf{F}_r$, with components $k_x x$ and $k_y y$, is not directed through the equilibrium position 0, since $k_x \neq k_y$. Therefore, the object will carry out a complicated type of motion in the x, y plane. If v_x/v_y is a rational number, the motion will repeat itself after a certain time, and the object will trace out a definite path. As an example, the path for $v_x/v_y = \frac{5}{3}$ is shown in Fig. 14.17. Although the motion is now a complicated one, the x and y components are simple harmonic; the position of the mass m is given by

$$x = x_0 \cos(2\pi v_x t + \phi_1)$$

$$y = y_0 \cos(2\pi v_y t + \phi_2),$$

where x_0 and y_0 are the coordinates of the point P. These simple motions into which the actual motion

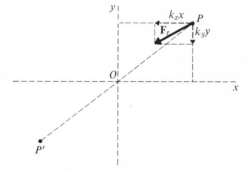

Fig. 14.16. The mass m is displaced from equilibrium in an arbitrary direction.

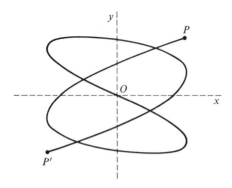

Fig. 14.17. Motion of the particle of mass m if $(v_x/v_y) = \frac{5}{3}$.

can be resolved, are called the **normal modes of vibration** of the mass m. The x and y coordinates are the **normal coordinates**.

Example. Deduce the form of the translational and rotational normal modes of the carbon-dioxide (CO_2) molecule.

Solution. For translational modes the molecule as a whole moves in the x, y or z direction. The form of the normal modes is readily deduced. A possible set of translational modes is sketched in Fig. 14.18.

For rotational modes the molecule as a whole rotates about axes through the center of mass. The form of the normal modes is readily deduced. A possible set of rotational modes is sketched in Fig. 14.19. Note that since CO_2 is a linear molecule, there is no angular momentum associated with rotation about the x axis. Therefore, such a rotation may be neglected.

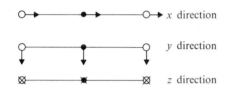

Fig. 14.18. A possible set of translational normal modes of a carbon-dioxide molecule.

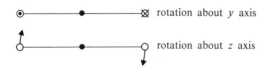

Fig. 14.19. A possible set of rotational normal modes of a carbon-dioxide molecule.

Now let us return to the vibrational motion of polyatomic molecules. If the N nuclei are displaced from their equilibrium positions in an arbitrary fashion, they will in general execute complicated motions. However, if the nuclei are displaced in certain definite ways simple motions result similar to the single-particle example. These normal modes of vibration are characterized by the following facts: 1) each nucleus carries out a simple harmonic motion, 2) all the nuclei have the same frequency of oscillation, and 3) in general, the nuclei oscillate in phase. An arbitrary displacement of the nuclei can be resolved into components, each of which is a normal mode of vibration. As an example, let us consider a water molecule, the equilibrium configuration of which is shown in Fig. 14.20. Since

$$N = 3, \qquad 3N - 6 = 3,$$

and the water molecule has three normal modes of vibration. These are shown in Fig. 14.21. The displacements characterizing the mode having frequency v_1 are designated Q_1, which is a normal coordinate. (Note that just one symbol Q_1 is used to denote the three displacements of the two

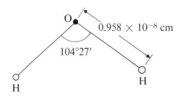

Fig. 14.20. Equilibrium configuration of a water molecule.

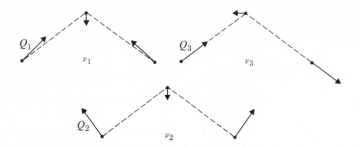

Fig. 14.21. The normal modes of vibration of a water molecule.

hydrogen atoms and one oxygen atom associated with each normal mode of vibration.)

As a second example, let us consider the linear CO_2 molecule. Again $N = 3$, but now

$$3N - 5 = 4,$$

and the CO_2 molecule has four normal modes of vibration. These are shown in Fig. 14.22. We note that the normal coordinates Q_3 and Q_4 describe modes of vibration that are identical but that take place in mutually perpendicular planes. They occur with the same frequency ν_3. The vibration is said to be **degenerate** and the **degree of degeneracy** is 2. For degenerate vibrations the normal modes are not uniquely defined by the geometric structure of the molecule. The modes of frequency ν_3 shown in Fig. 14.22 are only one possible pair.

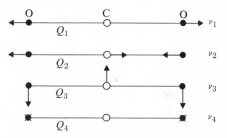

Fig. 14.22. The normal modes of vibration of a carbon-dioxide molecule.

Let us illustrate the mathematical formulation of the problem of determining the normal modes of vibration.

The N nuclei each exert forces on one another. In the equilibrium configuration the resultant of all forces acting on a given particle is zero. The ith nucleus is displaced from its equilibrium position by an amount Δs_i. Ordinarily we would write

$$F_{x_i} = -k_x \Delta x_i, \qquad F_{y_i} = -k_y \Delta y_i, \qquad F_{z_i} = -k_z \Delta z_i$$

for the components of the restoring force. However, in a many-nucleus system we cannot often make the assumption that movement of a nucleus in, say, the x direction is independent of the way in which it moves in the y or z directions. Therefore, to be completely general, we make allowance for the fact that the components of the restoring force may depend on all three components of the displacement by writing

$$F_{x_i} = -k_{xx}^{ii}\Delta x_i - k_{xy}^{ii}\Delta y_i - k_{xz}^{ii}\Delta z_i$$

$$F_{y_i} = -k_{yx}^{ii}\Delta x_i - k_{yy}^{ii}\Delta y_i - k_{yz}^{ii}\Delta z_i$$

$$F_{z_i} = -k_{zx}^{ii}\Delta x_i - k_{zy}^{ii}\Delta y_i - k_{zz}^{ii}\Delta z_i.$$

The meaning of, say, the force constant k_{yx} is as follows: the subscript yx tells us that we are considering the effect of the x component of the displacement on the y component of the restoring force; the superscript ii tells us that the restoring force on the ith nucleus is due to the displacement of the ith nucleus. The superscripts could be left off in the above equations, since they are all the same, but they are put in for completeness and as a preparation for the more general case.

If all N nuclei are displaced, then the force experienced by nucleus i will depend on the displacements of all the particles. The x component of the force experienced by nucleus i can be written

$$F_{x_i} = \sum_{j=1}^{N} (-k_{xx}^{ij}\Delta x_j - k_{xy}^{ij}\Delta y_j - k_{xz}^{ij}\Delta z_j).$$

The coefficient k_{xy}^{ij} determines how the x component of the force on nucleus i depends on the y component of the displacement of nucleus j. It can be shown that

$$k_{xy}^{ij} = k_{yx}^{ji}.$$

To find the normal vibrations, we must impose the condition for simultaneous simple harmonic motion for all the nuclei with the same frequency. For a particle i carrying out simple harmonic motion of frequency v, the displacement Δs_i is given by

$$\Delta s_i = \Delta s_i^0 \cos (2\pi vt + \phi)$$

where Δs_i^0 is the amplitude and ϕ a phase angle. The restoring force is then given by

$$F_i = m_i a_i = m_i \frac{d^2(\Delta s_i)}{dt^2}$$

$$= -4\pi^2 m_i v^2 \Delta s_i^0 \cos (2\pi vt + \phi)$$

$$= -4\pi^2 v^2 m_i \Delta s_i,$$

where m_i is the mass of the particle. This holds also for the component of motion in any direction. The normal vibrations are, therefore, given by the simultaneous solution of $3N$ equations of the form

$$4\pi^2 v^2 m_i \Delta x_i = \sum_{j=1}^{N} (k_{xx}^{ij}\Delta x_j + k_{xy}^{ij}\Delta y_i + k_{xz}^{ij}\Delta z_j).$$

The solution of these equations yields the form of the normal vibrations Q_i and their frequencies v_i. Six of the frequencies (five for a molecule that is linear in its equilibrium position) are zero. These correspond to **nongenuine normal vibrations** in which simply a translation along any one of the three coordinate axes or a rotation of the molecule as a whole about a principal axis takes place.

Note that a normal coordinate Q_i implies $3N$ displacements Δx_i, Δy_i, Δz_i corresponding to a normal frequency v_i. A normal coordinate may be written as

$$Q_i = Q_i^0 \cos(2\pi v_i t + \phi_i).$$

In a general vibration of the molecule, all of the normal vibrations are taking place simultaneously with different amplitudes and phases. This concept is illustrated in the film *Molecular Motions*[3].

PROBLEMS

1. (a) Find the position of the center of mass for the diatomic molecules listed in Table 14.1, assuming the internuclear distance to be 10^{-10} m in each case.
   (b) Find the reduced mass μ for each molecule. Plot a graph of μ versus M_2. Comment on the graph.

Table 14.1   Nuclear masses for some diatomic molecules

Molecule	M_1 (a.m.u.)	M_2 (a.m.u.)
H_2	1	1
HD	1	2
HF	1	19
HCl	1	35
HBr	1	79
HI	1	126

2. The ammonia molecule (NH_3) shown in Fig. 14.23 is a pyramid structure. The three hydrogen atoms define an equilateral triangle; the nitrogen atom is a distance $h = 0.38 \times 10^{-10}$ m above this plane and equidistant from each of the hydrogen atoms. Locate the center of mass of the molecule.

3. The ethylene molecule (C_2H_4), a planar molecule, is sketched in Fig. 14.24. The molecular parameters are:

$$CC = 1.35 \times 10^{-10} \text{ m}$$

$$CH = 1.07 \times 10^{-10} \text{ m}$$

$$\angle HCH = 120°.$$

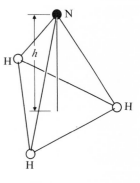

Fig. 14.23

[3] *Molecular Motions*, CHEM Study Film #4115. Distributed by Modern Learning Aids, 1212 Avenue of the Americas, N.Y. 10036.

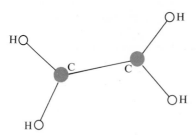

Fig. 14.24

The molecule is symmetric about the CC axis.

(a) Calculate the position of the center of mass of the molecule.

(b) Calculate the moments of inertia of the molecule about

(1) an axis through the center of mass parallel to the CC axis.

(2) an axis through the center of mass perpendicular to the CC axis in the plane of the molecule.

(3) an axis through the center of mass perpendicular to the CC axis and perpendicular to the plane of the molecule.

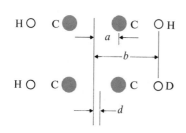

Fig. 14.25

*4. The acetylene (C_2H_2) molecule and its deuterated form (C_2HD) are shown in Fig. 14.25. The distance d represents the shift of the position of the center of mass caused by the replacement of an H atom by a D atom. The moments of inertia about axes through the centers of mass, perpendicular to the internuclear axes, are measured as 23.65×10^{-47} kg $\cdot$ m^2 for C_2H_2 and 28.04×10^{-47} kg $\cdot$ m^2 for C_2HD.

(a) Calculate d.

(b) Calculate the bond lengths r(C—H) and r(C=C) between the C and H and C and C atoms, respectively, in C_2H_2.

*5. A methane-type molecule is shown in Fig. 14.26. The carbon (C) atom is located at the center of the cube and the four other (X) atoms at the corners as shown. The CX distance is 1.09×10^{-10} m. Locate the center of mass of this structure for each of the following deuterated forms:

(a) $X_1 = X_2 = X_3 = X_4 = H$;   that is, CH_4.

(b) $X_1 = X_2 = X_3 = H$; $X_4 = D$;   that is, CH_3D.

(c) $X_1 = X_2 = H$; $X_3 = X_4 = D$;   that is, CH_2D_2.

(d) $X_1 = H$; $X_2 = X_3 = X_4 = D$;   that is, CHD_3.

(e) $X_1 = X_2 = X_3 = X_4 = D$;   that is, CD_4.

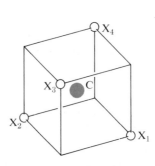

Fig. 14.26

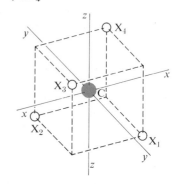

Fig. 14.27

*6. Calculate the moments of inertia I_x, I_y, and I_z about axes through the centers of mass for the CH_4, CH_2D_2, and CD_4 molecules introduced in Problem 5. Take x, y, and z axes to be perpendicular to the cube faces as indicated in Fig. 14.27.

7. A hydrogen-fluoride (HF) molecule is rotating at a frequency of 1.25×10^{12} Hz. Estimate the rotational frequency of a deuterated hydrogen-fluoride (DF) molecule rotating with the same rotational angular momentum.

8. A carbon monoxide (CO) molecule rotates at a frequency of 1.16×10^{11} Hz. The internuclear separation is 1.12×10^{-10} m. Calculate
(a) the reduced mass of the molecule.
(b) the rotational angular momentum.
(c) the rotational kinetic energy.

9. The reduced mass of a hydrogen (H_2) molecule is 8.3×10^{-28} kg. Calculate the mass of 6.02×10^{23} hydrogen molecules.

10. The carbon-dioxide (CO_2) molecule (see Fig. 14.28) is observed to rotate with a frequency of 2.68×10^{11} Hz and to have a moment of inertia of 71.1×10^{-47} kg·m². Assuming a rigid-rotator molecule,
(a) calculate the rotational angular momentum of the molecule.
(b) calculate the separation a between the carbon and oxygen atoms.

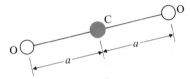

Fig. 14.28

11. A nitrogen (N_2) molecule, internuclear separation 1.09×10^{-10} m, rotates at a frequency of 1.21×10^{11} Hz. As a result of an interaction that lasts for 10^{-12} sec, the molecule jumps to a higher rotational state and rotates at a frequency of 3.63×10^{11} Hz. Calculate the average torque and average angular acceleration effective during the interaction.

12. A cyanide (CN) molecule vibrates at a frequency of 6.20×10^{13} Hz. Calculate the force constant of the molecule.

13. Estimate the ratio of the vibrational frequencies of a hydrogen iodide (HI) molecule to a deuterated hydrogen iodide (DI) molecule.

14. A diatomic molecule undergoes vibrational motion along the line joining the atoms with a frequency of 4×10^{13} Hz. One atom is of mass m, and its maximum displacement from equilibrium is 2.0×10^{-12} m.
(a) What is the maximum displacement of the second atom from its equilibrium position if its mass is $1.5m$?
(b) If the energy associated with the vibrational motion is 2.8×10^{-20} J and assuming simple harmonic motion of each atom about its equilibrium position, find the force constant.

15. Two spheres, each of mass 2 kg, are joined together by a spring of force constant 800 N·m^{-1} and unstretched length 0.20 m, as shown in Fig. 14.29.

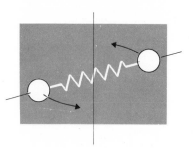

Fig. 14.29

(a) If the system is rotated about its center of mass at a frequency of 3 rev · sec^{-1}, calculate the extension of the spring.

(b) Determine the upper limit to the frequency of rotation.

16. A carbon-monoxide (CO) molecule is vibrating and rotating simultaneously. The rotational frequency is 1.16×10^{11} Hz, and the equilibrium internuclear separation is 1.12×10^{-10} m. Taking the change in internuclear separation resulting from the molecular rotation to be 3.5×10^{-16} m, estimate the vibrational frequency of the molecule.

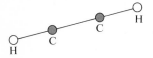

Fig. 14.30

*17. Write an expression for the rotational energy of a carbon tetrachloride (CCl$_4$) molecule in terms of the C—Cl distance. This is a spherical-top molecule like methane (CH$_4$).

18. Write an expression for the rotational energy of an acetylene (C$_2$H$_2$) molecule in terms of the internuclear distances shown in Fig. 14.30 and the mass of an H atom.

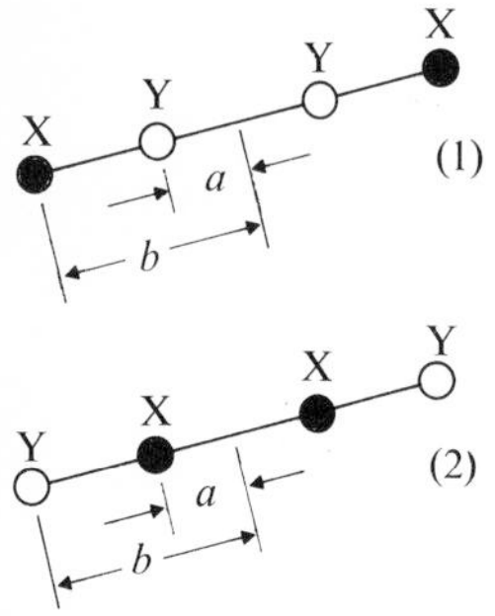

19. For the hypothetical molecule shown in Fig. 14.13, sketch three possible translational normal modes and three possible rotational normal modes.

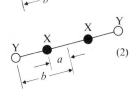

Fig. 14.31

20. Two possible configurations (1) and (2) of a symmetric, linear, four-atom molecule are shown in Fig. 14.31. Compare the rotational frequencies associated with the two configurations for a fixed rotational angular momentum. Consider the special case $2m_y = 3m_x$, $b = 2a$.

21. Which of the diagrams in Fig. 14.32 could represent a rotational normal mode of a planar molecule consisting of three identical atoms at the vertices of an equilateral triangle? Which of the diagrams could represent a translational normal mode?

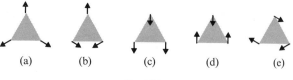

(a)   (b)   (c)   (d)   (e)

Fig. 14.32

22. State the number of normal modes of vibration associated with each of the following molecules: hydrogen sulfide (H$_2$S), hydrogen cyanide (HCN)— a linear molecule, ammonia (NH$_3$), and formamide (NH$_2$COH).

23. The nitrous oxide (N$_2$O) molecule has four normal modes of vibration. What can you deduce about its structure?

*24. Consider an object of mass m moving under the action of four springs as discussed in Section 14.7. Deduce the motion of m if $v_x/v_y = \frac{1}{2}$.

25. Four normal vibrational frequencies exist for the methane (CH_4) molecule. What are the degeneracies of these modes, assuming that one mode is non-degenerate and that no mode is more than 3-fold degenerate.

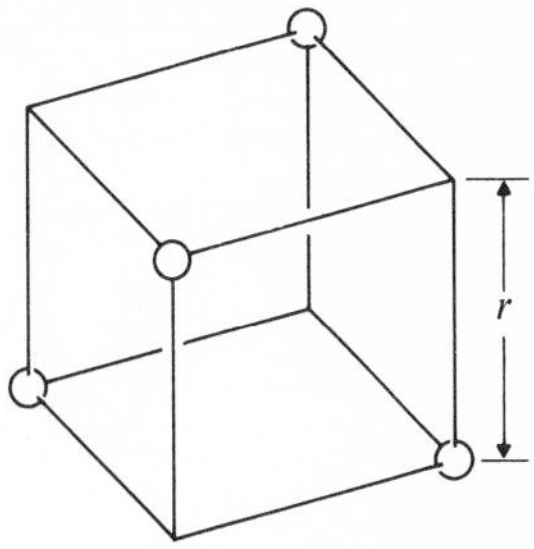

26. A tetrahedral molecule has four atoms of equal mass M situated at the corners of the cube, as shown in Fig. 14.33. Assume the edge of the cube to be of length r.
(a) Locate the center of mass of the molecule.
(b) Calculate the moment of inertia of the molecule about an axis drawn through the center of mass and perpendicular to one of the faces of the cube.
(c) About what axis can the molecule rotate so that a rotation through 120° would leave the configuration unaltered?
(d) State the number of degrees of vibrational freedom of the molecule.
(e) Sketch any two possible types of vibrational motion for the molecule. Note that the center of mass must remain stationary and that the molecule must not rotate as a whole.

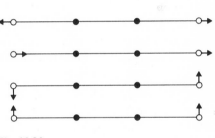

Fig. 14.33

27. Normal vibrations for a linear cyanogen (C_2N_2) molecule are partially indicated in Fig. 14.34; that is, the motion of the nitrogen atoms is shown.
(a) Deduce the corresponding motions of the carbon atoms for each of the normal vibrations.
(b) What can be said about the degeneracies of each of these modes?

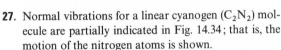

28. One of the normal modes of vibration of the linear hydrogen isocyanide (CHN) molecule is shown in Fig. 14.35. Deduce the corresponding normal vibration of

Fig. 14.34

the linear cyanogen chloride (CClN) molecule, the equilibrium configuration of which is shown in Fig. 14.36. All distances are given in units of 10^{-10} m.

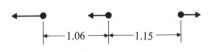

Fig. 14.35

Fig. 14.36

15 Introduction to wave motion

15.1 INTRODUCTION

We have all observed the disturbance resulting from a stone being dropped into quiet water. A characteristic pattern of concentric circles spreads out from the point of impact of the stone in the water. This pattern is an example of a wave. There are many types of waves other than water waves, such as mechanical waves, acoustical waves, and electromagnetic waves. All waves have certain similarities in their behavior. In this chapter we wish to consider the characteristic behavior of waves and then to describe this behavior in mathematical terms.

We shall find that waves are periodic disturbances that may be characterized by a small number of parameters, such as frequency, amplitude, wavelength, etc. One particularly simple type of periodic variation is a sinusoidal one. Not only is it relatively simple, but it is also extremely useful, since, as we shall see, it is always possible to write complex periodic variations as a sum of sinusoidal variations. We have seen in Section 11.4 that in simple harmonic motion the position of a particle varies sinusoidally with time. Therefore, sinusoidal waves can be generated by particles undergoing simple harmonic motion.

15.2 SOME SIMPLE EXAMPLES OF WAVE BEHAVIOR

A wave machine is a particularly useful device for the illustration of wave behavior[1]. A typical machine consists of a metal backbone with a number of crosspieces fastened to it at their centers and spaced equally along the backbone. This structure is supported in bearings that permit the structure to oscillate (see Figs. 15.1 and 15.2).

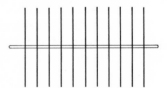

To generate a wave pulse we can raise the first crosspiece upward to any desired height and then return it to the rest position in one smooth continuous motion. Such a pulse will be seen to travel back and forth along the machine with a gradually decreasing amplitude, until eventually

Fig. 15.1. A metal backbone with 11 crosspieces fastened to it.

the pulse dies out completely. This gradual reduction of the pulse height is called **damping** and occurs as the wave gives up its energy in frictional losses to the air, to the bearing, and to the twisting and untwisting of the backbone itself. This decrease in amplitude is illustrated in Fig. 15.3.

Another basic property of waves is the **constancy of speed** with which they travel in a particular medium. By timing the travel of wave pulses

[1] J. N. Shive, *Similarities in Wave Behavior*, Bell Telephone Laboratories, Inc. Available from the Science Department of Allegri-Tech, Nutley, N.J.

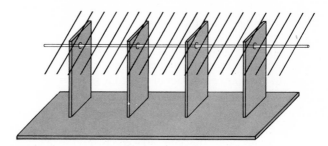

Fig. 15.2. A wave machine.

of different shapes and amplitudes along the machine, we can show that this is true. By using various wave machines that have backbones with different resistances to twisting and crosspieces of different masses, we can show that the speed depends upon the properties of the medium through which the wave is passing. The speed of the waves is found to increase if either (or both) the resistance of the backbone to twisting is increased or the mass of the crosspiece is decreased.

Fig. 15.3. Wave pulse traveling to the right with a gradually decreasing amplitude.

We have already noted that when a wave pulse reaches the end of the machine it is reflected back along the machine. **Reflection** is a common property of all waves. Let us consider reflection more closely. First we consider the transmission of a pulse generated at the left end along a

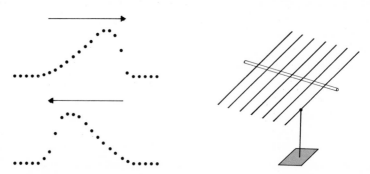

Fig. 15.4. The reflection of a wave pulse from a free end of a wave machine.

Fig. 15.5. The end of the wave machine rigidly fixed by a clamp.

machine with the right end unrestrained. The
reflected pulse is a reversed replica of the original
pulse (see Fig. 15.4). If, however, the right end of
the wave machine is rigidly fixed by a clamp, the
reflected pulse is an inverted replica of the original
(see Figs. 15.5 and 15.6).

We may study the reflection of a wave pulse
from the end of a wave machine that is neither
free nor rigidly clamped by connecting a device
called a dash pot (see Fig. 15.7) to one end of the
machine (see Fig. 15.8). The dash pot consists
of a piston suspended in a container of water.
The resistance of the water to the movement
of the piston through it prevents the crosspiece at
the end from moving freely, although it does

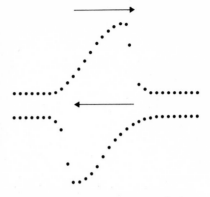

Fig. 15.6. The reflection of a wave pulse
from a fixed end of the wave machine.

not entirely prevent it from moving. By appropriately adjusting the
point of connection of the dash pot to the wave machine, we can eliminate
the reflected wave entirely. If we consider the wave machine as one system
and the dash pot as a second system and if we have adjusted the connection
between the two systems to eliminate reflection, we can say that we have
matched the impedance of the two systems. The energy that we manually
put into the wave machine to generate the pulse has traveled down the
machine, and the energy not lost by damping has been absorbed completely by the dash pot. We speak of the pulse as a **traveling wave disturbance**
that **transports energy** along the machine to the dash pot.

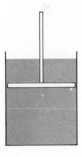

Fig. 15.7. A dash pot.

If instead of the wave pulse we were to generate a continuous wave,
for example, by connecting a motor and crank assembly to the wave
machine, as shown in Fig. 15.9, energy would be continually pumped into
one end of the machine by the motor, transported by the traveling wave
along the machine, and either absorbed by the dash pot or absorbed by
damping.

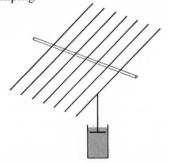

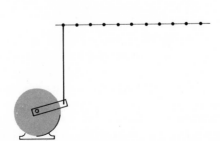

Fig. 15.8. The connection of the dash pot
may be adjusted to eliminate reflection.

Fig. 15.9. A motor and crank assembly
may be used to generate a continuous
wave.

Let us now consider the coupling together of two wave machines having identical backbones but crosspieces of different length. We terminate the second machine with a dash pot adjusted to eliminate reflection (see Fig. 15.10). If a pulse is started down the machine with the larger crosspieces, then part of the energy is reflected and part is transmitted when it reaches the junction between the two machines. Both the reflected and the transmitted pulses are upright; the transmitted pulse travels with increased speed and spreads out. This is shown in Fig. 15.11. The shorter the crosspieces of the second machine relative to those of the first, the more the junction looks like a free end and the larger the amplitude of the reflected pulse.

If the wave machines are interchanged and the pulse started down the machine with the shorter crosspieces, then the transmitted pulse is upright and the reflected wave is inverted; the reflected pulse travels more quickly than the transmitted pulse. This is illustrated in Fig. 15.12. The longer the crosspieces of the second machine, the more the junction looks like a fixed end and the larger the amplitude of the reflected pulse.

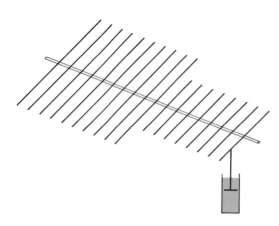

Fig. 15.10. Two machines with crosspieces of different length coupled together.

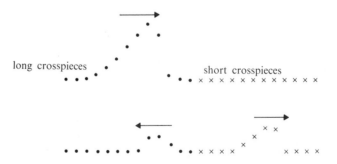

long crosspieces   short crosspieces

Fig. 15.11. Partial reflection of a wave pulse at a discontinuity. The pulse was started down a wave machine with long crosspieces.

These experiments illustrate the partial reflection of energy at a region of discontinuity in a transmitting medium. If one is interested in the efficient transmission of energy, it is undesirable to lose some of it by partial reflections in the transmission medium. Impedance matching devices are used to smooth over the discontinuities that result in partial reflections. In the present case, a tapered section of the machine can be used to mechanically match the impedance characteristics of two machines

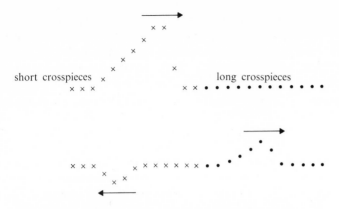

short crosspieces   long crosspieces

Fig. 15.12. Partial reflection of a wave pulse at a discontinuity. The pulse was started down a wave machine with short crosspieces.

Fig. 15.13. A tapered section of machine can be used to match the impedance characteristics of two machines having different length crosspieces.

having crosspieces of different length, thereby eliminating the partial reflections. The tapered section might be referred to as a **matching transformer**. Such a tapered section is shown in Fig. 15.13.

By introducing the tapered section, we replace a single sharp discontinuity with a series of small discontinuities. We have already pointed out that the amplitude of a partially reflected wave at a discontinuity depends on the sharpness of that discontinuity. By making the individual discontinuities sufficiently small, the reflected wave is eliminated, for all practical purposes.

15.3   THE EQUATION OF A SINUSOIDAL TRAVELING PLANE WAVE

Continuous sinusoidal plane waves are the easiest type of waves to describe mathematically. Sinusoidal plane waves will be generated in the wave machine if the motor and crank assembly (see Fig. 15.9) cause the end of the crosspieces to which they are connected to execute simple harmonic motion. Therefore, the displacement $\Delta y(t)$ of the end of the first crosspiece as a function of time will be of the form

$$\Delta y(t) = A \sin 2\pi\nu t,$$

where ν is the frequency of the simple harmonic motion in hertz and A is the maximum displacement of the crosspiece from its equilibrium position; the displacement is taken to be along the y axis, and it is assumed that at $t = 0$ the crosspiece is in its equilibrium position. As a result of the displacement of the first crosspiece by the motor and crank assembly, the other crosspieces are set into simple harmonic motion one after another. The motion is not passed on instantaneously to all crosspieces but rather each successive crosspiece is stimulated at a slightly later time than the previous one. That is, each successive crosspiece is slightly **out of phase** with the preceding one. Therefore, the displacement $\Delta y(x, t)$ at time t of a crosspiece situated a distance x from the one being initially stimulated will be of the form

$$\Delta y(x, t) = A \sin 2\pi v[t - \phi(x)],$$

where $\phi(x)$ is the time required for the stimulation to be propagated a distance x. Note that we have assumed that the maximum displacement at position x is also A; that is, we have neglected damping. Assuming the rate of propagation to be constant and equal to v, we have

$$\phi(x) = \frac{x}{v}$$

and

$$\Delta y(x, t) = A \sin 2\pi \left(vt - \frac{vx}{v} \right).$$

All points along the wave that have the same displacement and are moving in the same direction are said to be **in phase**. The points labeled a, b, c, and d in Fig. 15.14, for example, are in phase. The distance measured along the direction of the wave between adjacent points that are in phase is called the **wavelength** and is denoted by λ in Fig. 15.14. Since vt is just the number of oscillations occurring during time t, vx/v is the number of oscillations occurring in time x/v. When $x = \lambda$ this must correspond to one oscillation, and

$$\frac{v\lambda}{v} = 1,$$

thereby showing that the speed of wave propagation v is given by

$$v = \lambda v.$$

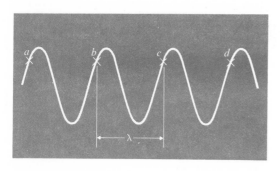

Fig. 15.14. Points a, b, c, and d are in phase.

Therefore, the equation of a **transverse sinusoidal traveling wave** being generated at the origin and traveling along the x axis may be written in the form

$$\Delta y(x, t) = A \sin 2\pi \left(vt - \frac{x}{\lambda} \right).$$

If the wave travels to the right, only positive values of x are allowed; if it travels to the left, only negative values of x are allowed. The maximum displacement at any point along the wave is A and is known as the **amplitude** of the wave. The quantity $2\pi(vt - x/\lambda)$ is called the **phase angle** of the wave at time t and position x.

It may readily be shown that, independent of the point of origin, any **sinusoidal plane wave** (that is, a wave traveling in one direction only) moving to the right may be written as

$$\Delta y(x, t) = A \sin 2\pi\left(vt - \frac{x}{\lambda} + \phi_0\right),$$

and any wave traveling to the left as

$$\Delta y(x, t) = A \sin 2\pi\left(vt + \frac{x}{\lambda} + \phi_0\right),$$

where ϕ_0 is an **initial phase angle** to account for the displacement at $x = t = 0$. The justification of this statement is left as an exercise for the reader.

The equation of a traveling plane wave may be written in a number of other forms by using one or more of the following relations:

$$\omega = 2\pi v, \tag{1}$$

where ω is the angular frequency in radians per second,

$$k = \frac{2\pi}{\lambda}, \tag{2}$$

where k is the **propagation constant** in meters^{-1}, and

$$T = \frac{1}{v}, \tag{3}$$

where v is the **period** of the wave in seconds.

We recall that we have been discussing undamped waves. Damping can be easily taken into account merely by replacing A with some appropriate function of time $A(t)$. However, we are concerned here primarily with the development of a wave disturbance in space and time. For that reason we shall neglect damping in all our discussions.

Example. A transverse sinusoidal wave of frequency 10 Hz and wavelength 1 m is traveling down a string.
(a) Plot a graph showing the wave at times $t = 0$, $t = T/4$, and $t = T/2$.
(b) Plot a graph showing the displacement of a particle at positions $x = 0$, $x = \lambda/4$, and $x = \lambda/2$ as a function of time.

Solution. The displacement is given by

$$\Delta y(x, t) = A \sin 2\pi\left(vt + \frac{x}{\lambda}\right).$$

Therefore,

$$\Delta y(x, t) = A \sin (20\pi t + 2\pi x).$$

(a) For $t = 0$, $y = A \sin 2\pi x.$

For $t = \dfrac{T}{4} = \dfrac{1}{4v} = \dfrac{1}{40}$ sec, $\Delta y = A \sin\left(\dfrac{\pi}{2} + 2\pi x\right).$

For $t = \dfrac{T}{2} = \dfrac{1}{2v} = \dfrac{1}{20}$ sec, $\Delta y = A \sin (\pi + 2\pi x).$

Since $\lambda = 1$ m, we should calculate Δy at intervals of 0.1 m in order to show the wave effectively. The calculations are summarized in Table 15.1. The displacements are plotted in Fig. 15.15 for all three times. Notice that the wave moves in the $-x$ direction as t increases.

Table 15.1   Displacement as a function of position and time

Position	$t = 0$	$t = \dfrac{T}{4}$	$t = \dfrac{T}{2}$
x (m)	Δy (m)	Δy (m)	Δy (m)
0.0	0	A	0
0.1	$0.57A$	$0.81A$	$-0.57A$
0.2	$0.95A$	$0.31A$	$-0.95A$
0.3	$0.95A$	$-0.31A$	$-0.95A$
0.4	$0.57A$	$-0.81A$	$-0.57A$
0.5	0	$-A$	0
0.6	$-0.57A$	$-0.81A$	$0.57A$
0.7	$-0.95A$	$-0.31A$	$0.95A$
0.8	$-0.95A$	$0.31A$	$0.95A$
0.9	$-0.57A$	$0.81A$	$0.57A$
1.0	0	A	0

(b) For $x = 0$, $\Delta y = A \sin 20\pi t.$

For $x = \dfrac{\lambda}{4} = \dfrac{1}{4}$ m, $\Delta y = A \sin\left(20\pi t + \dfrac{\pi}{2}\right).$

For $x = \dfrac{\lambda}{2} = \dfrac{1}{2}$ m, $\Delta y = A \sin (20\pi t + \pi).$

Since $T = 1/v = 0.1$ sec, we calculate Δy at intervals of 0.01 sec. A table of Δy values calculated at these intervals will be identical to Table 15.1, as you can readily verify. The displacements are plotted in Fig. 15.16 for all three positions. Notice that the peak displacement is reached at earlier times as x increases. This also indicates that the wave is traveling in the $-x$ direction.

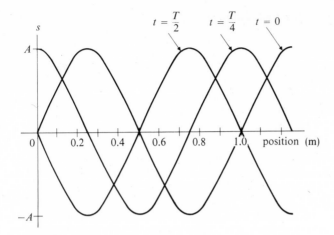

Fig. 15.15. The displacement as a function of position for times $t = 0, t = T/4, t = T/2$.

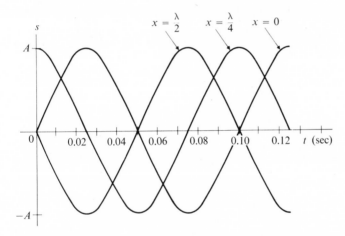

Fig. 15.16. The displacement as a function of time for positions $x = 0$, $x = \lambda/4, x = \lambda/2$.

15.4   THE DEPENDENCE OF THE SPEED OF TRAVELING MECHANICAL WAVES ON THE PROPERTIES OF THE TRANSMITTING MEDIUM

The manner in which the speed of propagation of mechanical waves depends upon the medium may be obtained directly by a mechanical analysis starting from Newton's second law for the motion of a particle in the medium through which the wave is passing. A second-order differential wave equation results which has as a possible solution the traveling wave that we have been discussing. The speed of propagation appears as a constant in the wave equation. Such an analysis for waves traveling along a string is carried out in the reference cited below[2]. We shall not reproduce such a detailed calculation here but shall rather use the powerful method of dimensional analysis to obtain the functional form of the dependence of the wave speed on the properties of the transmitting medium.

Let us digress for a moment and introduce the method of dimensional analysis by considering the motion of an object in the earth's gravitational field. It is known experimentally that the speed of a falling object near the earth's surface depends **only** on the distance it has fallen and the acceleration due to gravity. Therefore we can write

$$v = kg^x s^y,$$

where k is a dimensionless constant and x and y are to be determined by dimensional considerations. The condition to be satisfied by x and y is that the physical dimensions of the two sides of the equation must be the same.

The **dimensions of a physical quantity** give simply the functional dependence of that quantity on the basic dimensions of length L, mass M, and time T. For example, acceleration, which in the MKS system is measured in meters per second per second, has the dimensions of a length divided by the square of a time. We write

$$a \equiv (LT^{-2}).$$

Similarly, the reader can verify that

$$s \equiv (L)$$

$$v \equiv (LT^{-1}).$$

Therefore, in order that the above equation be dimensionally correct, it is necessary that

$$(LT^{-1}) \equiv (LT^{-2})^x (L)^y.$$

This equation is equivalent to the following two equations:

$$(L) \equiv (L^x)(L^y)$$

$$= (L^{x+y})$$

$$(T^{-1}) \equiv (T^{-2y}).$$

[2] G. Shortley and D. Williams, *Elements of Physics*, 4th Edition (Englewood Cliffs, N.J.: Prentice-Hall, Inc., 1965), Chapter 20.

That is,

$$x + y = 1,$$

$$2y = 1,$$

and

$$x = y = \tfrac{1}{2}.$$

Therefore,

$$v = k(gs)^{1/2}$$

or

$$v^2 = k^2 gs.$$

This may be compared with the theoretical expression

$$v^2 = 2gs.$$

We note that dimensional analysis does not provide a value for the dimensionless constant k.

Returning to the problem of the speed of the traveling waves, we must first decide which physical properties of the medium will be important for wave propagation. Let us consider a wave traveling along a string. As the wave moves, a distortion of the string is propagated along the string. The speed of propagation will clearly depend on the restoring forces brought to bear on the string by the distortion. The tension F in the string is a measure of these forces, and the wave speed v will depend on F. Also, the speed of propagation will depend on the inertia of the material in the medium that must be moved by the disturbance. That is, v will depend on the mass μ per unit length of the string. Assuming that F and μ are the only physical properties of the string upon which v depends, we can write

$$v = kF^x\mu^y,$$

where k is a constant of proportionality and x and y are to be determined by dimensional considerations. The condition to be satisfied by x and y is that the physical dimensions of the two sides of the equation must be the same. Since, as the reader can readily verify,

$$v \equiv (LT^{-1}),$$

$$\mu \equiv (ML^{-1}),$$

and

$$F \equiv (MLT^{-2}),$$

it is necessary that

$$(LT^{-1}) \equiv (MLT^{-2})^x(ML^{-1})^y,$$

in order that the above equations be dimensionally correct. This equation is equivalent to the following three equations:

$$(L) \equiv (L)^x(L^{-1})^y$$

$$(T)^{-1} \equiv (T^{-2})^x$$

$$(M)^0 \equiv (M)^x(M)^y.$$

Solving these equations gives $x = \frac{1}{2}$ and $y = -\frac{1}{2}$, so that

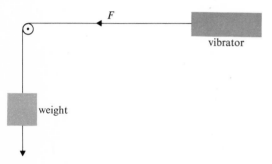

$$v = k\left(\frac{F}{\mu}\right)^{1/2}.$$

Fig. 15.17. Apparatus to study the speed of propagation of waves along a string.

This prediction may be readily checked using apparatus similar to that illustrated in Fig. 15.17. For a given string, the tension in the string may be varied by adding or subtracting weights; in this way the variation of wave speed with tension in the string may be investigated. For a given tension, the dependence of wave speed on the mass of the string may be studied by using strings of different thickness or composition. The apparatus of Fig. 15.17 may be used to generate **standing waves** (see Section 16.4), which in turn allow us to determine the wavelength λ very easily. Since the vibrator frequency is fixed, $v \propto \lambda$. Typical experimental results are shown in Fig. 15.18. The functional dependence of v on F and μ is verified by the results. The constant of proportionality as given by a detailed mechanical analysis is found to be unity.

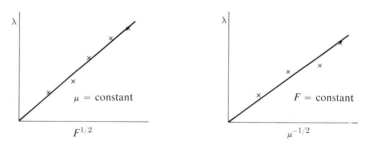

Fig. 15.18. Results obtained using the apparatus of Fig. 15.17 to show the dependence of wavelength (or speed, for a fixed frequency) on tension in the string and the mass per unit length of the string.

15.5　SOME OTHER FORMS OF TRAVELING WAVES

So far we have discussed only traveling plane waves in which oscillations take place in the plane perpendicular to the direction of propagation. Such waves are said to be **transverse**. Electromagnetic waves are an example of a transverse wave disturbance. It is also possible to produce waves in which the displacement is in the direction of propagation. Such waves are said to be **longitudinal**. Acoustical waves are an example of a longitudinal wave disturbance. A coil spring may be used to illustrate longitudinal waves. If one end of a coil spring is pulled out and then released, a longitudinal pulse of compression travels down the spring as illustrated in Fig. 15.19.

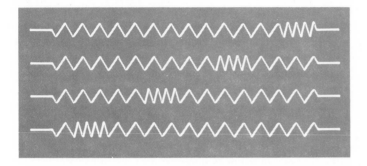

Fig. 15.19. A longitudinal pulse in a stretched spring.

The equations of sinusoidal longitudinal plane waves traveling in the positive and negative x directions, respectively, may be written in the forms

$$\Delta x(x, t) = A \sin 2\pi \left(vt - \frac{x}{\lambda} \right)$$

$$\Delta x(x, t) = A \sin 2\pi \left(vt + \frac{x}{\lambda} \right),$$

where $\Delta x(x, t)$ is the displacement from equilibrium position x at time t.

Example. A sound wave of frequency 200 Hz and of amplitude 10^{-3} m is generated by a loudspeaker. Write an equation representing the wave, assuming that the wave is traveling in one direction only.

Solution. Now $v = \nu\lambda$, where v is the speed of sound in air, which is 342 m · sec^{-1} at 18°C. Therefore,

$$\lambda = \frac{v}{\nu} = \frac{342}{200} = 1.71 \text{ m.}$$

The equation for the wave may be written as

$$\Delta x(x, t) = 0.001 \sin 2\pi \left(200t \pm \frac{x}{1.71} \right) \text{ m,}$$

where the plus sign is for a wave traveling in the $-x$ direction and the minus sign is for a wave traveling in the $+x$ direction.

The speed of propagation of longitudinal waves may be determined in the same manner as for transverse waves. For sound waves in a gas, the speed is

$$v = \left(\frac{\gamma P}{\rho} \right)^{1/2},$$

where P is the pressure, ρ the density, and γ a constant that depends on the gas (see Section 17.6). The reader can readily check the dimensional consistency of this equation. For **longitudinal waves in a solid bar**, the speed is

$$v = \left(\frac{Y}{\rho} \right)^{1/2},$$

where Y is the ratio of **stress** to **strain** in the bar or **Young's Modulus** (see Section 21.6) and ρ is the density of the solid. Strain is defined as the proportional change in a linear dimension of a solid in the direction in which a stress, or force per unit area, is applied.

Example. What is the wavelength of a sound wave produced in air at standard temperature and pressure by a source vibrating at 1000 Hz? Standard temperature and pressure are $273°K$ and $1.01 \times 10^5 \, N \cdot m^{-2}$, respectively.

Solution. The value of γ for air under these conditions is 1.402, and density is $1.293 \, kg \cdot m^{-3}$. Therefore, the speed of sound waves in air is

$$v = \left(\frac{\gamma P}{\rho} \right)^{1/2} = \left(\frac{1.402 \times 1.01 \times 10^5}{1.293} \right)^{1/2}$$

$$= 331 \, m \cdot sec^{-1}.$$

The wavelength of the sound wave is

$$\lambda = \frac{v}{v} = \frac{331}{1000}$$

$$= 0.331 \, m.$$

All waves are not plane waves. In Section 15.1, we noted that a stone dropped into quiet water gave rise to a disturbance that spread out in concentric circles about the point of impact of the stone. The simplest case of waves spreading out in all directions is the **spherical wave**. Longitudinal spherical acoustic waves can be produced by a sphere whose surface performs radial oscillations; transverse spherical light waves can be produced by a glowing source. Figure 15.20 is a plane section through a spherical wave disturbance, showing the position of the **wave front** (the initial portion or leading edge of the wave) at two instants of time t_1 and t_2, where the distances from the source S are r_1 and r_2, respectively. The energy carried by the waves is spread over the surfaces of spheres of ever increasing radii. It is clear that the energy transported through a unit area of space in the direction of propagation of the wave must decrease with distance from the source. The ratio of the energies per unit area at positions r_1 and r_2 must be inversely proportional to the ratio of the areas of the spherical wavefronts at these positions; that is,

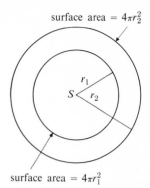

surface area $= 4\pi r_2^2$

surface area $= 4\pi r_1^2$

Fig. 15.20. A plane section through a spherical wave disturbance; the wave front is at a distance r_1 from the source at time t_1 and at a distance r_2 from the source at time t_2.

$$\frac{I_1}{I_2} = \frac{4\pi r_2^2}{4\pi r_1^2}$$

$$= \frac{r_2^2}{r_1^2}.$$

where I_1 and I_2 are the energies passing through a unit area per unit time. We see that the energy per unit area per unit time varies inversely as the square of the distance from the source.

In order to write an equation to describe a spherical wave, we require the relation between the energy transported by the wave and the amplitude of the wave. The energy transported per second is known as the **intensity**. We can demonstrate this relation for a medium that obeys **Hooke's law** (Sections 11.2 and 21.6), which says that the displacement of any particle in the medium is proportional to the force producing the displacement. The energy transported by a wave is just the energy expended in getting it started in the first place. A wave of a given amplitude, containing a given number of wavelengths, contains a definite amount of energy. If the force F required to generate a wave of amplitude A is

$$F = \kappa A,$$

then the work W expended in producing the displacement is

$$W = \int_0^A \kappa x \, dx$$

$$= \frac{\kappa A^2}{2}.$$

Therefore, the energy transported by a wave is proportional to the square of the wave amplitude. Hence, for a spherical wave, in order that the total energy over the wave front remains constant, the wave amplitude must vary inversely as the distance from the source. The equation of an outgoing longitudinal spherical wave may be written as

$$\Delta r(r, t) = \frac{A_0}{r} \sin 2\pi\left(vt - \frac{r}{\lambda}\right),$$

where $\Delta r(r, t)$ is the displacement at a radial distance r at time t and A_0 is the wave amplitude at unit distance from the source.

Example. The intensity of solar radiation at a point in space that is at the mean distance of the earth from the sun is 1.35×10^3 J $\cdot$ m$^{-2} \cdot$ sec^{-1}. This flux of energy is known as the **solar constant**. Determine the solar constant for each of the other planets in the solar system.

Solution. Mercury, the closest planet to the sun, is 36×10^6 miles from the sun. Since the earth is 93×10^6 miles from the sun, we have for the solar constant of Mercury

$$I_M = \left(\frac{r_E}{r_M}\right)^2 I_E,$$

where I_E is the solar constant for the earth. Therefore,

$$I_M = \left(\frac{93 \times 10^6}{36 \times 10^6}\right)^2 (1.35 \times 10^3)$$

$$= 9.0 \times 10^3 \text{ J} \cdot \text{m}^{-2} \cdot \text{sec}^{-1}.$$

The mean solar constants for all the planets are summarized in Table 15.2.

Table 15.2   Solar constants for the planets

Planet	Mean radius of orbit ($\times 10^6$ miles)	Solar constant) ($J \cdot m^{-2} \cdot sec^{-1}$)
Mercury	36	9000
Venus	67	2600
Earth	93	1350
Mars	142	580
Jupiter	485	50
Saturn	885	15
Uranus	1780	3.7
Neptune	2800	1.5
Pluto	~3700	~0.85

Our nearest neighbors, Venus and Mars, receive twice as much and less than half as much energy per unit area per second, respectively, than does the earth.

15.6   REFLECTION AND TRANSMISSION COEFFICIENTS

We have seen that when a traveling wave strikes a discontinuity in the transmitting medium, part of the energy is reflected. Since the total energy in the wave must be conserved at the discontinuity, the sum of the reflected energy E_r and the transmitted energy E_t must equal the incident energy E_i. That is,

$$E_r + E_t = E_i.$$

Therefore, for a plane wave

$$A_r^2 + A_t^2 = A_i^2,$$

where A_r, A_t, and A_i are the amplitudes of the reflected, transmitted, and incident waves, respectively. We can define a **reflection coefficient** r by

$$r = \left(\frac{A_r}{A_i}\right)^2$$

and a **transmission coefficient** t by

$$t = \left(\frac{A_t}{A_i}\right)^2.$$

Clearly

$$r + t = 1.$$

Example. Two wave machines (1) and (2) characterized by speeds of propagation v_1 and v_2 are coupled together. The reflection coefficient at the interface between the two machines is 0.05. Write equations for incident, reflected, and transmitted sinusoidal waves on the combined wave machine.

Solution. Assume that the discontinuity occurs at $x = 0$ and that the incident wave of amplitude A_0 travels from left to right on machine (1).

Since $r + t = 1$ and $r = 0.05$, then $t = 0.95$. Therefore,

$$A_r = (0.05)^{1/2} A_0 = 0.224 A_0$$

and

$$A_t = (0.95)^{1/2} A_0 = 0.975 A_0,$$

where A_r and A_t are the amplitudes of the reflected and transmitted waves, respectively. The wave equations are:
the **incident wave equation**

$$\Delta y_i(x, t) = A_0 \sin \left[v \left(t - \frac{x}{v_1} \right) + \phi \right], \qquad x < 0,$$

the **reflected wave equation**

$$\Delta y_r(x, t) = 0.224 A_0 \sin \left[v \left(t + \frac{x}{v_1} \right) + \phi \right], \qquad x < 0,$$

the **transmitted wave equation**

$$\Delta y_t(x, t) = 0.975 A_0 \sin \left[v \left(t - \frac{x}{v_2} \right) + \phi \right], \qquad x > 0.$$

15.7   THE DOPPLER EFFECT

A man is standing at the side of a road. An approaching automobile sounds its horn, and as it passes by the man notices a drop in the

pitch of the sound. (For a wave of a single frequency, the terms pitch and frequency are synonymous.) This is an example of the phenomenon known as the **Doppler effect**, after Christian Doppler. The same effect is noticed if the car sounding its horn is stationary and the man moves past the stationary car in a moving vehicle. The magnitude of the change in frequency is different, though; we shall consider the two cases separately, assuming for simplicity that the air is stationary.

Moving source

The source, such as a moving car, approaches the point of detection with speed u, whereas the sound waves spread out from the source at speed v fixed with respect to the air through which the waves travel (see Fig. 15.21). An observer, stationary with respect to the automobile, measures a frequency v_0 for the sound wave emitted by the horn. The associated wavelength is

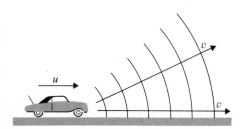

$$\lambda_0 = \frac{v}{v_0}.$$

Fig. 15.21. Sound waves travel at speed v after being emitted by a car moving at speed u.

Figure 15.22 shows the wave front from the moving source at a number of times; wave fronts 1, 2, 3, and 4 were emitted when the car was at positions S_1, S_2, S_3, and S_4, respectively. An observer at rest at P measures the apparent wavelength λ as shown. During a time interval Δt the source emits $v_0 \Delta t$ waves, and these waves must occupy a region of space $v\Delta t - u\Delta t$ in length, where $v\Delta t$ is the distance traveled by the wave emitted at the beginning of the time interval and $u\Delta t$ is the distance the source has traveled during the time interval. Since we have a total of $v_0 \Delta t$ waves in a distance $v\Delta t - u\Delta t$, the apparent wavelength is

$$\lambda = \frac{v - u}{v_0}.$$

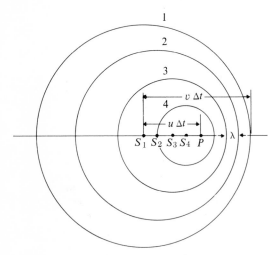

Fig. 15.22. Wave fronts 1, 2, 3 and 4 were emitted from a moving source when the source was at positions S_1, S_2, S_3 and S_4, respectively. After time Δt the source is at position P.

The ratio of the observed wavelength to the wavelength for a stationary source is

$$\frac{\lambda}{\lambda_0} = \left(\frac{v - u}{v_0} \right) \left(\frac{v_0}{v} \right) = \frac{v - u}{v},$$

and, since

$$\lambda = \frac{v}{\nu},$$

$$\frac{\nu}{\nu_0} = \frac{v}{v - u}$$

or

$$\nu = \nu_0 \frac{1}{1 - \dfrac{u}{v}}.$$

For an approaching source, u is positive, and the apparent frequency is greater than the frequency at emission; for a receding source, u is negative, and the apparent frequency is less than the frequency at emission. Therefore, when the car passes the man he observes the pitch of the sound coming from the horn to decrease rather suddenly to a lower value.

Moving observer

Let us suppose that the source is stationary at S (see Fig. 15.23) and the man is approaching the source with speed u. During a time interval Δt, the wave front advances a distance $v\Delta t$, and the man moves through a distance $u\Delta t$. The speed of the waves relative to the man is

$$v + u.$$

If ν_0 is the frequency of the emitted waves measured for a stationary source, the wavelength of the emitted waves will be

$$\lambda = \frac{v}{\nu_0}.$$

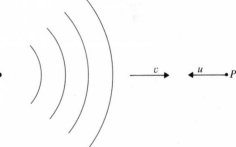

Fig. 15.23. Sound waves are spreading out at speed v from a stationary source S while an observer is moving toward the source at speed u.

The man will measure the same wavelength, since the source is stationary with respect to the air, which is the medium of propagation of the sound wave. The man, however, observes a speed $v + u$ for the waves relative to himself and concludes that the wavelength is given by

$$\lambda = \frac{v + u}{\nu'},$$

where v' is the observed frequency. Therefore,

$$\frac{v + u}{v'} = \frac{v}{v_0}$$

or

$$v' = v_0 \left(\frac{v + u}{v} \right)$$

$$= v_0 \left(1 + \frac{u}{v} \right).$$

As the man approaches the automobile, u is positive, and the apparent frequency is greater than the emitted frequency; as he recedes, u is negative, and the apparent frequency is less than the emitted frequency. As the man passes the automobile, he will observe the pitch of its horn to drop.

Although the effect is qualitatively the same for a moving observer and stationary source as for a stationary observer and moving source, there is a quantitative difference. The ratio of the apparent frequency v' for a observer moving at speed u_m to the frequency v for a source moving at speed u_A is

$$\frac{v'}{v} = \frac{v_0 \left(1 + \dfrac{u_m}{v} \right)}{v_0 \left(1 - \dfrac{u_A}{v} \right)^{-1}}$$

$$= \left(1 + \frac{u_m}{v} \right) \left(1 - \frac{u_A}{v} \right).$$

If we take $u_m = u_A = u$, which indicates the same relative speed of source and observer for both cases, we have

$$\frac{v'}{v} = 1 - \frac{u^2}{v^2},$$

which is less than unity. This should not bother us, particularly since a basic asymmetry exists between the two situations: sound is propagated at a fixed frequency relative to the air through which it is traveling, but in one case the source is stationary with respect to the air, whereas in the other case it is moving.

Example. An ambulance with siren screaming moves along a street at 18 $m \cdot sec^{-1}$. What is the ratio of the frequencies contributing to the sound wave from the siren that are observed when the ambulance is moving away from the man to the frequencies that are observed as the ambulance is approaching? The speed of sound in still air at 18°C is 342 $m \cdot sec^{-1}$.

Solution. We assume that when the ambulance is some distance from the observer, it can be considered to be coming directly at or receding directly away from the observer. Under this assumption, the relative speed of source and observer is just the speed of the ambulance. The observed frequency at approach for a particular frequency v_0 of the source is

$$v_a = v_0 \left(\frac{1}{1 - \dfrac{u}{v}} \right),$$

where $u = 18$ m · sec^{-1} and $v = 342$ m · sec^{-1}. The observed frequency for recession will be

$$v_r = v_0 \left(\frac{1}{1 + \dfrac{u}{v}} \right).$$

The ratio of these two frequencies is

$$\frac{v_r}{v_a} = \frac{v_0 \left(1 + \dfrac{u}{v} \right)^{-1}}{v_0 \left(1 - \dfrac{u}{v} \right)^{-1}} = \frac{1 - \dfrac{u}{v}}{1 + \dfrac{u}{v}} = \frac{v - u}{v + u}.$$

Note that this ratio is independent of the source frequency, so that all frequencies contained in the sound waves emitted by the source will change by the ratio

$$\frac{v_r}{v_a} = \frac{342 - 18}{342 + 18} = \frac{324}{360} = 0.90.$$

15.8 RELATIVISTIC DOPPLER EFFECT

If light waves are being observed rather than sound waves, our mathematical description of the observation expected when observer and source are in relative motion must be based upon the Lorentz transformation. Our preceding discussion has been based upon the Galilean transformation, which, as we discovered in Chapter 4, is valid only for speeds much less than the speed of light. Since all observers will measure a light signal to have the same speed independent of any relative motion of source and observer, the apparent frequency of a light signal for a given relative motion must be the same whether it is the source that moves or the observer. In the derivations of the equations describing the Doppler effect for sound waves, it was tacitly assumed that the frequency of a source is independent of the relative motion of source and observer. The apparent change in frequency was a result solely of the crowding together or spreading apart of the waves due to the relative motion of source and observer. Motion of the observer relative to the medium produced an apparent change in the speed of propagation.

Let us consider a light signal being emitted from time $t = 0$ to time $t = \tau$ by a source positioned at $x = 0$ in reference frame R (see Fig. 15.24). The leading portion of the light signal is received at time $t' = 0$ at the origin O' of a system R' moving in the x, x' direction with speed u. The trailing portion of the light signal is emitted in R at time $t = \tau$. The corresponding time in R' is obtained from the Lorentz transformation and is

$$t' = \gamma\left(t - \frac{ux}{c^2}\right)$$

$$= \gamma\tau,$$

since $x = 0$. At time $t = \tau$, the distance OO' as measured in R' is obtained from the Lorentz transformation and is given by

$$x' = \gamma(x - ut)$$

$$= -\gamma u\tau,$$

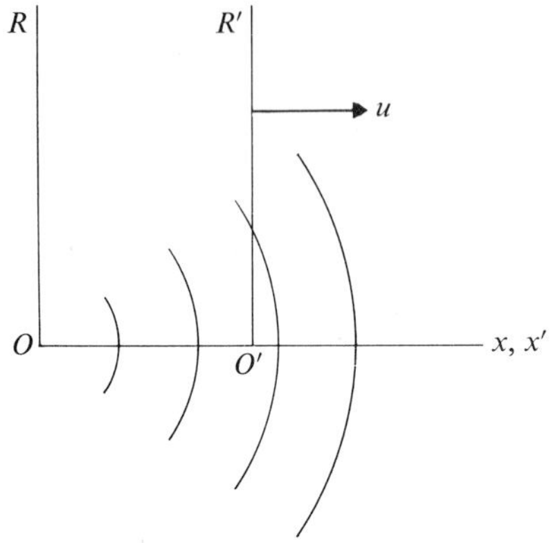

Fig. 15.24. A light signal emitted at time $t = 0$ at the position $x = 0$ in reference frame R is observed in reference frame R' moving at speed u with respect to frame R in the x, x' direction.

since $x = 0$. The trailing edge of the light signal emitted from the source at time $t = \tau$ must travel the distance $-\gamma u\tau$ to reach the origin O' of R'. The time taken will be

$$\Delta t' = \frac{|OO'|}{c} = \frac{\gamma u\tau}{c}.$$

The total time interval between the arrival at O' of the leading portion of the light wave emitted at $t = 0$ and the trailing portion emitted at $t = \tau$ will be

$$t' + \Delta t' = \gamma\tau + \frac{\gamma u\tau}{c}$$

$$= \gamma\tau\left(1 + \frac{u}{c}\right).$$

Since frequency is the reciprocal of a period of time

$$\frac{\nu}{\nu_0} = \frac{\tau}{t' + \Delta t'}$$

or

$$v = \frac{v_0\left(1 - \frac{u^2}{c^2}\right)^{1/2}}{1 + \frac{u}{c}}$$

$$= v_0\left(\frac{1 - \frac{u}{c}}{1 + \frac{u}{c}}\right)^{1/2},$$

where v_0 is the frequency of the light signal as measured in R and v is the frequency of the light signal as measured in R'. If the observer is receding from the source, u is positive and v is less than v_0. If the observer is approaching the source, u is negative and v is greater than v_0. Note that if we had solved the problem for the case of a stationary observer in R and a moving source in R' we would have obtained the same result. This is left as an exercise for the reader. In fact, this result was one of the requirements for a successful theory of relativity.

Frequencies corresponding to violet light are greater than those corresponding to red light. Therefore, the result of the Doppler effect is that a light source moving toward us appears more violet than normal, whereas a light source that is receding from us appears more red than normal. Astronomers have found that light from all stars is Doppler-shifted towards the red[3] and have concluded that the universe is expanding.

Example. In the early 1960's very strong radio sources in the sky were discovered to be connected with extremely distant, very faint stars. These objects were named **quasistellar radio sources** or **quasars**[4]. The most distant of them, a quasar numbered 3C9 in the star catalogs, was found to be approximately 8 to 10 billion light-years away from the earth and to be receding at a speed $u = 0.8c$. This large recession speed was measured by noting that a number of lines in the spectrum of light emitted from hydrogen gas in the quasar were Doppler shifted by a large amount from the normal frequency or wavelength. What is the expected red shift for a source receding at a speed of $u = 0.8c$?

Solution. The Doppler effect for a receding source in terms of frequency is

$$v = v_0\left(\frac{1 - \frac{u}{c}}{1 + \frac{u}{c}}\right)^{1/2}$$

[3] A. R. Sandage, "The Red Shift," *Scientific American*, September, 1956. Also available as *Scientific American Offprint 240* (San Francisco: W. H. Freeman and Co., Publishers).

[4] An interesting article on quasars is one by L. Lessing, "The Exploding Universe of Quasars," *The Physics Teacher*, IV, 3 (March, 1966).

and in terms of wavelength is

$$\lambda = \lambda_0 \left(\frac{1 + \dfrac{u}{c}}{1 - \dfrac{u}{c}} \right)^{1/2},$$

since wavelength is inversely proportional to frequency. For $u = 0.8c$

$$\lambda = \lambda_0 \left(\frac{1.8}{0.2} \right)^{1/2}$$

$$= 3\lambda_0.$$

Therefore, the wavelength of all light waves in the spectrum is increased by a factor of 3.

PROBLEMS

1. Write the equation for a longitudinal wave traveling in the $-x$ direction and having an amplitude of 0.03 m, a frequency of 30 Hz, and a wavelength of 0.18 m.

2. Write an equation to describe a transverse wave traveling in the $+y$ direction. Assume a wavelength of 0.65 m and a wave speed of 750 m · sec^{-1}. Draw graphs to represent the equation you write down.

3. A source of waves is oscillating at 500 Hz, and the waves travel at 15 m · sec^{-1}.
   (a) Determine the wavelength.
   (b) Find the phase angle of the source 1.5×10^{-3} sec after it passed through zero phase angle.
   (c) Find the phase angle of a particle 17 cm from the source at the instant that the source has zero phase angle.
   (d) If the displacement of the particle specified in part (c) is 0.9 mm at the instant specified, determine the amplitude of the wave at this point.

4. One end of a string such as in Fig. 15.17 is subject to a simple harmonic motion represented by $y = 0.10 \sin(4\pi t)$ m. Given that the wave so produced in the string will travel down the string at 1 m · sec^{-1}, plot on one graph the shape of the wave at times $t = 0, 0.1, 0.2, 0.3, 0.4,$ and 0.5 sec.

5. Write down an equation for the wave on the string of Problem 4. What will be the displacement of a point 0.5 m from the end of the string at $t = 0.75$ sec?

6. A wave of frequency 1000 Hz has a phase speed of 330 m · sec^{-1}.
   (a) How far apart are two points 60° out of phase?
   (b) What is the phase difference between displacements at a certain point at times 2.5×10^{-4} sec apart?

*7. What is the phase difference between the following transverse waves:

$$y = A \sin \left(\omega t + kx + \frac{\pi}{3} \right)$$

$$y = A \cos \left(\omega t + kx + \frac{\pi}{6} \right)?$$

8. A source A undergoing simple harmonic motion sends out plane waves that travel at $18 \text{ m} \cdot \text{sec}^{-1}$. The simple harmonic motion of the source is given by the relation $\Delta y = 1.5 \sin 50\pi t$ m.
(a) Determine the wavelength.
(b) At any instant find the phase difference between the motion of particles at two points B and C that are 2.52 and 2.34 m from the source, respectively.
(c) Find the displacement at B and C 2.08 sec after the source begins to vibrate.

9. A listener in a room is 5 m from one loudspeaker and 6.5 m from a second loudspeaker. The speed of sound in air at room temperature is $342 \text{ m} \cdot \text{sec}^{-1}$. Assuming that sound waves leave the speakers in phase, what will be the lowest frequency note for which the sound waves arrive at the listener's ear in phase?

10. Two simple harmonic oscillators A and B are vibrating with the same amplitude (0.10 m) and frequency (20 Hz). At a particular instant of time the displacements of the particles from their equilibrium positions are $+0.05$ m (A) and -0.03 m (B). Both particles are moving toward their equilibrium positions.
(a) What is the phase difference between A and B at this instant?
(b) How does this phase difference change with time?
(c) When the displacement of A is -0.08 m, what are the possible positions of B?

11. A fisherman sitting on a dock observes that his float makes ten complete oscillations in 4 sec and that the distance between crests is 2 m. What is the speed of the waves?

12. The speed of sound waves in a gas is
$$v = (\gamma P/\rho)^{1/2}$$
where P is the pressure, ρ the density, and γ a constant. Check the dimensional consistency of this equation.

13. Deduce the dimensions of G, the universal gravitational constant.

*14. A sinusoidal wave is traveling along a horizontal rope of **linear density** $0.28 \text{ kg} \cdot \text{m}^{-1}$ (that is, mass per unit length) held under a tension of 9 N.
(a) What is the speed of the wave?
(b) How much energy per meter length of rope is associated with the traveling wave if the wavelength is 5.7 m and the amplitude is 0.15 m?

15. A line source emits a cylindrical expanding wave. How do the amplitude and intensity of the wave depend on the distance from the source? What effect would an energy-absorbing medium have on the wave?

16. A satellite is designed to orbit the earth at distances less than 300 km. Communications with the earth are to be maintained by radio transmission at a carrier frequency of 3000 MHz ($\equiv$ 3 GHz). The radio waves are to be received by a radio telescope of diameter 10 m directly below the satellite. For detection, the power received must be at least 10^{-8} W ($\equiv$ J $\cdot$ sec^{-1}). If the satellite transmitter emits radiation equally in all directions, what must be its minimum power output? What is the ratio of wave amplitudes at ground level and 10 m from the transmitter?

*17. Electrical energy is transmitted along a transmission line from a power generating station to a large industrial complex. If there are 50 discontinuities along the line characterized by reflection coefficients 0.02 in value, estimate the fraction of the energy that arrives at its destination.

18. Three coil springs are joined together as shown in Fig. 15.25. The wavelength of compressional waves in the two end springs is λ and in the central spring is

Fig. 15.25

λ'. Write equations for sinusoidal longitudinal waves traveling along this system in terms of the reflection coefficient r at the junction of each pair of springs.

19. A train is approaching a level crossing at a speed of 30 m $\cdot$ sec^{-1}. A warning bell is sounding at the crossing. What is the ratio of the frequency of the sound coming from the bell to the frequency observed by a person on the train
(a) when the train is approaching the crossing?
(b) when the train is moving away from the crossing?

20. A train with its whistle sounding passes a man standing near the track. The frequency of the sound appears to drop as the train goes by, so that its value when receding is 0.75 of that when approaching. What is the speed of the train, assuming the speed of sound in air is 342 m $\cdot$ sec^{-1}?

21. A locomotive approaches a tunnel in the side of a large cliff. It emits a warning signal when 800 m from the tunnel entrance. The echo returns to the engine after 4.5 sec.
(a) Calculate the speed of the train.
(b) If the whistle emits a 500 Hz note, what will be the frequency of the echo as measured by an observer on the train when its speed of approach to the cliff is 16 m $\cdot$ sec^{-1} and the speed of sound in air is 342 m $\cdot$ sec^{-1}.

22. The frequency of a police-helicopter siren is observed to drop by a factor of two as it passes the observer, who is standing on a street corner. If the speed of sound is 342 m $\cdot$ sec^{-1}, what is the speed of the police helicopter?

23. A line observed in the spectrum of a certain star is displaced 5×10^{-10} m to the red from its normal position at 6.565×10^{-7} m. Calculate the relative velocity of the star.

24. Show that the following formulae for the Doppler shift in frequency for the case of source and observer moving closer together are identical to first order in u/c when $u \ll c$.

Moving source:

$$v = v_0 \frac{1}{1 - \dfrac{u}{c}}$$

Moving Observer:

$$v = v_0 \left(1 + \frac{u}{c} \right)$$

Relativistic:

$$v = v_0 \left(\frac{1 + \dfrac{u}{c}}{1 - \dfrac{u}{c}} \right)^{1/2}.$$

*25. A police radar set operates at a frequency of 3.00×10^9 Hz ($\equiv 3.00$ GHz). How accurately must the apparatus be able to measure frequency shifts in order that a speed of 30 m $\cdot$ sec^{-1} can be measured to within 0.5 m $\cdot$ sec^{-1}?

*26. The form of a wave as seen by an observer on the earth is $\cos(\omega t - kx)$. By using the Lorentz transformation, determine the form of the wave as seen by an observer traveling relative to the earth at speed u. Comment on your result.

27. Derive an expression for the nonrelativistic Doppler shift for the case of both a moving source and a moving observer, assuming that the air is stationary and that source and observer move in the same straight line.

28. Solve the problem of the relativistic Doppler effect for the case of a stationary observer in R and a source located in R' that is moving in the x, x' direction at speed u (see Section 15.8).

16 The superposition of waves

16.1   INTRODUCTION

Waves of a unique frequency and amplitude are rarely observed; more often a multiplicity of waves of various frequencies, amplitudes, and complex forms are present in the region under observation. The cumulative effect of a number of waves in a given region may give rise to new phenomena, such as **interference**, **diffraction**, and **beats**. All these phenomena can be understood in terms of the addition of two or more sine waves of the proper characteristics. Also of fundamental importance is the fact that any periodic disturbance, no matter how complex, can be written as a sum of a number of sinusoidal waves of different frequencies and of different amplitudes. Therefore, the very simple sinusoidal waves that we studied in the last chapter are sufficient to allow us to analyze the most complex waveforms.

When more than one wave disturbance occurs at any point in space at a given time, interesting and rather complicated effects can occur. Such effects can be explained by a very simple law known as the **principle of superposition**, proposed by Thomas Young (1802). It states that the resultant displacement at any point is merely the algebraic sum of the displacements due separately to the individual waves. For example, two complex wave forms of arbitrary shape may simply be added algebraically to give the resultant displacement. This has been done for two arbitrary wave forms in Fig. 16.1.

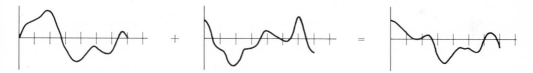

Fig. 16.1. Two complex wave forms can be added algebraically to give a resultant wave form.

16.2   A WAVE MACHINE DEMONSTRATION OF SUPERPOSITION

Suppose we simultaneously start two wave pulses of equal amplitude so that they travel in opposite directions from each end of a wave machine. When they meet in the middle of the machine, what happens? We readily observe that after the two pulses meet in the middle of the machine, two equal pulses continue to travel on toward the two ends of the machine. However, it is not easy to decide whether they collide with one another and rebound as particles would or whether they pass through each other. To resolve this ambiguity, we might start two more wave pulses toward each other from opposite ends of the machine but now make them of distinctly different amplitudes. The ambiguity is then resolved, as can be seen from Fig. 16.2. The two wave pulses pass through

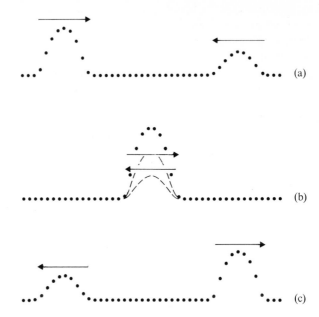

Fig. 16.2. A wave machine demonstration of the principle of superposition: (a) two waves approaching one another; (b) two waves passing through one another; (c) two waves moving away from one another.

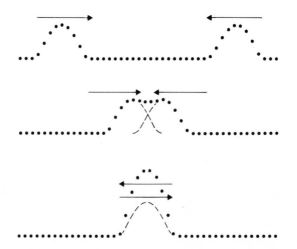

Fig. 16.3. Constructive interference of two equal pulses.

each other and continue on as if they had never met. At the instant that they occupy the same position on the wave machine, a resultant pulse is formed whose instantaneous amplitude is just the algebraic sum of the amplitudes of the individual component waves. This is a demonstration of the principle of superposition.

When the two pulses occupy the same portion of the wave machine, we observe an example of an interference phenomenon. If the two pulses are positive-going and of equal amplitude, the resultant pulse formed when the two pulses overlap one another will have an amplitude equal to twice that of the constituent pulses. This is an example of **constructive inter- ference** (see Fig. 16.3).

If the two pulses are of equal amplitude but one is positive-going and the other negative-going, the resultant pulse formed when the two pulses overlay one another will have zero amplitude. This is an example of **destructive interference** (see Fig. 16.4).

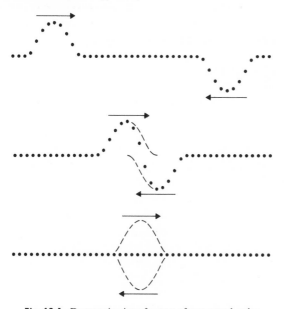

Fig. 16.4. Destructive interference of two equal pulses.

Let us try another experiment. We attach the motor and crank assembly described in Section 15.2 to one end of the wave machine and clamp the other end so that it cannot move. In this way we create a situation in which a train of continuous waves is generated, travels down the wave machine, and is then reflected and superposed upon itself. The resultant wave disturbance (Fig. 16.5) is called a **standing wave pattern**. That is, there are dead spots called **nodes** at which the cross- pieces of the wave machine are stationary and spots of maximum motion

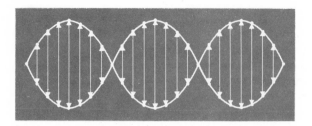

Fig. 16.5. A stationary wave pattern resulting when a continuous traveling wave is reflected and superposed upon itself.

called **loops** at which the crosspieces oscillate with maximum amplitude. Consecutive nodes or loops are separated by a half wavelength; this is shown in Section 16.4.

The effect can be enhanced by adjusting the physical dimensions of the machine by attaching the clamp to a different crosspiece. All solid objects possess a certain **natural frequency** at which they would oscillate if they were free to do so. This natural frequency is dependent upon the dimensions of the object (as well as upon other characteristics). When the machine is adjusted so that it exhibits a natural frequency of its own equal to the frequency of the disturbance to be transmitted on it, the standing wave pattern is enhanced, and the wave machine is said to be **tuned** to the frequency of the wave generated on the machine by external means. It is also said that the machine is in **resonance** with the externally impressed traveling wave. A more familiar example of resonance occurs in the tuning of a radio. When you turn the dial of a radio to select a particular station, you are bringing a selective circuit in the radio into resonance with radio waves of a particular frequency. A very dramatic illustration of resonance resulted in the collapse of the Tacoma Narrows Bridge[1] in 1940. Fluctuating winds excited torsional vibrations in the structure at one of its natural frequencies resulting in large amplitude oscillations (about 35° in each direction from the horizontal) and the ultimate collapse of the bridge.

16.3   A MATHEMATICAL DESCRIPTION OF INTERFERENCE

We shall restrict our consideration to the superposition of two transverse sinusoidal wave disturbances of the same frequency v and of the same amplitude A traveling in the direction of the positive x axis and vibrating in the x, y plane. Since the waves are of the same frequency and

---

[1] The Ealing Corporation, Film loop 80-218 (Cambridge, Mass).

are traveling in the same medium, they will travel with the same speed v and will have the same wavelength λ.

Figure 16.6(a) shows the two waves and the resultant wave after superposition for the case in which the two waves are in phase. Here the two waves reach peak value at the same time, and the resultant wave has the same frequency as the original waves but twice the amplitude. In Fig. 16.6(b) the original waves are taken to be one-half period or π rad (180°) out of phase. In this case they cancel each other, since peak displacement in a given direction for one wave corresponds to peak displacement in the opposite direction for the second wave. An intermediate situation is illustrated in Fig. 16.6(c). Here the two waves are one-quarter period or $\pi/2$ rad (90°) out of phase. The two waves add to give a wave of the same frequency as that of the original waves and an amplitude 1.4 times the original amplitude. Also, the maximum displacement occurs at a time different from the time of maximum displacement for either of the original waves.

Let us see if this behavior is predicted from our mathematical form for a traveling wave. We take the equations

$$\Delta y_1 = A \sin 2\pi\left(vt - \frac{x}{\lambda}\right)$$

$$\Delta y_2 = A \sin\left[2\pi\left(vt - \frac{x}{\lambda}\right) - \phi\right]$$

for the two waves, where ϕ is the phase difference between the waves. When the two waves are in phase, $\phi = 0$, and

$$\Delta y_1 = A \sin 2\pi\left(vt - \frac{x}{\lambda}\right)$$

$$\Delta y_2 = A \sin 2\pi\left(vt - \frac{x}{\lambda}\right),$$

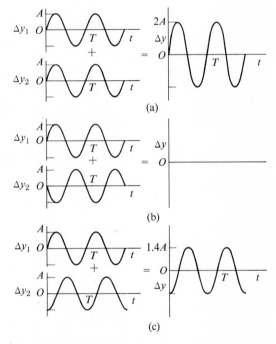

Fig. 16.6. The superposition of two wave disturbances of the same amplitude and frequency traveling in the same direction. The waves are in phase in (a), π rad out of phase in (b), and $\pi/2$ rad out of phase in (c).

and the resultant displacement, according to the principle of superposition, is

$$\Delta y = \Delta y_1 + \Delta y_2$$

$$= 2A \sin 2\pi\left(vt - \frac{x}{\lambda}\right).$$

The resultant wave has the same frequency and twice the amplitude of the original waves, in agreement with Fig. 16.6(a).

When the two waves are one-half period out of phase, $\phi = \pi$ rad, and

$$\Delta y_1 = A \sin 2\pi \left(vt - \frac{x}{\lambda} \right)$$

$$\Delta y_2 = A \sin \left[2\pi \left(vt - \frac{x}{\lambda} \right) - \pi \right]$$

$$= - A \sin 2\pi \left(vt - \frac{x}{\lambda} \right).$$

The resultant wave is given by

$$\Delta y = \Delta y_1 + \Delta y_2$$

$$= A \sin 2\pi \left(vt - \frac{x}{\lambda} \right) - A \sin 2\pi \left(vt - \frac{x}{\lambda} \right)$$

$$= 0.$$

This result is in agreement with Fig. 16.6(b).

When the two waves are one-quarter period out of phase, $\phi = \pi/2$ rad, and

$$\Delta y_1 = A \sin 2\pi \left(vt - \frac{x}{\lambda} \right)$$

$$\Delta y_2 = A \sin \left[2\pi \left(vt - \frac{x}{\lambda} \right) - \frac{\pi}{2} \right]$$

$$= A \cos 2\pi \left(vt - \frac{x}{\lambda} \right).$$

The resultant wave is given by

$$\Delta y = \Delta y_1 + \Delta y_2$$

$$= A \sin 2\pi \left(vt - \frac{x}{\lambda} \right) + A \cos 2\pi \left(vt - \frac{x}{\lambda} \right).$$

The maximum displacement is obtained when

$$\sin 2\pi \left(vt - \frac{x}{\lambda} \right) = \cos 2\pi \left(vt - \frac{x}{\lambda} \right),$$

which occurs for

$$2\pi \left(vt - \frac{x}{\lambda} \right) = \frac{\pi}{4} \text{ rad}$$

or 45°. Since $\sin \pi/4 = 0.707$, the maximum displacement is

$$\Delta y = 0.707A + 0.707A = 1.414A;$$

therefore, the amplitude of the resultant wave is 1.4 times the amplitude of the original wave in agreement with Fig. 16.6(c). Also, the maximum displacement occurs when $2\pi(vt - x/\lambda) = \pi/4$ rad or a multiple of $\pi/4$ rad, whereas the original waves have maximum displacement when $2\pi(vt - x/\lambda) = 0,\ \pi,\ 2\pi,\ldots$ rad and $2\pi(vt - x/\lambda) = \pi/2,\ 3\pi/2,\ldots$ rad, respectively (ignoring the sign of the displacement and considering magnitude only). The frequency of the resultant wave is the same as the frequency of the original waves.

An alternate analysis for the case of $\phi = \pi/2$ rad can be made by shifting the origin of time to the position indicated in Fig. 16.7. The first wave is now represented by

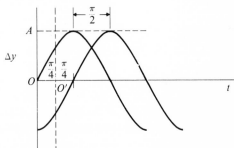

Fig. 16.7. When the origin is shifted from O to O', two waves which are $\pi/2$ rad out of phase have initial phases of $\pi/4$ and $-\pi/4$ rad, respectively.

$$\Delta y_1 = A \sin \left[2\pi\left(vt - \frac{x}{\lambda} \right) + \frac{\pi}{4} \right],$$

and the second wave is represented by

$$\Delta y_2 = A \sin \left[2\pi\left(vt - \frac{x}{\lambda} \right) - \frac{\pi}{4} \right].$$

The resultant displacement is (using the appropriate trigonometric identity in Appendix 2)

$$\Delta y = \Delta y_1 + \Delta y_2$$

$$= A \sin \left[2\pi\left(vt - \frac{x}{\lambda} \right) + \frac{\pi}{4} \right] + A \sin \left[2\pi\left(vt - \frac{x}{\lambda} \right) - \frac{\pi}{4} \right]$$

$$= 2A \cos \frac{\pi}{4} \sin 2\pi\left(vt - \frac{x}{\lambda} \right)$$

$$= 1.4A \sin 2\pi\left(vt - \frac{x}{\lambda} \right).$$

This approach can be immediately generalized for any phase difference ϕ to give a resultant represented by

$$\Delta y = 2A \cos \frac{\phi}{2} \sin 2\pi\left(vt - \frac{x}{\lambda} \right).$$

This example is a particularly simple case of the interference of waves. The various effects occurring in optical systems that are classified as diffraction or interference phenomena are in principle explained in this manner.

In order to observe interference effects, we must have a constant phase difference between two waves that remains constant over the period of observation. In many cases two separate sources of waves of the same frequency may be stable enough that the phase difference between the waves produced by them is constant. For other sources (such as sources of visible light) the waves from two different sources do not have constant phase differences over appreciable periods of time. In light sources countless numbers of atoms are radiating light for periods of the order of only 10^{-8} sec. However, interference effects can still occur between light waves if light waves from a single source are divided and sent over two different paths and are then recombined at a single point in space. Since different paths are traveled by the divided waves, different times are taken, resulting in a phase difference between the recombined waves. Since the recombined waves have come from the same source, this phase difference is constant, and interference can occur. Note that even when two sources have constant phase difference they are normally not at the same point in space, so that the phase difference between waves from the two sources interfering at some point in space will also depend upon the path difference. Note that the preceding discussion applies equally to longitudinal waves.

Example. Sound waves of frequency 688 Hz spread out at a speed of 344 m · sec^{-1} from two sources A and B separated by 2.0 m.

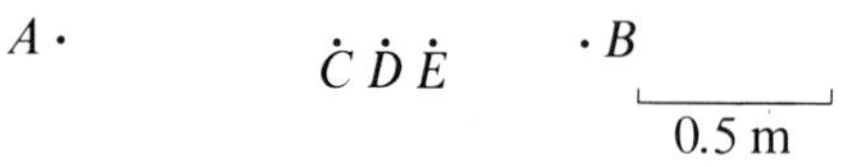

A · $\dot{C}\,\dot{D}\,\dot{E}$ · B

|_____|
   0.5 m

Fig. 16.8. Two sources A and B emit sound waves which interfere at points C, D, and E.

(a) Assuming that the two sources are in phase and emit waves of the same amplitude, determine the resultant amplitude at points C, D, E as shown in Fig. 16.8, where $AC = 1.0$ m, $AD = 1.125$ m, and $AE = 1.25$ m.
(b) If the sources had a constant phase difference of π rad, what would be the resultant amplitudes?

Solution
(a) Since $v = 344$ m · sec^{-1} and $v = 688$ Hz, the wavelength

$$\lambda = \frac{v}{v} = \frac{344}{688} = 0.5 \text{ m}.$$

Therefore, the sources are $2.0/0.5 = 4\lambda$ apart. Two sources of sound waves are positioned at A and B, and the resultant amplitude is measured at C, D, and E. Point C is 1.0 m $= 2\lambda$ from A and 1.0 m $= 2\lambda$ from B. Therefore, the path difference is $CA - CB = 2\lambda - 2\lambda = 0$, so that the waves are in phase at C and the resultant amplitude is twice the amplitude of the individual waves. Point D is 1.125 m $= 2.25\lambda$ from A and 0.875 m $= 1.75\lambda$ from B. Therefore, the path difference is $DA - DB = 2.25\lambda - 1.75\lambda = 0.5\lambda$, so that the waves are π rad out of phase and the resultant amplitude is zero. Point E is 1.25 m $= 2.5\lambda$ from A

and 0.75 m $= 1.5\lambda$ from B. Therefore, the path difference is $EA - EB = 1.0\lambda$, and the resultant amplitude is twice the amplitude of the individual waves.

(b) If there is a phase difference of π rad between the sources, the phase differences at C, D, and E would be π, 2π (or zero), and π rad, respectively. Therefore, the resultant amplitude would be zero at C and E and twice the amplitude of the individual waves at D.

Example. Two sources A and B, both emitting sound waves of frequency 3440 Hz, are placed 0.25 m apart along a line parallel to a wall that is 20 m distant, as shown in Fig. 16.9. (Note that the distance AB has been exaggerated for clarity.) What is the condition for the waves to interfere constructively at the point P on the wall?

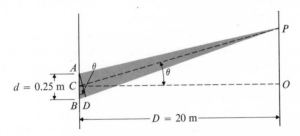

Fig. 16.9. Waves from A and B interfere at point P. Note that the distance $d = 0.25$ m has been exaggerated for clarity.

Solution. An arbitrary point P on the wall is a distance AP from source A and a distance BP from source B. Let us draw a line from A to point D on BP in such a way that $PD = PA$. The distance BD then represents the path difference between waves traveling from the two sources to P.

Since the distance $D = 20$ m to the wall is much greater than the separation $d = 0.25$ m between A and B, AD is almost perpendicular to both AP and BP. Therefore, angle BAD is almost equal to angle PCO, where C is the point midway between A and B. Both angles are marked θ in Fig. 16.9. Assuming A and B to be in phase, we see that the phase difference at P is due only to the path difference BD. The number of wavelengths contained in BD must be

$$BD = n\lambda, \qquad n = 0, 1, 2, \cdots$$

for constructive interference to occur at P. In triangle ABD

$$\frac{BD}{AD} = \frac{BD}{d} = \sin \theta$$

or

$$BD = d \sin \theta.$$

Therefore, for constructive interference at P, the condition is

$$d \sin \theta = n\lambda, \qquad n = 0, 1, 2, \cdots.$$

For $d = 0.25$ m and $\lambda = \dfrac{v}{v} = \dfrac{344}{3440} = 0.1$ m,

$$\sin \theta = \frac{n\lambda}{d} = 0.4n,$$

where $n = 0, 1, 2, \cdots$, or

$$\sin \theta = 0, 0.4, 0.8, 1.2, \cdots.$$

Therefore, $\theta = 0°$, $23.6°$ or $53.2°$ (the other solutions are inadmissable).

16.4 A MATHEMATICAL DESCRIPTION OF STANDING WAVES

We shall now consider the superposition of two sinusoidal wave disturbances of the same amplitude A and frequency v traveling in opposite directions and vibrating in the same plane. In Fig. 16.10 we show the spatial configuration of the two waves and their resultant at four particular times. Figure 16.10(a) shows the waves and their resultant at an instant of time when the waves are in phase. Figure 16.10(b) shows the waves and their resultant a short time later when each wave has traveled a distance of one-eighth of a wavelength. The peak displacements are now one-quarter wavelength apart, corresponding to a phase difference of $\pi/2$ rad. Figure 16.10(c) shows the waves and their resultant when the phase difference has increased to π rad and Fig. 16.10(d) shows the waves and their resultant when the phase difference has increased to 2π rad. The resultant waveforms for each case are superimposed in Fig. 16.11.

The amplitude of the resultant waves varies with position along the wave as well as with time. Positions of zero amplitude, called **nodes**, occur at points one-half wavelength apart. Between nodes, the amplitude varies periodically, with the maximum amplitude $(2A)$ occurring at points midway between the nodes. The regions between nodes are commonly called **loops**. Since the positions of the nodes and loops are independent of time, the wave appears to be fixed in space; this type of wave is known as a **standing wave**. The acoustical waves set up in an organ pipe are an example of longitudinal standing waves.

We shall write equations for the two original waves as

$$\Delta y_1 = A \sin 2\pi\left(vt + \frac{x}{\lambda}\right)$$

$$\Delta y_2 = A \sin 2\pi\left(vt - \frac{x}{\lambda}\right),$$

where the initial phase angle is taken to be zero. Parts (a), (b), (c), and (d) of Fig. 16.10 are for phase differences of 0, $\pi/4$, $\pi/2$, and 2π rad,

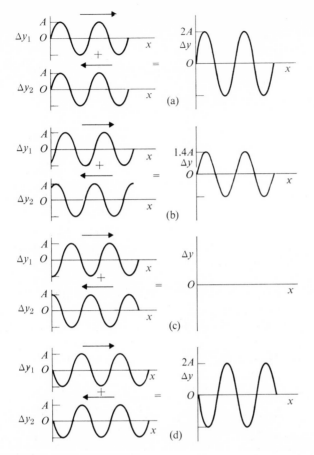

Fig. 16.10. The superposition of two waves of the same amplitude and frequency traveling in opposite directions. The phase difference is zero in (a), $\pi/2$ rad in (b), π rad in (c), and 2π rad in (d).

respectively. The resultant wave is represented by using the appropriate identity in Appendix 2.

$$\Delta y = \Delta y_1 + \Delta y_2$$

$$= A \sin 2\pi\left(vt + \frac{x}{\lambda}\right) + A \sin 2\pi\left(vt - \frac{x}{\lambda}\right)$$

$$= 2A \sin 2\pi vt \cos 2\pi\frac{x}{\lambda}$$

$$= \left(2A \cos 2\pi\frac{x}{\lambda}\right) \sin 2\pi vt.$$

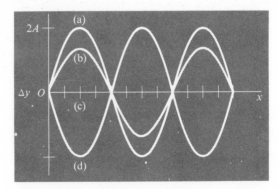

Fig. 16.11. The resultant wave forms of Fig. 16.10 are shown superimposed.

The amplitude, $2A \cos 2\pi x/\lambda$, is spatially dependent and has maximum value $2A$ or $-2A$ for values of x satisfying

$$2\pi \frac{x}{\lambda} = 0, \pi, 2\pi, \cdots \text{rad},$$

or when

$$2\pi \frac{x}{\lambda} = n\pi, \qquad n = 0, 1, 2, \cdots$$

or

$$x = \frac{n\lambda}{2}, \qquad n = 0, 1, 2, \cdots.$$

The minimum value of amplitude (zero) will then occur for values of x satisfying

$$2\pi \frac{x}{\lambda} = \frac{(2n + 1)\pi}{2}, \qquad n = 0, 1, 2, \cdots$$

or

$$x = \frac{(2n + 1)\lambda}{4}, \qquad n = 0, 1, 2, \cdots.$$

We now consider the standing wave pattern obtained when sinusoidal transverse mechanical waves are sent along a taut string that is held immovable at one end (as shown in Fig. 16.12). The traveling wave will be reflected and inverted at the fixed end and will travel back upon itself (see Section 15.2).

In this way two waves traveling in opposite directions and having the same frequency and amplitude are produced. Since the one end of the string is held fixed, a node in the resulting standing wave pattern must occur at this point. Since the string is of finite length, a node must occur at the other end also, or else the stationary wave pattern would not be stable under continued reflections of the waves at the ends of the string. If the wavelength is λ, nodes will occur at both ends of the string if the length of the string is $\lambda/2$, λ, $3\lambda/2$,

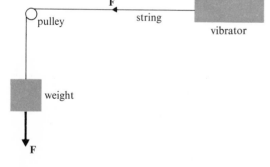

Fig. 16.12. A standing wave pattern will be obtained when a wave is sent along the stretched string in this apparatus.

$2\lambda, \cdots$, or if

$$l = \frac{n\lambda}{2}, \qquad n = 1, 2, 3, \cdots,$$

where l is the length of the string. This is the same relation as derived above from the mathematical expression for the wave. The first three modes of vibration are shown in Fig. 16.13. In each case a node is shown at the end of the vibrator. This is a good approximation, even though the end of the vibrator is clearly not at rest. Once the standing wave pattern has been established in the string, the vibrator need only supply sufficient energy per cycle to replace that dissipated in the string. The steady state amplitude of oscillations in the string is therefore much greater than the amplitude of the vibrator driving the string; the vibrator is near a node.

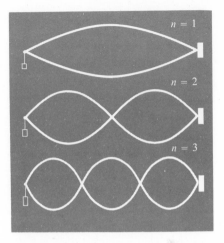

Fig. 16.13. The first three modes of vibration for a stretched string of fixed length (amplitude greatly exaggerated).

In laboratory experiments on standing waves, the length of the string is usually kept constant and the wavelength of the waves is varied by adding or removing weights from the weight holder, which changes the tension on the string. In this experiment l is constant and λ varies whereas in our discussion above we assumed λ constant and l variable. In either case $2l = n\lambda$ must be satisfied.

16.5   THE PHENOMENON OF BEATS

If two wave disturbances of slightly different frequencies traverse a medium, a stationary interference pattern will not be produced. If, however, the frequencies of the two waves are nearly equal, they can interfere in such a manner as to form a wave disturbance having a pulsating intensity at a given point in space. Such pulsations in intensity are known as **beats**. Tuning forks may be used to provide a simple demonstration of this phenomenon. If two tuning forks of frequencies 254 Hz and 256 Hz are sounded simultaneously, one will hear two beats every second. The reason for the production of beats follows from the principle of superposition. In Fig. 16.14 two sine curves of equal amplitude and frequencies 10 and 12 Hz are shown along with their resultant. The resultant is seen to be a wave of time-varying amplitude that attains maximum amplitude twice every second. The beat frequency corresponds to the difference frequency of the individual disturbances. We also might note from Fig. 16.14 that the resultant wave has a frequency of 11 Hz intermediate between the frequencies of the constituent waves.

Let us consider mathematically the superposition at $x = 0$ of two sine waves of frequencies v_1 and v_2 and equal amplitude A. That is, as the

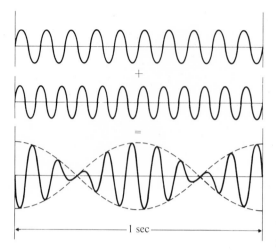

Fig. 16.14. The superposition of two waves of equal amplitude and frequencies 10 and 12 Hz.

original waves we take

$$\Delta x_1 = A \sin 2\pi v_1 t$$

$$\Delta x_2 = A \sin 2\pi v_2 t.$$

The resultant wave is

$$\Delta x = \Delta x_1 + \Delta x_2 = A(\sin 2\pi v_1 t + \sin 2\pi v_2 t).$$

This may be rewritten as

$$\Delta x = 2A \cos \frac{2\pi(v_1 - v_2)}{2} t \sin \frac{2\pi(v_1 + v_2)}{2} t.$$

This result follows immediately from the trigonometric identity (see Appendix 2)

$$2 \sin \frac{\theta + \phi}{2} \cos \frac{\theta - \phi}{2} = \sin \theta + \sin \phi$$

by substituting

$$\theta = 2\pi v_1 t, \qquad \phi = 2\pi v_2 t.$$

The result may be interpreted as a sinusoidal wave at the average frequency of the original waves, that is, of frequency

$$v = \frac{v_1 + v_2}{2}$$

and of time-dependent amplitude at the difference frequency $v_1 - v_2$. Note that the factor $\frac{1}{2}$ is omitted in the expression for the difference frequency. This is to take account of the fact that two maxima of the resultant wave occur for each oscillation at frequency $(v_1 - v_2)/2$ and these are detected as two beats.

16.6   THE AMPLITUDE MODULATION OF WAVES

Suppose we wish to use electromagnetic waves to transmit audio information through space from a transmitting station A to a receiving station B, as illustrated in Fig. 16.15. At the transmitter the audio signal is converted into an electromagnetic signal that is applied to an antenna that in turn radiates electromagnetic energy. An efficient transmitter must use an antenna that radiates a maximum amount of energy for a given amount of energy applied to it. A theoretical consideration of this problem shows that an antenna of physical dimensions of the order of magnitude of $\lambda/4$, where λ is the wavelength of the radiation being transmitted, should be used. A typical audio frequency might be taken as 10^3 Hz; this would correspond to a wavelength

$$\lambda = \frac{c}{1000} = 3 \times 10^5 \text{ m.}$$

Clearly, the construction of antennas with physical dimensions of the order of 7.5×10^4 m is not economically feasible. Suppose, however, that you were able to transmit the audio information at a typical radio frequency of, say, 10^6 Hz. Then the antenna dimensions would only have to be of the order of 25 m.

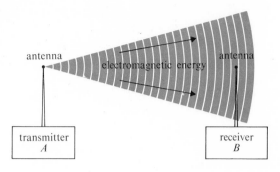

Fig. 16.15. The electromagnetic transmission of audio information.

Two schemes to transmit audio information at radio frequencies are in common use; they are known as **amplitude modulation** (AM) and **frequency modulation** (FM). In either case the transmission of energy occurs at a radio frequency v_c (or ω_c) called the **carrier frequency**. The audio information is impressed upon the carrier wave. For AM the amplitude of the carrier wave is varied with time at the audio frequency to be transmitted. For FM the amplitude of the carrier wave is kept constant, but the frequency is varied by a small amount about v_c at the audio frequency to be transmitted. An AM receiver is sensitive to the variations in the amplitude of the carrier wave picked out by a selective circuit coupled to the receiver's antenna; an FM receiver is sensitive to the variations in the frequency of the selected carrier wave. For TV transmission both AM and FM are used simultaneously.

We shall consider a mathematical description of amplitude modulation only. The carrier wave of fixed amplitude can be represented at point $x = 0$ by

$$\Delta y = A \sin 2\pi v_c t,$$

where v_c is the carrier frequency. If the amplitude is to be varied sinusoidally, we must add another term to this equation that will result in nonconstant values of amplitude. This can be done by adding a term including $\sin v_m t$, where v_m is the modulation frequency. The equation for the displacement becomes

$$\Delta y = A \sin 2\pi v_c t + Ab \sin 2\pi v_m t \sin 2\pi v_c t,$$

where b is a constant (normally ≤ 1). The equation may be rewritten as

$$\Delta y = A(1 + b \sin 2\pi v_m t) \sin 2\pi v_c t.$$

Note that the amplitude will vary between

$$A(1 + b) \quad \text{and} \quad A(1 - b)$$

at frequency v_m. The equation

$$\Delta y = A \sin 2\pi v_c t + Ab \sin 2\pi v_m t \sin 2\pi v_c t$$

may also be written as

$$\Delta y = A \sin 2\pi v_c t + \frac{Ab}{2}[\cos 2\pi(v_c - v_m)t - \cos 2\pi(v_c + v_m)t].$$

This result follows immediately from the trigonometric identity (see Appendix 2)

$$2 \sin \frac{\theta + \phi}{2} \sin \frac{\theta - \phi}{2} = \cos \phi - \cos \theta,$$

with

$$\theta = 2\pi(v_c + v_m)t, \qquad \phi = 2\pi(v_c - v_m)t.$$

The modulated wave may be considered to be the sum of three sinusoidal waves: one at the carrier frequency, one at the sum frequency $v_c + v_m$, and one at the difference frequency $v_c - v_m$. The sum and difference frequencies are often referred to as the **side bands**. A pictorial description of this summation is given in Fig. 16.16.

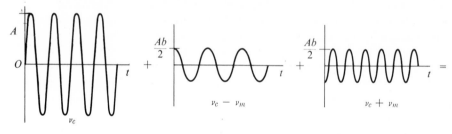

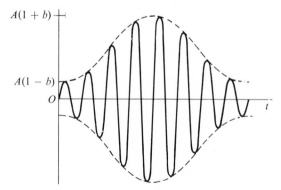

Fig. 16.16. A carrier wave of frequency v_c whose amplitude is modulated at frequency v_m can be thought of as the sum of a carrier wave, plus two additional waves, one of frequency $v_c - v_m$ and one of frequency $v_c + v_m$.

16.7   WAVE GROUPS AND GROUP VELOCITY

A point of constant phase on a traveling wave moves through space at a speed given by $v = v\lambda = \omega/k$. This speed should not be confused with the motion of the particle in the medium through which the wave travels. The speed of the individual particle depends on the amplitude and frequency of the wave. A point of constant phase moves at the **wave velocity** or the **phase velocity**. In the remainder of this section we shall use the term velocity where we would normally use speed (direction not indicated), in order to conform to the widely-used labels for the quantities that we are going to discuss.

In many cases, the properties of a medium through which a wave is passing are dependent upon the frequency of the wave. In such a case, the phase velocity is not a constant. That is, waves of different frequencies travel at different velocities; this is known as **dispersion**. The phase velocity now depends upon frequency. A common example of dispersion is the breakup of a beam of white light into a spectrum of colors when passing through a glass prism. The component waves have different velocities in the glass and consequently different directions upon emerging; they are said to be **dispersed**. The phenomenon of dispersion is closely related to the rate at which energy is carried by the wave. Energy does not necessarily travel with the wave at the phase velocity; the velocities are often very nearly equal, but they are rarely the same. The velocity with which the energy travels is called the **group velocity**, v_g.

We shall first consider the superposition of two waves of the same amplitude and in phase but differing in angular frequency by an amount $\Delta\omega$ and in propagation constant by Δk. The equations for the two waves may be taken as

$$\Delta y_1 = A \sin (\omega t - kx)$$

$$\Delta y_2 = A \sin [(\omega + \Delta\omega)t - (k + \Delta k)x].$$

The resultant displacement is

$$\Delta y = A \sin (\omega t - kx) + A \sin [(\omega + \Delta\omega)t - (k + \Delta k)x]$$

$$= 2A \cos \left[\Delta\omega\left(\frac{t}{2}\right) - \Delta k\left(\frac{x}{2}\right) \right] \sin (\omega t - kx).$$

The derivation is carried out on page 329. This equation may be described as a sinusoidal wave of angular frequency ω, propagation constant k, and time and space dependent amplitude

$$2A \cos \left(\frac{\Delta\omega}{2}t - \frac{\Delta k}{2}x \right).$$

The successive amplitude maxima are known as **wave groups** and are pictured in Fig. 16.17 at one instant of time. Wave groups are clearly closely related to the beats and amplitude modulation discussed in Section 16.5.

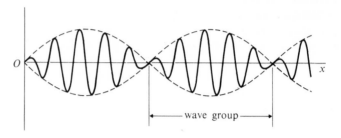

Fig. 16.17. A series of wave groups.

The resultant wave travels through space with the phase velocity

$$v = \frac{\omega}{k},$$

and the successive wave groups (or maxima in the modulation pattern) of Fig. 16.15 move through space, by analogy, with the group velocity

$$v_g = \frac{\dfrac{\Delta\omega}{2}}{\dfrac{\Delta k}{2}} = \frac{\Delta\omega}{\Delta k}.$$

In general, for waves differing infinitesimally in frequency and propagation constant,

$$v_g = \frac{d\omega}{dk}.$$

One important application of the wave group concept occurs in the quantum theory of matter[2]. According to this theory a moving particle has an associated wave nature that can be described as a wave group or a wave packet, as illustrated in Fig. 16.18. Now, however, we have only a single wave group to deal with rather than a whole series of wave groups such as resulted from the simple derivation used to introduce the concept of a wave group. It turns out that a single wave group may be mathematically described as a superposition of a very large number of sinusoidal waves each differing slightly in wavelength. The group velocity for a wave packet associated with a moving particle is identical to the

[2] E. Schrödinger, "What is Matter?" *Scientific American*, September, 1953. Available as *Scientific American Offprint 241* (San Francisco: W. H. Freeman and Co., Publishers).

particle's velocity through space[3] in agreement with the supposition made earlier that the group velocity is the velocity with which energy is transported by the wave. The special theory of relativity predicts that the magnitude of the group velocity can never exceed the speed of light c.

The interpretation of the wave group representation of a particle is as follows. Quantum mechanics denies the possibility of exactly knowing the position of a particle in space. The amplitude of the envelope of the wave packet at a given point is a measure of the probability of finding the particle at that point in an experimental measurement of its position. The narrower the wave packet, the better defined is the particle's position in space (see Chapter 18).

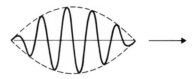

Fig. 16.18. According to quantum theory a moving particle has an associated wave nature which can be described as a wave group or wave packet.

Proof that

$$\Delta y = A \sin(\omega t - kx) + A \sin[(\omega + \Delta\omega)t - (k + \Delta k)x]$$
$$= 2A \cos\left(\frac{\Delta\omega}{2}t - \frac{\Delta k}{2}x\right) \sin(\omega t - kx).$$

Using the trigonometric identity (Appendix 2)

$$\sin a + \sin b = 2 \cos\left(\frac{b - a}{2}\right) \sin\left(\frac{b + a}{2}\right),$$

we obtain

$$\Delta y = A \sin(\omega t - kx) + A \sin[(\omega + \Delta\omega)t - (k + \Delta k)x]$$
$$= 2 \cos\left(\frac{\omega t + \Delta\omega\, t - kx - \Delta k\, x - \omega t + kx}{2}\right)$$
$$\times \sin\left(\frac{\omega t + \Delta\omega\, t - kx - \Delta k\, x + \omega t - kx}{2}\right)$$
$$= 2 \cos\left(\frac{\Delta\omega}{2}t - \frac{\Delta k}{2}x\right) \sin\left(\frac{2\omega t - 2kx + \Delta\omega\, t - \Delta k\, x}{2}\right).$$

If we assume that

$$\Delta\omega\, t - \Delta k\, x \ll 2(\omega t - kx),$$

we have

$$\Delta y = 2 \cos\left(\frac{\Delta\omega}{2}t - \frac{\Delta k}{2}x\right) \sin(\omega t - kx).$$

[3] A. Beiser, *Concepts of Modern Physics* (New York: McGraw-Hill Book Co., 1963), Chapter 3.

16.8   FOURIER ANALYSIS

Our discussions so far have been restricted to reasonably simple periodic disturbances that could be analyzed readily into a small number of sinusoidal disturbances. The exception was the wave packet, and then we simply stated without proof that a wave packet would be constructed from a very large number of sine waves. We now turn to a general periodic disturbance at a specific point in space, as for example the one shown in Fig. 16.19. This disturbance repeats itself in a time T. A mathematical theorem, known as Fourier's theorem after Jean Baptiste Fourier,

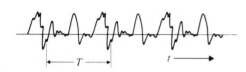

Fig. 16.19. A general periodic disturbance at $x = 0$.

states that it is possible to construct such a function from a series of sine functions of the form

$$C_n \sin\left(\frac{2\pi n t}{T} + \phi_n\right) = C_n \sin\left(2\pi n v_0 t + \phi_n\right),$$

where $n = 0, 1, 2, 3, \cdots$ by properly selecting the amplitudes C_n and the phase angles ϕ_n. The quantity $v_0 = 1/T$ is called the **fundamental frequency**; integral multiples of v_0 are called the **harmonic frequencies**.

Before discussing Fourier's theorem we shall discuss an example to illustrate why we might want to decompose a complicated wave form into sinusoidal components. Suppose we were confronted with the problem of designing an electronic amplifier for a hi-fi set that was to amplify a sawtooth wave without appreciable distortion. Distortion in an electronic amplifier can be measured quantitatively and is an indication of how faithfully the output signal resembles the input signal. A good hi-fi amplifier produces a distortion of much less than 1%. What amplification characteristics must the amplifier have?

To answer this question we analyze the input sawtooth wave (see Fig. 16.20) into component sine waves and then see, by using the principle of superposition, which of these components are essential to the construction of a resultant wave that reproduces the sawtooth within the desired limits of allowable distortion. Then we can say that the amplifier must have a uniform gain at all of the frequencies characterizing this group of components. For example, if the fundamental frequency of the input sawtooth wave is v_0 and if an analysis into component sine waves shows that the fundamental and four harmonics are required to adequately represent the sawtooth, then the amplifier would

Fig. 16.20. A sawtooth wave before and after electronic amplification.

have to be designed to have uniform amplification for frequencies from v_0 to $5v_0$.

Now let us return to Fourier's theorem. It states that any function $f(t)$, periodic in time T, can be expanded in the form

$$f(t) = \sum_n C_n \sin (2\pi n v_0 t + \phi_n)$$

or the mathematically equivalent form

$$f(t) = \sum_n A_n \cos 2\pi n v_0 t + \sum_n B_n \sin 2\pi n v_0 t.$$

The second expression follows from the first by using the trigonometric identity (see Appendix 2)

$$C_n \cos (\theta + \phi) = C_n \cos \theta \cos \phi - C_n \sin \theta \sin \phi$$

and by substituting

$$\theta = 2\pi n v_0 t, \qquad \phi = \phi_n$$

and by introducing the definitions

$$C_n \cos \phi_n = A_n, \qquad C_n \sin \phi_n = B_n.$$

We note that the average value of $\sin 2\pi v_0 t$ or $\cos 2\pi v_0 t$ or any harmonic of $\sin 2\pi v_0 t$ or $\cos 2\pi v_0 t$ over one complete period is zero. Therefore, averaging both sides of the second equation for $f(t)$ gives

$$\overline{f(t)} = A_0.$$

Using the method of averaging introduced in Chapter 8, we have

$$A_0 = \frac{1}{T} \int_0^T f(t) \, dt.$$

The remainder of the coefficients A_n, B_n may be obtained in the following rather ingenious manner[4]. To find a particular coefficient, say A_3, we multiply each side of the second equation for $f(t)$ by $\cos 6\pi v_0 t$ and average

[4] R. P. Feynman, R. B. Leighton, M. Sands, *The Feynman Lectures on Physics* (Reading, Mass.: Addison-Wesley Publishing Co., Inc., 1963).

over the periodic time T to give

$$
\begin{aligned}
\overline{f(t)\cos 6\pi v_0 t} =\ & A_0\,\overline{\cos 6\pi v_0 t} \\
& + A_1\,\overline{\cos 6\pi v_0 t \cos 2\pi v_0 t} + B_1\,\overline{\cos 6\pi v_0 t \sin 2\pi v_0 t} \\
& + A_2\,\overline{\cos 6\pi v_0 t \cos 4\pi v_0 t} + B_2\,\overline{\cos 6\pi v_0 t \sin 4\pi v_0 t} \\
& + A_3\,\overline{\cos 6\pi v_0 t \cos 6\pi v_0 t} + B_3\,\overline{\cos 6\pi v_0 t \sin 6\pi v_0 t} \\
& + A_4\,\overline{\cos 6\pi v_0 t \cos 8\pi v_0 t} + \cdots
\end{aligned}
$$

$$
\begin{aligned}
=\ & A_0\,\overline{\cos 6\pi v_0 t} \\[4pt]
& + \frac{A_1}{2}\,\overline{(\cos 8\pi v_0 t + \cos 4\pi v_0 t)} \\[4pt]
& + \frac{B_1}{2}\,\overline{(\sin 8\pi v_0 t - \sin 4\pi v_0 t)} \\[4pt]
& + \frac{A_2}{2}\,\overline{(\cos 10\pi v_0 t + \cos 2\pi v_0 t)} \\[4pt]
& + \frac{B_2}{2}\,\overline{(\sin 10\pi v_0 t - \sin 2\pi v_0 t)} \\[4pt]
& + \frac{A_3}{2}\,\overline{(\cos 12\pi v_0 t + 1)} + \frac{B_3}{2}\,\overline{(\sin 12\pi v_0 t)} \\[4pt]
& + \frac{A_4}{2}\,\overline{[\cos 14\pi v_0 t + \cos(-2\pi v_0 t)]} + \cdots \\[4pt]
=\ & \frac{A_3}{2}.
\end{aligned}
$$

Therefore,

$$
A_3 = \overline{2f(t)\cos 6\pi v_0 t} = \frac{2}{T}\int_0^T f(t)\cos 6\pi v_0 t\, dt.
$$

Any other coefficients may be determined in the same manner.

As an example of Fourier analysis we consider the sawtooth function shown in Fig. 16.21. By using the methods just described, we can evaluate the Fourier components; the function $f(t)$ can be written in the form

$$
f(t) = \sum_n \left(-\frac{2}{n}\cos n\pi \right) \sin 2\pi n v_0 t.
$$

The details of the calculation are given in the box below. Note that in this example only the coefficients of the sine terms are nonzero. Although an infinite number of sine terms are required to exactly reproduce the sawtooth wave, a reasonable approximation is obtained by considering only a small number of such terms. For example, suppose that we retain only the first four terms and represent $f(t)$ by

$$f(t) = 2 \sin 2\pi v_0 t - \sin 4\pi v_0 t$$
$$+ \tfrac{2}{3} \sin 6\pi v_0 t - \tfrac{1}{2} \sin 8\pi v_0 t.$$

The result is plotted in Fig. 16.22. It is clear that even four terms are enough to give the trend of the function.

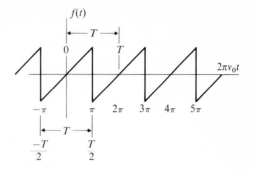

Fig. 16.21. A sawtooth function.

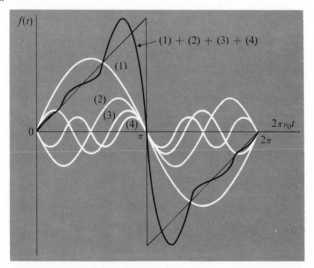

Fig. 16.22. The contribution of the first four terms of the Fourier series expansion of a sawtooth function.

Fourier analysis of a sawtooth function

We begin by considering the region $-\pi \leq 2\pi v_0 t \leq \pi$. Here the sawtooth function can be represented by

$$f(t) = 2\pi v_0 t.$$

Now,

$$A_0 = \frac{1}{T} \int_0^T f(t)\, dt$$

$$= v_0 \int_{-T/2}^{T/2} 2\pi v_0 t\, dt$$

$$= \frac{1}{2\pi} \int_{-\pi}^{\pi} x\, dx = 0.$$

In the second integral we replaced T by $1/v_0$, the period of the sawtooth, and changed the limits of the integral to $-T/2$ and $T/2$. We are still "averaging" over one period of the sawtooth; the same result is obtained whether we choose the period $-T/2$ to $T/2$ or 0 to T, but the former is more straightforward mathematically. In the third integral we replaced $2\pi v_0 t$ by x and changed the limits of integration to $-\pi$ and π. The coefficient A_1 is found to be equal to

$$A_1 = \overline{2f(t)\cos 2\pi v_0 t} = \frac{2}{T}\int_0^T f(t)\cos 2\pi v_0 t\, dt$$

in exactly the same way as A_3 was evaluated in the text. Making the same change of limits as for A_0, we have

$$A_1 = \frac{1}{\pi}\int_{-\pi}^{\pi} x\cos x\, dx$$

$$= \frac{1}{\pi}[\cos x + x\sin x]_{-\pi}^{\pi} = 0.$$

A similar calculation shows $A_3 = A_4 = \cdots = 0$. The coefficients of all cosine terms are zero. To find B_1 we multiply $f(t)$ by $\sin 2\pi v_0 t$ and average over one period to obtain

$$\overline{f(t)\sin 2\pi v_0 t} = A_0\,\overline{\sin 2\pi v_0 t}$$

$$+ A_1\,\overline{\sin 2\pi v_0 t\cos 2\pi v_0 t} + B_1\,\overline{\sin^2 2\pi v_0 t}$$

$$+ A_2\,\overline{\sin 2\pi v_0 t\cos 4\pi v_0 t} + \cdots$$

$$= B_1/2.$$

Therefore,

$$B_1 = \overline{2f(t)\sin 2\pi v_0 t} = \frac{2}{T}\int_0^T f(t)\sin 2\pi v_0 t\, dt.$$

Similarly,

$$B_2 = \overline{2f(t)\sin 4\pi v_0 t} = \frac{2}{T}\int_0^T f(t)\sin 4\pi v_0 t\, dt$$

and, in general,

$$B_n = \overline{2f(t)\sin 2n\pi v_0 t} = \frac{2}{T}\int_0^T f(t)\sin 2n\pi v_0 t\, dt.$$

Changing limits and putting $2\pi v_0 t = x$ yields

$$B_n = \frac{1}{\pi}\int_{-\pi}^{\pi} x\sin nx\, dx.$$

Therefore,

$$B_1 = \frac{1}{\pi}\int_{-\pi}^{\pi} x\sin x\, dx$$

$$= \frac{1}{\pi}(\sin x - x\cos x)_{-\pi}^{\pi} = 2.$$

Similarly,

$$B_2 = \frac{1}{\pi} \int_{-\pi}^{\pi} x \sin 2x \, dx = -1.$$

$$B_3 = \frac{1}{\pi} \int_{-\pi}^{\pi} x \sin 3x \, dx = -\tfrac{2}{3},$$

$$B_4 = \frac{1}{\pi} \int_{-\pi}^{\pi} x \sin 4x \, dx = \tfrac{1}{2},$$

etc. Therefore, we write

$$f(t) = 2 \sin 2\pi v_0 t - \sin 4\pi v_0 t + \tfrac{2}{3} \sin 6\pi v_0 t - \tfrac{1}{2} \sin 8\pi v_0 t + \cdots$$

PROBLEMS

1. Why do we not observe interference effects between the light beams emitted from two car headlights or between the sound waves emitted by the violins in the violin section of an orchestra?

2. Two plane waves of the same frequency but of amplitudes A and $0.5A$, respectively, are traveling in the same direction. Draw carefully a graph of the resultant wave when the phase difference between the two waves is (a) 0 and (b) $\pi/2$. Comment on the form of the resultant wave.

3. The time dependences of two square waves at a point in space are given by

$$f_1(t) = 3 \quad \text{if } (0 + 3n) < t < (1 + 3n) \quad \text{for } n = 0, 1, 2, \cdots$$
$$= 0 \quad \text{otherwise}$$
$$f_2(t) = 2 \quad \text{if } (0 + 3n) < t < (2 + 3n) \quad \text{for } n = 0, 1, 2, \cdots$$
$$= 0 \quad \text{otherwise.}$$

The planes of oscillation of the two waves coincide. Superpose the above two waves, giving both a mathematical expression and a scale drawing for the resultant wave.

4. (a) Write an equation to describe a longitudinal plane wave traveling in the z direction. Assume a wavelength of 0.22 m and a wave velocity of 331 m · sec^{-1}.
   (b) Write an equation to describe the superposition of the wave in part (a) with a wave of the same amplitude and frequency but traveling in the opposite direction.
   (c) Draw graphs to represent your answer to part (a). They should be clearly labeled and approximately to scale.

5. Two plane waves represented by the equations

$$\Delta y_1 = 0.10 \sin(4\pi t - 2\pi x) \, \text{m}, \qquad \Delta y_2 = 0.25 \sin(3\pi t - \pi x) \, \text{m},$$

respectively, are generated at points A and B, respectively. Assuming that the waves are initially in phase at A and B at $t = 0$, calculate the resultant displacement at a point 2 m from A and 3 m from B at time $t = 1.5$ sec.

*6. Spherical sound waves of frequency 1000 Hz traveling at a speed of 344 m $\cdot$ sec^{-1} spread out from a small source 5 m from a wall. There are two narrow apertures in the wall separated by 0.5 m and equidistant from the source. Each of these apertures may be considered themselves as new sources emitting spherical waves that interfere as they spread out toward a second wall 50 m distant. Calculate the intensity distribution of the sound at the position of the second wall.

7. Discuss the interference effects for sunlight passing through the two apertures described in Problem 6.

8. What is the shortest length of organ pipe in which a standing-wave sound pattern can be set up with a node at each end for a frequency of 30 Hz, 1000 Hz, and 15,000 Hz? Take the speed of sound to be 330 m $\cdot$ sec^{-1}.

9. In Western movies it is common to see the wheels of a stagecoach rotating in the wrong direction. Explain why this is and discuss its relation to the phenomenon of beats.

10. A standing-wave pattern is produced on a string 3 m long of linear density 3.0×10^{-3} kg $\cdot$ m^{-1}. Waves travel along the string at a speed of 70 m $\cdot$ sec^{-1}.
(a) For what frequency of vibration will four nodes occur in the standing wave pattern?
(b) What is the tension in the string?

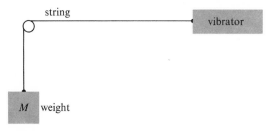

Fig. 16.23

11. A student uses the apparatus shown in Fig. 16.23 to study wave motion. For a selected weight and vibrator frequency the length of the string is varied to obtain a standing wave pattern. The distance between the nodes is measured with a ruler. Some of his results are given in Table 16.1. Use these data to deduce a relation between the wave speed and the tension in the string.

12. A student finds that the weight of a 90-cm sample of the string used in the experiment referred to in Problem 11 is 4.7 g. Using the data given in that problem deduce the frequency of the vibrator.

13. A student is given a disc having 12 equally spaced slits as shown in Fig. 16.24. He uniformly rotates the disc at a rate of 5 rev $\cdot$ sec^{-1}, and observes a vibrator through the disc. He notices what is in fact a beat frequency: the vibrator appears to vibrate slowly at 5 Hz. What can he deduce about the actual frequency of the vibrator?

Table 16.1

Tension (N)	Internodal distance (cm)
1.96	18.7, 17.4, 19.0, 18.7, 18.4
2.94	22.1, 22.1, 22.4, 22.6
3.92	25.2, 26.6, 26.2
4.90	27.6, 29.5, 29.4

14. Two strings held fixed at their ends have the same length and same mass. One string held under a certain tension T vibrates at frequency 500 Hz. If the tension on the other string is $6T/5$, what is the beat frequency between the two strings?

15. If a radio station wishes to broadcast a music program with frequencies up to 15,000 Hz, what frequency range will be required on the broadcast band at 540 kHz and at 120 MHz to carry this program? Express your answers as a fraction of the carrier frequency.

16. Two tuning forks are vibrating at 252 Hz and 256 Hz. What is the beat frequency resulting from the superposition of the two waves? At a certain point in the room in which the forks are vibrating, the sound amplitude from the first fork is A_1 and the amplitude from the second fork is $0.1A_1$. What will be the maximum and minimum sound amplitudes at that point?

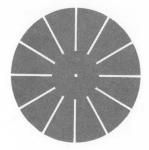

Fig. 16.24

*17. Amplitude modulation and the production of beats are closely related phenomena. Discuss.

*18. A modulated electromagnetic wave has the form

$$\Delta y = A(t) \cos 2\pi v_c t,$$

where the time dependent amplitude is given by

$$A(t) = A_0(1 + b \cos 2\pi v t + b' \cos 2\pi v' t).$$

This wave may alternatively be represented as a superposition of simple harmonic oscillations.
(a) What are the frequencies and amplitudes of the component oscillations?
(b) Sketch carefully and to scale the time dependent amplitude of the modulated wave and the modulated electromagnetic wave.
Take $b = b' = \frac{1}{2}$, $A_0 = 2$ units, $v' = 2v = 1000$ Hz and $v_c = 10^6$ Hz.

19. For the propagation of elastic waves along a linear lattice of atoms the dispersion relation (relation between ω and k) is

$$\omega = \pm \left(\frac{4\beta}{M}\right)^{1/2} \sin\left(\frac{ka}{2}\right),$$

where M is the mass of the atoms, β the force constant, and a the separation between atoms.
(a) Plot a graph of ω versus k for $0 < k < \pi/a$ for the positive branch of the dispersion relation.
(b) Write expressions for the phase velocity and group velocity of such elastic waves in terms of the constant $v_0 = (\beta a^2/M)^{1/2}$.

20. The phase velocity of ocean waves is given by $(g\lambda/2\pi)^{1/2}$, where g is the acceleration due to gravity. Deduce an expression for the group velocity of these waves.

21. It is wished to amplify a sawtooth wave by electronic means. To obtain the desired response requires linear amplification of the first ten terms of the Fourier series expansion of the function. If the periodic time of the sawtooth wave is 10 μsec (1 μsec = 10^{-6} sec), over what frequency range must the amplifier have uniform gain?

22. The Fourier series expansion of the wave shown in Fig. 16.25 is

$$f(t) = \frac{\pi}{2} + 2 \sum_{n=1}^{\infty} \frac{\sin(2n-1)t}{2n-1}.$$

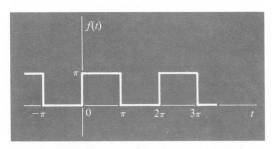

Fig. 16.25

Plot a graph of this function for values of t between 0 and 2π, using the first four terms of the series.

*23. Using Fourier analysis, evaluate the wave

$$f(t) = \cos^3 t.$$

Verify your result by direct trigonometric analysis.

The kinetic theory of gases 17

17.1 INTRODUCTION

The most familiar example of a gaseous substance is the earth's atmosphere. The atmosphere is not a pure gas but is a mixture of a large number of constituent gases, the major ones being nitrogen and oxygen. A 1 m^3 sample of the atmosphere at the earth's surface contains of the order of 10^{25} molecules. These molecules possess thermal energy, as a result of which they move about in a random manner. The gravitational attraction of the earth is responsible for the retention of the atmosphere by the earth. It is the balance between kinetic and potential energy that allows the lighter elements (such as hydrogen and helium) to escape from the atmosphere and the heavier ones to remain. The moving molecules exert forces on one another that are electrical in origin. Of course, gravitational forces of attraction also exist between the molecules, but they are infinitesimal in comparison with the forces with which we are concerned. Since the instantaneous force experienced by any molecule is determined by the instantaneous positions of all of the other molecules, the motions of the molecules are not independent. A detailed analysis of the molecular motions requires the simultaneous solution of 10^{25} coupled equations. Such an analysis is quite impossible even with high-speed electronic computers. In point of fact, we are not really interested in the detailed motions of the individual molecules but rather only in a relatively small number of macroscopic (large scale) parameters of the gas as a whole, such as the pressure, temperature, density, energy content, etc. We shall discuss two quite different approaches that attempt to relate these macroscopic parameters to the molecular motions. In the **kinetic theory** approach we make enough simplifying assumptions about the molecular motions so that we can solve the equations of motion. In the **statistical physics** approach, the detailed behavior of the individual molecules is ignored, and statistical methods are applied to the collection of molecules. This method will be discussed in Chapters 19 and 20.

When the kinetic theory and statistical physics were first developed it was generally believed that Newton's laws of mechanics were equally as valid in the microscopic world of the atom as in the macroscopic world of our daily experiences. The dramatic failure of classical mechanics in a number of instances led to the development of quantum mechanics. Some of these failures will be discussed in this and later chapters.

17.2 THE IDEAL GAS

In order to solve the equations of motion for the constituent molecules of a gas, we shall set up an idealized model for the gas that is amenable to mathematical analysis. To construct an adequate model we require some knowledge of the nature of the interaction force between molecules. The most important contribution to the force is electrical in origin[1]. For example, let us consider two hydrogen molecules separated by a distance R. Each molecule is composed of two positively charged nuclei and two negatively charged electrons that might better be pictured as a smeared-out distribution of negative charge (see Fig. 17.1). The interaction between the electrical charges constituting a molecule results in a stable molecular configuration with a characteristic internuclear spacing. The interaction between the electrical charges on different molecules results in a force that varies with intermolecular spacing R, as shown in Fig. 17.2. At large separations the force is attractive but of extremely small magnitude. As the molecules come closer, the attractive force increases, passes through a maximum, and decreases to zero at separation R_0; the force then becomes repulsive.

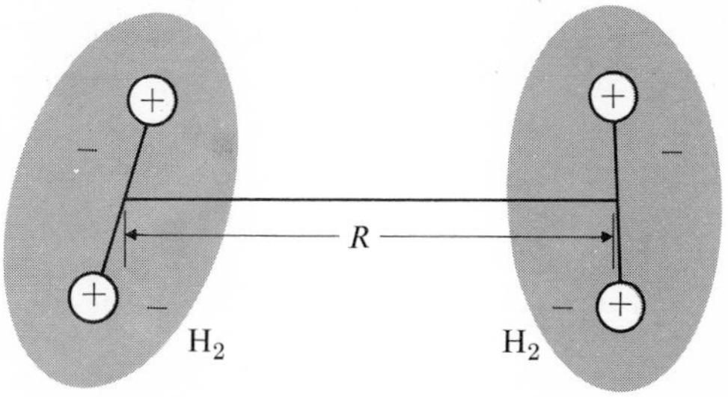

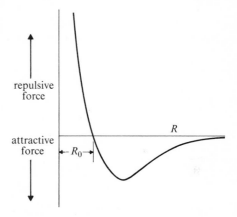

Fig. 17.1. Two hydrogen molecules separated by a distance R.

The separation R_0 is approximately equal to the diameter of the molecule; that is, the resultant force of interaction becomes zero when the two negative charge distributions slightly overlap. For a molecular substance in the solid state, where the kinetic energy of the molecules is relatively much less important than in the gas, the separation R_0 for adjacent pairs of molecules represents an equilibrium separation about which they oscillate.

The average separation between the molecules in the atmosphere at the earth's surface is about fifty molecular diameters, that is, about $50R_0$. Therefore, the forces between the molecules are extremely small most of the time. An **ideal gas** is defined as one whose molecules exert no force

Fig. 17.2. The force between two molecules is repulsive for separations less than a separation R_0 and attractive for separations greater than R_0; a maximum of the attractive force occurs just outside R_0 but the force approaches zero as $R \to \infty$.

on one another except when they physically collide. As a first approximation we adopt the **hard-sphere model** for a gas. The force-separation curve for this model is shown in Fig. 17.3. In this model the molecules are represented as impenetrable spheres of diameter R_0 between which there is no significant force of attraction. For this model we can conclude that the gas molecules will travel in straight-line paths at constant speed

---

[1] B. V. Derjaguin, "The Force Between Molecules," *Scientific American*, July, 1960. Available as *Scientific American Offprint 266* (San Francisco: W. H. Freeman and Co., Publishers).

between collisions with one another and with the walls of the container in which the gas is located.

A second assumption that we shall make is that the molecules are point particles; that is, they occupy a negligible volume. The justification for this assumption is that the molecular diameters are small compared to the average molecular separation. As a result of this assumption, only collisions with the walls of the container are significant. It should be noted that collisions between the molecules are essential for establishing a state of equilibrium within the gas but that such collisions may be disregarded when discussing the equilibrium behavior of the gas.

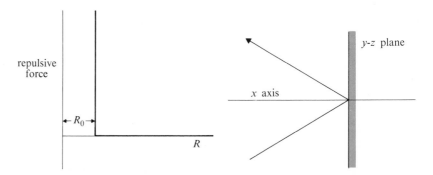

Fig. 17.3. A force vs. separation curve for a "hard sphere" model of a molecule.

Fig. 17.4. A molecule striking a wall undergoes a change in direction of motion which results in a change in its momentum.

In addition we shall make the reasonable assumption that all directions of motion for the molecules are equally probable. That is, if we select a molecule at random it is just as likely to be traveling in any one direction as in any other. No direction is preferred in the absence of an external force.

We would not expect the nature of the interaction of the molecules with the walls of the container to play any essential role in the determination of the macroscopic properties of the gas. We shall therefore assume for simplicity that collisions with the walls are perfectly elastic and that no change in the tangential component of the molecular velocity occurs upon collision. That is, the molecules travel with the same speed before and after they collide with a wall of the container, and the molecular trajectories make equal angles with the container wall before and after a collision.

We are now ready to carry out the mathematical analysis for our model gas. If a molecule collides with a wall lying in the y, z plane, as shown in Fig. 17.4, the change in the x component of its momentum is just

$$\Delta p = 2mv_x,$$

where v_x is the magnitude of the x component of its velocity. The ith molecule will collide with the wall shown in Fig. 17.4 within a time interval of Δt sec if it is within a distance $v_{x_i}\Delta t$ of the wall at the beginning of the interval and moving toward it. The probability of this occurring is $(v_{x_i}\Delta t A)/2V$, where V is the volume of the container, A is the cross-sectional area of the container perpendicular to the x direction, $v_{x_i}\Delta t A$ is that part of the volume in which the molecule must be at the start of the interval if it is to make a collision within the specified interval (see Fig. 17.5), and the factor $\frac{1}{2}$ takes account of the fact that $\frac{1}{2}$ of the molecules in the specified volume will be traveling in the wrong direction to make a collision.

The total force exerted by the molecules in the gas on the area A of the wall is

$$F_x = \sum_i \frac{\Delta p_i}{\Delta t}$$

$$= \sum_i \left(\frac{2mv_{x_i}}{\Delta t}\right)\frac{1}{2}\left(\frac{v_{x_i}\Delta t A}{V}\right)$$

$$= \frac{mA}{V}\sum_i v_{x_i}^2.$$

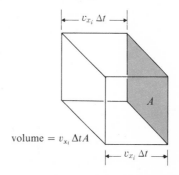

volume $= v_{x_i}\Delta t A$

Fig. 17.5. One half of the molecules in a volume $v_{x_i}\Delta t A$ of a container will make a collision with the wall within time t.

Defining the **pressure** P as the force per unit area, we can write

$$P = \frac{F_x}{A}$$

$$= \frac{m}{V}\sum_i v_{x_i}^2.$$

Defining the **root mean square value** $(v_x)_{\text{rms}}$ of the x component of the molecular velocity by

$$(v_x)_{\text{rms}} = \left(\frac{1}{N}\sum_i v_{x_i}^2\right)^{1/2},$$

we can write

$$P = \frac{mN}{V}(v_x)_{\text{rms}}^2.$$

Since we are assuming that the molecular motion is random, we can assume that

$$(v_x)_{\text{rms}}^2 = (v_y)_{\text{rms}}^2 = (v_z)_{\text{rms}}^2.$$

Since

$$v_{rms}^2 = (v_x)_{rms}^2 + (v_y)_{rms}^2 + (v_z)_{rms}^2,$$

it follows that

$$v_{rms}^2 = 3(v_x)_{rms}^2.$$

Therefore, we have

$$PV = \frac{2N}{3}\left(\frac{mv_{rms}^2}{2}\right)$$

$$= \frac{2N}{3}\left(\frac{\overline{mv^2}}{2}\right),$$

where $\overline{mv^2}/2$ is the **average translational kinetic energy per molecule** and $N\overline{mv^2}/2$ is the **total energy** U of the gas of point molecules (usually called the **total internal energy** of the gas). Therefore,

$$PV = \frac{2U}{3}.$$

This we call the **thermal equilibrium equation of state** of an ideal gas. (A discussion of the concept of **thermal equilibrium** is given in Section 23.2.)

This is as far as kinetic theory can take us. It has succeeded in providing a relation between the macroscopic quantities, P, V, and U and of giving molecular interpretations to P and U. The pressure associated with a gas is not a static push but is the average effect of many tiny impulsive blows resulting from the collisions of the molecules with the walls of the container. The energy content of the gas is the sum of the translational kinetic energies of the constituent molecules.

Example. A beam of protons of 1 MeV energy is incident normally upon a metal target in a small accelerator. If the proton **current** is 1 μA (10^{-6} A), what is the force exerted by the proton beam on the target, assuming all protons are absorbed in the target?

Solution. A current of 1 A is defined as one coulomb (1 C) of charge passing a given point in a circuit per second. Therefore,

$$1 \ \mu A = 10^{-6} \ C \cdot sec^{-1}.$$

The charge on a proton is 1.6×10^{-19} C. Therefore, in a 1 μA beam of protons there are

$$N = \frac{10^{-6}}{1.6 \times 10^{-19}}$$

$$= 6.3 \times 10^{12} \text{ protons}$$

striking the target per second.

Let us assume, for the moment, that 1 MeV protons are moving slowly enough that we may use Newtonian physics to calculate their speed. Since 1 MeV = 1.6 × 10⁻¹³ J and the mass of the proton is 1.67×10^{-27} kg, we have

$$K = \tfrac{1}{2}mv^2$$

$$v^2 = \frac{2K}{m}$$

$$= \frac{2 \times 1.6 \times 10^{-13}}{1.67 \times 10^{-27}}$$

$$= 1.92 \times 10^{14}$$

$$v = 1.4 \times 10^7 \, \text{m} \cdot \text{sec}^{-1}.$$

Since

$$\left(\frac{v}{c}\right)^2 = \left(\frac{1.4 \times 10^7}{3 \times 10^8}\right)^2$$

$$= 0.002 \ll 1,$$

the Newtonian calculation is sufficient. The total momentum imparted to the target per second will be

$$Nmv = 6.3 \times 10^{12} \times 1.67 \times 10^{-27} \times 1.4 \times 10^7$$

$$= 1.5 \times 10^{-7} \, (\text{kg} \cdot \text{m} \cdot \text{sec}^{-1}) \cdot \text{sec}^{-1},$$

which is equivalent to a force of 1.5×10^{-7} N.

17.3   TEMPERATURE AND THE IDEAL GAS LAW

Temperature is a fundamental quantity in physics. We know from our own experience that some objects are hotter than others. If heat flows from one object to a second object we describe the first object as being hotter than the second or, conversely, that the second is colder than the first. The relative "hotness" of different objects (or substances) is noted by assigning a **temperature** to them, with higher values of temperature denoting increasingly hotter objects. Temperature is a scalar quantity. A more complete discussion of the concepts of temperature, and of hot and cold, will be given in Chapter 23.

Like length, mass, and time, temperature must be defined operationally. Therefore, we take temperature as that quantity measured by a **thermometer**. A difficulty arises, however, in trying to choose a standard thermometer. Consider the common thermometer made by sealing alcohol in glass (Fig. 17.6). The temperature scale of such a thermometer can be readily established with the help of two reference baths. Convenient reference baths are provided by melting ice and boiling water.

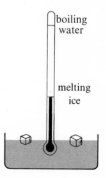

Fig. 17.6. A common alcohol in glass thermometer.

The difference in the height of the column of alcohol when placed in the two baths can be subdivided arbitrarily into a number of equal parts. On the **centigrade (Celsius) scale**, the temperature of melting ice is taken as 0°C and the interval between the temperatures of melting ice and boiling water as 100°C.

If we now choose some **thermometric substance** other than alcohol, such as mercury, and construct a second thermometer in the same manner, the two thermometers must agree when compared at the 0 and 100°C points because of the procedure used in establishing the scales. But, and this is the important point, they do not necessarily agree at intermediate temperatures. Discrepancies result from the slightly different rates of expansion of different liquids in different temperature regions.

Many physical properties other than the expansion of a liquid are temperature dependent and can form the basis of a thermometer. The expansion of a gas, the differential expansion of two metals, the equilibrium pressure of the vapor above a liquid, the electrical resistance of a wire, the voltage developed at the junction of two dissimilar metals, and the color of a hot object are all examples of temperature-dependent physical properties commonly used in the design of thermometers. Any such thermometers will give results that, for temperatures other than the fixed points used for establishing their scales, depend more or less on the thermometric substance of the thermometer.

Experiments show that the best agreement is found among **constant-volume gas thermometers** of the type illustrated in Fig. 17.7. The gas G is

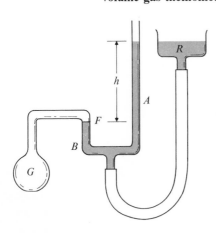

Fig. 17.7. A constant volume gas thermometer.

contained in the bulb, and the pressure exerted by it is measured using the mercury manometer to which it is attached. As the temperature increases, the gas expands, tending to force the mercury down tube B and up tube A. Tubes A and B are coupled through a flexible tube to a mercury reservoir R. By adjusting the height of R, we can always return the mercury level in B back to a fiducial mark F, thereby keeping the gas at constant volume. By measuring the difference in heights h of the mercury columns A and B, we can readily determine the pressure P on the gas G. We define the ratio of two temperatures to be the ratio of the corresponding gas pressures. Therefore,

$$\frac{T}{T_s} = \frac{P}{P_s},$$

where P_s and T_s are the pressure and temperature of a standard fixed point. In modern thermometry the only standard fixed point is the **triple point of water**. The triple point is the temperature at which the solid, liquid, and vapor phases coexist in equilibrium. Unlike the temperature of melting ice or the temperature of boiling water, which vary with atmospheric pressure, the triple-point temperature is well defined, since coexistence of the three phases in equilibrium can occur only at one definite pressure.

It is a fact of experiment that all constant-volume gas thermometers at the same temperature approach the same reading as the pressure of the gas in them becomes vanishingly small. Figure 17.8 shows the typical behavior of three gas thermometers as the gas pressure approaches zero. Therefore, we can define a **gas temperature scale** by the equation

$$\frac{T}{T_s} = \lim_{P_s \to 0} \left(\frac{P}{P_s}\right).$$

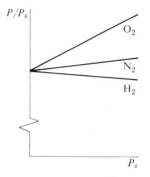

Fig. 17.8. All constant volume gas thermometers approach the same value for P/P_s as $P_s \to 0$.

One might argue that although this is obviously a very good temperature scale, it still depends upon the properties of gases in general and therefore is not the best possible one. In Section 23.4 we shall introduce the **Kelvin temperature scale** (often called the absolute temperature scale), which indeed is independent of the properties of any substance or group of substances. In fact, it can be shown that the gas temperature scale and the Kelvin scale are identical in the temperature region in which gas thermometers can be used. The Kelvin scale has an added advantage in as much as it defines an absolute zero of temperature. The temperature T_s of the triple point of water on the Kelvin scale is

$$T_s = 273.16°K.$$

The temperature in degrees Kelvin is related to the temperature in degrees centigrade simply by a shift in origin. That is, the temperature of the triple point of water is 0.01°C. The gas temperature scale is defined as

$$T = (273.16) \lim_{P_s \to 0} \left(\frac{P}{P_s}\right)_{\text{constant volume}} °K.$$

Now that we have some feeling for the concept of temperature, let us return to our kinetic theory equation of state

$$PV = \frac{2U}{3}.$$

Experimental measurements on gases at low pressures are consistent with the equation of state

$$PV = nRT,$$

where n is the number of kilogram-moles (kg-mole) of gas, R is the universal constant $8.314 \times 10^3 \, \text{J} \cdot (\text{kg-mole})^{-1} \cdot (°K)^{-1}$ and T is the absolute

(Kelvin) temperature. (Experimentally, equations of state are often quoted in CGS units, in which case n is in gram-moles (g-mole) and R is $8.314 \times 10^7 \, \text{erg} \cdot (\text{g-mole})^{-1} \cdot (°K)^{-1}$. We shall use CGS units later in this chapter.) This equation can be checked experimentally in a student laboratory making use of very simple apparatus (see Problem 5 at the end of this chapter). Experiments performed at a constant temperature yield

$$PV = \text{constant.}$$

This is known as **Boyle's law**. Experiments performed with a constant volume of gas yield

$$\frac{P}{T} = \text{constant.}$$

This is known as **Charles' law.** The theoretical and experimental laws will be in complete agreement if we set

$$\frac{2U}{3} = nRT.$$

The average translational kinetic energy of an individual molecule is then

$$\frac{U}{N} = \frac{1}{2} m \overline{v^2} = \frac{3}{2} \frac{nRT}{N}.$$

The number of moles n is given by

$$n = \frac{N}{N_0},$$

where $N_0 = 6.02 \times 10^{26}$ is the number of molecules in a kilogram-mole and is known as **Avogadro's number.** Therefore,

$$\frac{1}{2} m \overline{v^2} = \frac{3}{2} \frac{RT}{N_0} = \frac{3}{2} kT,$$

where $k = R/N_0 = 1.38 \times 10^{-23} \, \text{J} \cdot (°K)^{-1}$ and is known as **Boltzmann's constant.** That is, the average kinetic energy per molecule for an ideal gas depends only on the temperature of the gas and is independent of the pressure, the volume, and the species of molecule.

The mean square speed of molecules in an ideal gas is given by

$$\overline{v^2} = \frac{3kT}{m}.$$

and the square root of the mean square speed by

$$v_{rms} = (\overline{v^2})^{1/2}$$

$$= \left(\frac{3kT}{m}\right)^{1/2}.$$

A more detailed kinetic theory analysis can also make a prediction about the spread in the actual speeds of individual molecules about the average value. It can be shown that the number of molecules dN having speeds within the interval bounded by speeds v and $v + dv$ is given by

$$dN = \frac{4N}{\pi^{1/2}}\left(\frac{m}{2kT}\right)^{3/2} v^2 \exp\left(\frac{-mv^2}{2kT}\right) dv.$$

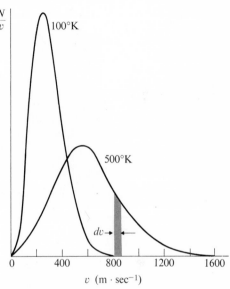

(See Chapters 19 and 20 for the development of this equation by statistical methods.) This result, known as the **Maxwell–Boltzmann distribution of molecular speeds**, is shown in Fig. 17.9 for two temperatures. Collisions between molecules are essential for establishing and maintaining the Maxwell–Boltzmann distribution. The area of the shaded strip is

$$\left(\frac{dN}{dv}\right) dv = dN\,;$$

the area under the curve is

$$\int_0^N dN = N.$$

Fig. 17.9. The Maxwell–Boltzmann distribution of molecular speeds of hydrogen molecules at 100°K and 500°K.

For hydrogen gas at 300°K, the mean molecular speed is 1.96×10^3 m·sec^{-1}, and about 99.9% of the molecules have speeds less than 4.0×10^3 m·sec^{-1}. Since the molecules of all other gases are heavier, the molecular speeds will on the average be less at the same temperature; that is, the Maxwell–Boltzmann distributions for other gases will peak at smaller values of v. From this discussion we can conclude that, except at extremely high temperatures, relativistic effects will be quite negligible.

Since the mean velocity of hydrogen molecules is about ten times smaller than the escape velocity (see Section 12.8) from the earth's surface, only the most rapidly moving hydrogen molecules are lost to the earth's atmosphere. Once they are lost, however, the Maxwell–Boltzmann distribution is reestablished by collisions and again the most rapidly moving ones escape. For any given initial population of hydrogen in the

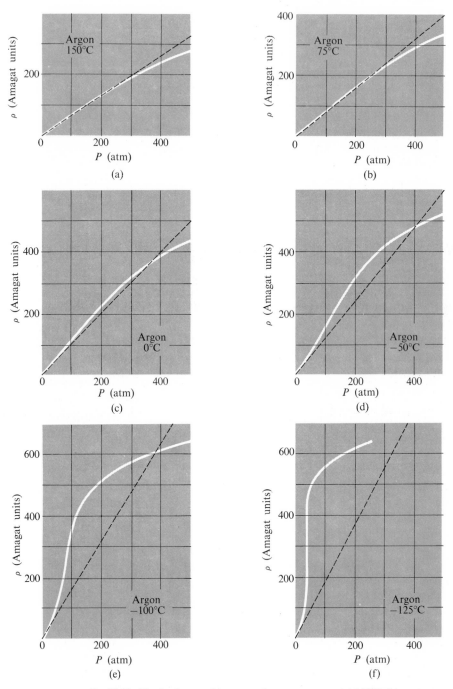

Fig. 17.10. The isotherms of argon at six temperatures: (a) 150°C, (b) 75°C, (c) 0°C, (d) −50°C, (e) −100°C, (f) −125°C.

earth's atmosphere, most of the hydrogen would eventually escape. In general, the time for this to occur depends on the difference between the escape velocity and the mean molecular velocity. A reasonable criterion for the retention of a specific molecular constituent as a "permanent" part of the atmosphere of a celestial body is that the mean molecular speed be less than 20% of the escape velocity.

17.4   THE BEHAVIOR OF REAL GASES

The behavior of gases over a wide range of temperatures and pressures deviates substantially from the ideal gas law. In Fig. 17.10 the behavior of argon (Ar) at six different temperatures is shown. Fig. 17.11 shows the behavior of ethylene (C_2H_4), methane (CH_4), and hydrogen (H_2) at 0°C. These data were obtained by Dutch scientists working at the van der Waals Laboratory in Amsterdam. In these graphs the density ρ is plotted as a function of the gas pressure P. To compare these results with the ideal gas law we shall rewrite it in the form

$$P\tilde{V} = RT,$$

where $\tilde{V}$ is the volume per kilogram-mole, and define a **number density**

$$\rho = \frac{1}{\tilde{V}}$$

so that

$$P = \rho RT.$$

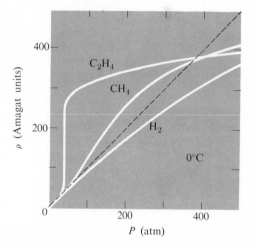

Fig. 17.11. The isotherms of ethylene, methane, and hydrogen at 0°C.

In these graphs the density is given as a multiple of the density at 1 atm pressure and 0°C, that is, in so-called **Amagat units**. In these units the ideal gas law becomes

$$P = \rho\left(\frac{T}{273}\right),$$

where T is the absolute temperature in degrees Kelvin. The dashed lines represent ideal gas behavior.

From graphs of this sort it is possible to make two general observations concerning the behavior of real gases. From Fig. 17.10 we see that the **isotherms** (curves at constant temperature) of argon undergo a marked change in shape with temperature, tending to become more and more ideal as the temperature increases. A comparison of the isotherms of a number of gases suggests that the shape of the isotherms of all gases varies in a similar fashion with temperature but that a particular shape of isotherm occurs at different temperatures for different gases. Let us consider Figs. 17.10 and 17.11. The isotherm of ethylene at 0°C behaves very much like that of argon at −125°C; the isotherm of methane at 0°C behaves very much like that of argon at −50°C; and the isotherm of hydrogen at 0°C behaves very much like that of argon at 75°C. That is, hydrogen approaches ideal behavior at a lower temperature than argon, which approaches ideal behavior at a lower temperature than methane, which approaches ideal behavior at a lower temperature than ethylene. Let us compare this behavior with the temperature of liquefaction of these gases; hydrogen liquefies at −252.8°C, argon at −186°C, methane at −164.5°C, and ethylene at −103°C. This illustrates the general result that gases tend to approximate ideal behavior as their temperature becomes high relative to the temperature at which they liquefy.

It is possible to fit the experimental isotherms with empirical relations of the form

$$\frac{P}{\rho RT} = 1 + B(T)\rho + C(T)\rho^2 + \cdots.$$

This form of equation is known as a **virial equation of state**. The temperature-dependent functions $B(T)$, $C(T)$, etc., are referred to as the second, third, etc., **virial coefficients**. In order to understand the existence of the second and higher order virial coefficients, let us reexamine the two essential assumptions of the ideal gas theory, that the molecules exert no force on one another except during collisions and that the molecules can be treated as point particles.

Neither assumption can be strictly valid. The effect of intermolecular attractions results in a modification of the pressure exerted by the gas. It has been shown[2] that the required modification of the pressure cannot be accounted for by any reduction in molecular speeds or by mere alterations in directions of motion. Rather, a reduction in pressure results from a reduced frequency of collisions, which in turn results from time delays caused by the curved trajectories of the interacting molecules. That is, real molecules will not actually collide physically but will have their directions of motion altered continuously when near each other. Indeed, their orbits will be very much like the hyperbolic orbits for central force motions as described in Section 13.6. Since the molecules follow curved trajectories instead of straight-line paths, the average time between collisions with the container is longer and the pressure lower

[2] W. C. Thorburn, "Role of Intermolecular Attractions in Liquids and Gases" and "Dynamic Consequences of Intermolecular Attraction in Gases," *American Journal of Physics*, **34** (February 1966), 132, 136.

than that for ideal point molecules. To a first approximation we may take account of this effect by replacing the pressure P in the ideal gas law by $P + (an^2/V^2)$, where a is an empirical constant representing the strength of the intermolecular forces and n the number of moles of gas. The finite size of the molecules means that the volume available for the motion of the molecules (the free volume) is less than the total volume. Since pressure changes can cause a change in the free volume only, the total volume V in the ideal gas law should be replaced by the free volume $V - nb$, where b is an empirical constant representing the volume of 1 kg-mole of molecules.

With these two changes we obtain as an approximate equation of state for a real gas

$$\left(P + \frac{an^2}{V^2}\right)(V - nb) = nRT.$$

This is known as the **van der Waals equation of state**. The constants a and b for a particular gas are obtained by fitting this equation to experimental isotherms. Since the van der Waals equation represents only a rather crude attempt to overcome the limitations of the ideal gas equation, it is not surprising that it holds for a particular gas over only a relatively small temperature range. Nonetheless, it is quite adequate to account for the gross features of real gas behavior that have been discussed with reference to Figs. 17.10 and 17.11. To see this we first relate the van der Waals equation to the virial equation. Since $V = n\tilde{V}$, the van der Waals equation may be written

$$\left(P + \frac{a}{\tilde{V}^2}\right)(\tilde{V} - b) = RT;$$

therefore,

$$P = \frac{RT}{\tilde{V} - b} - \frac{a}{\tilde{V}^2}$$

$$= \frac{\rho RT}{1 - b\rho} - a\rho^2.$$

Using a binomial expansion for $(1 - b\rho)^{-1}$, we obtain

$$\frac{P}{\rho RT} = 1 + \left(b - \frac{a}{RT}\right)\rho + b^2\rho^2 + \cdots.$$

By comparison with the virial equation we have

$$B(T) = \left(b - \frac{a}{RT}\right), \qquad C(T) = b^2.$$

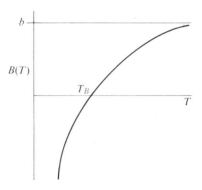

Fig. 17.12. The second virial coefficient $B(T)$ for a van der Waals gas is negative at low temperatures and positive at high temperatures. It is zero at the Boyle temperature, T_B.

The temperature dependence of $B(T)$ is shown in Fig. 17.12. For low enough temperatures $B(T)$ is negative, and for high enough temperatures it is positive. At a particular temperature T_B known as the **Boyle temperature**, $B(T) = 0$. The third virial coefficient and all higher order ones are positive and independent of temperature. Therefore, at low densities, where the term linear in ρ will dominate, the pressure P required to maintain a given ρ may either be greater than or less than the ideal gas prediction, depending on the temperature. At high densities where quadratic terms of higher power in ρ become dominant, the pressure P required to maintain a given ρ will always exceed the ideal gas prediction. Figure 17.13 shows the shape of the isotherms predicted by the van der Waals equation. Compare Fig. 17.13 with Fig. 17.11.

It is an experimental fact that a substance cannot exist in the liquid state at a temperature above that of its **critical point** (see Section 23.7), regardless of how high a pressure is applied to it. The critical point is characterized by a **critical pressure** P_c, a **critical volume** $\tilde{V}_c$, and a **critical temperature** T_c that satisfy the mathematical relations

$$\left(\frac{\partial P}{\partial V}\right)_T = \left(\frac{\partial^2 P}{\partial V^2}\right)_T = 0$$

for the equation of state of the substance. For the van der Waals equation, the critical constants may be shown to be

$$P_c = \frac{a}{27b^2}, \qquad \tilde{V}_c = 3b, \qquad T_c = \frac{8a}{27bR}.$$

(The proof of these relations is rather difficult as you will find if you attempt Problem 20.) Once the critical constants are known, the van der Waals parameters can be readily deduced.

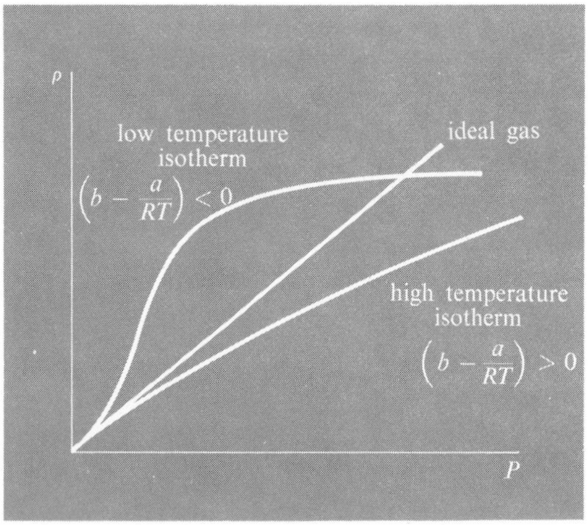

Fig. 17.13. High and low temperature isotherms as predicted by the van der Waals equation of state.

(Figure labels: low temperature isotherm $\left(b - \frac{a}{RT}\right) < 0$; ideal gas; high temperature isotherm $\left(b - \frac{a}{RT}\right) > 0$)

Example. The critical constants for argon have been measured as $T_c = 151°K$, $\tilde{V}_c = 75.2 \text{ cm}^3$ and $P_c = 48$ atm. Deduce the van der Waals parameters. Compare the van der Waals equation of state at $-50°C$ with the measured equation of state as depicted in Fig. 17.10. (Note: 1 g-mole $\equiv$ one gram-mole $\equiv$ mass in grams equal to the molecular weight of the gas.)

Solution. The constants b and a are

$$b = \frac{\tilde{V}_c}{3}$$

$$= \frac{75.2}{3}$$

$$= 25.1 \text{ cm}^3 \cdot \text{g-mole}^{-1}$$

$$= 0.0251 \; l \cdot \text{g-mole}^{-1}$$

$$a = \frac{27bRT_c}{8}$$

$$= \frac{27 \times 0.0251 \times 0.08207 \times 151}{8}$$

$$= 1.05 \; l^2 \cdot \text{atm} \cdot \text{g-mole}^{-2},$$

where $1 \; l \equiv \text{one liter} \equiv 10^3 \text{ cm}^3$. The substitution of these parameters into the van der Waals equation gives

$$P = \frac{\rho RT}{1 - 0.0251\rho} - 1.05\rho^2.$$

For

$$R = 0.08207 \; l \cdot \text{atm} \cdot \text{g-mole}^{-1} \cdot (^\circ K)^{-1} \quad \text{and} \quad T = 223^\circ K,$$

$$P = \frac{18.3\rho}{1 - 0.0251\rho} - 1.05\rho^2$$

with ρ in the units $(\text{gram-moles})^{-1}$. By selecting convenient values of ρ, the corresponding values of P may be calculated. The densities may be converted to Amagat units by multiplying them by $22.4 \; l$, the volume occupied by 1 mole of gas at 1 atm and $273^\circ K$. Some results are given in Table 17.1. The experimental data and the deduced van der Waals equation are plotted in Fig. 17.14. Although the van der Waals curve does not exactly reproduce the experimental curve at the higher densities it certainly does reproduce the general trend.

Table 17.1 P–ρ data for argon at $223^\circ K$

ρ ($g\text{-}mole \cdot l^{-1}$)	P (atm)	ρ (Amagat units)
10	139	224
15	206	336
20	315	448
23	440	516

The van der Waals parameters for hydrogen, argon, methane, and ethylene as deduced from critical point data are quoted in Table 17.2. As we go down the series of gases we see that a increases faster than b. Therefore, with a lot of reflection, we can see that a certain shape of isotherm will occur at a lower temperature for hydrogen than for argon, than in turn for methane, than in turn for ethylene. This agrees with our earlier conclusions drawn from a consideration of Figs. 17.10 and 17.11.

Table 17.2   van der Waals parameters deduced from critical point data

Gas	b $l \cdot (g\text{-}mole)^{-1}$	a $l^2 \cdot atm \cdot (g\text{-}mole)^{-2}$
Hydrogen (H_2)	0.0217	0.20
Argon (Ar)	0.0251	1.05
Methane (CH_4)	0.0330	1.74
Ethylene (C_2H_4)	0.0410	3.21

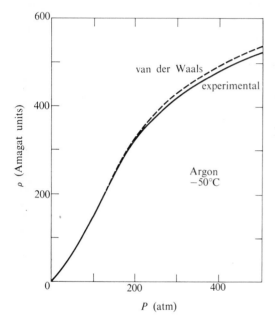

Fig. 17.14. Experimental and van der Waals isotherms of argon at $-50°C$.

17.5   EQUIPARTITION OF ENERGY

We saw in Section 17.3 that the average translational kinetic energy of a molecule of an ideal gas at thermal equilibrium at temperature T is $\frac{3}{2}kT$. If we add thermal energy (heat) to the gas and allow it to come to equilibrium at a higher temperature $T + \Delta T$, the average kinetic energy per molecule will increase by $\frac{3}{2}k\Delta T$.

Now let us consider a mixture of gases in thermal equilibrium. We assume that the gases are at a temperature and pressure that are consistent with ideal gas behavior and that the gases do not react chemically with one another. In such situations it is found experimentally that the total pressure of the mixture is equal to the sum of the pressures that each component gas would exert if it alone occupied the entire volume containing the mixture. That is, for each gas,

$$P_i V = N_i k T,$$

where P_i is the partial pressure due to the N_i molecules of type i in the mixture and V and T are the common volume and temperature, respectively. This is known as **Dalton's law of partial pressures**.

Following the development in Section 17.2 we can write

$$P_i V = \frac{2}{3} N_i \left(\frac{m_i \overline{v_i^2}}{2} \right).$$

Combining these two equations, we have

$$\tfrac{1}{2} m_i \overline{v_i^2} = \tfrac{3}{2} k T.$$

That is, the mean translational kinetic energy of all the molecules is the same despite the differences in molecular weight. For example, in a mixture of hydrogen and oxygen molecules at temperature T, the average energy of the hydrogen molecules is the same as that of the oxygen molecules. Since oxygen molecules are 16 times heavier than hydrogen molecules, this implies that the average speed of the oxygen molecules is only $\frac{1}{4}$ that of the hydrogen molecules.

Although the average energy of monatomic molecules such as helium and argon in a gaseous sample at temperature T is indeed $\frac{3}{2}kT$, such is not the case for diatomic and polyatomic molecules. We have seen in Chapter 14 that such molecules have structure and that, as well as having energy associated with their translational motion, they have energy associated with internal rotational and vibrational motions. If thermal energy is added to a gas of diatomic or polyatomic molecules, how is that energy shared among the translational, rotational, and vibrational motions? The principle of equipartition of energy has something to say about this problem.

Before stating the principle it is convenient to introduce the concept of a **degree of freedom**. One degree of freedom is attributed to each independent coordinate required for the complete specification of the energy of a system of particles. As an example, we shall consider point molecules. The thermal energy associated with a point molecule is entirely in the form of translational kinetic energy. Since the molecules move randomly about in a three-dimensional container, a specification of their energy requires a knowledge of three velocity (or momentum) components with respect to some coordinate system. That is, the energy of a point molecule can be written as

$$U = \frac{m}{2}(v_x^2 + v_y^2 + v_z^2)$$

or

$$U = \frac{1}{2m}(p_x^2 + p_y^2 + p_z^2).$$

A point molecule has three degrees of freedom[3] associated with its motion.

---

[3] There is another definition of the number of degrees of freedom that is often used. It associates one degree of freedom with each position coordinate necessary to specify the positions of all the particles in the system. Unfortunately, there is not a one-to-one correspondence between position coordinates and coordinates necessary for the specification of the energy. For example, consider a one-dimensional harmonic oscillator. The position of the oscillator can be specified by a single coordinate x. The energy of the oscillator is given by an expression of the form

$$E = \frac{kx^2}{2} + \frac{mv^2}{2},$$

so that the energy requires both a position and a velocity coordinate for its specification.

As a second example we shall consider the water molecule. It was shown in Section 14.7 that the energy of a water molecule can be expressed in the form

$$U = \frac{m}{2} \sum_{i=x,y,z} v_i^2 + \frac{1}{2} \sum_{i=x,y,z} I_i \omega_i^2 + \frac{1}{2} \sum_{i=1,2,3} \left[a_i \left(\frac{dQ_i}{dt} \right)^2 + b_i Q_i^2 \right].$$

The first three terms represent the translational energy of the center of mass of the molecule: there are three degrees of freedom associated with the translational motion specified by the variables v_x, v_y, and v_z. The next three terms represent the rotational energy of the molecule rotating as a rigid body about the center of mass: there are three degrees of freedom associated with the rotational motion specified by the variables ω_x, ω_y, and ω_z. The last six terms represent the vibrational energy of the normal modes of vibration relative to the center of mass: there are six degrees of freedom associated with the vibrational motion specified by the variables dQ_1/dt, dQ_2/dt, dQ_3/dt, Q_1, Q_2, and Q_3. Note that the energy expression contains only squared terms. The energy of any molecule can be expressed in a similar form.

A statement of the **principle of equipartition of energy** is the following: if the energy associated with any degree of freedom is proportional to the square of the variable specifying the degree of freedom, then the average value of its contribution to the total thermal energy is $kT/2$. According to this principle, the average thermal energy per molecule for point molecules at temperature T is

$$\overline{U} = \frac{3kT}{2}$$

as we have already seen. Also, the average thermal energy per molecule for water molecules at temperature T executing rotational and vibrational motion is

$$\overline{U} = 12 \cdot \left(\frac{kT}{2} \right) = 6kT.$$

In general it may be shown that the average energy per molecule for nonlinear x-atom molecules for which all rotational and vibrational modes of motion are excited is given by

$$\overline{U} = 3(x - 1)kT$$

(see Problem 23).

A further discussion of the principle of equipartition of energy is contained in Section 20.3.

17.6 THE SPECIFIC HEATS OF GASES

An acceptable theory of gases must be able to account for all of the macroscopic parameters of gases that can be measured by experiment. As an example we shall consider in detail one such quantity, the specific heat. If an amount of heat ΔQ is added to a sample, the temperature T of the sample increases by an amount ΔT unless special provisions are made to keep the temperature constant (for example, allowing the gas to expand and do work rather than to increase its temperature). Over certain ranges of temperature ΔT is found experimentally to be approximately proportional to ΔQ. The constant of proportionality, which is just the ratio $\Delta Q/\Delta T$, is called the **heat capacity** of the substance. The heat capacity depends both upon the amount of matter making up the sample and the physical nature of the particular type of matter involved. Of particular physical interest is the heat capacity per unit mass or the **specific heat**. We define the average specific heat $\bar{c}$ over the temperature range from T to $T + \Delta T$ by

$$\bar{c} = \frac{\Delta Q}{m\Delta T},$$

where m is the mass of the sample. The units[4] for specific heat are $\text{J} \cdot \text{kg}^{-1} \cdot (^\circ\text{K})^{-1}$. The specific heat at a particular temperature T can then be defined by

$$c = \frac{dQ}{mdT}.$$

We also define the molar specific heat

$$C = \frac{dQ}{ndT},$$

where n is the number of kilogram-moles of sample. The units of C are $\text{J} \cdot (\text{kg-mole})^{-1} \cdot (^\circ\text{K})^{-1}$.

Experimental measurements of the specific heats of gases may be performed subject to a variety of constraints. The measured values[5] depend markedly on the manner in which the system is constrained during the measurement. In particular, we designate the specific heat as measured for a sample maintained at **constant pressure** by C_p and the specific heat as measured for a sample maintained at **constant volume** by C_v.

---

[4] For historical reasons heat is often expressed in calories (cal); 1 cal = 4.186 J. The specific heat is then given in the units $\text{cal} \cdot \text{kg}^{-1} \cdot (^\circ\text{K})^{-1}$.

[5] Methods of Experimental Physics, Vol. 1 (New York: Academic Press Inc., 1959).

Let us calculate C_p and C_v for an ideal gas. Suppose that we have a sample of gas in a cylinder fitted with a movable and essentially friction-less piston of cross-sectional area A. For the piston to remain stationary against the influence of molecular bombardment an external force must be applied to it, as illustrated in Fig. 17.15. The magnitude of the force is

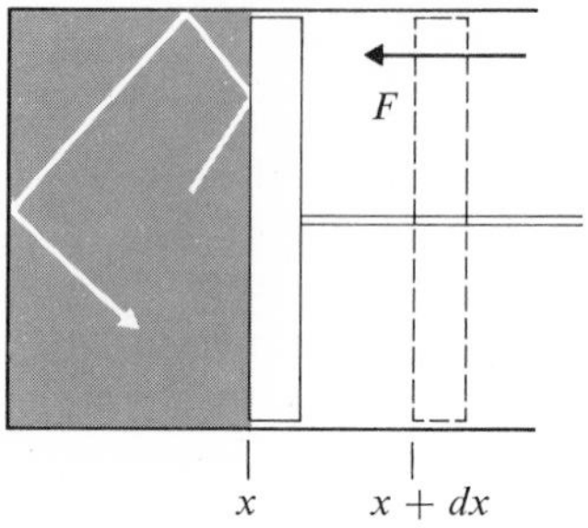

$$F = PA$$

$$= \frac{nRTA}{V}.$$

Fig. 17.15. An external force F must be applied to the piston of a cylinder containing a gas to keep the piston stationary. When heat is added the gas must do work in order to push back the piston against the external force.

If the volume occupied by the gas is to be kept constant as heat is added, then the force must be increased in proportion to the increase in temperature. If the external force is kept constant as heat is added to the gas, the position of the piston adjusts itself to maintain a constant gas pressure. That is, part of the heat added to the gas is converted to work to push back the piston against the opposition of the external force. The amount of work that must be provided by the gas to move the piston an infinitesimal amount dx is

$$dW = F \, dx$$

$$= PA \, dx$$

$$= P \, dV,$$

where dV is the change in volume of the gas. For a finite expansion from a volume V_1 to a volume V_2, the work done by the gas is

$$W = \int_{V_1}^{V_2} P \, dV.$$

If the pressure is kept constant,

$$W = P(V_2 - V_1).$$

Note that if $V_2 < V_1$ the work done by the gas is negative. That is, work is being done on the gas.

Now we are ready to calculate C_p and C_v. In general, as the result of adding an amount of heat dQ to a gas, an increase dU in the internal energy of the gas occurs, and an amount of work dW is done to change the volume occupied by the gas. That is,

$$dQ = dU + dW.$$

If the pressure of the gas is kept constant, then, using the definition of C_p, we can write

$$dQ = nC_p\, dT.$$

The principle of equipartition of energy tells us that the change in internal energy associated with a change in temperature dT for a gas of N point molecules is

$$dU = \tfrac{3}{2}Nk\, dT.$$

Since the number of kilogram-moles $n = N/N_0$ and the gas constant $R = N_0 k$, where N_0 is Avogadro's number, we can write

$$dU = \tfrac{3}{2}nR\, dT.$$

The work done by the gas to maintain constant pressure for a change in temperature dT is

$$dW = P\, dV.$$

Assuming the ideal gas equation of state

$$PV = nRT,$$

we have by differentiation

$$P\, dV = nR\, dT$$

and

$$dW = nR\, dT.$$

By substitution in $dQ = dU + dW$, we obtain

$$nC_p\, dT = \tfrac{3}{2}nR\, dT + nR\, dT$$

and

$$C_p = \tfrac{5}{2}R.$$

If the volume of the gas is kept constant then, using the definition of C_v we can write

$$dQ = nC_v\, dT.$$

As before

$$dU = \tfrac{3}{2}nR\, dT.$$

Since the volume remains constant no work is done by the gas and

$$dW = 0.$$

By substitution in $dQ = dU + dW$, we obtain

$$nC_v \, dT = \tfrac{3}{2} nR \, dT$$

and

$$C_v = \tfrac{3}{2} R.$$

For an ideal gas of point particles $C_p = \tfrac{5}{2} R$ and $C_v = \tfrac{3}{2} R$, and $C_p - C_v = R$. The **ratio of specific heats** is

$$\gamma = \frac{C_p}{C_v} = 1.67.$$

We would expect this theoretical result to be in agreement with experimental measurements of γ for monatomic gases. From Table 17.3 we see that experiment and theory are in good agreement.

With this success we are encouraged to go on to consider diatomic molecules. Now there will be contributions to the internal energy U from rotational and vibrational motions of the nuclei with respect to the center of mass of the molecule. For diatomic molecules we have already seen (Chapter 14) that two position coordinates are required to specify the rotation and one to specify the vibration. That is, there are two degrees of rotational freedom and two degrees of vibrational freedom. Using the principle of equipartition of energy, we obtain for the average energy per molecule

$$\overline{U} = 7\left(\frac{kT}{2}\right);$$

the change in internal energy associated with a change in temperature dT is

$$dU = \tfrac{7}{2} nR \, dT.$$

It follows that $C_p = \tfrac{9}{2} R$, $C_v = \tfrac{7}{2} R$, and $\gamma = 1.29$ for a diatomic gas. Experimental measurements of γ for various diatomic molecules are given in Table 17.4. Except for the heavy bromine and iodine molecules, the diatomic molecules listed have values of $\gamma \simeq 1.40$ rather than the predicted value of $\gamma \simeq 1.29$. Similar discrepancies occur for most polyatomic molecules.

Table 17.3 Ratio of specific heats for monatomic gases at 15°C

Gas	γ
Helium (He)	1.66
Neon (Ne)	1.64
Argon (Ar)	1.67
Krypton (Kr)	1.68
Xenon (Xe)	1.66

Table 17.4 Ratio of specific heats for diatomic gases at 15°C

Gas	γ
Hydrogen (H$_2$)	1.41
Oxygen (O$_2$)	1.40
Hydrogen chloride (HCl)	1.41
Carbon monoxide (CO)	1.40
Nitric oxide (NO)	1.40
Nitrogen (N$_2$)	1.40
Chlorine (Cl$_2$)	1.36
Bromine (Br$_2$)	1.32
Iodine (I$_2$)	1.30

A further discrepancy appears when one studies the temperature dependence of γ. The theory that has been presented predicts a temperature-independent γ. Although this prediction is borne out for monatomic gases, such is not the case in general. As an example we consider hydrogen gas. As shown in Fig. 17.16, γ varies from about 1.67 at $-185°C$ to 1.29 at $2200°C$. Above $2200°C$ the experimental value of γ agrees with the theoretical value. As the temperature decreases the experimental value of γ twice increases quite suddenly within the temperature ranges labeled A and B. The accepted explanation is that in crossing the region A toward lower temperatures the vibrational motion of the molecules is **frozen out** and in crossing the region B the rotational motion is frozen out. Most of the experimental values of γ for diatomic molecules as

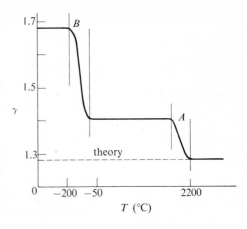

Fig. 17.16. Experimental values of γ for hydrogen as a function of temperature. The temperature is plotted on a logarithmic scale.

listed in Table 17.5 can be brought into agreement with the theory provided that we are justified in assuming that no energy goes into the molecular vibrations. If this is the case then

$$dU = \tfrac{5}{2}nR \, dT$$

instead of

$$dU = \tfrac{7}{2}nR \, dT,$$

so that $C_p = \tfrac{7}{2}R$, $C_v = \tfrac{5}{2}R$, and $\gamma = 1.40$. For the bromine and iodine molecules the molecular vibrations are apparently not frozen out at room temperature. A fundamental understanding of the freezing out of various molecular motions with decreasing temperatures requires a knowledge of the basic concepts of statistical physics and the quantum theory. More will be said about this in Section 20.5. For the moment we shall simply point out that the need for arbitrary rules about the freezing out of certain degrees of freedom was one of the serious defects of classical mechanics and kinetic theory that prepared the way for the acceptance of quantum theory.

Finally, we might note that no account has been taken of any contribution to the specific heat due to the extranuclear electrons. That their presence should indeed be neglected cannot be justified on the basis of classical mechanics. If such a contribution is included, the agreement between theory and experiment becomes even worse. Statistical physics and the quantum theory also provide the justification for neglecting any contribution of the extranuclear electrons to the specific heat.

PROBLEMS

1. In our discussion of collisions of point molecules with the wall of a container as carried out in Section 17.2, we stated that one-half of the molecules in a volume $v_{x_i} \Delta t A$ of a container will make a collision with the wall within time Δt. This statement is not strictly true. Discuss the reasons for this. Is the conclusion reached concerning the pressure exerted by the gas still valid?

2. Repeat the calculation in the example in Section 17.2 for 10 MeV protons; calculate the momentum both relativistically and with Newtonian physics and determine the error introduced by a nonrelativistic calculation.

3. A Ping-Pong ball (mass 10^{-3} kg, radius 10^{-2} m) bounces up and down between the table and a bat held horizontally 0.04 m above the table. Assuming that the speed of the ball remains constant at $2 \, \text{m} \cdot \text{sec}^{-1}$ (that is, neglecting energy loss and gravitational acceleration), what average upward force does the ball exert on the bat?

*4. A beam of molecules, each of mass 5.4×10^{-26} kg and speed $460 \, \text{m} \cdot \text{sec}^{-1}$, strikes a wall at an angle of 45°. Each molecule bounces off the wall with the same speed and at an angle of 45° to the surface. What is the pressure on the wall if the beam contains 1.5×10^{20} molecules $\cdot \, \text{m}^{-3}$?

5. The amount of energy incident on a unit area per unit time at right angles to the sun's rays at the earth's mean distance of 1.50×10^{11} m is 333 cal $\cdot \, \text{m}^{-2} \cdot \text{sec}^{-1}$.
(a) Calculate the mass that the sun loses each minute.
(b) Assume that all of the radiation is of wavelength 500 nm and consists of energy packets, called photons, of energy

$$E = h\nu$$

and

$$\text{momentum} = \frac{h\nu}{c},$$

where $h = 6.63 \times 10^{-34} \, \text{J} \cdot \text{sec}$, ν is the frequency of the radiation, and c is the speed of light. Calculate how many photons strike a surface of area $1.0 \, \text{m}^2$ each second on the earth.
(c) From the result of (b), estimate the pressure of the sun's radiation on a perfect reflector.

6. Using whatever references you can find, write short essays on thermometers based on the temperature variation of each of the following physical properties:
(a) the electrical resistance of a wire,
(b) the voltage developed at the junction of two dissimilar metals,
(c) the equilibrium pressure of a vapor above a liquid,
(d) the color of a hot object.

*7. Suppose a thermometer measures a temperature T when immersed in an ideal gas. Now let the thermometer move through the gas with a speed v_0 in the $+x$ direction. What temperature does the thermometer now read?

*8. Plot the distribution function in molecular speeds as a function of the molecular speed for hydrogen gas at 200°K. Determine the speed of maximum probability and compare it with the root-mean-square speed. Estimate the fraction of the molecules having speeds greater than twice the root-mean-square speed.

*9. Titan is the largest satellite of the planet Saturn and is unique in that it is the only satellite in the solar system known to have an atmosphere. The mass and diameter of Titan are 1.37×10^{23} kg and 4800 km, respectively. The mean surface temperature is 94°K. Would methane and/or hydrogen be possible constituents of the atmosphere of Titan?

10. A box of volume 10^{-3} m^3 contains helium at standard pressure.
   (a) What is the total average kinetic energy of the atoms in the box?
   (b) If the box contains 10^{-5} kg-mole of the gas, what would the temperature of the gas be?
   (c) At approximately what temperature would one have to consider relativistic modifications to the perfect gas law?

11. Two ideal monatomic gases have the same pressure. The density of the first gas is $2 \text{ kg} \cdot \text{m}^{-3}$, its atoms are five times as heavy as the second gas, and it is at a temperature of 400°K.
   (a) What is the density of the second gas if it is at a temperature of 200°K?
   (b) If one had a room filled with the first gas and in the room a balloon with a volume of 10^{-3} m^3 filled with the second gas, how much weight could the balloon lift?

12. (a) Calculate the square root of the mean-square molecular speed in nitrogen at 0°C.
   (b) Calculate the ratio of dN/dv at the speed in (a) to the value of dN/dv at ten times this speed.

13. The best modern vacuum pumps are able to reduce the pressure in a container to about 10^{-13} Torr (1 Torr $\equiv$ 1 torricelli $\equiv 10^{-3}$ m of Hg).
   (a) At such a pressure, how many molecules are in 10^{-6} m^3 at 0°C?
   (b) Estimate the average distance between the molecules in the gas.

14. A student uses the apparatus shown in Fig. 17.17 to study the relation between the volume occupied by a fixed quantity of gas and the pressure exerted on it

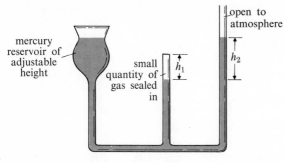

Fig. 17.17

Table 17.5

$h_1(cm)$	$h_2(cm)$
12.1	-10.8
11.0	-4.6
10.1	1.5
9.3	8.3
8.6	14.7

when the temperature is kept constant. Some of his measurements of the heights h_1 and h_2 are given in Table 17.5 (a negative h_2 means that the mercury column in the open tube is lower than that in the closed tube). Atmospheric pressure was recorded as 75.9 cm on a standard mercury manometer. The temperature was recorded as 29°C. Discuss the validity of Boyle's law in terms of these measurements.

Table 17.6

$h(cm)$	$T(°C)$
−9.0	0
−3.3	24.5
0.4	41
5.5	60
9.0	80
14.8	99.5

15. A student uses the apparatus shown in Fig. 17.18 to study the relation between the pressure exerted on a given mass of gas and its temperature when its volume is kept constant. Some of his measurements are given in Table 17.6 (a negative h means that the mercury column in the open tube is lower than the arbitrary fixed level). Atmospheric pressure was recorded as 74.7 cm on a standard mercury manometer. Discuss the validity of Charles' law in term of these measurements.

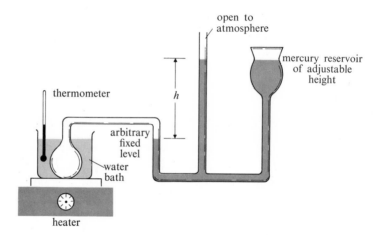

Fig. 17.18

16. Show on one graph the temperature dependence of the van der Waals second virial coefficients of hydrogen, argon, methane, and ethylene. The van der Waals parameters are given in Table 17.2.

17. Draw the P-ρ graphs for helium, neon, argon, krypton, and xenon at 0°C that you think would approximate the experimental curves.
Hint: look up the liquefaction temperatures of these gases.

*18. The critical constants for hydrogen have been measured as $T_c = 33.2°K$, $\tilde{V}_c = 0.065 \text{ m}^3 \cdot (\text{kg-mole})^{-1}$ and $P_c = 12.8 \text{ atm}$.
(a) Deduce the van der Waals parameters.
(b) Compare the van der Waals equation of state at 0°C with the measured equation of state given in Fig. 17.11.

19. Draw a P-ρ graph for argon at 150°C assuming van der Waals parameters $a = 1.05\, l^2 \cdot \text{atm} \cdot (\text{g-mole})^{-2}$ and $b = 0.0251\, l \cdot (\text{g-mole})^{-1}$. Compare your graph with the experimental equation of state given in Fig. 17.10.

*20. Show that for the van der Waals equation of state the critical constants are given by

$$P_c = \frac{a}{27b^2}, \qquad \tilde{V}_c = 3b, \qquad T_c = \frac{8a}{27bR}.$$

21. A gas consists of a mixture of two monatomic gases whose atoms have masses of 4 and 30 amu, respectively. The average value of the velocity squared of the heavier atoms is $4 \times 10^6\, \text{m}^2 \cdot \text{sec}^{-2}$. What is the average kinetic energy of the lighter atoms? What is the temperature of the gas?

22. Avogadro's number is derived from **Avogadro's law**, which states that equal volumes of gas contain equal numbers of molecules if they have the same temperature and pressure. Show that this is so.

23. Show that the average energy per molecule for nonlinear x atom molecules for which all the rotational and vibrational modes of motion are excited is given by

$$\overline{U} = 3(x - 1)kT.$$

24. The specific heat at constant pressure of a certain gas is $2.5 \times 10^2\, \text{cal} \cdot \text{kg}^{-1} \cdot (°\text{C})$. The molecular weight of the gas is 28.
(a) What is the specific heat at constant volume in units of $\text{cal} \cdot \text{kg}^{-1} \cdot (°\text{C})^{-1}$?
(b) What is γ for this gas?
(c) What can you tell about the molecular structure of this gas?

25. One cubic meter of air at 20°C and atmospheric pressure is contained in a cylinder fitted with a frictionless piston. The air is heated to 60°C, the piston moving so as to keep the pressure constant. Calculate the work done by the expanding air and the increase in internal energy of the air. (Specific heat of air at constant pressure $= 2.37 \times 10^2\, \text{cal} \cdot \text{kg}^{-1} \cdot (°\text{C})^{-1}$.)

26. (a) Hydrogen sulfide (H_2S) at room temperature may be assumed to have 3 degrees of translational freedom and 3 degrees of rotational freedom. Calculate γ.
(b) The H—S bond length is determined experimentally to be $1.33 \times 10^{-10}\, \text{m}$ and the H—S—H bond angle to be about 90°. Taking x and y axes in the plane of the molecule through its center of mass as shown in Fig. 17.19 and the z axis perpendicular to the x and y axes, calculate the moments of inertia of the molecule about each of the three axes, assuming point nuclei.
(c) If the molecules of the gas at $T = 300°\text{K}$ rotate about the x axis, calculate their mean angular frequency of rotation.

27. The hydrogen cyanide molecule (HCN) is a linear molecule (see Fig. 17.20).
(a) Locate the center of mass of the molecule, assuming that the masses of N, C, and H are 14, 12, and 1 amu.

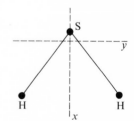

Fig. 17.19

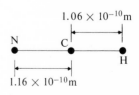

Fig. 17.20

(b) Calculate the moment of inertia of the molecule about an axis through the center of mass perpendicular to the axis of the molecule.

(c) How many vibrational degrees of freedom does the molecule have?

(d) Sketch any two possible modes of vibrational motion for the molecule.

(e) Calculate the specific heat at constant volume for a gas of these molecules. Assume that no energy is taken up by vibrational modes of motion ($1\ \text{amu} = 1.67 \times 10^{-27}\ \text{kg}$).

Elementary quantum concepts

18

18.1 INTRODUCTION

Newtonian physics is able to adequately account for the macroscopic motions of objects familiar to us from our day-to-day experiences. In previous chapters, for example, we have considered the motion of projectiles over the earth's surface and the orbital paths of earth satellites and have found Newtonian physics adequate for a detailed analysis of both types of motion.

However, in the microscopic world of atomic motions Newtonian physics fails to provide an explanation for many features of observed behavior. For example, we mentioned in Section 14.3 that molecular rotation could only occur for certain definite values of the rotational angular momentum, and in Section 17.6 we saw that the specific heat of polyatomic gases is temperature dependent. Neither of these results can be reconciled with the predictions of Newtonian mechanics.

In this chapter we wish to introduce some of the fundamental concepts of a more recent formulation of mechanics known as **quantum mechanics**[1].

18.2 PLANCK'S RADIATION LAW

The first important step leading to the development of quantum mechanics was taken by Max Planck in 1900. Planck was considering the variation of the intensity of radiation within a hot cavity (such as a furnace) with the frequency of the radiation. Experimental measurements on this type of radiation had been carried out as the radiation escaped from a small hole in a metal cavity maintained at a constant temperature T (see Fig. 18.1). The variation of the intensity of radiation per unit frequency interval as a function of frequency is shown in Fig. 18.2; the same curve is obtained for a given temperature T regardless of the material of which the cavity is constructed.

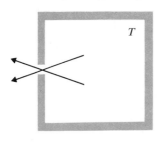

It is convenient to define an **ideal solid object** as one which absorbs all radiation falling upon it. Such an object would appear black if viewed in reflected light (unless its temperature were high enough for it to be self-luminous). Such an ideal object is called a **blackbody**. The cavity considered by Planck behaves in much the same manner as a blackbody in the sense that any radiation

Fig. 18.1. A cavity maintained at temperature T with a small hole through which radiation can escape.

[1] K. K. Darrow, "The Quantum Theory," *Scientific American*, March, 1952. Available as *Scientific American Offprint 205* (San Francisco: W. H. Freeman and Co., Publishers).

entering the cavity by the small hole is trapped inside for many reflections. The radiation within the cavity is therefore called **blackbody radiation**.

The radiation within the cavity might be thought of as consisting of a series of standing electromagnetic waves. We can picture the waves in terms of a three-dimensional generalization of the type of standing waves considered in Section 16.4. In the present case the condition for standing waves is that the path length from wall to wall in any direction be an integral multiple of a half wavelength. A number of possible standing waves for a two-dimensional cavity are illustrated in Fig. 18.3.

Owing to the multiplicity of possible paths there will be a great number of different standing waves that are possible for any value of the wavelength. In fact, it can be shown that the number of waves per unit volume having wavelengths between λ and $\lambda + \Delta\lambda$ is given by

$$\Delta n_\lambda = \frac{8\pi \, \Delta\lambda}{\lambda^4}.$$

Since $c = \nu\lambda$, it readily follows that the number of waves per unit volume having frequencies between ν and $\nu + \Delta\nu$ is given by

$$\Delta n_\nu = \frac{8\pi\nu^2 \, \Delta\nu}{c^3}.$$

Now we must decide how much energy to associate with each of these waves. In order to arrive at a theory that would account for the observed distribution of blackbody radiation, Planck introduced the radical concept of **energy quantization**.

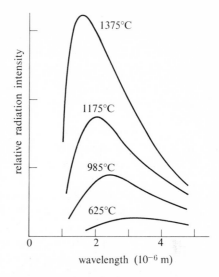

Fig. 18.2. Frequency distribution of the radiation intensity within a cavity for four different temperatures.

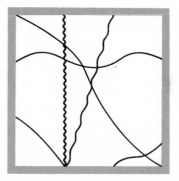

Fig. 18.3. Some of the possible standing waves in a two-dimensional cavity.

He postulated that, **through its interaction with matter, radiation is emitted and absorbed only in discrete amounts.** He called these amounts **quanta**, each quantum having energy

$$E = h\nu,$$

where $h = 6.63 \times 10^{-34}$ J·sec $= 4.14 \times 10^{-15}$ eV·sec. h is now known as **Planck's constant**. Planck believed, however, that this discontinuity applied only to the interaction of radiation with matter and that the radiation within the cavity itself behaved classically as if it consisted of continuous waves.

With the assumption of energy quantization, Planck was able to deduce that the average energy associated with a standing wave of frequency v in the cavity is

$$\bar{E} = \frac{hv}{\exp\left(\dfrac{hv}{kT}\right) - 1}$$

where k is Boltzmann's constant and T the temperature in °K. This result can be explained as follows. Since the formation of high-frequency quanta requires more energy than the formation of low-frequency quanta, one might expect an excess number of low-frequency quanta to be present in the cavity. Therefore, the average energy associated with high-frequency standing waves in the cavity will decrease with increasing frequency for sufficiently high frequencies simply because there are relatively few high-energy quanta available.

Now we are ready to deduce Planck's formula for the total energy per unit volume in the frequency interval between v and $v + \Delta v$. That is,

$$E_v \, \Delta v = \bar{E} \, \Delta n_v$$

$$= \frac{\left(\dfrac{8\pi h}{c^3}\right) v^3 \, \Delta v}{\exp\left(\dfrac{hv}{kT}\right) - 1}.$$

This equation is in excellent agreement with the experimental measurements.

Example. Deduce the limiting low-frequency value of the average energy associated with a standing wave.

Solution. To study the case of low frequencies, we consider the series expansion of e^x, namely,

$$e^x = 1 + x + \frac{x^2}{2!} + \cdots.$$

When x is small we may approximate e^x by

$$e^x = 1 + x.$$

Therefore, when

$$hv \ll kT,$$

we may write

$$\bar{E} = \frac{hv}{\exp\left(\dfrac{hv}{kT}\right) - 1}$$

$$= \frac{hv}{1 + \dfrac{hv}{kT} - 1}$$

$$= kT.$$

The limiting low-frequency value of the average energy is kT. That is, electromagnetic waves are transverse waves and therefore have two degrees of freedom associated with their motion, each of which contributes $kT/2$ to $\bar{E}$.

18.3 THE PHOTOELECTRIC EFFECT

In 1905 Albert Einstein took the next important step toward the development of quantum mechanics when he provided an explanation of the photoelectric effect. Experiments to study the emission of electrons from a metal surface by an incident light beam defied any explanation in terms of classical physics. The emitted electrons are called **photoelectrons** and the phenomenon the **photoelectric effect**. The characteristics of the photoelectron production are as follows:

1. The energy distribution of the photoelectrons is independent of the intensity of the incident light. That is, although an intense light yields more photoelectrons than does a weak light, the average energy per photoelectron remains the same.

2. The energy of the photoelectrons depends on the frequency of the incident light. At frequencies below some critical value v_0 characteristic of the metal surface, no electrons are emitted regardless of the intensity of the incident light. Above this threshold frequency the photoelectrons exhibit a range of energies from 0 to a maximum value E_{max} determined by the frequency v of the incident light. This is illustrated in Fig. 18.4 for two different metals. The relation between E_{max} and v is given simply by

$$E_{max} = h(v - v_0),$$

where v_0 depends on the metal and h is Planck's constant. That is, a faint blue light will produce

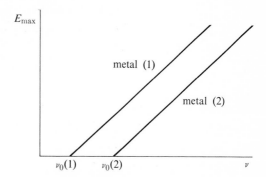

Fig. 18.4. The variation of the maximum energy of photoelectrons with incident light frequency for two different metals.

electrons with a greater average energy than those produced by an intense red light.

3. The time lag between the arrival of light at the metal surface and the emission of photoelectrons, if indeed any lag exists, is too small to be readily measured. That is, the time lag is less than 3×10^{-9} sec.

None of these observations can be understood on the basis of classical physics. In particular, let us consider the question of a time lag between the arrival of light at the metal surface and the ejection of a photoelectron. To be specific, we suppose violet light is incident on a sodium surface. Experiments show that $10^{-6} \, \text{J} \cdot \text{m}^{-2} \cdot \text{sec}^{-1}$ of energy are sufficient to observe a photoelectric current; the individual electrons have energies of the order of 1 eV. If we assume that the energy is uniformly absorbed by the electrons in the first few layers of the surface, then a simple calculation (see Problem 8) shows that each electron will receive energy at an average rate of less than $10^{-7} \, \text{eV} \cdot \text{sec}^{-1}$. Therefore, it should take more than 10^{7} sec, or about 1 year, for any single electron to accumulate the 1 eV of energy of the photoelectrons. That is, this type of classical argument predicts an appreciable time lag.

Einstein was able to provide a simple explanation of the photoelectric effect by assuming that the energy carried by a beam of light of frequency v is concentrated into packages called **photons** of energy hv. That is, he extended Planck's idea: radiation is not only emitted and absorbed in discrete amounts but also propagates through space in discrete bundles. He proposed that an entire photon could be instantaneously absorbed by a single electron and wrote down the equation

$$hv = E_{\text{max}} + \phi.$$

In this equation ϕ represents the minimum energy necessary to dislodge an electron from the metal surface. This minimum energy is known as the **work function** of the surface; it represents the energy required to push an electron through the metal surface from just beneath it against the attractive force of the surface atoms tending to draw the electron back into the surface. If we identify ϕ with hv_0, this equation agrees with the experimental results, that is,

$$E_{\text{max}} = hv - hv_0$$
$$= h(v - v_0).$$

Einstein's equation is simply a statement of energy conservation. That portion of the energy of the incident photon not required to remove the electron from the surface appears as kinetic energy. The equation applies only to electrons originating just beneath the surface. Since many photoelectrons originate deeper within the metal, these photoelectrons will tend to lose energy by collisions with atoms in the metal, and therefore they are ejected with less kinetic energy.

Following Einstein's work, the concept of the dual character of electromagnetic radiation in its interaction with matter found ever increasing support. There was ample experimental evidence for a wave theory (for example, interference experiments); no one could deny that. But now evidence was growing to support a particle theory (for example, the photoelectric effect). Our present-day view of light is that the energy is localized in particles (photons) whose paths through space are governed by a wave equation.

Example. In an experiment on the photoelectric effect it is determined that, when light of wavelength 300 nm falls on the metal surface in question, the maximum kinetic energy of the emitted photoelectrons is 0.51 eV, while for light of wavelength 200 nm the maximum kinetic energy is 2.6 eV. Determine the work function for the surface and the value of Planck's constant from these data.

Solution. Let us put

$$\lambda_1 = 300 \text{ nm} \quad \text{and} \quad \lambda_2 = 200 \text{ nm}.$$

Therefore,

$$v_1 = \frac{c}{\lambda_1} = \frac{3 \times 10^8}{3 \times 10^{-7}} = 10^{15} \text{ Hz}$$

$$v_2 = \frac{c}{\lambda_2} = \frac{3 \times 10^8}{2 \times 10^{-7}} = 1.5 \times 10^{15} \text{ Hz}.$$

The maximum kinetic energy is given by

$$E_{\max}(1) = hv_1 - \phi$$

$$E_{\max}(2) = hv_2 - \phi.$$

Subtracting gives

$$E_{\max}(2) - E_{\max}(1) = h(v_2 - v_1)$$

$$h = \frac{E_{\max}(2) - E_{\max}(1)}{v_2 - v_1}$$

$$= \frac{2.6 - 0.51}{0.5 \times 10^{15}}$$

$$= \frac{2.1}{0.5} \times 10^{-15}$$

$$= 4.2 \times 10^{-15} \text{ eV} \cdot \text{sec}.$$

This compares well with the accepted value of $4.14 \times 10^{-15}\,\text{eV} \cdot \text{sec}$. The work function is given by

$$\phi = h\nu_2 - E_{max}(2)$$

$$= 4.2 \times 10^{-15} \times 1.5 \times 10^{15} - 2.6$$

$$= 6.3 - 2.6$$

$$= 3.7\,\text{eV}.$$

This is close to the value of the work function for zinc, which is $3.63\,\text{eV}$.

18.4   ATOMIC SPECTRA AND THE BOHR THEORY

In Section 18.2 we discussed blackbody radiation. The radiation emitted by a solid heated to incandescence approximates blackbody radiation. The radiation is essentially independent of the nature of the individual atoms of the solid and is a product of the collective behavior of the enormous number of interacting atoms that make up the solid.

In contrast, we might expect the radiation emitted from the atoms or molecules of a suitably excited gas at low pressure to be characteristic of the constituent atoms or molecules, since these atoms or molecules act essentially independently of each other. That is, the atoms or molecules of a gas are on the average separated by many molecular diameters and effectively interact only during collisions. An atomic gas may be excited to emit radiation by passing an electrical current through it. It is found experimentally that the resultant radiation consists of discrete frequencies characteristic of the atoms that are emitting. The frequencies may be classified into well-defined sets called **spectral series** that may be specified by empirical formulas.

Let us turn our attention to the simplest of all atomic systems, hydrogen. A spectral series known as the **Balmer series** appears in the visible region of the spectrum (that is, the frequencies in the series are in the range to which the eye is sensitive). The frequencies of the various members of the series are given by

$$\nu = C\left(\frac{1}{2^2} - \frac{1}{n^2}\right), \qquad n = 3, 4, 5, \cdots,$$

with C a constant. The visible spectrum of hydrogen is illustrated in Fig. 18.5. The spectrum consists of a series of parallel lines that are just images of a slit that is used in the experimental apparatus. Starting with the line of lowest frequency, the lines are labeled H_α, H_β, H_γ, $\cdots$. As the frequency increases the lines become closer together and weaker in intensity, until the **series limit** is reached when $n = \infty$. No lines occur beyond the series limit.

A number of other series have been observed, all of which may be accounted for by the empirical formula

$$v = C\left(\frac{1}{n_f^2} - \frac{1}{n_i^2}\right),$$

with

$$n_i = n_f + 1, \quad n_f + 2, \cdots$$

and

$$n_f = 1, 2, 3, \cdots.$$

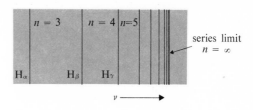

Fig. 18.5. The Balmer series of spectral lines from hydrogen gas.

For the Balmer series $n_f = 2$.

How can the preceding observations be explained? First, we require an atomic model. The model we presently accept was put forward by Ernest Rutherford in 1911. In this model most of the mass of the atom is concentrated in a tiny positively charged nucleus surrounded by an appropriate number of negatively charged electrons. In particular, the hydrogen atom depicted in Fig. 18.6 consists of a nucleus with one unit of positive charge and a single extranuclear electron with one unit of negative charge. For such a structure to be in stable equilibrium against the attractive electrostatic force, the electron must carry out an accelerated motion. The simplest such motion is a circular motion about the nucleus with the electrostatic force

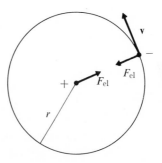

Fig. 18.6. The classical picture of the hydrogen atom.

$$F_{\text{electrostatic}} = k'\frac{e^2}{r^2}$$

providing the centripetal acceleration

$$a_{\text{centripetal}} = \frac{v^2}{r}$$

to keep the electron moving in the circular path. That is,

$$F_{\text{electrostatic}} = k'\frac{e^2}{r^2}$$

$$= \frac{mv^2}{r},$$

where m is the mass of the electron.

The electron has both kinetic energy and potential energy. The total energy is (see Section 7.8)

$$E = E_{kin} + E_{pot}$$

$$= \frac{1}{2}mv^2 - k'\frac{e'^2}{r}.$$

But

$$mv^2 = k'\frac{e^2}{r},$$

so that

$$E = \frac{k'}{2}\frac{e^2}{r} - k'\frac{e^2}{r}$$

$$= -\frac{k'}{2}\frac{e^2}{r}.$$

(Recall from Section 7.8 that a negative energy is required in order that the electron be bound to the nucleus.)

It has been shown experimentally that an energy of 13.6 eV or 2.2×10^{-18} J is required to ionize a hydrogen atom, that is, to separate it into a proton and an electron. Therefore, the binding energy of the hydrogen atom is -2.2×10^{-18} J, and we can write

$$E = -\frac{k'e^2}{2r}$$

$$= -2.2 \times 10^{-18}$$

or

$$r = \frac{k'e^2}{2(2.2 \times 10^{-18})}.$$

The constant $k' = 8.99 \times 10^9$ MKS units. Therefore,

$$r = \frac{8.99 \times 10^9 (1.60 \times 10^{-19})^2}{2 \times 2.2 \times 10^{-18}}$$

$$= 5.3 \times 10^{-11} \text{ m}.$$

That is, the Rutherford model of the hydrogen atom leads to a dynamically stable system with the electron revolving in a circular orbit about the nucleus at a distance of 5.3×10^{-11} m.

Classical physics can in no way account for the characteristics of hydrogen atoms that we have just described. It is a well-known result of electromagnetic theory that an electron moving in a circular orbit emits radiation and therefore loses energy. The loss of energy would be accompanied by a decrease in the orbital radius. Since the electron would be expected to radiate continuously, it would gradually spiral into the proton. The frequency of the radiation is just the frequency of revolution (or a multiple of it), which varies as $r^{-3/2}$, so that one would not expect a number of discrete frequencies. Also, the only stable configuration would be that with the proton and the electron in contact.

The Bohr model of the hydrogen atom

Niels Bohr adopted the concept of the quantization of radiation as presented by Planck and Einstein and in 1913 put forward an explanation of the emission of radiation by the excited atoms of hydrogen gas. Bohr's explanation was based on the following two general postulates.

1. An atomic system can exist in certain **stationary states**, each one corresponding to a definite value of the energy E of the system. Transitions between stationary states are accompanied by the emission (or absorption) of an amount of energy equal to the difference in energy of the two states.
2. The frequency of the radiation emitted as the result of a transition from an initial state of energy E_i to a final state of lower energy E_f is given by
$$E_i - E_f = h\nu.$$

From the experimental frequencies the energies of some of the stationary states of hydrogen could be deduced. Bohr found that he could account for these stationary states by restricting the allowed angular momentum of circular electron orbits to be integral multiples of $h/2\pi$. That is, he used the condition

$$mvr_n = \frac{nh}{2\pi}, \qquad n = 1, 2, 3, \cdots$$

to deduce the allowed values of r, denoted by r_n. The integer n is known as a **quantum number**. From our earlier considerations we see that

$$(mvr_n)^2 = (mv^2)mr_n^2$$
$$= k'me^2r_n$$
$$= \frac{n^2h^2}{4\pi^2}.$$

Therefore,

$$r_n = \left(\frac{h^2}{4\pi^2me^2}\right)n^2.$$

But

$$E = -\frac{k'e^2}{2r},$$

so that the stationary states have energy

$$E_n = -\frac{k'e^2}{2} 4\pi^2 \frac{me^2}{h^2 n^2}$$

$$= -\left(\frac{2\pi^2 k' me^4}{h^2}\right)\frac{1}{n^2}.$$

The first six of these stationary **electronic energy states** are shown in an **energy level diagram** in Fig. 18.7. A vertical energy scale is selected, and a horizontal line is drawn at the energy corresponding to each of the stationary states. The lowest energy state is called the **ground electronic state**. The predicted frequencies of emission according to Bohr's second postulate are given by

$$v = \frac{E_i - E_f}{h}$$

$$= -\frac{2\pi k' me^4}{h^3}\left(\frac{1}{n_i^2} - \frac{1}{n_f^2}\right)$$

$$= C\left(\frac{1}{n_f^2} - \frac{1}{n_i^2}\right)$$

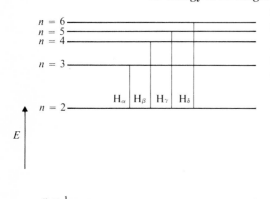

Fig. 18.7. The energy level diagram of atomic hydrogen.

in agreement with the empirical formula. The transitions giving rise to the first few members of the Balmer series are shown on the energy level diagram in Fig. 18.7. Atoms in the $n = 2$ stationary state return to the lowest energy state or ground state corresponding to $n = 1$ with the emission of ultraviolet radiation.

In spite of the success of the Bohr theory in accounting for the frequencies of the spectral lines in atomic hydrogen, the theory cannot be extended to the treatment of more complex atoms having two or more extranuclear electrons. Nonetheless the two postulates of Bohr retain, to a great extent, their validity in the quantum mechanics that has replaced the Bohr theory.

Example. The wavelengths of the H_α and H_β lines of hydrogen have been measured experimentally to be 656.279 nm and 486.133 nm, respectively. Compare the ratio of the measured wavelengths to the ratio predicted by the Bohr theory.

Solution. For the Balmer series $n_f = 2$ and $n_i = 3$ for H_α and $n_i = 4$ for H_β. Since $v \propto \lambda^{-1}$, we have

$$\lambda_\alpha = \frac{1}{C}\left(\frac{1}{2^2} - \frac{1}{3^2}\right)^{-1} = \frac{1}{C}\left(\frac{5}{36}\right)^{-1}$$

$$\lambda_\beta = \frac{1}{C}\left(\frac{1}{2^2} - \frac{1}{4^2}\right)^{-1} = \frac{1}{C}\left(\frac{3}{16}\right)^{-1}.$$

Therefore,

$$\frac{\lambda_\alpha}{\lambda_\beta} = \left(\frac{3}{16}\right)\left(\frac{36}{5}\right) = 1.3500000.$$

The experimental ratio is

$$\frac{\lambda_\alpha}{\lambda_\beta} = \frac{656.279}{486.133} = 1.34999(9).$$

The theoretical and experimental ratios are identical to approximately one part in 10^7.

18.5 DE BROGLIE WAVES AND THE UNCERTAINTY PRINCIPLE

In 1924 Louis de Broglie speculated that particles might exhibit wave properties under certain circumstances. After all, electromagnetic waves had been shown to behave in a manner identical to that of particles in certain cases. Why should the opposite situation not be true?

In order to arrive at the de Broglie formula, let us reconsider the particle nature of electromagnetic waves. The photon has an energy

$$E = hv.$$

If we consider the photon as a relativistic "particle" traveling with speed c, then its energy is

$$E = m_p c^2,$$

where m_p, the relativistic mass of the photon, may be obtained by equating the two expressions for the energy of a photon to give

$$hv = m_p c^2$$

and

$$m_p = \frac{hv}{c^2}.$$

The photon momentum is given by

$$p = m_p c = \frac{h\nu}{c} = \frac{h}{\lambda}.$$

That is,

$$\lambda = \frac{h}{p}.$$

De Broglie suggested that this expression was a general expression, applying to material particles as well as to photons. For a material particle

$$p = mv,$$

and the de Broglie wavelength is

$$\lambda = \frac{h}{mv},$$

with m the relativistic mass and v the speed.

In Section 12.5 we noted that for any particle,

$$E^2 = p^2 c^2 + m_0^2 c^4,$$

where m_0 is the rest mass. Since, for a photon,

$$p = \frac{h\nu}{c}$$

and

$$E = h\nu,$$

we have

$$E = pc,$$

which implies that $m_0 = 0$ for a photon.

The de Broglie hypothesis was experimentally verified several years later by the observation of the scattering of electrons from a large single crystal. The experiment is illustrated in Fig. 18.8. In the experiment, the energy of the incident electrons, the angle of incidence θ of the incident beam, and the scattering angle ϕ seen by the detector were all varied. If electrons are classical particles, we would expect the scattered electrons to emerge in all directions with only a slight dependence of their intensity on the scattering angle and essentially no dependence on the energy of the incident beam. However, the intensity of the

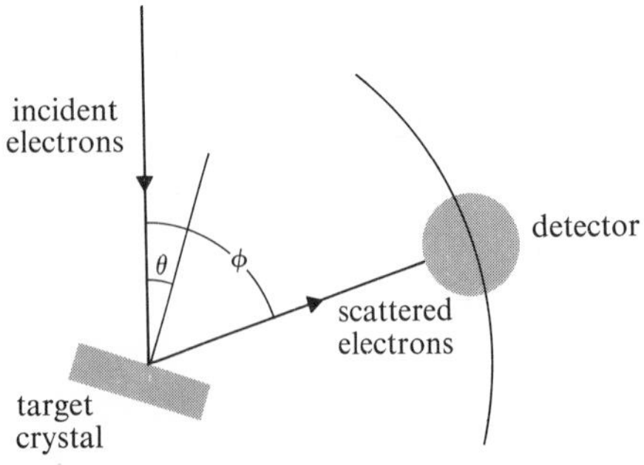

Fig. 18.8. An experiment to study the wave nature of material particles.

scattered electron beam was observed to be a definite function of the scattering angle with positions of distinct maxima and minima; these positions were found to vary with the incident electron energy.

These observations can be explained in terms of interference effects similar to those described for waves in Chapter 16. That is, the electrons may be considered as waves that are scattered by atoms in the crystal; the scattered waves interfere constructively in directions corresponding to a difference in path length of an integral number of wavelengths for electrons scattered from adjacent atoms, and they interfere destructively in directions corresponding to a difference in path length of a half-integral number of wavelengths (see Fig. 18.9 and Section 16.3). Such

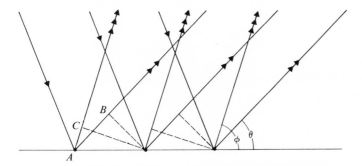

Fig. 18.9. When electrons are scattered from the atoms of a crystal constructive interference occurs at angle θ (AB = integral number of wavelengths) and destructive interference occurs at angle ϕ (AC = half-integral number of wavelengths).

interference effects are discussed in detail for x-rays in Section 21.4. Note that interference effects are strictly wave phenomena and cannot be explained using a particle picture.

In order that the wave nature of material particles be observable, it is necessary that the associated de Broglie wavelength be of the same order as the dimensions of the particle. For example, for interference effects to be observable between waves scattered by atoms in a crystal, the spacing between atoms must be of the order of the wavelength of the waves. If the spacing is too small, path differences will be much less than the wavelength, and constructive interference cannot occur; if the spacing is too large, directions of constructive interference become so close together that it is not possible to observe them separately. Therefore, particles must have de Broglie wavelengths of the order of atomic dimensions for interference effects to be observable when they are scattered from a crystal.

For an electron traveling at a speed of $5 \times 10^6 \, \mathrm{m \cdot sec^{-1}}$,

$$\lambda = \frac{h}{mv}$$

$$= \frac{6.63 \times 10^{-34}}{9.1 \times 10^{-31} \times 5 \times 10^6}$$

$$= 1.5 \times 10^{-10} \, \mathrm{m},$$

which is of the order of magnitude of atomic dimensions. In contrast, let us consider a macroscopic particle such as a golf ball of mass 0.10 kg and traveling with a speed of $50 \, \mathrm{m \cdot sec^{-1}}$. For this particle

$$\lambda = \frac{h}{mv}$$

$$= \frac{6.63 \times 10^{-34}}{0.10 \times 50}$$

$$= 1.3 \times 10^{-34} \, \mathrm{m},$$

which is negligible compared to its dimensions. We would not expect to observe any wave aspects in the motion of a golf ball. This is an example of the **correspondence principle**, which asserts that the classical and quantum pictures of nature must make identical predictions in the limit of the macroscopic world of our daily experiences.

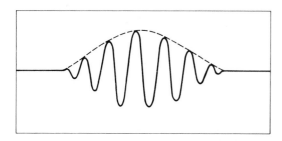

Fig. 18.10. A wave packet associated with a moving particle.

What is the nature of the waves associated with material particles? It may be argued that the amplitude of a de Broglie wave at a specific time at a particular position in space is a measure of the probability that the associated particle be found at that position at that time. If we adopt this interpretation, the wave associated with a moving object might reasonably be expected to correspond to a wave packet or wave group, as illustrated in Fig. 18.10 and introduced in Section 16.7. The envelope of the wave group (dashed line in Fig. 18.10) is proportional to the probability of finding the particle at position x. The wave group travels with the velocity of the particle; the shape of the wave group will in general be a function of the time.

A wave packet may be mathematically decomposed into a series of sinusoidal waves, each with constant amplitude and with a slightly different frequency (see Section 16.8). A fundamental property of this mathematical process is that the shorter the wave group is, the greater the frequency spread of the component sine waves that are required to reproduce the wave group.

What are the consequences of the necessity to regard a moving particle as a de Broglie wave group? First of all, it imposes a fundamental limit to the accuracy with which the position of the particle can be specified. The shorter the wave packet associated with a particle, the more accurately its position in space is defined. Second, it imposes a fundamental limit to the accuracy with which the momentum of the particle can be specified. From the de Broglie formula we see that the momentum of a particle is proportional to the de Broglie frequency, so that if the frequency is not well-defined, neither is the momentum. Therefore, the longer the wave packet associated with a particle, the smaller is the range of frequencies required to describe it, and the better defined is the particle's momentum. That is, a long wave packet describes a particle whose momentum is well-defined but whose position is ill-defined; a short wave packet describes a particle whose momentum is ill-defined but whose position is well-defined.

The position x of a particle and its momentum p are referred to as **conjugate variables**. The fundamental uncertainty Δx in x and the fundamental uncertainty Δp in p may be shown to be related by

$$\Delta x \, \Delta p = \frac{h}{4\pi}.$$

This is an example of the **uncertainty principle**[2] put forward by Werner Heisenberg. It states that it is conceptually impossible to simultaneously measure values of conjugate variables precisely.

Another important form of the uncertainty principle involves the energy E emitted in some process and the time τ at which the energy is emitted. The fundamental uncertainty ΔE in E and the fundamental uncertainty $\Delta \tau$ in τ are related by

$$\Delta E \, \Delta \tau = \frac{h}{4\pi}.$$

This result is quite easy to illustrate for the case of the emission of electromagnetic radiation by an atomic system. The time duration of the emission $\Delta \tau$ limits the number of waves in the emitted wave train and therefore limits the accuracy to which we can determine the frequency. That is, since

$$\nu = \frac{\text{number of waves}}{\Delta \tau}$$

[2] G. Gamow, "The Principle of Uncertainty," *Scientific American*, January, 1958. Available as *Scientific American Offprint 212* (San Francisco: W. H. Freeman and Co., Publishers).

and since the uncertainty in the number of waves that we can count in a wave train is about one,

$$\Delta v \sim \frac{1}{\Delta \tau}.$$

But

$$E = hv,$$

so that

$$\Delta E = h \, \Delta v$$

and

$$\Delta E \, \Delta \tau \sim h.$$

A more rigorous derivation is required to obtain the factor 4π.

18.6 SCHROEDINGER'S FORMULATION OF QUANTUM MECHANICS

Stimulated by de Broglie's attribution of a wave character to material particles, Erwin Schroedinger published in 1926 a form of quantum mechanics known as **wave mechanics**. About the same time Werner Heisenberg formulated an independent version of quantum mechanics known as **matrix mechanics**. Subsequently these two forms of quantum mechanics were shown to be identical. We shall restrict our discussion to Schroedinger's formulation.

Schroedinger's dynamics differs entirely from Newton's dynamics. Whereas a solution of Newton's equation yields a prediction of the exact positions and momenta of the particles of a system in a given state of motion, a solution of the fundamental equation of wave mechanics (known as Schroedinger's equation) yields only a prediction of the **probable positions and momenta of the particles**.

A solution of the Schroedinger equation subject to its auxiliary postulates yields certain functions Ψ of the position coordinates and the time. These are called **wave functions** and the square of their absolute value $|\Psi|^2$ may be interpreted as a **probability distribution function**. For example, the solution of the Schroedinger equation for the normal state of a hydrogen atom yields the probability distribution function shown in Fig. 18.11. The ordinate $P(r)$ is proportional to the probability of finding the electron at a distance r from the nucleus. The most probable distance of the electron from the nucleus is 5.3×10^{-11} m, which is equal to the radius of the lowest energy Bohr orbit for a hydrogen atom. This is an example of the general result that the Bohr orbits correspond to the most probable distances of finding the electron from the nucleus.

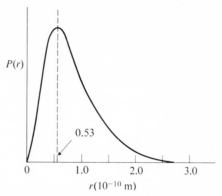

The Schroedinger formalism also provides a means of calculating the energies of the stationary states of an atomic system, the existence of which was postulated by Niels Bohr as we have discussed earlier. Now, however, no arbitrary quantization conditions have to be introduced. Instead, quantum numbers enter automatically in the process of finding acceptable solutions for Schroedinger's equation. For example, the solution of Schroedinger's equation for the hydrogen atom yields three quantum numbers designated

Fig. 18.11. Probability distribution function for the normal state of the hydrogen atom.

n, l, and m. The quantum number n, which can take values

$$n = 1, 2, 3, \cdots,$$

is called the **principal quantum number**. The energy of the electron depends only on the value of n according to

$$E_n = -\left(\frac{2\pi^2 k'me^4}{h^2}\right)\frac{1}{n^2},$$

which is precisely the Bohr formula. The quantum number l, which can take values

$$l = 0, 1, 2, \cdots, (n - 1),$$

is called the **orbital quantum number**. The physical interpretation of l is that it specifies the orbital angular momentum of the electron. In particular, the magnitude of the electronic orbital angular momentum L is related to l through the equation

$$L = [l(l + 1)]^{1/2}\left(\frac{h}{2\pi}\right).$$

This may be compared with Bohr's quantization condition on the orbital angular momentum, namely

$$L = l\left(\frac{h}{2\pi}\right),$$

with

$$l = 1, 2, 3, \cdots.$$

Note that when $l = 0$, $L = 0$, and quantum mechanics predicts stationary states of the electron with zero angular momentum. In terms of the Bohr model this is inconceivable: how can an orbiting electron have zero orbital angular momentum? The quantum number m, which can take values

$$m = -l, -(l - 1), \cdots, 0, \cdots, (l + 1), l,$$

is called the **magnetic quantum number**. It describes the quantization of the direction of the electron's orbital angular momentum with respect to a unique direction in space as provided by an external magnetic field. (See Sections 7.6 and 12.6 for a discussion of magnetic forces applied to moving charged particles.) The allowed components of the angular momentum in this direction are

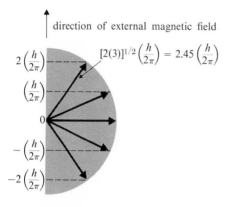

$$l\left(\frac{h}{2\pi}\right), (l - 1)\left(\frac{h}{2\pi}\right), \cdots, -l\left(\frac{h}{2\pi}\right).$$

These are illustrated in Fig. 18.12 for $l = 2$. Note that there are $2l + 1$ allowed components.

Unlike the Bohr theory, the Schroedinger formalism may be applied to atoms with more that one electron. In such cases the energy states depend on both the n and l quantum numbers. This is understandable if we consider the nature of the electron probability distribution for different values of l and take account of the variation of the electron-nuclear interaction for a particular electron due to the presence of the other electrons. For $l = n - 1$, a cross-section through the probability distribution has an approximately circular contour; for small values of l, the cross-section has an approximately elliptical contour, the eccentricity increasing with decreasing l. This is shown symbolically in Fig. 18.13 for the outer electron of a many-electron atom. An electron in an energy

Fig. 18.12. Vector diagram to illustrate the space quantization of an orbital angular momentum described by $l = 2$.

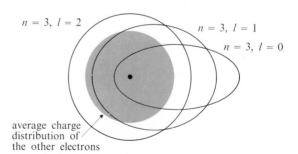

Fig. 18.13. Symbolic representation of the electron "orbits" specified by $(n = 3, l = 2)$ $(n = 3, l = 1)$ and $(n = 3, l = 0)$ in a many-electron atom.

state corresponding to a small value of l spends part of its time close to the nucleus, where it is poorly shielded by the average charge distribution of the other electrons as compared to an electron in a state corresponding to $l = n - 1$. That is, the electron energy for a given value of n decreases with decreasing values of l due to the increased interaction of these negative electrons with the positive nucleus. The energy level diagram of the lithium atom, shown in Fig. 18.14, illustrates this result. Energy states corresponding to different values of l are displaced horizontally for clarity.

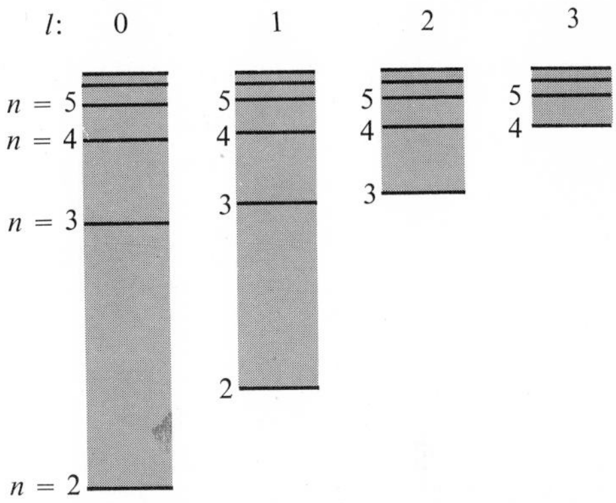

Paul Dirac in 1926 developed a method allowing the calculation of the probability of the transition of an atomic system from one stationary state to another in the presence of some time-dependent interaction, for example, that provided by the presence of electromagnetic radiation. This method allows the calculation of the relative intensities of the spectral components emitted by an atomic system.

Fig. **18.14**. Energy level diagram of the Li atom.

We should think of the Schroedinger equation and its auxiliary restrictions as the starting point of quantum mechanics, just as Newton's second law is the starting point of Newtonian mechanics. It is not derived from other physical laws nor obtained as a necessary consequence of any experiment, although it has much experimental experience built into it. The only justification of the Schroedinger equation is its success in predicting the outcome of a wide variety of experiments in the microscopic domain.

18.7   ELECTRON SPIN AND THE PAULI EXCLUSION PRINCIPLE

The quantum theory, as we have so far presented it, was still unable to account for some of the properties of the hydrogen atom. For example, the first line of the Balmer series, which arises from transitions between the $n = 3$ and $n = 2$ energy states, appears as two closely spaced lines under high resolution. A small effect, but nonetheless a failure of the theory. The explanation of this effect requires the inclusion into the theory of an idea proposed by Samuel Goudsmit and George Uhlenbeck. They pictured the electron as a classical charged sphere which, if it is rotated about an axis, would possess an angular momentum. In 1928 Paul Dirac showed that electrons must possess such a **spin angular**

momentum of magnitude

$$L_s = \left[\frac{1}{2}\left(\frac{3}{2}\right)\right]^{1/2}\left(\frac{h}{2\pi}\right)$$

$$= \frac{3^{1/2}}{2}\left(\frac{h}{2\pi}\right).$$

The allowed directions of the spin angular momentum with respect to an external magnetic field are quantized, there being only two allowed orientations, as specified by the quantum number m_s, having the values $\pm\frac{1}{2}$. The allowed components of the spin angular momentum in the direction of the magnetic field are

$$\frac{1}{2}\left(\frac{h}{2\pi}\right) \quad \text{and} \quad -\frac{1}{2}\left(\frac{h}{2\pi}\right).$$

It should be noted that although one can visualize an electron's spin as arising from the rotation of a charge cloud, such an explanation is physically incorrect. This can be seen if one calculates the speed of rotation necessary to produce the observed angular momentum.

The energy states of the extranuclear electrons of a complex atom may be labeled by stating values for the four quantum numbers n, l, m, and m_s of each electron. It is a basic law of nature that an atomic system left to itself will tend to the state of lowest energy available to it. For example, for a hydrogen atom in its normal state the electron has an energy of -13.6 eV, the lowest value available to it. What then are the lowest energy states available to the electrons of other atoms? To answer this question it is necessary to invoke the **Pauli exclusion principle**[3]. The stationary states of an atom can be determined from its spectrum, and the appropriate quantum numbers of the electrons can be inferred. Wolfgang Pauli, from the observation of atomic spectra, found that certain expected spectral lines did not appear and noted that the missing lines correspond to postulated transitions between stationary states involving two or more electrons with the same set of quantum numbers. This empirical finding led to the Pauli exclusion principle, which states that the existence of stationary states in which two or more electrons would have the same set of quantum numbers is forbidden.

All electrons with the same n value are on the average about the same distance from the nucleus and therefore have about the same energy. We say that they occupy the same **atomic shell** and designate the shells as follows:

n value	1	2	3	4	$\cdots$
shell designation	K	L	M	N	$\cdots$

[3] G. Gamow, "The Exclusion Principle," *Scientific American*, July, 1959. Available as *Scientific American Offprint 264* (San Francisco: W. H. Freeman and Co., Publishers).

The letters used to designate these shells (and the subshells mentioned below) were assigned by physicists who studied atomic spectra long before quantum theory was developed. They were used to designate certain lines or series of lines in atomic spectra. Their use has been extended to the quantum theory merely due to convenience.

The electron energy depends to a lesser degree on the value of the l quantum number, as we have discussed previously. The smaller the l value, the lower the energy. We say that electrons with the same n and l values occupy the same **atomic subshell** and designate the subshells as follows:

l value	0	1	2	3	$\cdots$
subshell designation	s	p	d	f	$\cdots$

The occupancy of a subshell is conventionally expressed by identifying the subshell by its numerical n value followed by the alphabetic designation of its l value. The number of electrons in the subshell is indicated by a superscript after the letter. For example, the **electron configuration** of a lithium atom in its lowest energy state is specified as $1s^2 2s$, which says that the $1s$ ($n = 1$, $l = 0$) subshell contains two electrons and the $2s$ ($n = 2$, $l = 0$) subshell contains one electron.

According to the Pauli exclusion principle, the number of electrons in any subshell is limited. That is, for a given value of l there are $2l + 1$ possible values of m, and for each value of m there are two possible values of m_s. Each subshell therefore contains a maximum of $2(2l + 1)$ electrons. For example, a p subshell can contain a maximum of $2(2 \times 1 + 1) = 6$ electrons. The six electrons of the $2p$ subshell have the quantum numbers given in Table 18.1. It is left as an exercise (Problem 22) for the reader to show that each shell contains a maximum of $2n^2$ electrons. The energy states characterized by the quantum numbers n, l, m, and m_s are called **electronic energy states**.

Table 18.1   Quantum numbers in the $2p$ subshell

n	l	m	m_s
2	1	1	$\frac{1}{2}$
2	1	1	$-\frac{1}{2}$
2	1	0	$\frac{1}{2}$
2	1	0	$-\frac{1}{2}$
2	1	-1	$\frac{1}{2}$
2	1	-1	$-\frac{1}{2}$

The energies of electrons in the various possible subshells have been shown experimentally to increase in the following order:

$$1s, 2s, 2p, 3s, 3p, 4s, 3d, 4p, 5s, 4d,$$

and so on. Using this result in conjunction with the Pauli exclusion principle allows the electron configuration of lowest energy to be specified. As an example, we consider the sodium atom with its eleven extranuclear electrons. The $1s$ and $2s$ subshells can contain a maximum of two electrons each, so that they will be completely filled. The $2p$ subshell can contain a maximum of six electrons, so that it too will be completely filled. The final electron will occupy the $3s$ subshell. That is, the electron configuration of the ground state of the sodium atom is

$$1s^2, 2s^2, 2p^6, 3s.$$

It is convenient to associate with each element an **atomic number** that is just equal to the number of protons in the nucleus of an atom of the element. Dmitri Mendeleev first noted that when the elements are listed in order of their atomic numbers, elements with similar physical and chemical properties recur at regular intervals. A tabular arrangement of the elements that exhibits this behavior is called a **periodic table**. A portion of one form of periodic table is given in Table 18.2. The elements in any vertical column exhibit similar properties. For example, the elements He, Ne, and Ar are very inert chemically and almost never form compounds with other elements. They are known as inert gases.

Table 18.2   A periodic table of some of the elements

H 1							He 2
Li 3	Be 4	B 5	C 6	N 7	O 8	F 9	Ne 10
Na 11	Mg 12	Al 13	Si 14	P 15	S 16	Cl 17	Ar 18

The notation of electron shells and subshells fits perfectly into the pattern of the periodic table, and the Pauli exclusion principle provides us with an understanding of it. For example, we consider the electron configurations of the inert gases mentioned above:

$$\begin{array}{ll} \text{He} & 1s^2 \\ \text{Ne} & 1s^2, 2s^2, 2p^6 \\ \text{Ar} & 1s^2, 2s^2, 2p^6, 3s^2, 3p^6. \end{array}$$

Note that in each case all subshells contain their maximum number of electrons. It follows that the effective charge distributions of these atoms are spherically symmetric and that they neither strongly attract other electrons nor readily give up electrons. Such atoms might well be expected to be chemically inert.

18.8   VIBRATIONS AND ROTATIONS OF DIATOMIC MOLECULES

The solution of the Schroedinger equation for a harmonic oscillator representation of the vibrational motion of a diatomic molecule yields **vibrational energy states** of energy

$$E_v = (v + \tfrac{1}{2})hv,$$

where the **vibrational quantum number** v can take the integral values

$$v = 0, 1, 2, \cdots$$

and v is a characteristic oscillator frequency. The first few vibrational energy states are plotted on an energy level diagram in Fig. 18.15. The lowest energy state is called the **ground vibrational state**; the energy $\tfrac{1}{2}hv$ is called the **zero point energy** of the oscillator.

Transitions are allowed only between energy states whose vibrational quantum numbers differ by one. Since the energy states are equally spaced, the predicted vibrational spectrum consists of a single line of frequency

$$\frac{E_{v+1} - E_v}{h} = \left[(v + 1) + \frac{1}{2}\right]v - \left(v + \frac{1}{2}\right)v$$

$$= v.$$

Fig. 18.15. The first few vibrational energy states of a quantized oscillator.

This frequency is experimentally observed to occur in the near infrared region of the electromagnetic spectrum (that is, at frequencies of the order of 10^{14} Hz).

The solution of the Schroedinger equation for a rigid rotator representation of the rotational motion of a diatomic molecule yields **rotational energy states** of energy.

$$E_J = \frac{J(J + 1)h^2}{8\pi^2 I},$$

where the **rotational quantum number** J can take the integral values

$$J = 0, 1, 2, \cdots$$

and I is the moment of inertia of the molecule about an axis perpendicular to the internuclear axis and passing through the center of mass. The first few rotational energy states are shown in the energy level diagram of Fig. 18.16.

Transitions are allowed only between energy states whose rotational quantum numbers differ by one. Since the spacing of the rotational energy states increases with increasing J, the predicted rotational spectrum consists of a series of lines

Fig. 18.16. The first few rotational energy states of a quantized rotator.

with frequencies

$$\frac{E_{J+1} - E_J}{h} = \frac{[(J+1)(J+2) - J(J+1)]h}{8\pi^2 I}$$

$$= \frac{(J+1)h}{4\pi^2 I},$$

that is, a series of equally spaced lines. These frequencies are experimentally observed to occur in the far infrared region of the electromagnetic spectrum (that is, at frequencies of the order of 10^{13} Hz).

If vibrational and rotational motions occur simultaneously and are assumed to be independent of one another, the solution of the appropriate Schroedinger equation yields vibration-rotation energy states

$$E_{J,v} = (v + \tfrac{1}{2})hv + \frac{J(J+1)h^2}{8\pi^2 I}$$

$$= (v + \tfrac{1}{2})hv + J(J+1)Bh,$$

where

$$v = 0, 1, 2, \cdots,$$

$$J = 0, 1, 2, \cdots,$$

and $B = h/8\pi^2 I$ is called the **rotational constant**. The energy level diagram is shown in Fig. 18.17. Note that the separation between consecutive vibrational energy states is of the order of ten times the separation between consecutive rotational energy states.

Transitions are allowed between energy states whose vibrational and rotational quantum numbers each differ by one. That is, a change in the vibrational energy is accompanied by a change in the rotational energy. The near infrared spectrum is now predicted to consist of a series of closely spaced lines (a **vibration-rotation band**) centered about the characteristic vibrational frequency, as illustrated in Fig. 18.18(a). Note that no line occurs at the position of the characteristic vibrational frequency. The transitions giving rise to the spectrum shown in Fig. 18.18(a) are shown in Fig. 18.18(b).

Finally, if electronic motions are superimposed on the vibrational and rotational motions, the nature of the **electronic band spectra** of diatomic molecules may be predicted. Since the spacing

$v = 2, J = 3$

$v = 2, J = 2$
$v = 2, J = 1$
$v = 2, J = 0$

$v = 1, J = 3$

$v = 1, J = 2$
$v = 1, J = 1$
$v = 1, J = 0$

$v = 0, J = 3$

$v = 0, J = 2$
$v = 0, J = 1$
$v = 0, J = 0$

Fig. 18.17. The superposition of the vibrational and rotational energy states of a diatomic molecule.

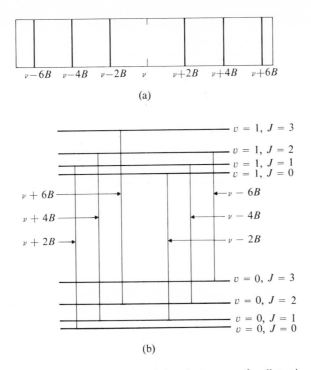

Fig. 18.18. (a) The predicted near infrared spectrum of a diatomic molecule. (b) The transitions giving rise to the near infrared spectrum shown in (a).

of the electronic energy states is an order of magnitude greater than that of the vibrational energy states, the electronic spectra appear in the visible region of the electromagnetic spectrum (that is, at frequencies of the order of 10^{15} Hz). An electronic band spectrum consists of a series of closely spaced lines resulting from the simultaneous changes of electronic, vibrational, and rotational energies.

PROBLEMS

1. Discuss briefly the justification that astronomers have for referring to stars appearing reddish in color as "cool" stars and those appearing bluish as "hot" stars.

2. Plot a graph to show the variation of the energy of a standing electromagnetic wave in a cavity at temperature T with the frequency of the wave.

3. Starting from Planck's formula for blackbody radiation expressed in terms of frequency, deduce the corresponding formula expressed in terms of wavelength.

4. What energy is carried by each photon in
(a) a beam of blue light of wavelength 470 nm?
(b) a beam of red light of wavelength 660 nm?

5. A certain radioactive substance emits γ-rays of 1.46 MeV energy. What is the wavelength of the γ-rays?

6. The photoelectric threshold of tungsten occurs for a wavelength of 273 nm. Calculate its work function in electron volts.

7. The work function for a zinc surface is 3.63 eV. Calculate the maximum kinetic energy of electrons ejected from the surface by incident light of wavelength 300 nm. Will visible light cause the emission of photoelectrons from such a surface?

8. The neighboring conduction electrons in a metallic solid are separated by $\sim 10^{-10}$m. If radiation of intensity $10^{-6}\,\mathrm{J}\cdot\mathrm{m}^{-2}\cdot\mathrm{sec}^{-1}$ is incident on the surface of the metal and is entirely and uniformly absorbed by the first layer of electrons, estimate the time it will take for an individual electron to accumulate an energy of 1 eV.

9. The elastic scattering of x-rays by electrons, or the **Compton effect** as it is called, can be understood on the basis of a "billiard-ball collision" between an incident photon and a stationary electron. Assuming that the photon is scattered through an angle θ, write equations for conservation of energy and momentum for the collision including possible relativistic effects. *Hint:* The scattered photon has a frequency different from that of the incident photon.

10. The inverse photoelectric effect results in the emission of x-ray photons by a metal when electrons are incident on it. In such cases the work function is small compared to the photon energy and can be neglected. Deduce the minimum wavelength of x-rays emitted from a surface if 3000-eV electrons are incident on a metal surface.

11. A series of lines in the spectrum of hydrogen gas known as the **Lyman series** has wavelengths of 121.6 nm, 102.6 nm, 97.0 nm, 94.9 nm, 94.0 nm, etc. Compare these experimental wavelengths with the predictions of the Bohr theory as given by

$$v = C\left(\frac{1}{n_i^2} - \frac{1}{n_f^2}\right).$$

Obtain a value of C. Show the transitions on an energy level diagram.

*12. The Bohr theory calculations presented in this chapter assumed that the nucleus is of infinite mass. Repeat the calculation for a nucleus of finite mass. Compare the predicted energy states for a deuterium atom with those for a hydrogen atom. *Hint:* The nucleus and the electron now revolve about a common center of mass.

13. The Bohr theory is also applicable to ions containing single electrons, as for example singly ionized helium and doubly ionized lithium. Such ions are called hydrogenic atoms. Assuming a nucleus of infinite mass, compare the energy levels for the hydrogenic atoms with the energy levels for the hydrogen atom.

14. Calculate the de Broglie wavelength of
    (a) a baseball of mass 125 g traveling at a speed of $25 \text{ m} \cdot \sec^{-1}$.
    (b) a bullet of mass 1 g traveling at $1000 \text{ m} \cdot \sec^{-1}$.
    (c) a neutron of mass $1.67 \times 10^{-27} \text{ kg}$ traveling at $3300 \text{ m} \cdot \sec^{-1}$.

15. Calculate the de Broglie wavelengths of electrons of energies 30 MeV and 1000 MeV. At what energy is the wavelength comparable to the size of a proton ($\sim 10^{-15} \text{ m}$)?

16. A particle is confined to a cubic container 10^{-8} m to a side. What is the uncertainty in the particle's speed
    (a) if the particle is an electron?
    (b) if the particle is a neutron?

17. If an atom makes a transition from one stationary state to another in $10^{-8} \sec$, what is the fundamental uncertainty in the energy of the emitted radiation? Comment on the effect that such an uncertainty will have on the nature of the energy levels of the atom.

18. The existence of a zero-point energy associated with a harmonic oscillator is a consequence of the uncertainty principle. Comment on this statement.

19. Draw a diagram to illustrate space quantization for a system whose orbital quantum number is $l = 3$.

20. Specify the lowest energy electron configuration for the following atoms:
    (a) the boron atom having 5 extranuclear electrons.
    (b) the aluminum atom having 13 extranuclear electrons.
    (c) the chlorine atom having 17 extranuclear electrons.

*21. It has been said that if we lived in a universe in which Planck's constant h was very much larger than in our own, quantum effects would be everyday occurrences. Comment on this hypothesis and indicate approximately how large h would have to be.

*22. Use the Pauli exclusion principle to show that an atomic shell can contain a maximum of $2n^2$ electrons.

23. The two atoms in the hydrogen molecule are $7.42 \times 10^{-11} \text{ m}$ apart and have a mass of $1.67 \times 10^{-27} \text{ kg}$ each.
    (a) Calculate the moment of inertia of the molecule about an axis through the center of mass and perpendicular to the nuclear axis.
    (b) Calculate the energies of the first five rotational states and plot them on an energy level diagram.
    (c) Plot the first five vibrational energy levels, assuming the characteristic frequency to be $1.32 \times 10^{12} \text{ Hz}$ and compare the resulting diagram with part (b).

19 Introductory statistical physics

19.1   INTRODUCTION

The aim of statistical physics is the same as that of the kinetic theory: to relate the macroscopic behavior of a system of particles to the microscopic behavior of the individual particles. Statistical physics is concerned with the most probable behavior of the particles rather than with the actual motions and interactions of the particles themselves. The basic assumption is that a calculation of the thermal equilibrium properties of a system of particles requires only a knowledge of the most probable behavior of the particles. The fundamental problem of statistical physics is the calculation of the most probable distribution of an amount of energy E among the various members of a system of N identical particles.

19.2   BASIC PROBABILITY IDEAS[1]

When we flip a coin, it is possible for it to land in either of two ways: heads up or tails up. If the coin is symmetrical and uniform in constitution, it is reasonable to assume that the appearance of a head is just as likely as that of a tail. The same is true of a uniform die; when it is rolled, it is reasonable to assume that it is equally probable for the die to stop with any given one of its six numbered sides facing upward. In reality, we have no certain way of knowing if a head or a tail is equally likely to occur on the flip of a coin. Our best justification for making this assumption is that mathematical models based on it predict the behavior of real coins. Mathematical models in probability theory are based upon the expected behavior of ideal objects such as ideal dice, ideal coins, ideal colored balls, etc. If objects in our real world correspond closely enough to their ideal counterparts, we find that the mathematical models predict their behavior quite well.

Suppose that we are to deal with an experiment that can result in N different equally probable outcomes, in which N may be any positive integral number. We suppose further that, of these N equally probable outcomes, we can consider n of them to be "favorable" in that they result in the outcome in which we are interested. If we denote this favorable outcome to be the **event** E, we may define the **probability** $W(E)$ of an event E to be

$$W(E) = \frac{n}{N}.$$

[1] W. Weaver, *Lady Luck: the Theory of Probability* (Garden City, N.Y.: Doubleday and Company, Inc., 1963). J. N. P. Hume and D. G. Ivey, "Random Events," a film available from Educational Services Incorporated, Watertown, Mass.

Note that the event E may arise in n different ways out of a total of N outcomes. As an example, let us consider a deck of cards. If we were to choose one card from the deck at random (say by shuffling and then cutting the deck), there are 52 possible outcomes corresponding to the 52 different cards in the deck. If we are interested only in the color of the card chosen, the probability of picking a black card will be

$$W(\text{black}) = \tfrac{26}{52} = 0.5,$$

since there are $n = 26$ black cards in the deck of $N = 52$ cards. On the other hand, if we are interested in diamonds, the probability of picking a diamond is given by

$$W(\text{diamond}) = \tfrac{13}{52} = 0.25,$$

since there are $n = 13$ diamonds in the deck. Again, if we are interested in sevens, the probability of picking a seven is given by

$$W(\text{seven}) = \tfrac{4}{52} = 0.077,$$

since there are $n = 4$ sevens in the deck. If $n = N$ for a particular experiment, $W(E) = 1$, and the event is certain to happen; on the other hand, if $n = 0$, $W(E) = 0$, and the event will certainly not happen. These probabilities are often given the name **a priori probabilities** because they are derived from basic or first principles.

We are very often concerned with events composed of two or more component events. These are known as **compound events**. The roll of two dice is a compound event composed of two component events: the roll of one die and the roll of the second die. When a die is rolled there are six possible outcomes. Suppose that the first die stops with the number 3 up. If the roll of the second die is assumed to be independent of the first, then it is equally likely to stop with any one of its six sides up. Therefore, for each of six possible outcomes for the first die, there are six possible outcomes for the second die or a total of $6 \times 6 = 36$ possible outcomes in all for the two dice. The reader may wish to tabulate the possible outcomes starting with $(1, 1), (1, 2), \cdots$ and ending with $\cdots, (6, 5),$ $(6, 6)$ to verify this result. For the roll of five dice, similar reasoning indicates the number of outcomes for this compound event to be

$$6 \times 6 \times 6 \times 6 \times 6 = 6^5 = 7776.$$

The following general rule may be formulated: if a compound event is composed of several component events, one of which has α outcomes, a second of which has β outcomes, etc., then the number of outcomes of the compound event is given by the product $\alpha \times \beta \times \cdots$.

Permutations

Suppose we take the first three letters of the alphabet A, B, and C and ask how many different ways there are to write them down. We have three choices for writing down the first letter; once this is done we have only two choices for the second letter and, finally, one choice for the third letter. This is a triply compound event, and the total number of outcomes is then $3 \times 2 \times 1 = 6$, the product of the outcome for each component event. These outcomes are ABC, ACB, BAC, BCA, CAB, CBA. In each of the six cases, the letters A, B, and C appear in a different order; each arrangement is said to be a **permutation** of the three basic letters. If we have n objects, the total number of permutations denoted $_nP_n$ will be

$$_nP_n = n(n - 1)(n - 2) \cdots (2)(1).$$

The symbol $_nP_n$ simply means "the number of ways in which n objects chosen from a group of n objects may be ordered." If we now ask the question "how many ways may we order r objects taken out of a group of n objects," the answer is

$$_nP_r = n(n - 1)(n - 2) \cdots (n - r + 1).$$

This product stops short after the rth term.

In order to avoid writing down long products when dealing with permutations, we introduce the notation

$$n! = n(n - 1) \cdots (2)(1),$$

where $n!$ is read **factorial** n. It simply means "the product of all the integers from n down to 1." In terms of this notation we may write

$$_nP_n = n!$$

$$_nP_r = \frac{n!}{(n - r)!}.$$

The reader should verify for himself that the $(n - r)!$ in the denominator of $_nP_r$ cancels the unwanted tail of the $n!$ in the numerator. The quantity $n!$ increases very rapidly: $5! = 120$, $10! = 3,628,800 = 3.63 \times 10^6$ to three significant figures, and $15! = 1.31 \times 10^{12}$.

Example. In how many ways can we select two books from a group of four?

Solution. Using the formula we obtain

$$_4P_2 = \frac{4!}{(4 - 2)!} = \frac{24}{2} = 12.$$

From first principles, we may reason as follows. There are four different choices for the first book and three different choices for the second book for a total of $4 \times 3 = 12$ permutations.

Combinations

Often we are not concerned about the order of the objects in a group but only in the constitution of the group in question. Suppose, for example, we have four cards and we are interested in how many different groups or **combinations** of two cards we can make out of the group of four. By labeling the cards a, b, c, and d and writing down pairs systematically, you will readily find that there are six possible combinations. They are ab, ac, ad, bc, bd, and cd. We desire a general mathematical expression that will allow us to evaluate the number of combinations in a given situation.

This problem is very similar to the problem of finding $_nP_r$, the number of permutations of r objects taken from a group of n objects, except that we are no longer interested in the order of the r objects within their group but merely desire to know how many combinations of r objects can be made from a group containing n objects. Each group of r objects chosen out of the group of n objects is counted as only one combination; however, there are $r!$ possible permutations of the objects within the group. If we denote the number of combinations of n things taken r at a time by $_nC_r$, we can write

$$_nC_r = \frac{_nP_r}{r!} = \frac{n!}{r!(n-r)!}.$$

Applied to our example of four cards above, we see that the number of combinations of two cards taken from four cards is

$$_4C_2 = \frac{4!}{2!2!} = \frac{24}{4} = 6$$

in agreement with earlier conclusions.

Suppose we have n objects and are interested in how many ways we can choose from them a group or combination of α objects, a second group of β objects, and a third group of γ objects. The answer is

$$_nC_{\alpha,\beta,\gamma} = \frac{n!}{\alpha!\beta!\gamma![n-(\alpha+\beta+\gamma)]!},$$

where the last factorial in the denominator includes all those objects of the original n that have not fallen in one of the chosen groups. This result can be readily expanded to any arbitrary number of groups.

19.3   INTRODUCTORY STATISTICS FOR MOLECULES

We introduce the statistical approach with a simple example. We consider an ideal gas of N molecules in a container, assume an imaginary partition dividing the volume into equal parts, and ask for the probability of finding N_1 molecules on the left side of the partition and N_2 molecules on the right side, where $N_1 + N_2 = N$. That is, we ask for the probability of occurrence of a certain **distribution** of the molecules N_1 and N_2. The situation is shown in Fig. 19.1.

We begin by assuming that we are equally likely to find a particular molecule in either section, since we have no reason to assume anything else. Therefore, the a priori probability for finding a specific molecule in either compartment is 0.5. The probability of finding N_1 specific molecules on the left side and N_2 specific molecules on the right side is then

$$(0.5)^{N_1}(0.5)^{N_2}.$$

This is called the **a priori probability of the distribution** (N_1, N_2). It is the same for all distributions, since for each molecule in turn we ask for the probability that it will be found on one side or the other. This is also obvious from the mathematical expression for the a priori probability, since

$$(0.5)^{N_1}(0.5)^{N_2} = (0.5)^{N_1 + N_2}$$

$$= 0.5^N$$

independent of the particular values for N_1 and N_2. Note that we have implicitly assumed that the molecules are distinguishable, that is, that we could tag each molecule in some way and know where it is at all times.

However, we are interested only in the total number of molecules that appear on each side and must therefore add up the number of distinct ways in which the molecules can be placed into the box to give the distribution (N_1, N_2). For example, if we have three molecules labeled A, B, and C, the distribution $(2, 1)$ can be realized in three different ways, as illustrated in Fig. 19.2. In general, for N molecules, the distribution (N_1, N_2) can be realized in

$$\frac{N!}{N_1!(N - N_1)!} = \frac{N!}{N_1!N_2!}$$

ways, since we are just looking for the number of possible combinations in which N_1 molecules out of a group of N may be put into one side of the box with the remaining $N - N_1 = N_2$ molecules going into the other side. In statistical physics the quantity

$$\frac{N!}{N_1!N_2!}$$

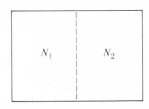

Fig. 19.1. For an ideal gas of N molecules, N_1 are found to the left and N_2 to the right of an imaginary partition.

$A\ B$	C
$B\ C$	A
$A\ C$	B

Fig. 19.2. There are only three ways of realizing the distribution $(2, 1)$ of the three letters A, B, and C.

is called the **statistical weight**, or **thermodynamic probability** of the distribution (N_1, N_2). For the distribution $(2, 1)$, the statistical weight is

$$\frac{3!}{2!1!} = 3$$

in agreement with our direct counting analysis above.

The total probability $W(N_1, N_2)$ is the product of the a priori probability and the thermodynamic probability, that is,

$$W(N_1, N_2) = (0.5)^{N_1}(0.5)^{N_2}\frac{N!}{N_1!N_2!}.$$

W is tabulated for all possible distributions for $N = 4$ and $N = 5$ in Table 19.1. Clearly, the distribution of maximum probability is that for which there are equal numbers of molecules on the two sides for N even, and for N_2 and N_1 differing by only one molecule for N odd.

Table 19.1   Values of $W(N_1, N_2)$ for $N = 4$ and $N = 5$

N_1	$N = 4$ N_2	W	N_1	$N = 5$ N_2	W
4	0	$\frac{1}{16}$	5	0	$\frac{1}{32}$
3	1	$\frac{4}{16}$	4	1	$\frac{5}{32}$
2	2	$\frac{6}{16}$	3	2	$\frac{10}{32}$
1	3	$\frac{4}{16}$	2	3	$\frac{10}{32}$
0	4	$\frac{1}{16}$	1	4	$\frac{5}{32}$
			0	5	$\frac{1}{32}$

If we define

$$N_1 = \frac{N}{2} + n, \qquad N_2 = \frac{N}{2} - n,$$

where n is a measure of the deviation from the most probable value, then

$$W(n) = (0.5)^N \frac{N!}{\left(\dfrac{N}{2} + n\right)!\left(\dfrac{N}{2} - n\right)!}.$$

For very large N and for $N \gg n$ this expression may be shown to reduce to

$$W(n) = \left(\frac{2}{\pi}\right)^{1/2}\left(\frac{1}{N}\right)^{1/2}\exp\left(-\frac{2n^2}{N}\right).$$

The mathematics is rather too involved to pursue here but may be found in most texts on statistical physics. This probability distribution is of the general form

$$W(n) = a \exp(-bn^2),$$

where a and b are constants, and is known as **the Gaussian or normal distribution**. In Fig. 19.3, the total probability $W(N_1, N_2)$ for $N = 4$ is plotted as a function of N_1; a normal distribution of the same area is plotted for comparison. It is easy to see that $W(N_1, N_2)$ should approach a normal distribution as N_1 becomes large.

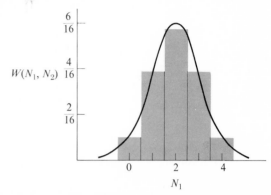

The width of the normal distribution at one-half the peak value may be shown to be $b^{1/2}$; in the present case

$$b^{1/2} = \left(\frac{2}{N}\right)^{1/2} \simeq \left(\frac{1}{N}\right)^{1/2}$$

for a large N. For a real system of molecules N is usually extremely large, so that the normal curve is extremely narrow and it is overwhelmingly more probable to find equal numbers of molecules in the two sections of the container. Even a volume of gas so small as to be 10^{-3} mm^3 (a cube 0.1 mm on a side) has more than 10^{13} molecules

Fig. 19.3. The total probability $W(N_1, N_2)$ for $N = 4$ as a function of N; a normal distribution of the same area is plotted for comparison.

in it. Here the value of $(1/N)^{1/2}$ is about 10^{-7}, so that the probability for the system to fluctuate very far from its most probable state is extremely small.

Example. There are 10^{22} molecules in a container. What is the probability that we should at some instant find 5.000001×10^{21} molecules in one half of the container and 4.999999×10^{21} molecules in the second half?

Solution. Here $n = N_1 - N/2 = 10^{15}$. Therefore,

$$W(n) = \left(\frac{2}{\pi N}\right)^{1/2} \exp\left(-\frac{2n^2}{N}\right)$$

$$= \frac{1}{10^{11}} \exp(-2 \times 10^8)$$

$$= 10^{-108}$$

$$= 10^{-100000000},$$

which is an extraordinarily small number! Even fluctuations of one part in one million are extremely unlikely.

The state of a real, macroscopic system in thermal equilibrium is determined by specifying a very small number of parameters (such as pressure, volume, energy, etc.) derived from a study of the statistical properties of the microscopic behavior. Experimentally it is found that any system left to itself will approach a state of thermal equilibrium, and, for a given initial total energy, the same state is reached independent of any other initial conditions imposed upon the constituent particles. We see from our preceding discussion that this results from the fact that there is a large number of particles contained in any macroscopic system. Organized motions tend to become randomized, and the probability that a macroscopic parameter will be found to have a value different from the most probable value is negligible.

As a final example let us consider again the container of Fig. 19.1 with initially N_1 molecules of one kind of gas on one side of the container and N_2 molecules of a different kind of gas on the second side, with a partition separating them. Our preceding discussions suggest that when the partition is removed the most probable state of the gas in the container will consist of equal numbers of molecules of each kind in each half of the container. This means that the gases should become uniformly mixed, which indeed is observed to occur. Moreover, as time passes and the system is checked periodically, it is found that the gases remain uniformly mixed. It should be emphasized that it is not impossible that at some instant in the future we might discover that the molecules had become separated again into their initial state; however, it is so extremely improbable that this should occur that it may for all practical purposes be disregarded[2].

19.4   THE MAXWELL–BOLTZMANN DISTRIBUTION

The basic problem of statistical physics is the calculation of the most probable distribution of an amount of energy E over the N constituent "particles" constituting a physical system.

For the present we shall consider a system containing N identical but distinguishable "particles," for example, the N molecules of a low-density gas. We wish to calculate the most probable distribution of the total energy E of this system among the N particles. The distribution that we shall obtain is known as the Maxwell–Boltzmann distribution. In general, there are a number of possible energies that the individual particles in the system may have. However, the following discussion does not depend on the number of energy states available to each particle of a system but only on the probability of a particle having that energy.

Let us assume that each particle can have the possible energies E_1, E_2, E_3, $\cdots$, E_i, $\cdots$, and that the corresponding a priori probabilities are g_1,

---

[2] See the film by B. J. Adler and F. Reif, "Irreversibility and Fluctuations" (New York: McGraw-Hill Book Company, College Division).

$g_2, \cdots, g_i, \cdots$. We assume that our system is isolated from its surroundings so that the total energy E is constant. If there are N_i particles in the ith energy state, the contribution to the total energy of the system by these particles will be $N_i E_i$, where E_i is the energy of the ith state. The total energy is the sum of the energies associated with each of the energy states or

$$E = N_1 E_1 + N_2 E_2 + \cdots + N_i E_i + \cdots$$
$$= \sum_i N_i E_i,$$

where the sum is over all the energy states available to the system of particles. The total number of particles N must be

$$N = N_1 + N_2 + \cdots + N_i + \cdots$$
$$= \sum_i N_i.$$

These two sums express conservation of energy and conservation of the number of particles, respectively. They are conditions that the system must satisfy at all times.

Since the a priori probabilities for each state are $g_1, g_2, \cdots$, the a priori probability for the distribution will be

$$(g_1)^{N_1}(g_2)^{N_2} \cdots (g_i)^{N_i} \cdots$$

following the reasoning of Sections 19.2 and 19.3. The statistical weight will be

$$\frac{N!}{N_1! N_2! \cdots N_i! \cdots}.$$

These are combined to give for the total probability $W(N_1, N_2, \cdots)$ of the distribution $(N_1, N_2, \cdots)$

$$W(N_1, N_2, \cdots) = \frac{(g_1)^{N_1}(g_2)^{N_2} \cdots N!}{N_1! N_2! \cdots}.$$

Since we associate the thermal equilibrium properties of the system with the distribution of maximum probability, we only have to discover which particular distribution $(N_1, N_2, \cdots, N_i, \cdots)$ has the maximum probability subject to the conditions of conservation of energy and conservation of the number of particles. Let us assume that all the g quantities are the same and set them equal to unity. That is, we assume each state to have equal a priori probability. The total probability then becomes

$$W(N_1, N_2, \cdots) = \frac{N!}{N_1! N_2! \cdots}.$$

The process of finding the distribution of maximum probability subject to the conservation conditions is purely mathematical in nature and makes use of the method of Lagrangian multipliers. The analysis, carried out in the box, yields the **Maxwell–Boltzmann distribution**

$$N_i = \left(\frac{N}{Z}\right) \exp\left(-\frac{E_i}{kT}\right),$$

where N_i is the number of particles in the ith energy state, k is Boltzmann's constant, and T is the absolute temperature of the system. The quantity Z, known as the **partition function** or the **sum over states** is

$$Z = \sum_i \exp\left(-\frac{E_i}{kT}\right).$$

It depends only on the temperature and the particle state energies E_i.

Derivation of the Maxwell–Boltzmann Distribution

Suppose we have a function $f(x_1, \cdots, x_N)$ of the variables $x_1, \cdots, x_N$ that is not subject to any restrictive conditions. Then, according to the calculus, if the function possesses a maximum value for a certain set of variables $x_1', \cdots, x_N'$, these variables may be obtained by the solutions of the set of independent equations

$$\frac{\partial f(x_1, \cdots, x_N)}{\partial x_i} = 0,$$

where i assumes the values $1, 2, \cdots, N$. If however, the function $f(x_1, \cdots, x_n)$ is subject to a restrictive condition $g(x_1, \cdots, x_N) = 0$, then the set of variables $x_1', \cdots, x_N'$ are most readily obtained by the solution of the set of independent equations

$$\frac{\partial f(x_1, \cdots, x_N)}{\partial x_i} - \frac{\lambda \partial g(x_1, \cdots, x_N)}{\partial x_i} = 0$$

where λ is known as a Lagrangian multiplier.

For the present case there are two restrictive conditions on the function $W(N_1, N_2, \cdots)$:

$$\sum_i N_i = N \quad \text{and} \quad \sum_i N_i E_i = E.$$

Both the functions $W(N_1, N_2, \cdots, N_N)$ and $\ln W(N_1, N_2, \cdots)$ will attain their maximum values for the same set of variables $N_1, N_2, \cdots$. We choose to maximize $\ln W(N_1, N_2, \cdots)$ and therefore are led to consider equations of the form

$$\frac{\partial}{\partial N_i}\left[\ln W(N_1, N_2, \cdots) - \alpha \sum_j N_j - \beta \sum_j N_j E_j\right] = 0$$

where α and β are the appropriate Lagrangian multipliers. The values of N_1

and N_2 are fixed by choosing α and β so that

$$\frac{\partial}{\partial N_1}\left[\ln W - \alpha \sum_j N_j - \beta \sum_j N_j E_j\right] = 0$$

$$\frac{\partial}{\partial N_2}\left[\ln W - \alpha \sum_j N_j - \beta \sum_j N_j E_j\right] = 0.$$

Now all the remaining equations are independent of each other, since the only connections between the various possible N_i's are through the restrictions of the conservation of mass and the conservation of energy. N_1 and N_2 have been chosen to satisfy these conservation laws while all other N_i's remain as independent variables. Therefore, we may write for all N_i

$$\frac{\partial}{\partial N_i}\left[\ln W(N_1, N_2, \cdots) - \alpha \sum_j N_j - \beta \sum_j N_j E_j\right] = 0.$$

Now

$$W(N_1, N_2, \cdots) = \frac{N!}{N_1!N_2!\cdots}.$$

Taking natural logarithms of both sides of this equation gives

$$\ln W(N_1, N_2, \cdots) = \ln N! - \sum_j \ln N_j!$$

For large N the function $N!$ may be approximated by Stirling's formula[3] as

$$\ln N! = N \ln N - N$$

Therefore,

$$\ln W(N_1, N_2, \cdots) = N \ln N - N - \sum_j N_j \ln N_j + \sum_j N_j$$

$$= N \ln N - \sum_j N_j \ln N_j.$$

The condition for a maximum then becomes

$$\frac{\partial}{\partial N_i}\left[N \ln N - \sum_j N_j \ln N_j - \alpha \sum_j N_j - \beta \sum_j N_j E_j\right] = 0.$$

Differentiating gives, for the ith term,

$$-\ln N_i - 1 - \alpha - \beta E_i = 0$$

$$\ln N_i = -1 - \alpha - \beta E_i$$

$$N_i = \exp\left(-1 - \alpha - \beta E_i\right).$$

The quantity α may be determined by substituting for N_i in the equation

$$\sum_i N_i = N.$$

[3] I. S. and E. S. Sokolnikoff, *Higher Mathematics for Engineers and Physicists* (New York: McGraw-Hill Book Company, 1941).

Therefore,

$$\sum_i \exp\left(-1 - \alpha - \beta E_i\right) = N$$

and

$$\frac{N_i{}'}{N} = \frac{\exp\left(-1 - \alpha - \beta E_i\right)}{\sum_i \exp\left(-1 - \alpha - \beta E_i\right)}$$

$$= \frac{\exp\left(-\beta E_i\right)}{\sum_i \exp\left(-\beta E_i\right)}$$

or

$$N_i = \left(\frac{N}{Z}\right) \exp\left(-\beta E_i\right),$$

where

$$Z = \sum_i \exp\left(-\beta E_i\right).$$

Thermodynamic arguments are required to show that

$$\beta = \frac{1}{kT},$$

thereby leading to the final expression

$$N_i = \left(\frac{N}{Z}\right) \exp\left(-\frac{E_i}{kT}\right)$$

for the Maxwell–Boltzmann distribution of energy.

The Maxwell–Boltzmann distribution of energy is plotted in Fig. 19.4 for temperatures T_1 and $T_2 = 2T_1$. The curves are normalized to contain the same area under the curve from $E_i = 0$ to $E_i = \infty$.

The most characteristic feature of the distribution is that the number of particles in a given state decreases exponentially with increasing energy of the state for a given temperature. As the temperature increases E_i/kT becomes smaller and $\exp\left(-E_i/kT\right)$ increases; Z also increases. The net result is that more particles are found in higher energy states when the temperature is increased. This is consistent with our intuitive identification of "hotter" systems with higher energies.

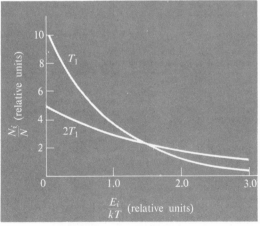

Fig. 19.4. The Maxwell–Boltzmann energy distribution for two temperatures T_1 and $2T_1$.

Example. What are the relative numbers of molecules in a gas at 300°K having energies of 10^{-2} eV, 5×10^{-2} eV, and 10^{-1} eV?

Solution. Boltzmann's constant k is 1.38×10^{-23} J $\cdot (°\text{K})^{-1}$ or 8.63×10^{-5} eV $\cdot (°\text{K})^{-1}$. Since $T = 300°\text{K}$,

$$kT = 8.63 \times 10^{-5} \times 300$$
$$= 2.59 \times 10^{-2} \text{ eV}.$$

For $E_1 = 10^{-2}$ eV,

$$N_1 = \frac{N}{Z} \exp\left(-\frac{E_1}{kT}\right)$$

$$= \frac{N}{Z} \exp\left(-\frac{10^{-2}}{2.59 \times 10^{-2}}\right)$$

$$= \frac{N}{Z} \exp(-0.386)$$

$$= 0.680 \frac{N}{Z}.$$

For $E_2 = 5 \times 10^{-2}$ eV,

$$N_2 = \frac{N}{Z} \exp\left(-\frac{5 \times 10^{-2}}{2.59 \times 10^{-2}}\right)$$

$$= \frac{N}{Z} \exp(-1.93)$$

$$= 0.145 \frac{N}{Z}.$$

For $E_3 = 10^{-1}$ eV,

$$N_3 = \frac{N}{Z} \exp\left(-\frac{10^{-1}}{2.59 \times 10^{-2}}\right)$$

$$= 0.0211 \frac{N}{Z}.$$

Comparing N_2 and N_3 to N_1, we have

$$\frac{N_2}{N_1} = \frac{0.145}{0.680} = 0.213,$$

$$\frac{N_3}{N_1} = \frac{0.0211}{0.680} = 0.031.$$

The number of molecules having 10^{-1} eV energy is only 3.1 % of the number having an energy of 10^{-2} eV. For an energy of 1 eV, the number drops to 1.7×10^{-15} % of the number of molecules having an energy of 10^{-2} eV. The verification of this last number is left as an exercise for the reader.

It is readily shown that, if the g values are not all the same, the Maxwell–Boltzmann distribution takes the form

$$N_i = \frac{N}{Z} g_i \exp\left(-\frac{E_i}{kT}\right),$$

with

$$Z = \sum_i g_i \exp\left(-\frac{E_i}{kT}\right).$$

Example. Suppose that E_i is proportional to a quantity $J_i(J_i + 1)$ and g_i is proportional to a quantity $(2J_i + 1)$. (See Sections 18.6 and 18.8.) What form does the Maxwell–Boltzmann distribution take?

Solution. The Maxwell–Boltzmann distribution may be written as

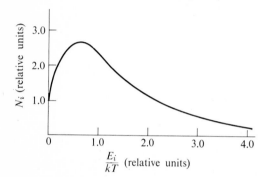

$$N_i = C(2J_i + 1) \exp\left[-\frac{BJ_i(J_i + 1)}{kT}\right],$$

where C and B are constants. To illustrate the form of this curve we set

$$\frac{B}{kT} = 0.1 \quad \text{and} \quad C = 1,$$

so that

$$N_i = (2J_i + 1) \exp[-0.1J_i(J_i + 1)].$$

Calculated values of N_i are given in Table 19.2 for several values of J_i; the calculated values are plotted in Fig. 19.5.

Fig. 19.5. The form of the Maxwell–Boltzmann distribution for $E_i \propto J_i(J_i + 1)$ and $g_i \propto (2J_i + 1)$.

Table 19.2   Values of N_i as a function of J_i

J_i	$(2J_i + 1)$	$J_i(J_i + 1)$	$\exp[-BJ_i(J_i + 1)]$	N_i
0	1	0	1.000	1.000
1	3	2	0.819	2.457
2	5	6	0.549	2.745
3	7	12	0.301	2.107
4	9	20	0.135	1.215
5	11	30	0.050	0.550
6	13	42	0.015	0.195

PROBLEMS

1. If you take a deck of cards and begin to turn them up one at a time, what is the probability that nine cards be turned up and still one suit not have appeared?

2. Suppose you wish to telephone a friend but cannot remember his number other than that it consists of seven digits. If you start dialing numbers at random, at

the rate of one per minute, what is the probability that you will get the correct number
(a) within 3 hours?
(b) within 3 weeks working 8 hours per day?
(c) within 3 years working 8 hours per day and 5 days per week?

3. Five pennies are dropped to the floor. What is the probability that all pennies will land heads up?

4. Calculate the factorials of the numbers from one to ten and plot the results on a graph.

5. Evaluate $_5P_3$, $_6P_2$, and $_4P_3$.

6. Evaluate $_5C_3$, $_6C_2$, and $_4C_3$.

7. What is the probability that a player will be dealt a bridge hand of 13 spades? What is the probability that a player will be dealt the very ordinary hand of:

Spades (King, 10, 6)
Hearts (Queen, 9, 2)
Diamonds (10, 9, 5, 4)
Clubs (Ace, 8, 5)?

8. Calculate $W(N_1, N_2)$ for $N = 6, 7$, and 8.

9. In how many ways can eight golfing physicists divide themselves into two teams of four players if:
(a) the two best players must be on opposite teams, and
(b) two of the four novices must be on each team.

*10. In a random group of 25 persons, what is the probability that at least two of them will have common birthdays?

*11. Calculate the value of n for which $W(n) = W(0)/2$. Evaluate the resulting expression for n for the following values of N: 10^{22}, 10^{10}, 10^6, 10^4, and 10. List the corresponding values of n/N. Comment on your results.

12. Plot $W(N_1, N_2)$ as a function of N_1 for $N = 8$; plot a normal curve of approximately the same area on the same graph for comparison.

13. Plot, on one graph, normal distributions all having the same value of the constant a and having values for the constant b of 1, 0.5, 0.1, and 0.01. Comment.

14. What is the probability for a fluctuation of one part in 10^{10} in the number of molecules on one side of the container described in Section 19.3, assuming that there are 10^{22} molecules in the container?

15. What is the relative number of molecules in a gas at 300°K having energies of 10^{-3} eV and 10^{-1} eV?

16. The particle energy states available to a hydrogen atom have energies of 0, 10.2, 12.1, 12.8, $\cdots$ eV. Calculate the ratio of the number of hydrogen atoms in the 10.2-eV state to the number in the zero-energy state at 300°K, using the Maxwell–Boltzmann distribution.

17. Using the data of Problem 16, calculate the temperature of a gas of hydrogen atoms required to have 1% of the hydrogen atoms in the 10.2-eV energy state, given that $Z \simeq 1$.

18. The lowest two energy states available to a vibrating molecule differ in energy by ΔE J. Calculate the ratio of the number of molecules in these two vibrational levels for $T = 1000°K$ for the gases listed in Table 19.3.

Table 19.3

Gas	$\Delta E(J)$
Hydrogen	8.28×10^{-20}
Nitrogen	4.64×10^{-20}
Oxygen	3.09×10^{-20}
Chlorine	1.01×10^{-20}
Iodine	0.424×10^{-20}

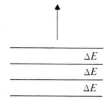

ΔE

ΔE

ΔE

Fig. 19.6

19. There are an infinite number of energy states available to a vibrating molecule, all equally spaced by ΔE J as shown in Fig. 19.6. Using the data given in Problem 18, deduce the number of vibrational states that need to be considered for each gas at 1000°K so as to include only those levels having populations greater than 1% of that of the ground state.

20. Evaluate the partition function for a system having only two energy states 0 and E, assuming equal a priori probabilities for the two states.

21. Evaluate the partition function for a system having energy states

$$E_n = (n + \tfrac{1}{2})h\nu,$$

where $n = 0, 1, 2, \cdots$, assuming equal a priori probabilities for the states.

22. Consider an atomic system having only two energy states E_a and E_b. In some instances such a system is describable by a **negative temperature**. What can you say about the relative number of particles in the two energy states in such a case?

*23. You were introduced to the concept of a negative temperature in Problem 22. Discuss the following statements dealing with negative temperatures.
   (a) A negative temperature is hotter than any positive temperature.
   (b) Only those physical systems having a finite upper limit of their allowed energies can ever be in a state describable by a negative temperature.

*24. Evaluate the partition function for a system having energy states

$$E_J = -BJ(J + 1),$$

where $J = 0, 1, 2, \cdots$, assuming a priori probabilities g_J for the states

$$g_J = 2J + 1$$

and that T is large, B is small. *Hint:* For sufficiently large T and small B the sum over states can to a good approximation be replaced by an integral.

***25.** Suppose we have a bag filled with colored balls and that there are K different colors present in fractional amounts $f, f_2, \cdots, f_K$. We select N balls from the bag in a random manner.

(a) Write an expression for the probability of occurrence of the various possible distributions.

(b) By applying the method of Lagrange multipliers, determine the most probable distribution.

(c) Does your answer to (b) agree with your intuitive feelings?

20 Some applications of statistical and quantum concepts

20.1 INTRODUCTION

In Chapters 18 and 19 we developed the basic concepts of quantum and statistical physics. In this chapter we treat a selection of topics using these concepts. For example, when we were dealing with specific heats of gases from the kinetic theory point of view (Section 17.6), we uncovered certain discrepancies between theory and experiment. By incorporating appropriate quantum concepts into the theory and by applying a statistical approach, we shall be able to progress much further toward a satisfactory description of the specific heats of gases.

20.2 A MONATOMIC IDEAL GAS

The allowed energy states for the molecules in a gas may be determined from quantum-mechanical considerations. The quantized energy states for translational motion are found to be so close together that they are virtually inseparable. This is not surprising, since translational motion takes place over distances very large compared to the size of an atom and we noted in Chapter 18 that quantum effects are not noticeable for large scale phenomena. Since (as we saw in Chapter 17) only translational motion contributes to the energy of a monatomic gas, classical mechanics adequately accounts for the behavior of such gases. In particular we saw that classical mechanics could account for the observed specific heat values of a monatomic gas.

Now we shall apply the Maxwell–Boltzmann distribution to determine the distribution of molecular speeds. The energy E_j of the jth particle in a monatomic gas is given by

$$E_j = \frac{p_j^2}{2m},$$

where m is the mass of the particle and p_j is the magnitude of its momentum. For a gas of N monatomic molecules, the total energy E is

$$E = \sum_{j=1}^{N} E_j.$$

We assume that any value of energy E_j can be associated with the jth molecule consistent with the total energy available to the gas. We define $N(E)$ as the number of molecules per unit energy range having energy E. It is now more convenient to speak of the number of molecules $N(E)\,dE$ having energies lying in the range between E and $E + dE$ rather than the number of molecules N_j in the jth energy state. That is, we write

$$E = \int EN(E)\, dE.$$

The Maxwell–Boltzmann distribution is now written in the form

$$N(E)\, dE = \frac{N}{Z} \exp\left(-\frac{E}{kT}\right) E^{1/2}\, dE.$$

The factor $E^{1/2}$ arises from the fact that we are now talking about a range of energies rather than a specific energy; in this case the a priori probability is no longer unity but can be shown to be equal to $E^{1/2}$.

At this point it is convenient to define a **probability distribution function** $f(E)$ as

$$f(E) = \frac{N(E)}{N},$$

where $f(E)$ is **normalized** so that

$$\int_0^\infty f(E)\, dE = \int_0^\infty \frac{N(E)}{N}\, dE$$

$$= \frac{N}{N}$$

$$= 1.$$

This is analogous to the probability $W(E)$ defined in Section 19.2. In terms of $f(E)$ the Maxwell–Boltzmann distribution becomes

$$f(E)\, dE = \frac{1}{Z} \exp\left(-\frac{E}{kT}\right) E^{1/2}\, dE.$$

Note that $f(E)\, dE$ is the probability that the energy of an arbitrarily selected molecule lies within the range E to $E + dE$. The partition function can be evaluated (it is an integral over the energy), and hence $f(E)$ can be determined.

We may also write the distribution in terms of momentum by noting that

$$E = \frac{p^2}{2m},$$

so that

$$dE = \frac{2p\, dp}{2m} = p\frac{dp}{m}.$$

Substituting this into the Maxwell–Boltzmann distribution gives

$$f(p)\,dp = \frac{1}{Z}\exp\left(-\frac{p^2}{2mkT}\right)\frac{p}{(2m)^{1/2}}\frac{p\,dp}{m}$$

$$= \frac{1}{Z}\frac{p^2}{2^{1/2}m^{3/2}}\exp\left(-\frac{p^2}{2mkT}\right)dp.$$

The E values on the left-hand side of the expression have been changed to p values, since the probability distribution is now considered to be a function of the momentum p rather than the energy E. The value of Z can now be shown to be

$$Z = \frac{\pi^{1/2}(kT)^{3/2}}{2},$$

so that the momentum distribution is given by

$$f(p)\,dp = \left(\frac{2}{\pi}\right)^{1/2}\left(\frac{1}{mkT}\right)^{3/2}p^2\exp\left(-\frac{p^2}{2mkT}\right)dp.$$

The calculations are carried out in the box.

Finally, we may write the distribution in terms of speed since $p = mv$, where v is the molecular speed. It is easily seen that

$$f(v)\,dv = \frac{1}{Z}\exp\left(-\frac{mv^2}{2kT}\right)\frac{m^{3/2}}{2^{1/2}}v^2\,dv,$$

where $f(v)$ means that the probability distribution is now a function of the molecular speed. The partition function Z is now an integral over v rather than an integral over p but has the same value as for the momentum distribution, as shown in the box.

The Maxwell–Boltzmann distribution of molecular speeds is

$$f(v)\,dv = \left(\frac{2}{\pi}\right)^{1/2}\left(\frac{m}{kT}\right)^{3/2}v^2\exp\left(-\frac{mv^2}{2kT}\right)dv;$$

this distribution was introduced in Section 17.3, and its properties are discussed there.

The Maxwell–Boltzmann distribution of momentum

The distribution is given by the expression

$$f(p)\,dp = \frac{1}{Z}\frac{p^2}{2^{1/2}m^{3/2}}\exp\left(-\frac{p^2}{2mkT}\right)dp,$$

where Z is the partition function. For a finite number of available energy states,

Z is given by

$$Z = \sum_i \exp\left(-\frac{E_i}{kT}\right).$$

For a continuous range of energies, Z becomes

$$Z = \int_0^\infty \exp\left(-\frac{E}{kT}\right) E^{1/2} \, dE.$$

For translational motion

$$E = \frac{p^2}{2m}, \qquad dE = \frac{p \, dp}{m},$$

so that

$$Z = \int_0^\infty \exp\left(-\frac{p^2}{2mkT}\right) \frac{p}{(2m)^{1/2}} \frac{p \, dp}{m}$$

$$= \frac{1}{2^{1/2} m^{3/2}} \int_0^\infty p^2 \exp\left(-\frac{p^2}{2mkT}\right) dp.$$

The integral is of a standard form given in most tables of integrals. The value of the standard integral is given by

$$\int_0^\infty x^2 \exp\left(-ax^2\right) dx = \frac{1}{4}\left(\frac{\pi}{a^3}\right)^{1/2},$$

where a is a constant. In our example, $a = (2mkT)^{-1}$, and

$$Z = \frac{1}{2^{1/2} m^{3/2}} \frac{1}{4} \frac{\pi^{1/2}}{(2mkT)^{-3/2}}$$

$$= \frac{\pi^{1/2}(kT)^{3/2}}{2}.$$

The Maxwell–Boltzmann distribution of momentum is then given by

$$f(p)\,dp = \frac{2}{\pi^{1/2}(kT)^{3/2}} \frac{p^2}{2^{1/2} m^{3/2}} \exp\left(-\frac{p^2}{2mkT}\right) dp$$

$$= \left(\frac{2}{\pi}\right)^{1/2} \frac{1}{(mkT)^{3/2}} p^2 \exp\left(-\frac{p^2}{2mkT}\right) dp.$$

The Maxwell–Boltzmann distribution of molecular speeds can be determined by substituting

$$p = mv$$

in the above calculations. The partition function becomes

$$Z = \frac{1}{2^{1/2}m^{3/2}} \int_0^\infty m^2 v^2 \exp\left(-\frac{mv^2}{2kT}\right) m\, dv$$

$$= \frac{m^{3/2}}{2^{1/2}} \int_0^\infty v^2 \exp\left(-\frac{mv^2}{2kT}\right) dv$$

$$= \frac{m^{3/2}}{2^{1/2}} \frac{1}{4} \frac{\pi^{1/2}}{\left(\frac{2kT}{m}\right)^{-3/2}}$$

$$= \frac{\pi^{1/2}}{2}(kT)^{3/2}.$$

The Maxwell–Boltzmann distribution of molecular speeds is

$$f(v)\, dv = \left(\frac{2}{\pi}\right)^{1/2} \frac{1}{(mkT)^{3/2}} m^2 v^2 \exp\left(-\frac{mv^2}{2kT}\right) m\, dv$$

$$= \left(\frac{2}{\pi}\right)^{1/2} \left(\frac{m}{kT}\right)^{3/2} v^2 \exp\left(-\frac{mv^2}{2kT}\right) dv.$$

20.3 THE DISTRIBUTION OF ENERGY IN A CLASSICAL SYSTEM

The energy of a system of classical particles may be written as a sum of terms involving either position coordinates (x, y, z) or momentum coordinates (p_x, p_y, p_z). In all the examples that we have discussed so far in this text, the energy terms depend on the square of a position coordinate ($\frac{1}{2}kx^2$ is the potential energy in simple harmonic motion) or on the square of a momentum coordinate ($p_x^2/2m$ is the x component of the translational kinetic energy of a particle). Let us suppose then that, in general, the energy of a system in thermal equilibrium can be written in the form

$$E = \sum_{i=1}^{3N} (a_i q_i^2 + b_i p_i^2),$$

where the q_i are appropriately selected position coordinates and the p_i are the corresponding momentum coordinates of the N particles in the system. Note that the summation is over $3N$ terms, since we live in a three-dimensional world that requires three coordinates to specify the position (or momentum) of a single particle. For a monatomic ideal gas this expansion reduces to

$$E = \sum_{i=1}^{3N} \frac{p_i^2}{2m} = \sum_{i=1}^{N} \frac{1}{2m}(p_{x_i}^2 + p_{y_i}^2 + p_{z_i}^2);$$

there are no coordinate terms.

For a gas of classical diatomic molecules, as we have seen in Chapter 14, the energy expression is

$$E = \sum_{i=1}^{3N} \frac{p_i^2}{2m} + \sum_{i=1}^{3N} I_i \omega_i^2 + \sum_{i=1}^{N} \left[\mu \left(\frac{dr_i}{dt} \right)^2 + \frac{1}{2} kr_i^2 \right],$$

where the first term is due to translational motion, the second term is the contribution from rotational motion, and the third term is the contribution from vibrational motion. We can put the rotational term into the proper form by noting that $I = \mu r^2$ and $\omega = v/r$, so that

$$I\omega^2 = \mu r^2 \left(\frac{v^2}{r} \right) = \frac{1}{\mu} (p')^2,$$

where $p' = \mu v$ is a momentum that may be associated with the rotating reduced particle. Therefore, the rotational energy is of the form

$$\sum_{i=1}^{3N} \frac{1}{\mu} (p_i')^2.$$

The energy expression for a diatomic molecule contains both position coordinate and momentum coordinate terms.

We can define the **average value** $\bar{g}$ of any quantity $g(x)$ over a distribution $f(x)$ to be (see Section 9.5)

$$\bar{g} \int f(x)\,dx = \int g(x) f(x)\,dx$$

or

$$\bar{g} = \frac{\int g(x) f(x)\,dx}{\int f(x)\,dx}.$$

We may now determine the average value of the energy associated with the value of the ith coordinate q_i or the ith momentum p_i of our general system, assuming a Maxwell–Boltzmann distribution of energy for the particles. For the ith coordinate we have

$$\overline{a_i q_i^2} = \frac{\int_0^\infty a_i q_i^2 \exp \left(-\frac{a_i q_i^2}{kT} \right) dq_i}{\int_0^\infty \exp \left(-\frac{a_i q_i^2}{kT} \right) dq_i},$$

where $\overline{a_i q_i^2}$ replaces g and the Boltzmann distribution replaces $f(x)$ in the general expression for the average value. The integration over all other coordinates is common to both numerator and denominator and therefore does not enter our consideration. The integrals are standard forms found in most tables of integrals. When the integrals are evaluated we find that

$$\overline{a_i q_i^2} = \tfrac{1}{2}kT.$$

Similarly, for the momentum coordinate p_i, the average value of the energy is found to be

$$\overline{b_i p_i^2} = \tfrac{1}{2}kT.$$

We see that an amount of energy $\tfrac{1}{2}kT$ is associated with each term in the energy expression that is quadratic in a position coordinate or in a momentum coordinate. This is the **principle of equipartition of energy**. Its validity depends upon only two assumptions:

1. that the Maxwell–Boltzmann distribution is applicable;
2. that the energy of the system can be expressed as a continuous function of position coordinates and momentum coordinates in the form

$$E = \sum_{i=1}^{3N} (a_i q_i^2 + b_i p_i^2).$$

According to the principle of equipartition of energy, a one-dimensional classical harmonic oscillator in thermal equilibrium at temperature T has an average energy kT independent of its frequency. In contrast, the average energy of a quantum-mechanical oscillator (which is identical to the energy attributed to a standing wave in a cavity by Planck) at temperature T is

$$E = \frac{h\nu}{\exp\left(\dfrac{h\nu}{kT}\right) - 1},$$

which depends on the frequency as was discussed in Section 17.2. In the example at the end of that section we saw that when $h\nu$ is small compared to kT this expression reduces to kT; that is, the results of quantum physics reduce to those of classical physics. This is another example of the correspondence principle mentioned in Section 18.5.

20.4   ENTROPY AND PROBABILITY

When a physical system undergoes a change in its state we mean that it exchanges some energy with its surroundings. A cold object placed in a warm atmosphere receives energy in the form of heat from its surroundings; a warm object placed in a cold environment loses energy in the form of heat to its surroundings; a solid when it melts requires energy in the form of heat; a gas when it condenses liberates energy in the form of heat; the foregoing processes are all examples of changes of state of simple systems. The energy exchange ceases when the object attains thermal equilibrium with its surroundings (see Section 23.2 for a further discussion of thermal equilibrium). In speaking of energy changes in a system on a macroscopic scale, it is convenient to define a special parameter called the **entropy** of the system to aid in the discussion of energy changes. If a system gains an amount of heat dQ when its temperature is T, the entropy S of the system by definition changes by an amount

$$dS = \frac{dQ}{T}.$$

Of course, adding heat to a system will increase its temperature, so that the total change in entropy is found by integrating all the contributions dS from an infinite number of infinitesimal exchanges of heat that make up the finite change.

When the entropy change is calculated for any physical process taking place in an isolated system, it is always found that the total entropy has increased during the process if we add together the entropy changes in all parts of the system. This is known as the **principle of the increase in entropy**. Therefore, an isolated system that is not initially in thermal equilibrium will tend toward a state of increased entropy as it approaches thermal equilibrium. The entropy at thermal equilibrium will have its maximum value.

We have seen in Sections 19.2 and 19.3 that the equilibrium state of a system from the statistical point of view is the state of maximum probability. Therefore, in the equilibrium state both the entropy and the probability have their maximum values; we should therefore expect some connection between them. It turns out that the relation between entropy and probability is of the form

$$S = k \ln W,$$

where k is Boltzmann's constant. This logarithmic relation was first established by Ludwig Boltzmann as a proportionality relation. Subsequently, Max Planck determined that the proportionality constant was k. A state of high probability is more disordered or "mixed-up" than a state of low probability, as we saw in Section 19.3. We therefore conclude that isolated systems tend to become more disordered or more random in their microscopic motions (that is, motions on the molecular scale). A more detailed discussion of entropy will be given in Section 23.4.

20.5 THE SPECIFIC HEATS OF A GAS OF DIATOMIC MOLECULES

In Section 17.6 we saw that the kinetic theory, applied to a gas of diatomic molecules, led to a constant value of $7R/2$ for C_v. This prediction is in poor agreement with experiment. For example, C_v for hydrogen gas is not a constant but varies with temperature in the manner illustrated in Fig. 20.1.

How can we understand this behavior? First we recall that a diatomic molecule, unlike a monatomic molecule, can have appreciable energy associated with rotational and vibrational motion about the center of mass (Chapter 14) as well as translational energy of the center of mass. The allowed energy states for the rotational and vibrational motion may be obtained by solving an appropriate form of Schroedinger's equation. The solutions were discussed in Section 18.8. Unlike the translational motion that takes place over distances that are very large relative to the molecular dimensions, the rotational and vibrational motions take place on a molecular scale of dimensions. Therefore, we should expect quantum effects to be important for the rotational and vibrational motions but not for the translational motion.

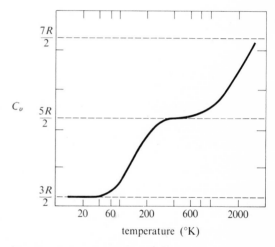

Fig. 20.1. The experimental measurements of C_v for hydrogen gas as a function of temperature.

To determine the distribution of the molecules among the possible energy states, we use the Maxwell–Boltzmann distribution in the form

$$N_i = \frac{N}{Z}\exp\left(-\frac{E_i}{kT}\right).$$

For vibrational energies we have

$$N_v = \frac{N}{Z}\exp\left[-\left(v+\frac{1}{2}\right)\frac{hv}{kT}\right].$$

We define a **characteristic temperature for vibration** to be

$$\theta_v = \frac{hv}{k}.$$

The partition function then becomes

$$Z = \sum_v \exp\left[-\frac{(v + \frac{1}{2})h\nu}{kT} \right]$$

$$= \exp\left(-\frac{h\nu}{2kT} \right) + \exp\left(-\frac{3h\nu}{2kT} \right) + \exp\left(-\frac{5h\nu}{2kT} \right) + \cdots$$

$$= \exp\left(-\frac{\theta_v}{2T} \right) + \exp\left(-\frac{3\theta_v}{2T} \right) + \exp\left(-\frac{5\theta_v}{2T} \right) + \cdots.$$

This is a **geometric** series of the form

$$Z = a + ar + ar^2 + \cdots$$

where

$$a = \exp\left(-\frac{\theta_v}{2T} \right) \quad \text{and} \quad r = \exp\left(-\frac{\theta_v}{T} \right).$$

The sum of such a series is

$$\frac{a}{1 - r}.$$

Therefore,

$$Z = \frac{\exp\left(-\dfrac{\theta_v}{2T} \right)}{1 - \exp\left(-\dfrac{\theta_v}{T} \right)}.$$

The distribution of molecules among the vibrational energy states is then given by

$$N_v = \frac{N\left[1 - \exp\left(-\dfrac{\theta_v}{T} \right) \right]}{\exp\left(-\dfrac{\theta_v}{2T} \right)} \exp\left[-\left(v + \frac{1}{2} \right)\frac{\theta_v}{T} \right]$$

$$= N \exp\left(-\frac{v\theta_v}{T} \right) \left[\exp\left(\frac{\theta_v}{2T} \right) - \exp\left(-\frac{\theta_v}{2T} \right) \right] \exp\left(-\frac{\theta_v}{2T} \right)$$

$$= N \exp\left(-\frac{v\theta_v}{T} \right) \left[1 - \exp\left(-\frac{\theta_v}{T} \right) \right].$$

N_v gives the number of molecules in the vibrational state characterized by the quantum number v. As $T \to 0°\text{K}$,

$$N_v(v = 0) \to N$$

$$N_v(v \neq 0) \to 0,$$

and all molecules are in their vibrational ground state when $T = 0$. Note that the total vibrational energy at $T = 0$ is not zero but rather $Nh\nu/2$; this is called the **zero point energy**.

The total vibrational energy at any temperature is found by multiplying the energy of each state by the number of molecules in that energy state and summing over all possible states. When this is done it is found that the mean energy per molecule is given by

$$\bar{E}_v = k\theta_v \left[\frac{1}{2} + \frac{1}{\exp\left(\dfrac{\theta_v}{T}\right) - 1} \right].$$

The calculation is carried out below.

The Mean Vibrational Energy Per Molecule

The total vibrational energy is given by

$$\sum_{v=0}^{\infty} E_v N_v = \sum_{v=0}^{\infty} \left(v + \frac{1}{2}\right) hvN \exp\left(-\frac{v\theta_v}{T}\right)\left[1 - \exp\left(-\frac{\theta_v}{T}\right)\right].$$

The mean energy per molecule is

$$\bar{E}_v = \frac{\displaystyle\sum_{v=0}^{\infty} E_v N_v}{N} = \frac{\displaystyle\sum_{v=0}^{\infty} E_v N_v}{\displaystyle\sum_{v=0}^{\infty} N_v}$$

$$= \frac{\displaystyle\sum_{v=0}^{\infty} \left(v + \frac{1}{2}\right) hvN \exp\left(-\dfrac{v\theta_v}{T}\right)\left[1 - \exp\left(-\dfrac{\theta_v}{T}\right)\right]}{\displaystyle\sum_{v=0}^{\infty} N \exp\left(-\dfrac{v\theta_v}{T}\right)\left[1 - \exp\left(-\dfrac{\theta_v}{T}\right)\right]}$$

$$= \frac{\displaystyle\sum_{v=0}^{\infty} \left(v + \frac{1}{2}\right) hv \exp\left(-\dfrac{v\theta_v}{T}\right)}{\displaystyle\sum_{v=0}^{\infty} \exp\left(-\dfrac{v\theta_v}{T}\right)}$$

$$= \frac{\dfrac{hv}{2} \displaystyle\sum_{v=0}^{\infty} \exp\left(-\dfrac{v\theta_v}{T}\right) + \displaystyle\sum_{v=0}^{\infty} vhv \exp\left(-\dfrac{v\theta_v}{T}\right)}{\displaystyle\sum_{v=0}^{\infty} \exp\left(-\dfrac{v\theta_v}{T}\right)}$$

$$= \frac{hv}{2} + \frac{hv \displaystyle\sum_{v=0}^{\infty} v \exp\left(-\dfrac{v\theta_v}{T}\right)}{\displaystyle\sum_{v=0}^{\infty} \exp\left(-\dfrac{v\theta_v}{T}\right)}$$

$$= \frac{hv}{2} + \frac{hv\left[\exp\left(-\dfrac{\theta_v}{T}\right) + 2\exp\left(-\dfrac{2\theta_v}{T}\right) + 3\exp\left(-\dfrac{3\theta_v}{T}\right)\cdots\right]}{\left[1 + \exp\left(-\dfrac{\theta_v}{T}\right) + \exp\left(-\dfrac{2\theta_v}{T}\right)\cdots\right]}.$$

Putting $x = \exp(-\theta_v/T)$, we see that the series in the numerator is

$$x(1 + 2x + 3x^2 + \cdots) = x(1 - x)^{-2}.$$

The series in the denominator is

$$1 + x + x^2 + \cdots = (1 - x)^{-1}.$$

Therefore,

$$\bar{E}_v = \frac{h\nu}{2} + h\nu \exp\left(-\frac{\theta_v}{T}\right)\left[1 - \exp\left(-\frac{\theta_v}{T}\right)\right]^{-1}$$

$$= k\theta_v\left[\frac{1}{2} + \frac{1}{\exp\left(\dfrac{\theta_v}{T}\right) - 1}\right].$$

The mean molecular energy for vibration is always less than the classical value of kT but approaches kT as $T \to \infty$; the quantum-mechanical value approaches the classical value as T becomes large. The specific heat at constant volume is found by taking the derivative of the energy with respect to temperature. The variation of specific heat due to vibrational motion is plotted against T/θ_v in Fig. 20.2. Note that for

$$\frac{T}{\theta_v} \ll 1, \qquad C_v \to 0$$

and for

$$\frac{T}{\theta_v} > 1, \qquad C_v \to R.$$

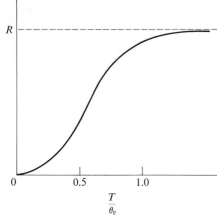

Fig. 20.2. The variation of specific heat due to vibrational motion as a function of T/θ_v.

The vibrational motion starts to be "frozen out" when T/θ_v becomes less than one.

A physical explanation for the "freezing out" of the vibrational motion at low temperatures is as follows. Suppose the system of molecules starts out at $T = 0$ with all of the molecules in their lowest vibrational energy state. If a small amount of heat is added to the system, causing the temperature to increase to T, then on the average each molecule receives an amount of energy of the order of kT. If kT is much less than the energy spacing $h\nu$ between consecutive vibrational levels (see Fig. 20.3), then it is very unlikely that a molecule will receive enough energy to change its vibrational energy state. That is, essentially none of the heat added is used to increase the vibrational energy of the molecules. As more and more heat is added and kT becomes of the same order of magnitude as $h\nu$, it becomes easy for a molecule to change its vibrational energy state.

Part of the heat added is now used to increase the vibrational energy of the molecules.

A similar analysis may be carried out for the rotational motion. The analysis is somewhat more difficult since there are different distinguishable molecular rotational states with the same energy. (This is known as **degeneracy**.) Therefore, the a priori probability weights are not all the same and depend on J. The appropriate form for the Maxwell–Boltzmann distribution is discussed in an example at the end of Section 18.4. A **characteristic temperature for rotation** is defined by

$$\theta_r = \frac{h^2}{8\pi^2 kI}.$$

Values of θ_r are much less than θ_v for all molecules, as is shown in Table 20.1. Therefore, the mean rotational energy per molecule approaches the classical value at much lower temperature than does the mean vibrational energy; the rotational motion does not begin to be "frozen out" until T/θ_r becomes less than one. The general characteristics of the curve in Fig. 20.1 are explained by the quantum theory of a diatomic molecule. At low temperature only translational motion occurs, and $C_v = 3R/2$. As the temperature increases, rotational motion begins to occur, until all molecules are rotating and $C_v = 5R/2$; this occurs when $T/\theta_r > 1$. When the temperature is raised to a much higher value, vibrational motion occurs, and $C_v = 7R/2$ when $T/\theta_v > 1$. Similar considerations may be applied to polyatomic molecules and to their specific heats.

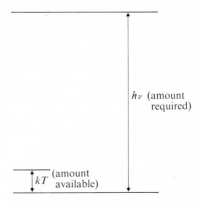

Fig. 20.3. If the mean energy per molecule kT available for vibration is much less than $h\nu$, it is very unlikely that any one molecule will receive enough energy to change its vibrational energy state.

Table 20.1   Values of θ_v and θ_r for various diatomic gases

Gas	θ_v	θ_r
Hydrogen (H_2)	6140°K	85.5°K
Hydrogen chloride (HCl)	4300°K	15.3°K
Carbon monoxide (CO)	3120°K	2.77°K
Oxygen (O_2)	2260°K	2.09°K

20.6   INDISTINGUISHABLE PARTICLES: THE FERMI–DIRAC AND BOSE–EINSTEIN DISTRIBUTIONS

In our calculations of the specific heats of gases we have ignored any contribution due to the electrons. That this is reasonable follows from a consideration of the allowed electron energy states according to quantum mechanics. Just as we saw that the separation between consecutive molecular vibrational levels is of the order of ten times greater than the separation between consecutive molecular rotational levels, quantum mechanics also shows that the separation between consecutive molecular electronic energy states (see Section 18.7) is of the order of ten times greater than that between the vibrational levels. Therefore, kT becomes of the order of the spacing between the electronic energy states only when T becomes on the order of $10^{5\circ}$K.

For metallic solids (see Chapters 21 and 22), however, such is not the case. A reasonable model for a metal is a regular arrangement of positively charged ions permeated by a "gas" of weakly interacting electrons. The forces exerted by the positively charged nuclei contain this gas within the metal. If the free electron gas is treated as a monatomic ideal gas, it should make a contribution of $3R/2$ to C_v. Experimentally, however, such a contribution is not observed. The conclusion is that the Maxwell–Boltzman distribution does not apply to the electrons in a metal. We shall see that the Maxwell–Boltzmann distribution represents the high-temperature limit of two other distributions, namely the **Fermi–Dirac distribution** and the **Bose–Einstein distribution**. The Fermi–Dirac distribution is able to account for the behavior of the electrons in a metal. The resulting Fermi–Dirac electron gas has some quite interesting properties. For example, even at the absolute zero of temperature, the pressure of the gas is of the order of 500,000 times atmospheric pressure. In contrast, a Maxwell–Boltzmann electron gas would exert zero pressure at the absolute zero of temperature.

In our derivation of the Maxwell–Boltzmann distribution, we assumed that the particles in the system were distinguishable from each other. In reality this is not the case. We might then ask why the Maxwell–Boltzmann distribution works so well for describing the macroscopic behavior of gases, since it is derived on the basis of the distinguishability of particles. For one thing, for macroscopic distributions the allowed energy states for the particles of the system are so numerous and so close together that most energy states have no particles in them and the rest have one only. In this case, one can correct for indistinguishability by dividing the constant (the partition function Z) in the Maxwell–Boltzmann distribution by the factor $N!$, which is just the number of ways in which the N particles can be rearranged in the N occupied states.

As an example, in Fig. 20.4 we show the number of different ways in which we can put three particles into three particular cells of a rectangular grid, assuming (a) distinguishable particles, and (b) indistinguishable particles. The distinguishable particles are distinguished by the labels 1, 2, and 3; the indistinguishable particles are each labeled x. Clearly, there are $3! = 6$ different ways of placing the distinguishable particles and

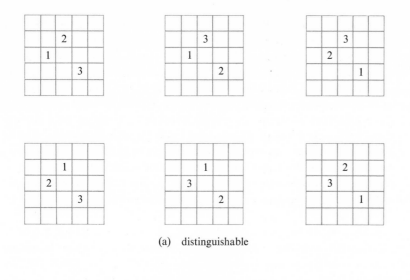

(a)　distinguishable

(b)　indistinguishable

Fig. 20.4. The possible arrangements of three particles into three particular cells of a rectangular grid assuming (a) distinguishable particles, (b) indistinguishable particles.

only one way of placing the indistinguishable particles into the three cells in question. We should point out here that the thermodynamic probability

$$\frac{N!}{N_1!N_2!\cdots}$$

we used earlier assumes distinguishability of particles and therefore badly overestimates the number of different distributions for indistinguishable particles. The form of the Maxwell–Boltzmann distribution does not change if the particles are indistinguishable, as long as the assumption of only one particle per energy state is a good one. If the number of available energy states for the particles in a system is small enough, then the probability that more than one particle will occupy the same state becomes nonnegligible. Under these circumstances we must take account of the fact that on the atomic scale we must use quantum theory rather than classical theory.

Actual particles (those found in nature) are of two types, both of which are indistinguishable. For one set of particles as many can be packed into a given energy state as possible up to the total N of the number of particles present. Such particles are called **bosons** and are governed by the Bose–Einstein distribution. Photons and helium nuclei (α-particles) are examples of bosons. The other kind of particle found in nature has the very different behavior of refusing to occupy an energy state already occupied by another similar particle. Each particle energy state in a system can have zero particles or one particle in it but no more. These unsocial particles obey the Pauli exclusion principle (see Section 18.7). They are called **fermions** and are governed by the Fermi–Dirac distribution. Electrons and protons are examples of fermions.

The Bose–Einstein and Fermi–Dirac distributions are obtained in a manner similar to that which leads to the Maxwell–Boltzmann distribution, with the important difference that the particles are treated as indistinguishable from the beginning and allowances are made for the number of particles that can occupy one energy state for the two types of particles. The Maxwell–Boltzmann (MB), Bose–Einstein (BE), and Fermi–Dirac (FD) distributions are tabulated together here for comparison. They are

$$\frac{N_i}{N} = \frac{1}{Z \exp\left(\dfrac{E_i}{kT}\right)} \qquad \text{(MB distribution)}$$

$$\frac{N_i}{N} = \frac{1}{Z\left[\exp\left(\dfrac{E_i}{kT}\right) - 1\right]} \qquad \text{(BE distribution)}$$

$$\frac{N_i}{N} = \frac{1}{Z\left[\exp\left(\dfrac{E_i}{kT}\right) + 1\right]} \qquad \text{(FD distribution)}$$

Note that for E_i values large compared to kT, the -1 in the BE distribution and the $+1$ in the FD distribution are negligible compared to the exponential term and can be ignored. In this correspondence principle limit, the BE and FD distributions are identical to the MB distribution. It is for this reason that the MB distribution is so useful.

As the temperature $T \rightarrow 0$, the MB distribution becomes a poorer and poorer approximation to the real distribution. The BE and FD distributions behave quite differently from each other, however. As the temperature falls, the particles in the system fall into lower and lower energy states. For fermions, only one particle per energy state is allowed; that is, fermions obey the Pauli exclusion principle. The probability that a state will be occupied is plotted in Fig. 20.5 for $T = 0$ and for a few other temperatures $T_3 > T_2 > T_1 > 0$. When $T = 0$ each state up to and including the state having energy E_F (known as the Fermi energy) has one particle in it, whereas all higher energy states contain no particles. It is this behavior of fermions that accounts for the failure of the MB distribution to provide a correct prediction for the specific heat of a metal (see Section 22.5).

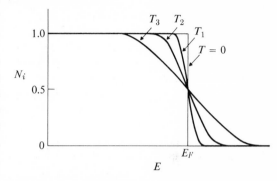

Fig. 20.5. The Fermi–Dirac distribution for temperature $T = 0$ and $T_3 > T_2 > T_1 > 0$.

On the other hand there is no limit to the number of bosons that can occupy a given particle energy state. As $T \to 0$, the lower energy states fill up rapidly, and at $T = 0$ all particles are in the lowest energy or ground state. This collapse of the particles into ground states as $T \to 0$ is believed to be responsible for the low-temperature behavior of helium, since helium atoms are bosons. Remarkable physical properties[1] described as **superfluidity** are exhibited by liquid helium at temperatures below 2.18°K. There are many other boson liquids (all even-mass-number nuclei are bosons), but they all solidify at much higher temperatures than those for which they would be likely to exhibit superfluidity. The rapid increase in ground state population for a system of bosons as $T \to 0$ is known as an **Einstein condensation**.

PROBLEMS

1. Plot the Maxwell–Boltzmann momentum distribution for two different temperatures T and $5T$ on the same graph. Where does the maximum for each curve occur? Explain the reason for this behavior.

2. The Maxwell–Boltzmann distribution of molecular speeds may be checked experimentally using an apparatus as shown schematically in Fig. 20.6. A beam of atoms emerges from oven slit S, passes through a collimating slit S', and, if the effect of gravity were negligible, would impinge on a hot wire W, causing ionization and resulting in an ion current proportional to the number of ions impinging on the wire per second. Actually, each atom passing through S' travels in a parabolic path determined by its speed; as the detector is moved down from W to W' and W'', those atoms with speeds characteristic of the dashed trajectories will be collected at W' and W'', respectively.

[1] E. M. Lifshitz, "Superfluidity," *Scientific American*, June, 1958, and F. Reif, "Quasi-particles," *Scientific American*, November, 1960. Both articles are available as *Scientific American Offprints 224* and *272*, respectively (San Francisco: W. H. Freeman and Co., Publishers).

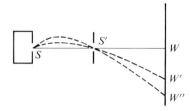

(a) If $SS' = S'W = 1.0$ m, calculate the distance of the detector below the central position W for cesium atoms having a speed equal to the root-mean-square speed in a beam emerging from an oven at a temperature of 475°K.

(b) Calculate the angle of elevation of the trajectory of these atoms.

Fig. 20.6

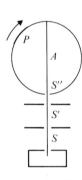

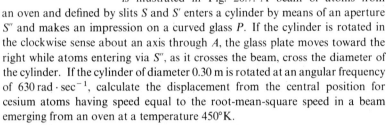

3. An alternative method of experimentally checking the Maxwell–Boltzmann distribution of molecular speeds is illustrated in Fig. 20.7. A beam of atoms from an oven and defined by slits S and S' enters a cylinder by means of an aperture S'' and makes an impression on a curved glass P. If the cylinder is rotated in the clockwise sense about an axis through A, the glass plate moves toward the right while atoms entering via S'', as it crosses the beam, cross the diameter of the cylinder. If the cylinder of diameter 0.30 m is rotated at an angular frequency of 630 rad $\cdot$ sec^{-1}, calculate the displacement from the central position for cesium atoms having speed equal to the root-mean-square speed in a beam emerging from an oven at a temperature 450°K.

Fig. 20.7

4. Two methods for experimentally checking the Maxwell–Boltzmann distribution of molecular speeds are described in Problems 2 and 3. Comment on the inherent accuracy of the two methods.

*5. A flux of 10^{14} neutrons $\cdot$ m$^{-2} \cdot$ sec^{-1} emerges from a port in a nuclear reactor. Calculate the density of neutrons in the beam, assuming them to have a Maxwell–Boltzmann speed distribution corresponding to $T = 300$°K.

6. Consult a table of integrals and verify that $\overline{a_i q_i^2} = \frac{1}{2}kT$ by evaluating the integrals given in Section 20.3.

*7. The thermodynamic probability for the distribution (N_1, N_2) of distinguishable molecules into two halves of a container was given in Section 19.3 as

$$\frac{N!}{\left(\dfrac{N}{2} + n\right)!\left(\dfrac{N}{2} - n\right)!},$$

where n is a measure of the deviation from the most probable value and

$$N = N_1 + N_2.$$

Calculate the entropy for $n = 0$ and $n = N/100$ and compare.

*8. Plot a graph to show the variation of entropy with n/N, assuming a large number of particles.

9. Why would you expect θ_r to be larger for hydrogen (H_2) than for oxygen (O_2)?

10. The fundamental vibration frequency of carbon monoxide (CO) is 6.53×10^{13} Hz. Calculate the ratio of the number of molecules in the first to that in the zeroth vibrational level for 300°K and 1000°K.

11. Using the data of Table 20.1, determine the moments of inertia of the H_2, HCl, CO, and O_2 molecules. Estimate the separation between the two atoms of each molecule.

12. The temperature dependences of C_p for H_2 and D_2 are shown in the graph in Fig. 20.8 for a certain range of temperatures. Comment on the nature of the two curves.

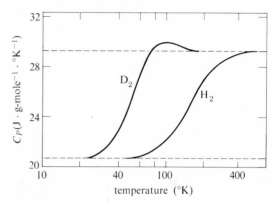

Fig. 20.8

13. The energy of a certain system of molecules may be expressed in the form

$$E = E_{trans} + E_{int}.$$

Assume that the translational motion can be treated classically and that only two internal energy states exist differing in energy by an amount ε. Use the definition

$$C_v = \frac{dE}{dT}$$

to deduce an expression for C_v.

14. Show that the number of ways in which A indistinguishable particles can be placed in B numbered cells with no more than one particle per cell is

$$\frac{B!}{A!(B-A)!}.$$

15. Plot on the same graph N_i/N versus E_i/kT for the BE, MB, and FD distributions. Assume a fixed value of T.

21 The symmetry of crystalline solids

21.1   INTRODUCTION

The study of the physics of the solid state has developed only within the present century. One of the major breakthroughs came at the turn of the century when it was realized that, with the exception of a small group of solids such as glasses, all solids exist in crystalline form. That is, the atoms of which solids are composed are arranged in regular patterns with the regularity of order extending over distances large compared to the atomic diameters. The characteristic crystal structure of a solid can be deduced by observing the directions in which x-rays, electrons, and neutrons are preferentially scattered from the regular planes of atoms that make up the solid. Most naturally occurring solids are **polycrystalline**. They are composed of very large numbers of small crystals of various sizes, packed together in an irregular way.

21.2   THE CRYSTALLOGRAPHIC SPACE GROUPS

Starting from an arbitrary point O, let us generate a spatial array of points (that is, a **lattice**) in the following manner. Through O we draw three axes x, y, and z that make angles α, β, and γ with one another as shown in Fig. 21.1. We next define lattice parameters a, b, and c along the x, y, and z axes, respectively. If we now generate an array of points such that any point $P(x, y, z)$ has coordinates

$$P(x, y, z) = (ma, nb, pc),$$

with m, n, and p positive or negative integers, then the array of points defines a **space lattice**. Such a lattice possesses **translational symmetry**. That is, if the lattice is displaced parallel to the x, y, or z axes, an integral multiple of the lattice parameters a, b, or c, the lattice so obtained is identical to the original one. The volume element bounded by a, b, and c is the smallest cell that when repeated in space will span the entire lattice. Such a cell is known as a **primitive unit cell**. We shall find occasion later to introduce unit cells that are not primitive.

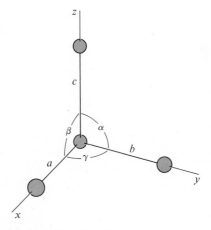

Fig. 21.1. The definition of the lattice parameters a, b, c.

Space lattices may be conveniently categorized according to the **degree of symmetry** that they possess. Consider, for example, the portion of the lattice shown in Fig. 21.2(a). For the lattice shown, $\alpha = \beta = \gamma = 90°$, and $a = b \neq c$. For clarity, we show only the origin and the six nearest neighbor lattice points. A rotation of this lattice about the z axis through 90° or any multiple of 90° will provide an array of lattice points identical with the original arrangement. The structure is said to possess **rotational symmetry** of order four about the z axis. The z axis is known as a **symmetry element**, and a rotation of the structure through an integral multiple of $\pi/2$ is known as a **symmetry operation**. For example, for a rotation through 90°, site 2 is replaced by site 1, site 3 by site 2, site 4 by site 3, and site 1 by site 4 according to the numbering of Fig. 21.2. Similarly, a rotation about the x axis through 180° or a multiple of 180° will provide an identical arrangement of lattice points. The structure possesses rotational symmetry of order two about the x axis.

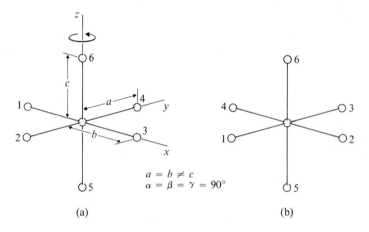

Fig. 21.2. (a) A portion of a lattice. (b) The lattice after rotation about the z axis through 90°.

Now consider the x, y plane as a mirror plane in which lattice points can be reflected. For a reflection in the x, y plane, sites 2 and 4 are interchanged. The x, y plane is a **plane of symmetry**; it is also a **symmetry element**. The associated symmetry operation is reflection in the plane. In all there are 16 different symmetry elements for this lattice. Each of the associated symmetry operations may be described as a rotation about an appropriate axis, as a reflection in an appropriate plane, or as a rotation about an appropriate axis followed by a reflection in an appropriate plane. Included in the 16 are the **identity element** and the **center of inversion**. The corresponding symmetry operations correspond to leaving the structure unaltered and reflecting each point in the center of inversion, respectively.

All lattices may be shown to belong to one of seven **crystal systems** according to the symmetry elements that they possess. These are listed in Table 21.1 with the relations between the lattice parameters and the

Table 21.1   The seven crystal systems

Crystal system	Lattice parameters	Angles between axes
Triclinic	$a \neq b \neq c$	$\alpha \neq \beta \neq \gamma$
Monoclinic	$a \neq b \neq c$	$\alpha = \beta = \pi/2, \qquad \gamma \neq \pi/2, 2\pi/3$
Orthorhombic	$a \neq b \neq c$	$\alpha = \beta = \gamma = \pi/2$
Tetragonal	$a = b \neq c$	$\alpha = \beta = \gamma = \pi/2$
Rhombohedral	$a = b = c$	$\alpha = \beta = \gamma \neq \pi/2, 2\pi/3, \text{ or } \cos^{-1}\left(-\frac{1}{3}\right)$
Hexagonal	$a = b \neq c$	$\alpha = \beta = \pi/2, \qquad \gamma = 2\pi/3$
Cubic	$a = b = c$	$\alpha = \beta = \gamma = \pi/2$

angles between the axes. The crystal systems are listed in order of increasing symmetry.

From the seven basic crystal systems 14 so-called **Bravais lattices** may be deduced by considering the possibility of locating additional lattice

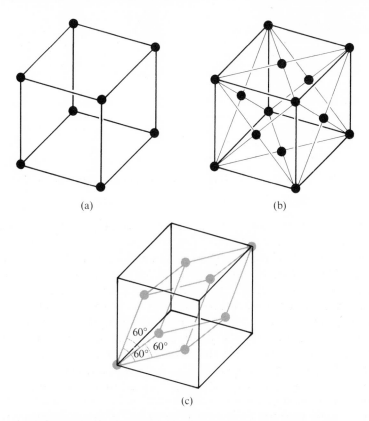

Fig. 21.3. (a) The primitive unit cell of a simple cubic lattice. (b) The conventional unit cell of the face-centered cubic lattice. (c) The primitive unit cell of the face-centered cubic lattice.

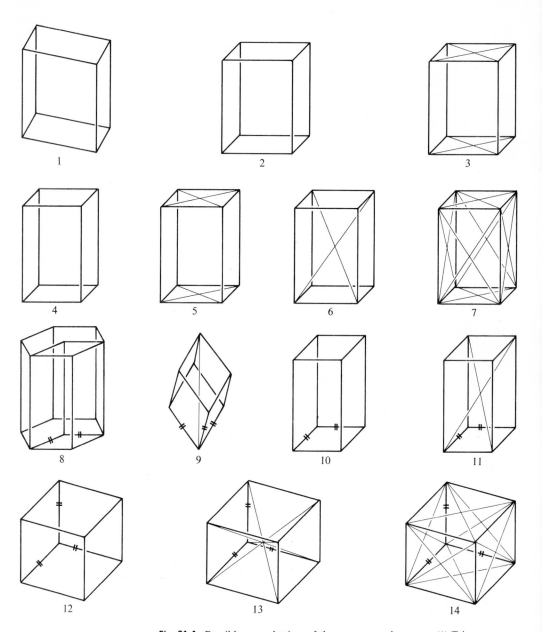

Fig. 21.4. Possible space lattices of the seven crystal systems. (1) Triclinic, simple; (2) monoclinic, simple; (3) monoclinic, base-centered; (4) orthorhombic, simple; (5) orthorhombic, base-centered; (6) orthorhombic, body-centered; (7) orthorhombic, face-centered; (8) hexagonal; (9) rhombohedral; (10) tetragonal, simple; (11) tetragonal, body-centered; (12) cubic, simple; (13) cubic, body-centered; (14) cubic, face-centered.

points at suitable locations such as face centers or cell centers. Such a procedure is allowed only if all of the lattice points so obtained, including the original ones, can be fitted into a single lattice by a suitable choice of new axes and lattice parameters, the lattice itself not suffering a deterioration in symmetry.

As an example we consider the face-centered cubic lattice formed from the simple cubic lattice shown in Fig. 21.3(a) by locating a lattice point at the center of each face of the unit cube as shown in Fig. 21.3(b). The face-centered cubic lattice may be regarded as a special rhombohedral lattice in which $a = b = c$ and $\alpha = \beta = \gamma = 60°$. That is, in Fig. 21.3(c) the light lines show the face-centered cubic **conventional unit cell** and the heavy lines the special rhombohedral primitive unit cell. The 14 Bravais lattices are pictured in Fig. 21.4.

Now any finite group of atoms possesses definite symmetry elements with respect to a fixed point relative to the group. For example, consider the symmetry elements of the atoms situated at the vertices of an equilateral triangle about the center of the triangle. The five symmetry elements are the identity element, the rotation axis of order three through the center of the triangle and perpendicular to the plane of the triangle, and the three planes perpendicular to the plane of the triangle and containing the center of the triangle and one of the vertices (see Fig. 21.5).

Associated with the five symmetry elements are six unique symmetry operations. For a rotation axis of order three, rotations through $2\pi/3$ and $4\pi/3$ are different operations. Such a group of symmetry operations characterizes a specific **point group**. There are 32 different point groups.

Starting with each of the 14 Bravais lattices and locating a group of atoms possessing either the same symmetry as the lattice or a lower symmetry at every lattice point, we can specify 32 unique **crystal classes**.

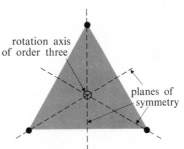

Fig. 21.5. The symmetry elements of the atoms situated at the vertices of an equilateral triangle about the center of the triangle.

As an example let us consider the triclinic lattice ($a \neq b \neq c$; $\alpha \neq \beta \neq \gamma$). This lattice possesses only two symmetry elements, namely the identity element and the center of inversion. Two unique crystal classes are obtained by locating at the lattice points groups of atoms having as symmetry elements about the lattice points either the identity element and the center of inversion or just the identity element. The two situations are analogous to the two-dimensional illustrations of Fig. 21.6.

The 32 crystal classes in turn lead to the various possible **space groups**. These arise by combining the symmetry of the point groups with the basic translational symmetry of the Bravais lattices. Two new symmetry ele-

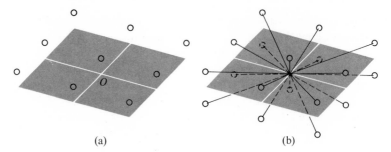

Fig. 21.6. Two-dimensional lattice possessing identity and center of inversion as symmetry elements. (a) Each atom has only the identity as a symmetry element with respect to the neighboring lattice point. (b) Each pair of atoms has the symmetry elements with respect to the neighboring lattice point.

ments are now possible. The symmetry operation associated with the **glide reflection plane** is a reflection followed by a translation or a glide parallel to the plane of reflection, as illustrated in Fig. 21.7(a). The symmetry operation associated with the **screw axis of rotation** is a rotation about some axis followed by a translation parallel to the axis. The

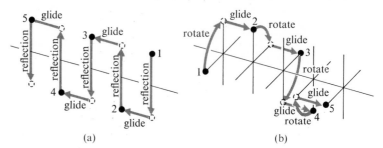

Fig. 21.7. (a) Illustration of a glide reflection plane. (b) Illustration of a screw axis of rotation.

various space groups are obtained from the crystal classes by substituting glide reflection planes for ordinary reflection planes and screw axes of rotation for ordinary rotation axes. In this manner the 230 unique space groups may be obtained. Any crystalline solid must be characterized by one of the space groups.

21.3   SOME COMMON CRYSTAL STRUCTURES

Face-centered cubic (fcc) structure

The fcc structure is shown in Fig. 21.8. The space lattice is fcc. At each lattice site there is located a single atom. Both the conventional and the primitive unit cells are shown in Fig. 21.8. The eight corner atoms of the conventional unit cell are shared by the eight neighboring unit cells; the six face-centered atoms are each shared by two unit cells. Therefore, there are

$$\tfrac{8}{8} + \tfrac{6}{2} = 4$$

atoms per conventional unit cell. Copper, silver, gold, argon, calcium, and neon are examples of solids having the fcc structure.

Body-centered cubic (bcc) structure

The bcc structure is shown in Fig. 21.9. The space lattice is obtained from the simple cubic lattice by adding a site at the center of each cube. A single atom is located at each lattice site. There are

$$\tfrac{8}{8} + \tfrac{1}{1} = 2$$

atoms per conventional unit cell. The edges of the primitive cell are shown by heavy lines in Fig. 21.9. Lithium, sodium, barium, and potassium are examples of solids having the bcc structure.

Hexagonal close-packed (hcp) structure

An hcp structure may be constructed in the following manner. Suppose spheres are packed in a two-dimensional array to form a close-packed layer as shown in Fig. 21.10(a). A second close-packed layer can then be placed on top of the first. One of the two possible and essentially identical ways in which this can be done is shown by the open circles in Fig. 21.10(b). When we come to placing the third layer we find that two nonidentical choices exist. We may either select the hollows in the second layer that will place the spheres of the third level directly above those of the first layer or we may select the other set of hollows (indicated by the shaded regions) for which the spheres of the third layer are not directly above those of the first layer. In the former case the spheres define an hcp structure; in the latter an fcc structure. The conventional

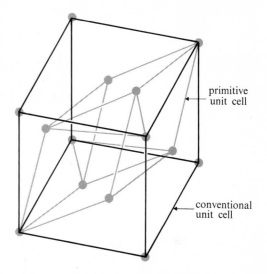

Fig. 21.8. The conventional and primitive unit cells for the fcc structure.

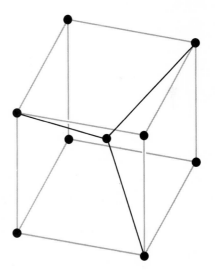

Fig. 21.9. The conventional unit cell for the bcc structure. The heavy lines show the edges of the primitive cell.

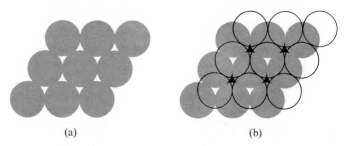

(a)   (b)

Fig. 21.10. (a) A two-dimensional array of close packed spheres. (b) A second layer of spheres can be added as shown by the open circles; a third layer can be placed either over the first layer or over the black regions indicated.

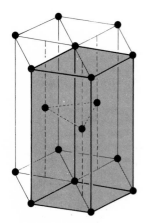

Fig. 21.11. The conventional and primitive unit cells of the hexagonal close-packed structure.

unit cell of the hcp structure is shown in Fig. 21.11. It consists of two hexagons with atoms at each corner and one in the center, separated by three atoms filling the allowed spaces between the top and bottom layers. The primitive unit cell is shown by the heavy lines. It has sides of length a and c. Beryllium, cadmium, magnesium, titanium, and tin are examples of solids having the hcp structure.

Cesium chloride (CsCl) structure

The CsCl structure is shown in Fig. 21.12. The space lattice is simple cubic. At first sight the structure looks like bcc if the distinction between cesium and chlorine ions is ignored. Actually, at each lattice site there occurs a pair of different atoms with separation one-half the body diagonal of the unit cube. Each atom is at the center of a cube of atoms of the opposite kind. Beryllium, copper, zinc, rubidium chloride, and cesium chloride are examples of solids having the CsCl structure.

Sodium chloride (NaCl) structure

The NaCl structure is shown in Fig. 21.13. The space lattice is fcc. At each lattice site there occurs a pair of different atoms, with separation one-half the body diagonal of the unit cube. The structure may be thought of as the interpenetration of two fcc lattices. Lithium hydride, potassium bromide, lead sulfide, manganese oxide, and sodium chloride are examples of solids having the NaCl structure.

Diamond structure

The space lattice is fcc. At each lattice site there appears a pair of like atoms with separation one-quarter the body diagonal of the unit

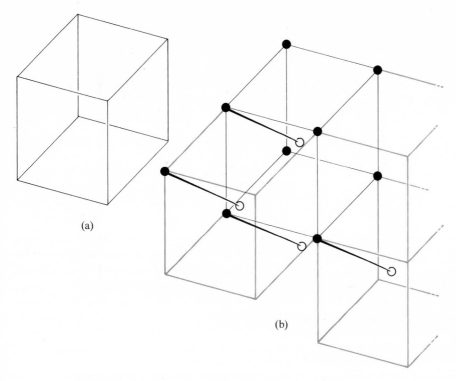

Fig. 21.12. (a) A simple cubic unit cell. (b) The CsCl structure.

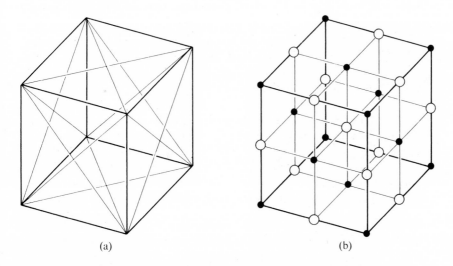

Fig. 21.13. (a) A face-centered cubic unit cell. (b) The NaCl structure.

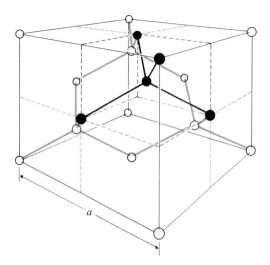

Fig. 21.14. The diamond structure.

cube. The structure is shown in Fig. 21.14. Each atom has about it four equally distant atoms arranged at the corners of a regular tetrahedron. Carbon, silicon, germanium, and gray tin are examples of solids having the diamond structure.

21.4   SCATTERING OF WAVES BY CRYSTALS

The present-day method for the determination of the crystal structure of solids had its origin in the 1912 experiments of Max von Laue with crystals and x-rays. He found that x-rays incident on a crystal from a particular direction were scattered or diffracted by the crystal in a number of well-defined directions. Sir William Bragg found that the directions of the scattered radiation could be readily accounted for by means of a simple model that assumed that x-rays are reflected specularly from the various planes of atoms within a crystal. The scattered radiation is observed only for those special directions for which the reflected radiation from parallel planes of atoms interferes constructively.

In Fig. 21.15 we consider one possible set of atomic planes separated by a distance d for a two-dimensional array of atoms. If radiation of wavelength λ is incident upon these planes at an angle θ, then the path difference for rays reflected from adjacent planes (shown by the heavy lines) is $2d \sin \theta$. In order for constructive interference (see Section 16.2) between such rays to occur, this path difference must be an integral multiple of λ. That is,

$$2d \sin \theta = n\lambda.$$

This is the **Bragg law.** If Bragg reflection is to occur, then wavelengths

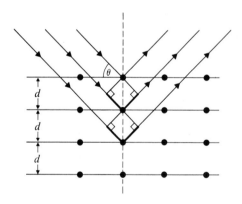

Fig. 21.15. A set of atomic planes separated by distance d. The path difference for rays reflected from adjacent planes is $2d \sin \theta$.

$$\lambda \leqslant 2d$$

are required. These wavelengths correspond to radiation in the x-ray region of wavelengths. In Fig. 21.16 the same array of atoms is shown, with two other sets of atomic planes indicated. We see that the same crystal can be regarded as made up of many different sets of planes with different spacings and atomic densities. The intensity of radiation scattered in a specific direction will depend upon the number and type of scattering centers situated in the planes responsible for that particular reflection. The atomic arrangement must be inferred from the nature of the scattered radiation. A considerable amount of ingenuity is required in order to do this.

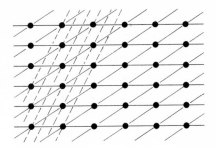

Clearly it is desirable to have a simple way of designating the various crystallographic planes. **Miller indices** are used for this purpose, and they may be deduced in the following manner.

Fig. 21.16. Three different sets of atomic planes for the same two-dimensional array of atoms.

1. Obtain the intercept lengths x, y, and z at which the plane of interest intersects the crystallographic axes.
2. Express the intercepts as fractions x/a, y/b, and z/c of the lattice parameters a, b, and c along the respective axes.
3. Take the reciprocals of these fractions and reduce them to the set of smallest integers h, k, and l having the same ratios.
4. Write (hkl) as the Miller indices of the plane.

Some important examples are shown in Fig. 21.17.

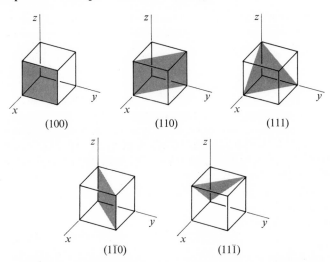

Fig. 21.17. The Miller indices of some important crystallographic planes. A bar over one of the indices indicates that the corresponding intercept is negative.

Example. The (100) set of atomic planes for a bcc structure is shown in Fig. 21.18. Show under what conditions the intensity of the $n = 1$ scattered radiation from these planes will be zero.

Solution. If we consider only the first and third planes, then for reinforcement the radiation contributions must be out of phase by 2π rad. In this case, however, the radiation contributions from the second and fourth planes will be out of phase with respect to that from the first by π and 3π, respectively. Therefore, if the various planes show equal scattering intensity, as will be the case if they are identical in composition, the radiation contributions from the second and fourth planes will just cancel the contributions from the first and third planes. The scattered intensity will then be zero.

Because of the wave nature of particles, as discussed in Section 18.5, electron or neutron beams of appropriate energy can also be used to probe crystal structures. Electron and neutron scattering are useful supplementary tools to x-ray scattering for probing crystal structures. Although the penetration depth of electrons in most solids is small, the

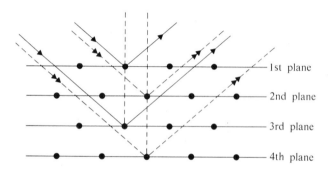

Fig. 21.18. The (100) set of atomic planes for a bcc structure.

scattering efficiency of electrons by atoms is considerably greater than that of x-rays. Electron scattering is therefore useful for studying thin solid films. The scattering of neutrons by light elements is, in contrast to x-rays, quite often relatively strong. Neutron scattering has been particularly useful for locating the positions of hydrogen and carbon atoms in organic crystals.

In reality, almost all crystals have various kinds of imperfections occurring in various places throughout their structure. These imperfections or **defects** are of several different types and must be taken into account when discussing the properties of crystalline solids in detail. For the purposes of our present discussion, however, it is quite sufficient to ignore their existence. Readers wishing a more detailed introduction to crystallography are referred to the book *Crystals and Light*[1].

[1] Elizabeth A. Wood, *Crystals and Light*, (Princeton, N.J.: D. Van Nostrand Co., Inc., 1964).

21.5   THE NATURE OF THE BINDING FORCES IN SOLIDS

Electrostatic forces (Section 7.6) are almost entirely responsible for binding atoms and molecules in all solids. However, due to qualitative differences in the manner in which the electrons are distributed around the atoms and molecules, it is possible and indeed useful to distinguish between a number of types of **chemical binding**.

Ionic solids

In an ionic solid, electrons are transferred from atoms of one type to atoms of another type, so that the solid is composed of positive and negative ions. The ionic bond is essentially the electrostatic interaction of oppositely charged ions. Solids having the CsCl or NaCl structure in general exhibit strong **ionic bonding**.

Covalent solids

In a covalent solid, electrons are shared by pairs of atoms. A covalent bond is usually formed from two electrons, one contributed by each of the two participating atoms. Solids of the diamond structure in general exhibit strong **covalent bonding**.

Metallic solids

Metals are solids possessing electrons that are free to move about the lattice. Such electrons are called **conduction electrons** because they are responsible for the high degree of electrical conductivity characterizing metallic solids. In metals it is often the interaction of the positively charged ions fixed at the lattice positions with the relatively free conduction electrons that provides the bonding.

Molecular solids

Inert gas atoms such as neon, argon, and krypton are weakly bound in the solid phase by electrostatic forces known as **van der Waals forces**. If one calculates (using the methods of wave mechanics) the time-average value of the interaction associated with the instantaneous positions of the electrons in the atoms, a net weak attractive force results. The inert gas solids crystallize at relatively low temperatures into cubic close-packed structures.

21.6   THE ELASTIC PROPERTIES OF SOLIDS

All of the physical properties of a crystal are related to the nature and arrangement of the atoms comprising the crystal. Once the symmetry of the crystal structure is known, the symmetry of its physical properties is also known.

As an example, let us consider the elastic behavior of solids. Elasticity may be defined as the response of a material to an applied force. An elastic change specifically refers to a change that exists only during the application of the force. The equilibrium arrangement of atoms in a crystal corresponds to a balance of the repulsive and attractive electromagnetic forces on each atom exerted by its neighbors. When the crystal is subjected to an external force, the internal electromagnetic force balance is disturbed, the atoms change their positions in response, and a resulting dimensional change of the crystal occurs. If the displacements from equilibrium are small we should expect the atoms to experience restoring forces proportional to their displacements.

Consider the sodium chloride structure. A two-dimensional view of the structure is shown in Fig. 21.19. If a force of compression is applied

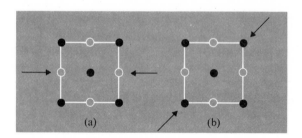

Fig. 21.19. A two-dimensional view of the sodium chloride structure showing external forces applied in two different directions.

in the direction shown in (a), then the effect of the compression is predominantly to reduce the distances between unlike atoms. If a force of compression is applied in the direction shown in (b), then the effect of the compression is predominantly to reduce the distances between like atoms. It is therefore not surprising that the elastic properties of a crystal are in general different in different directions.

The solid state theory of elasticity attempts to relate the atomic force constants to macroscopic parameters describing the deformation. In order to obtain some feeling for the manner in which the microscopic and macroscopic parameters are related, let us assume for the remainder of this section that the elastic properties are independent of direction. That is, let us ignore the details of the crystal structure. Since solids are usually encountered in the polycrystalline form with no preferred orientations of the constituent crystallites, most solid specimens closely follow the elastic laws for an isotropic crystal.

Let us consider a uniform solid bar of cross-sectional area A and length L in its equilibrium state. If a force F is applied to the ends of the bar normal to a cross-section, as shown in Fig. 21.20, the length changes

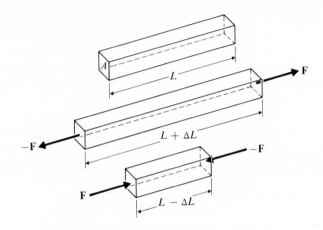

Fig. 21.20. Strain in a metal bar under stress.

by an amount ΔL; if ΔL is positive the bar is under **tension**, and if ΔL is negative the bar is under **compression**. The **stress** σ in the bar is defined as the force per unit area; that is,

$$\sigma = \frac{F}{A}.$$

The **strain** ε resulting from the application of stress σ is defined as the relative change in length of the bar; that is,

$$\varepsilon = \frac{\Delta L}{L}.$$

Experiments show that for sufficiently small stresses the strain is proportional to the stress. Therefore, we can write

$$\sigma = Y\varepsilon,$$

where the constant Y is called **Young's modulus**. This equation is known as **Hooke's law** after Robert Hooke, who discovered it. Substituting for the stress and strain, we have

$$\frac{F}{A} = Y\frac{\Delta L}{L}.$$

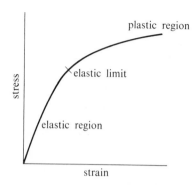

Fig. 21.21. A typical stress-strain curve for a solid.

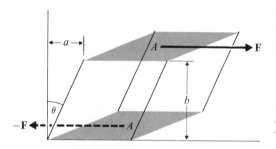

Fig. 21.22. A shearing stress applied to a simple cube.

Figure 21.21 is a plot of strain as a function of stress for a typical crystalline metal. The initial portion of the curve is linear in agreement with Hooke's law. The behavior in this region is **elastic** if the stress is removed, the atoms return to their former equilibrium positions, and the solid returns to its original length. Above a certain value of the stress, the strain increases rapidly, and the solid exhibits **plastic** behavior; the position of the elastic limit is indicated in Fig. 21.21. If the stress carries the solid into the plastic region a permanent deformation results; the strain does not return to zero when the stress is removed, as some permanent damage has been done to the lattice structure. Now when the stress is removed not all atoms return to their former equilibrium positions, and a permanent strain results. If a further increase of stress occurs the solid will fracture.

If a force is applied tangentially to the surface of a solid cube as shown in Fig. 21.22 a lateral displacement of adjacent atomic planes results. The **shearing stress** is defined as

$$S = \frac{F}{A}$$

and the **shearing strain** as

$$\gamma = \frac{a}{b} = \tan \theta.$$

Experiments show that, provided the shearing stress is sufficiently small, the shearing strain is proportional to the shearing stress. Therefore, we can write

$$S = E_s \theta,$$

where E_s is the **shear modulus** or the **modulus of rigidity** and we have replaced $\tan \theta$ by θ.

Let us proceed in a very elementary fashion[2] to try to relate Young's modulus for the bar of Fig. 21.20 to the atomic force constant for the constituent atoms. We suppose that the bar is isotropic and composed of atoms spaced a distance d apart when no force is applied. Since the

[2] K. G. McNeill, "Interrelationship of Physical Quantities," *American Journal of Physics*, **28** (April 1960), 375.

length of the bar is initially L, there are L/d atoms along the length of the bar. Also, the number of atoms in a cross-sectional area is A/d^2. Therefore, when a force of magnitude F is applied uniformly over an area A at each end of the bar, each atom on the end surface feels a force Fd^2/A, which is then transmitted along a "chain" of L/d atoms, each of which is displaced a distance x from equilibrium. The change in length must be

$$\Delta L = \left(\frac{L}{d}\right)x.$$

Assuming the force on each atom to be proportional to the displacement, we have

$$\frac{Fd^2}{A} = k_A x = k_A \Delta L \frac{d}{L},$$

where k_A is the **force constant of the atom**. Therefore,

$$k_A = \left(\frac{F}{A}\right)\left(\frac{L}{\Delta L}\right)d$$

$$= Yd$$

so that

$$Y = \frac{k_A}{d}.$$

Example. Given that the speed of sound v in a copper bar is 3.5×10^3 m $\cdot$ sec^{-1} and that the density ρ of copper is 9.0×10^3 kg $\cdot$ m^{-3}, estimate the atomic force constant of the copper atoms. *Hint:* Recall from Section 15.5 that the speed of sound in a solid bar is related to Young's modulus for the material by

$$v = \left(\frac{Y}{\rho}\right)^{1/2}.$$

Solution. Young's modulus for the copper bar is given by

$$Y = \rho v^2$$

$$= 9.0 \times 10^3 \times (3.5)^2 \times 10^6$$

$$= 1.1 \times 10^{11} \text{ N} \cdot \text{m}^{-2}.$$

Now let us estimate the distance between the copper atoms. To do so, we must recall that there are 6.0×10^{26} atoms (Avogadro's number) in every kilogram-mole of the solid. That is, for copper there are $N = 6.0 \times 10^{26}$ atoms in every 64 kg. The volume V occupied by the amount of copper is

$$V = \frac{64}{9.0 \times 10^3} \text{ m}^3.$$

An estimate of the distance between the atoms is

$$d = \left(\frac{V}{N}\right)^{1/3}$$

$$= \left(\frac{64}{9.0 \times 10^3} \times \frac{1}{6.0 \times 10^{26}}\right)^{1/3}$$

$$= 2.3 \times 10^{-10} \, \text{m}.$$

Now we can estimate the atomic force constant of the copper atoms as

$$k_A = Yd$$

$$= 1.1 \times 10^{11} \times 2.3 \times 10^{-10}$$

$$= 25 \, \text{N} \cdot \text{m}^{-1}.$$

PROBLEMS

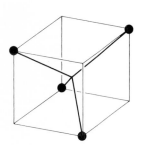

Fig. 21.23

1. Identify the six symmetry operations associated with the three atoms situated at the corners of an equilateral triangle with the six permutations of three atoms over three sites.

2. Express the symmetry operation of reflection in a center of inversion as a rotation about an appropriate axis followed by a reflection in an appropriate plane.

*3. For a tetrahedral arrangement of identical atoms (as shown in Fig. 21.23), list the symmetry elements about the center of the structure.

*4. Show that the complete group of unique symmetry operations associated with the symmetry elements deduced in the previous problem correspond to the 24 permutations of four atoms over four allowed sites.

5. Copper is an fcc structure. Given that the molecular weight of copper is 63.5 and the lattice parameter is 3.61×10^{-10} m, calculate the density of copper and compare your answer with the experimental value.

6. Zinc is an hcp structure. Given that the molecular weight of zinc is 65.37 and the lattice parameters are $a = 2.66 \times 10^{-10}$ m and $c = 4.93 \times 10^{-10}$ m, calculate the density of zinc and compare your answer with the experimental value.

7. For the simple body-centered and face-centered cubic lattices make a table of the following quantities:
   (a) the volume of the conventional unit cell,
   (b) the number of lattice points per unit volume,
   (c) the nearest neighbor distance,
   (d) the second nearest neighbor distance,
   (e) the number of nearest neighbors,
   (f) the number of second nearest neighbors.

8. Calculate the ratio of the lattice parameters for an ideal hexagonal close-packed structure.

*9. Solid spheres of radius R are packed together to form sc, bcc, and fcc structures. Find the radius r of the largest sphere that can fit into the free space within each structure.

10. Calculate the maximum fraction of the total volume that may be occupied by solid spheres arranged in each of the following structures:
(a) simple cubic,
(b) body-centered cubic,
(c) face-centered cubic,
(d) hexagonal close-packed,
(e) diamond.

11. Find the minimum kinetic energy in keV of a beam of electrons that is suitable for the observation of Bragg reflection from a set of atomic planes with separation 3.0×10^{-10} m.

12. A beam of thermal neutrons is incident on a set of atomic planes with separation 1.5×10^{-10} m. Can Bragg reflection occur? *Hint:* Take the effective temperature of the neutrons as 300°K.

13. Deduce the Miller indices for the cube faces of a simple cubic crystal.

14. Show in separate diagrams planes having Miller indices (210), (211), ($2\bar{1}1$), and (310).

15. Compare the number of atoms per square meter in a (100), a (110), and a (111) plane in gold. Gold is an fcc structure with lattice parameter 4.07×10^{-10} m.

*16. For a simple cubic lattice of lattice parameter a, deduce an expression for the perpendicular distance between adjacent planes of a set of planes described by Miller indices (hkl).

17. The speed of sound in aluminum is 5.0×10^3 m · sec^{-1}. Calculate Young's modulus for aluminum.

18. A 1-kg mass is suspended from the lower end of a vertical aluminum wire of length 2 m and diameter 0.75×10^{-3} m. The resultant strain is measured to be 3.05×10^{-4}. Calculate Young's modulus for aluminum. Compare your result with that obtained in Problem 17.

19. Using the value of Young's modulus obtained in Problem 17, estimate the atomic force constant for aluminum.

20. Young's modulus for a certain metal is measured as 1.83×10^{10} N · m^{-2} and the speed of sound in the material as 1.28×10^3 m · sec^{-1}. Estimate the inter-atomic spacing in this metal. *Hint:* Identify the metal from the density.

21. When a cylindrical metallic rod experiences a longitudinal strain, there is an accompanying transverse, or radial, strain. It is found that the radial strain ε_r is given by $\varepsilon_r = -\mu\varepsilon_l$, where ε_l is the relative longitudinal extension, ε_r is the relative lateral contraction, and μ is a constant known as the **Poisson constant**. If a rod of metal with a Poisson constant of 0.2 suffers a longitudinal strain of 5×10^{-2}, find the change in diameter of the rod if the original diameter is 0.005 m.

22. A cube of metal lies on a smooth, frictionless table. Describe one method by which this cube may be subjected to a shear stress. Is it possible to apply a shear stress to the cube that does not cause it to rotate? Justify your answer.

23. To measure Young's modulus for a metal two wires of the metal are hung side by side. One carries a millimeter scale, the other a vernier and a scale pan. As 500-g weights are added to the scale pan and taken from it, readings on the vernier are recorded. Some typical data are given in Table 21.2. The average diameter of the wire was measured as 6.32×10^{-2} cm and the length as 2.10 m.
(a) Determine Young's modulus for the wire.
(b) Comment on the use of two identical wires to support the vernier and the scale.
(c) Of what material is the wire probably made?
(d) Estimate the speed of sound in the material from which the wire is constructed.

Table 21.2

Load (g)	Reading: increasing wt. (cm)	Reading: decreasing wt. (cm)
0	1.83	1.81
500	1.84	1.85
1000	1.89	1.90
1500	1.95	1.95
2000	1.99	2.00
2500	2.04	2.04
3000	2.09	2.09
3500	2.14	2.14
4000	2.19	2.20
4500	2.24	2.24

The dynamics of 22
crystalline solids

22.1 INTRODUCTION

In this chapter we discuss some aspects of the motions of the atoms in a solid about their equilibrium positions. Such motions, called **lattice vibrations**, constitute elastic (acoustical) waves in the solid. The periodic nature of crystalline solids determines the allowed modes of vibration. We shall see how the lattice vibrations are important for an understanding of such thermal properties of solids as specific heats, thermal conductivity, and thermal expansion.

22.2 THE DYNAMICS OF A ONE-DIMENSIONAL ARRAY OF SIMILAR ATOMS

We shall first consider an infinite line of identical and equally spaced atoms of mass M. We take a for the lattice parameter and Δx_n for the displacement of atom n from its equilibrium position, as shown in Fig. 22.1. We shall assume that the restoring force F_n acting on atom

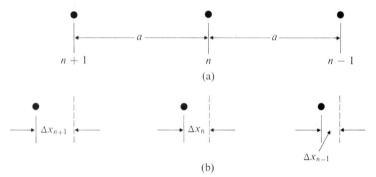

Fig. 22.1. (a) The atoms $n - 1$, n, and $n + 1$ in their equilibrium positions. (b) The displacements from equilibrium of atoms $n - 1$, n, and $n + 1$.

n to return it to its equilibrium position results entirely from the interaction of atom n with its nearest neighbors, atoms $n - 1$ and $n + 1$. The quantity $\Delta x_{n+1} - \Delta x_n$ represents the change of separation of atoms $n + 1$ and n from a; the quantity $\Delta x_n - \Delta x_{n-1}$ represents the change of the separation of atoms n and $n - 1$ from a. We shall assume that the forces on atom n due to these changes in equilibrium separation are proportional to the change in equilibrium separation. Taking k_A as the atomic force constant, we can write

$$F_n = k_A(\Delta x_{n+1} - \Delta x_n) - k_A(\Delta x_n - \Delta x_{n-1})$$
$$= k_A(\Delta x_{n+1} + \Delta x_{n-1} - 2\Delta x_n).$$

If

$$\Delta x_{n+1} > \Delta x_n > \Delta x_{n-1},$$

then the two forces acting on atom n will be oppositely directed.

The equation of motion of atom n as given by Newton's second law is

$$M \frac{d^2(\Delta x_n)}{dt^2} = k_A(\Delta x_{n+1} + \Delta x_{n-1} - 2\Delta x_n).$$

We shall see that this equation has a solution of the form

$$\Delta x_n = \xi \cos(\omega t + kna + \phi),$$

which is essentially a traveling wave (see Section 15.3) of amplitude ξ, angular frequency ω, and propagation constant k. The quantity na represents the equilibrium position of atom n and replaces the usual variable x because of the discrete nature of the lattice. Now, taking the second derivative of the trial solution for Δx_n, we see that

$$\frac{d^2(\Delta x_n)}{dt^2} = -\omega^2 \Delta x_n;$$

also,

$$\begin{aligned}
\Delta x_{n+1} &= \xi \cos[\omega t + k(n+1)a + \phi] \\
&= \xi \cos[(\omega t + kna + \phi) + ka] \\
&= \xi \cos(\omega t + kna + \phi)\cos ka - \xi \sin(\omega t + kna + \phi)\sin ka \\
\Delta x_{n-1} &= \xi \cos[\omega t + k(n-1)a + \phi] \\
&= \xi \cos(\omega t + kna + \phi)\cos ka + \xi \sin(\omega t + kna + \phi)\sin ka
\end{aligned}$$

and

$$\begin{aligned}
\Delta x_{n+1} + \Delta x_{n-1} &= 2\xi \cos(\omega t + kna + \phi)\cos ka \\
&= 2\Delta x_n \cos ka.
\end{aligned}$$

Upon substitution into the equation of motion, we obtain

$$\begin{aligned}
-M\omega^2 \Delta x_n &= 2k_A \Delta x_n(\cos ka - 1) \\
&= -4k_A \Delta x_n \sin^2\left(\frac{ka}{2}\right)
\end{aligned}$$

and

$$M\omega^2 = 4k_A \sin^2\left(\frac{ka}{2}\right).$$

Therefore,

$$\Delta x_n = \xi \cos(\omega t + kna + \phi)$$

is indeed a solution of the equation of motion for atom n, provided that

$$\omega = \pm \left(\frac{4k_A}{M}\right)^{1/2} \sin\left(\frac{ka}{2}\right).$$

This relation between ω and k is known as a **dispersion relation**. The positive portion of the curve is plotted in Fig. 22.2. A maximum frequency of propagation ω_m is predicted for a propagation constant k_m given by

$$\frac{k_m a}{2} = \pm \frac{\pi}{2}$$

or

$$k_m = \pm \frac{\pi}{a};$$

the maximum frequency is

$$\omega_m = \left(\frac{4k_A}{M}\right)^{1/2}.$$

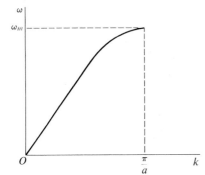

Fig. 22.2. The dispersion relation for a one-dimensional solid with one kind of atom.

Values of k outside of the range

$$-\frac{\pi}{a} \leqslant k \leqslant \frac{\pi}{a}$$

give us nothing new but merely repeat motions already described. This region is known as the first **Brillouin zone**.

We might note that, when $\omega = \omega_m$,

$$\Delta x_n = \xi \cos\left(\omega_m t + k_m na + \phi\right)$$

$$= \xi \cos\left(\omega_m t \pm n\pi + \phi\right)$$

$$= \xi \cos\left(\omega_m t + \phi\right) \cos n\pi,$$

which represents a standing wave (see Section 15.4). Further, we might note that the value $k_m = \pm \pi/a$ satisfies Bragg's law

$$2d \sin \theta = n\lambda$$

for $n = \pm 1$, where θ is the angle that the waves make with the line of atoms from which the waves are reflected (see Fig. 21.14). In this case $\theta = \pi/2$, $a = d$, and $\lambda = 2\pi/k$, so that

$$2d \sin \theta = 2a$$

and

$$n\lambda = n\left(\frac{2\pi}{k_m}\right)$$

$$= \pm 2na$$

$$= 2a$$

for $n = \pm 1$. That is, when the Bragg condition is satisfied a traveling elastic wave cannot propagate along the lattice. Rather, through successive Bragg reflections a standing wave is set up.

The velocity of propagation of the lattice vibrations is

$$v = v\lambda = \frac{\omega}{2\pi}\left(\frac{2\pi}{k}\right) = \frac{\omega}{k} = \left(\frac{4k_A}{M}\right)^{1/2}\frac{\sin\left(\frac{ka}{2}\right)}{k} = v_0\frac{\sin\left(\frac{ka}{2}\right)}{\left(\frac{ka}{2}\right)},$$

with

$$v_0 = \frac{a}{2}\left(\frac{4k_A}{M}\right)^{1/2}.$$

The maximum or cutoff frequency ω_m is therefore

$$\omega_m = v_m k_m$$

$$= \frac{v_0}{\left(\frac{k_m a}{2}\right)}k_m$$

$$= \frac{2v_0}{a}.$$

Since the speed of sound in crystals is of the order of $3 \times 10^3 \, \text{m} \cdot \text{sec}^{-1}$ and a is of the order of 3×10^{-10} m,

$$\omega_m \sim \frac{2 \times 3 \times 10^3}{3 \times 10^{-10}}$$

$$= 2 \times 10^{13} \, \text{rad} \cdot \text{sec}^{-1},$$

which is a frequency in the infrared region of the electromagnetic spectrum.

22.3   THE DYNAMICS OF A ONE-DIMENSIONAL ARRAY COMPOSED OF TWO KINDS OF ATOMS

We now consider an infinite one-dimensional lattice characterized by lattice parameter a, with atoms of mass m located at odd-numbered lattice sites and atoms of mass M at even-numbered lattice sites as shown in Fig. 22.3. Making the same assumption as in Section 22.2, namely, that forces proportional to the change in equilibrium separation act between nearest neighbors, we can easily deduce the equations of motion

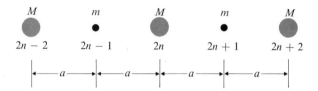

Fig. 22.3. A one-dimensional lattice with two kinds of atoms.

for the two types of atoms; they are

$$M\frac{d^2(\Delta x_{2n})}{dt^2} = k_A(\Delta x_{2n+1} + \Delta x_{2n-1} - 2\Delta x_{2n})$$

$$m\frac{d^2(\Delta x_{2n+1})}{dt^2} = k_A(\Delta x_{2n+2} + \Delta x_n - 2\Delta x_{2n+1}).$$

We shall verify that this coupled set of equations has traveling wave solutions of the form

$$\Delta x_{2n} = \xi \cos[\omega t + 2nka + \phi];$$

$$\Delta x_{2n+1} = \eta \cos[\omega t + (2n+1)ka + \phi].$$

We leave it as an exercise for the reader (see Problem 1) to show that the substitution of solutions of this form into the equation of motion yields the following conditions.

$$(2k_A - \omega^2 M)\xi = (2k_A \cos ka)\eta,$$

$$(2k_A \cos ka)\xi = (2k_A - \omega^2 m)\eta.$$

Equating the ratios of the coefficients of ξ and η as given by these equations yields

$$\frac{2k_A - \omega^2 M}{2k_A \cos ka} = \frac{2k_A \cos ka}{2k_A - \omega^2 m}$$

or

$$(2k_A - \omega^2 M)(2k_A - \omega^2 m) - (2k_A \cos ka)^2 = 0$$

or

$$mM\omega^4 - 2k_A(M + m)\omega^2 + (4k_A^2 - 4k_A^2 \cos^2 ka) = 0$$

or

$$mM\omega^4 - 2k_A(M + m)\omega^2 + 4k_A^2 \sin^2 ka = 0.$$

It is necessary that the traveling wave solutions to the equations of motion satisfy this relation for all values ξ and η. Solving this quadratic equation in ω^2 yields

$$\omega^2 = k_A\left(\frac{M+m}{Mm}\right) \pm \frac{[4k_A^2(M+m)^2 - 4mM(4k_A^2\sin^2 ka)]^{1/2}}{2mM}$$

$$= k_A\left(\frac{1}{m} + \frac{1}{M}\right) \pm k_A\left[\left(\frac{1}{m} + \frac{1}{M}\right)^2 - \frac{4\sin^2 ka}{mM}\right]^{1/2}.$$

This dispersion relation is plotted in Fig. 22.4 for $M > m$. It has two branches, one called the **acoustical branch**, the other the **optical branch**. The nature of the vibrations for the two branches is illustrated for transverse waves in Fig. 22.5. For the optical branch the two types of atoms vibrate against one another in such a way that the centers of mass of the atom pairs remain fixed. If the solid is ionic, a light (electromagnetic) wave incident on the solid can excite this type of vibration. Therefore, such vibrations are called **optical vibrations**. For the acoustical branch the two types of atoms and the centers of mass of atom pairs move together as in an acoustical disturbance.

We note from Fig. 22.4 that wave-like solutions do not exist for frequencies in the range

$$\left(\frac{2k_A}{M}\right)^{1/2} < \omega < \left(\frac{2k_A}{m}\right)^{1/2}.$$

This is an example of a **forbidden band**. Sometimes it is referred to as a **frequency gap** at the boundary of the first Brillouin zone. The consideration of Sections 22.2 and 22.3 can be extended to three-dimensional lattices. In so doing no new physical concepts are introduced.

22.4 THE FREQUENCY DISTRIBUTION OF LATTICE VIBRATIONS

In Sections 22.2 and 22.3 we considered infinite one-dimensional arrays of atoms. As a one-dimensional model for a real solid it would be more realistic to consider a finite though very large one-dimensional array. Let us consider $N + 1$ atoms with separation a defining a line of length

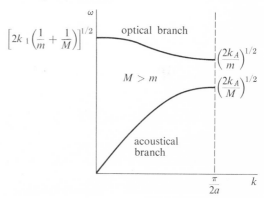

Fig. 22.4. The dispersion relation for a one-dimensional solid with two kinds of atoms.

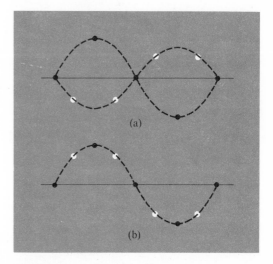

Fig. 22.5. (a) An illustration using transverse waves of the nature of the optical modes in a solid. (b) An illustration using transverse waves of the nature of the acoustical modes in a solid.

L (see Fig. 22.6). Then

$$Na = L.$$

For such a finite array we must take special account of the end points by introducing some sort of **boundary conditions**. It turns out, as is indeed reasonable, that the actual form of the boundary conditions is not important for the determination of bulk properties of the solid. We shall

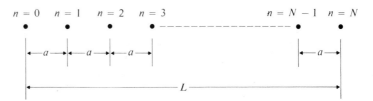

Fig. 22.6. A finite one-dimensional array of atoms.

assume for simplicity that the particles $n = 0$ and $n = N$ at the ends of the line are held fixed. Then the only acceptable wave solutions must be such that $\Delta x_n = 0$ for $n = 0$ or N. We can assure this by choosing

$$\Delta x_n = \zeta \sin(\omega t + \phi) \sin kna,$$

with k taking only those values for which

$$\sin kNa = \sin kL = 0.$$

That is, the boundary conditions restrict the allowed values of k to

$$k = \frac{\pi}{L}, \frac{2\pi}{L}, \frac{3\pi}{L}, \dots, \frac{(N-1)\pi}{L}.$$

Note that if

$$k = \frac{N\pi}{L} = \frac{\pi}{a},$$

then

$$\sin kna = \sin n\pi,$$

which vanishes for all n so that no motion of the lattice occurs. This set of solutions, each of which describes a standing wave, gives the **normal modes of vibration** of the finite lattice. The first four normal modes are shown in Fig. 22.7.

The standing wave or normal mode solutions may be constructed from the traveling wave solutions obtained in Section 22.2 by taking appropriate combinations of them. That is, if we combine the functions

$$\cos(\omega t + kna + \phi) \quad \text{and} \quad \cos(\omega t - kna + \phi),$$

assuming them to have equal amplitudes ξ but to be out of phase by π, we obtain

$$\Delta x_n = \xi \cos(\omega t - kna + \phi) - \xi \cos(\omega t + kna + \phi)$$

$$= \xi[\cos(\omega t + \phi)\cos kna + \sin(\omega t + \phi)\sin kna]$$

$$\quad - \xi[\cos(\omega t + \phi)\cos kna - \sin(\omega t + \phi)\sin kna]$$

$$= 2\xi \sin(\omega t + \phi)\sin kna$$

$$= \zeta \sin(\omega t + \phi)\sin kna.$$

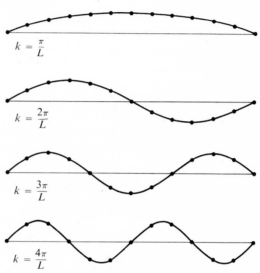

Note that there are $N - 1$ allowed values for k. This number equals the number of particles that are allowed to move. It follows that for a three-dimensional solid containing N atoms, all of which are allowed to move, the number of normal modes equals $3N$.

The **density of states in k space** $\Omega(k)$ is defined as the number of normal modes per unit propagation constant range. For the one-dimensional array there is one normal mode for each interval

$$\Delta k = \frac{\pi}{L}.$$

Therefore,

$$\Omega(k) = \frac{1}{\left(\dfrac{\pi}{L}\right)} = \frac{L}{\pi}.$$

Fig. 22.7. The first four normal modes of a finite one-dimensional array of atoms.

The **frequency distribution of normal modes** $C(\omega)$ is defined as the number of normal modes per unit angular frequency range. Since the total number of normal modes is $N - 1$, it follows that

$$\int_{\text{vol}} C(\omega)\, d\omega = \int_{\text{vol}} \Omega(k)\, d\mathbf{k} = N - 1,$$

from which it may be shown that

$$C(\omega)\, d\omega = \Omega(k)\left(\frac{dk}{d\omega}\right) d\omega.$$

In Section 22.2 we saw that

$$\omega = \omega_m \sin\left(\frac{ka}{2}\right)$$

or

$$k = \frac{2}{a} \sin^{-1}\left(\frac{\omega}{\omega_m}\right),$$

so that

$$\frac{dk}{d\omega} = \frac{2}{a(\omega_m^2 - \omega^2)^{1/2}}.$$

Therefore,

$$C(\omega) = \frac{L}{2\pi} \frac{2}{a(\omega_m^2 - \omega^2)^{1/2}}$$

$$= \frac{L}{\pi a}(\omega_m^2 - \omega^2)^{-1/2}.$$

The extension to three dimensions is quite difficult.

22.5   THE SPECIFIC HEATS OF A SOLID

For solids the specific heats measured at constant pressure, C_p, and at constant volume, C_v, differ by only 3–5% at room temperature. Figure 22.8 shows the variation with temperature of C_p and C_v for copper. We note that C_v approaches a constant value of ~$3R$ at high temperatures but that C_p continues to increase with temperature. Similar results obtain for most solids; C_v always approaches the same constant value. We confine our discussion to the behavior of C_v.

For nonmetallic solids heat added at constant volume goes entirely toward an increase in the energy associated with the lattice vibrations. For metallic solids part of the heat goes to an increase of the kinetic energy of the conduction electrons.

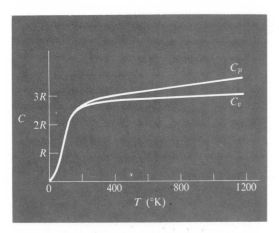

Fig. 22.8. The variation with temperature of C_p and C_v for copper.

Dulong and Petit specific heat

Since the lattice vibrations take place on an atomic scale of dimensions, we should expect quantum effects to be important. Nonetheless, let us first consider the classical prediction of C_v, which according to the correspondence principle should be the limiting value at high temperature of the quantum prediction.

In Section 22.4 we suggested that an N-atom solid has associated with it $3N$ independent lattice oscillations. These oscillations are associated with the coupled motions of the N atoms about their equilibrium positions. For restoring forces that are proportional to the change in separation, the atoms move in simple harmonic motion. Since the atomic motions can always be resolved into components in three mutually perpendicular directions, it is possible to picture an atomic oscillator as consisting of three independent harmonic oscillators. The lattice vibrations can therefore be thought of as the coupled motions of $3N$ harmonic oscillators. Classically, according to the principle of equipartition of energy, the total energy associated with the lattice of a solid at temperature T is $3NkT$, since a linear harmonic oscillator has two degrees of freedom. The internal energy per kilogram-mole is

$$E = \frac{3NkT}{n} = 3RT,$$

since

$$nR = Nk.$$

Therefore,

$$C_v = \left(\frac{dE}{dT}\right)_V = 3R.$$

(The subscript V outside the parentheses means that the quantity inside the parentheses is to be evaluated under conditions of constant volume.) This is known as the Dulong and Petit law, and it indeed agrees with the high-temperature limit of the experimental results.

Einstein specific heat

An explanation for the decrease in C_v at low temperatures was first given by Einstein, who considered $3N$ quantum mechanical oscillators of some average frequency ν_E called the **Einstein frequency**. As our analysis of Section 22.3 has shown, it is an oversimplification to assign all $3N$ oscillators the same frequency.

The average energy of the $3N$ quantum mechanical oscillators of frequency ν_E is, according to the result of Section 20.5,

$$\bar{E} = 3Nk\theta_E \left[\frac{1}{2} + \frac{1}{\exp\left(\dfrac{\theta_E}{T}\right) - 1}\right],$$

where

$$\theta_E = \frac{h\nu_E}{T}$$

is the **Einstein temperature**.

The molar specific heat is given by

$$C_v = \frac{1}{n}\frac{dE}{dT}$$

$$= \frac{3Nk\theta_E}{n}\frac{-1}{\left[\exp\left(\dfrac{\theta_E}{T}\right) - 1\right]^2}\left(-\frac{\theta_E}{T^2}\right)\exp\left(\frac{\theta_E}{T}\right)$$

$$= 3R\left(\frac{\theta_E}{T}\right)^2\frac{\exp\left(\dfrac{\theta_E}{T}\right)}{\left[\exp\left(\dfrac{\theta_E}{T}\right) - 1\right]^2}.$$

As $T \to 0$, $C_v \to 0$, as

$$C_v \simeq 3R\left(\frac{\theta_E}{T}\right)^2\exp\left(-\frac{\theta_E}{T}\right).$$

At very low temperatures the exponential factor is dominant, and C_v is predicted to vary exponentially with temperature. For an appropriate choice of the Einstein temperature θ_E, the equation gives a reasonable fit to the experimental data. For example, in Fig. 22.9 we show a comparison of experimental values and values calculated on the Einstein model for the heat capacity of diamond, using $\theta_E = 1320°\text{K}$.

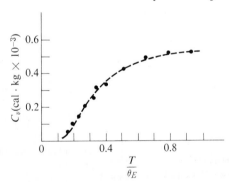

Fig. 22.9. A comparison of experimental values of the specific heat of diamond and values calculated on the Einstein model.

Debye specific heat

Experiments show that the variation of C_v at low temperatures obeys a T^3 law and not an exponential one. Such a low-temperature T^3 law was predicted later by the work of Peter Debye, who represented the $3N$ lattice vibrations with a set of elastic waves having a continuous range of frequencies up to a maximum value ν_D called the **Debye frequency**. Mathematically the Debye theory is considerably more complicated than the Einstein theory, and we state only the limiting low-temperature result

$$C_v = 3R\frac{4\pi^4}{5}\left(\frac{T}{\theta_D}\right)^3, \qquad T \ll \theta_D,$$

where

$$\theta_D = \frac{h\nu_D}{k}$$

is the **Debye temperature**. θ_D is a parameter chosen to make the equation fit the experimental data.

Specific heat of conduction electrons in a metal

The conduction electrons in a metal are in a certain sense free to move through the metal. It is possible to understand a number of the physical properties of metals by assuming that the conduction electrons constitute a free electron gas within the metal. In this model the interactions of the conduction electrons with themselves and with the ion cores of the original atoms are neglected.

If the free electron gas were a gas of classical particles, one contributed by each atom making up the solid, the electron gas would contribute an amount

$$E(\text{el}) = \tfrac{3}{2}NkT$$

to the internal energy of the metal and a contribution

$$C_v(\text{el}) = \frac{3}{2}\frac{Nk}{n} = \frac{3}{2}R$$

per kilogram-mole to the specific heat. Therefore, at temperatures high compared to the Debye temperature θ_D, one would expect a total molar specific heat for metals of

$$C_v = C_v(\text{lattice}) + C_v(\text{el})$$

$$= 3R + \tfrac{3}{2}R.$$

However, the high-temperature specific heats for metals are not significantly larger than for nonmetals. That is, there is no evidence that the electrons contribute to the specific heat according to the prediction of classical statistics.

The observed thermal behavior of the free electron gas is that of a Fermi–Dirac gas (see Section 20.6). At $T = 0$ the lowest energy states available to the electrons are filled up in turn; as soon as the number of electrons is exhausted, the probability of the next highest state having electrons in it drops to zero. The electrons therefore occupy the lowest states allowed by the Pauli exclusion principle. The energy E_F of the highest occupied level at absolute zero is called the **Fermi energy**. When $T > 0$ the lower energy states begin to be vacated, since more energy is available and the probability for higher energy states to be populated increases. For temperatures that are low compared to temperatures of the order of $3 \times 10^{4°}$K, only a small fraction of the electrons can be thermally excited. This number increases in proportion to the absolute temperature, so that

$$E(\text{el}) \propto T^2$$

and

$$C_v(\text{el}) = \gamma T,$$

where γ is a constant characteristic of the particular metal. For typical values of γ the electron contribution at room temperature is only of the order of 10^{-2} (or 1%) of the lattice contribution to the specific heat.

22.6 THERMAL CONDUCTIVITY

Suppose that a long solid bar has its axis in the z direction and that a temperature difference is maintained between the ends of the bar. Thermal energy will flow from the hot end to the cold end. We shall assume that a steady state has been established and define the **thermal conductivity** K of the bar by the equation

$$Q = -K \frac{dT}{dz},$$

where Q is the thermal energy crossing a unit area per unit time and dT/dz, called the **temperature gradient**, is the rate of change of temperature with distance in the z direction. In pure metals most of the heat is transported by the conduction electrons. That the temperature gradient enters the expression is a result of the importance of the interaction of the electrons with the lattice vibrations. If the electrons passed freely through the metal, the rate of heat transfer would depend upon the temperature difference rather than the temperature gradient.

We shall carry out a kinetic theory calculation of the thermal conductivity of a classical electron gas. In particular we consider the transfer of energy by the electrons crossing a particular x, y plane with z coordinate z_0 (see Fig. 22.10) and assign a value $\bar{E}$ to the mean energy per electron for electrons in the plane. According to the kinetic theory all directions of motion of the electrons are equally probable, and at equilibrium the number of electrons per unit volume is the same throughout

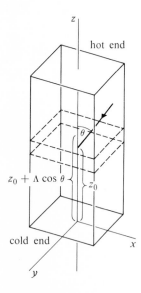

Fig. 22.10. The geometry for the calculation of the thermal conductivity of a classical electron gas.

the whole volume of gas. If there are N electrons per kilogram-mole having an average speed $\bar{v}$, then there will be $N/2$ electrons with a component of velocity in the $-z$ direction. All electrons within a distance $\bar{v}$ of the plane $z = z_0$ would cross the plane in 1 sec if they were all traveling in the $-z$ direction. However, the electrons are moving randomly in all directions, and the number traveling in directions making angles between 0 and $0 + d\theta$ with the z axis is

$$\frac{N}{2} \sin\theta \, d\theta$$

per kilogram-mole (the reader may check this expression by integrating over θ). These electrons have a component of velocity $\bar{v} \cos\theta$ normal to the $z = z_0$ plane, so that the number of electrons crossing a unit area of the plane per second from above and traveling in directions making angles between 0 and $0 + d\theta$ with the z axis is

$$\frac{N}{2} \bar{v} \cos\theta \sin\theta \, d\theta.$$

If Λ is the mean distance that the electrons travel between collisions (called the **electron mean free path**), then on the average these electrons must have originated at a plane whose z coordinate is $z_0 + \Lambda \cos\theta$. The mean energy of these electrons is

$$\bar{E} + \Lambda \cos\theta \left(\frac{d\bar{E}}{dz}\right)$$

where $d\bar{E}/dz$ is the rate of change of mean electron energy with z. The energy carried per second across the plane z_0 by these electrons is therefore

$$-\left(\bar{E} + \Lambda \cos\theta \frac{d\bar{E}}{dz}\right) \frac{N}{2} \bar{v} \cos\theta \sin\theta \, d\theta,$$

the minus sign indicating that the direction of energy flow is the direction of decreasing z. Integration over all values of θ from 0 to π yields for the total flow of energy across a unit area of the plane $z = z_0$

$$-\frac{N}{3} \Lambda \bar{v} \frac{d\bar{E}}{dz}.$$

By the definition of the thermal conductivity

$$-K \frac{dT}{dz} = -\frac{N}{3} \Lambda \bar{v} \frac{d\bar{E}}{dz}.$$

Now

$$N\frac{d\bar{E}}{dz} = N\frac{d\bar{E}}{dT}\frac{dT}{dz},$$

since $\bar{E}$ is a function of the temperature only. Therefore,

$$N\frac{d\bar{E}}{dz} = C_v\frac{dT}{dz},$$

where C_v is the molar specific heat at constant volume. Finally

$$K\frac{dT}{dz} = \frac{1}{3}\Lambda\bar{v}C_v\frac{dT}{dz}$$

and

$$K = \tfrac{1}{3}C_v\bar{v}\Lambda.$$

The corresponding result for a Fermi–Dirac gas is obtained by writing

$$C_v = \gamma T$$

and substituting for $\bar{v}$ the **Fermi speed** v_F as given by

$$\tfrac{1}{2}mv_F^2 = kT_F = E_F,$$

where T_F is the **Fermi temperature** and E_F the **Fermi energy**. The Fermi speed is the speed of the electrons at the top of the Fermi–Dirac distribution, and it is these electrons that are free to contribute to the thermal conductivity.

The measured temperature dependence of the thermal conductivity of a typical metal, namely copper, at low temperatures is shown in Fig. 22.11. To give a physical explanation of this behavior let us look at the temperature dependence of each of the quantities appearing in the kinetic theory expression for the thermal conductivity that we have developed.

In Section 22.5 we saw that the contribution to the specific heat C_v associated with the conduction electrons in a metal increases linearly with temperature. That is,

$$C_v \propto T.$$

The Fermi speed of the electrons is a constant independent of temperature. Now, what about the temperature dependence of Λ, the electron

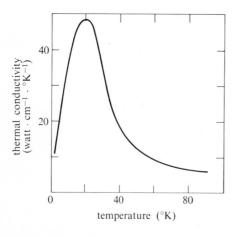

Fig. 22.11. The experimental thermal conductivity of copper at low temperatures.

mean free path? It turns out that one may distinguish two contributions to Λ (which we shall label Λ_d and Λ_l) that have different temperature dependences.

No real metal can be represented by a perfect crystal lattice. For example, in every real metal lattice there are sites at which no atoms occur (called **vacancy sites**), and extra atoms (called **interstitial atoms**) appear between sites designated in the corresponding perfect lattice. As a result of such **static defects** the conduction electrons are inhibited in their motion through the lattice. The effect of electron interactions with such defects may be characterized by a mean free path Λ_d independent of temperature. For this contribution to Λ, it follows that the temperature dependence of the corresponding contribution K_d to the thermal conductivity is given by

$$K_d \propto T.$$

In Fig. 22.11 we see that, at the very lowest temperatures, K increases linearly with increasing temperature. In this region the rate of heat conduction is limited by the presence of impurities.

Even for a perfect crystal lattice the electrons cannot move unhindered through the metal because of the thermal vibrations of the crystalline lattice. Since the number of elastic waves set up in the lattice increases with the temperature, it is to be expected that the contribution Λ_l to the electron mean free path resulting from the interaction of the conduction electrons with the lattice vibrations will decrease with increasing temperature. If a Debye spectrum of lattice vibrations is assumed, it may be shown that for temperatures much less than the Debye temperature

$$\Lambda_l \propto T^{-3}.$$

For this contribution to Λ, it follows that the temperature dependence of the corresponding contribution K_l to the thermal conductivity is given by

$$K_l \propto T^{-2}.$$

In Fig. 22.11 we see that, at the highest temperatures, K decreases (in an approximately quadratic fashion) with increasing temperature. In this region the rate of heat conduction is limited by the lattice vibrations.

The two effects may be taken into account simultaneously by defining the total thermal conductivity K by

$$\frac{1}{K} = \frac{1}{K_d} + \frac{1}{K_l}$$

$$= \frac{\alpha}{T} + \beta T^2.$$

In Fig. 22.12(a) we see the individual contributions to $1/K$ (called the thermal resistivity) due to impurities and to lattice vibrations. In Fig. 22.12(b) we see the predicted theoretical behavior of the total electronic

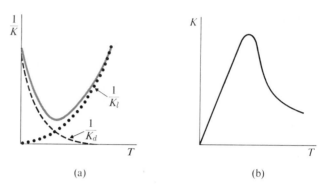

(a)   (b)

Fig. 22.12. (a) The thermal resistivity as a function of temperature showing the individual contributions due to impurities and to lattice vibrations. (b) The theoretical electronic thermal conductivity as a function of temperature.

contribution to the thermal conductivity. From a visual comparison of Figs. 22.11 and 22.12(b) we see that the theory is certainly able to account for the qualitative behavior of the experimental measurements.

22.7   THERMAL EXPANSION

It is well known that when a solid is heated the dimensions of the solid increase. This is the phenomenon known as **thermal expansion**. Let us see if we can account for it in terms of the thermal energy of the lattice vibrations.

We consider a one-dimensional solid and first assume a linear restoring force for the atomic vibrations about their equilibrium positions at $0°K$. Therefore, the potential energy of atom n when its displacement is Δx_n is

$$V(\Delta x_n) = \frac{k_A}{2}(\Delta x_n)^2,$$

where k_A is the atomic force constant. We now calculate the average displacement $\overline{\Delta x_n}$ at temperature T, weighting the possible values of Δx_n

according to the Maxwell–Boltzmann distribution. That is,

$$
\overline{\Delta x_n} = \frac{\displaystyle\int_{-\infty}^{\infty} \Delta x_n \exp\left[-\frac{V(\Delta x_n)}{kT}\right] d(\Delta x_n)}{\displaystyle\int_{-\infty}^{\infty} \exp\left[-\frac{V(\Delta x_n)}{kT}\right] d(\Delta x_n)}
$$

$$
= \frac{\displaystyle\int_{-\infty}^{\infty} \Delta x_n \exp\left[-\frac{k_A(\Delta x_n)^2}{2kT}\right] d(\Delta x_n)}{\displaystyle\int_{-\infty}^{\infty} \exp\left[-\frac{k_A(\Delta x_n)^2}{2kT}\right] d(\Delta x_n)}
$$

$$
= \frac{\displaystyle\int_{-\infty}^{\infty} x \exp(-ax^2)\, dx}{\displaystyle\int_{-\infty}^{\infty} \exp(-ax^2)\, dx},
$$

with

$$
a = \frac{k_A}{2kT}, \qquad x = \Delta x_n.
$$

These integrals may be easily evaluated with the help of a table of integrals to give

$$
\int_{-\infty}^{\infty} \exp(-ax^2)\, dx = \left(\frac{\pi}{a}\right)^{1/2},
$$

$$
\int_{-\infty}^{\infty} x \exp(-ax^2)\, dx = 0.
$$

Therefore,

$$
\overline{\Delta x_n} = 0.
$$

For expansion there must be a net increase in the distance between atom 1 and atom N. In order to affect this distance, the quantity Δx_n must increase for the majority of the atoms. Such increases add to yield a macroscopic increase in the dimension of the solid. If, however, $\overline{\Delta x_n} = 0$, the quantities Δx_n do *not* increase on the average, and the solid does not expand.

In fact, because of the mutual repulsion of the atoms the restoring force is not really proportional to the displacement. Figure 22.13 shows diagrammatically the potential energy variation between atoms in a crystal lattice. Superposed on the atomic potential energy curve is a potential energy curve for a harmonic oscillator. We see that the assumption that

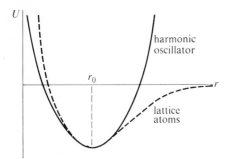

Fig. 22.13. The variation in potential energy for a harmonic oscillator compared with a typical variation in potential energy for lattice atoms as a function of the separation of the atoms.

the atoms undergo harmonic oscillations about their equilibrium positions is an approximation that is good only for small amplitudes of oscillation for which the harmonic oscillator potential energy curve and the atomic potential energy curve coincide.

Therefore, let us introduce an **anharmonic term** $g_A(\Delta x_n)^2$ into the expression for the restoring force experienced by atom n, where g_A is the **anharmonic atomic force constant**. The potential energy of atom n then becomes

$$V(\Delta x_n) = \frac{k_A}{2}(\Delta x_n)^2 - \frac{g_A}{3}(\Delta x_n)^3.$$

Let us once again calculate the average displacement $\overline{\Delta x_n}$. Now

$$\overline{\Delta x_n} = \frac{\displaystyle\int_{-\infty}^{\infty} \Delta x_n \exp\left\{\frac{\left[-\frac{k_A}{2}(\Delta x_n)^2 + \frac{g_A}{3}(\Delta x_n)^3\right]}{kT}\right\} d(\Delta x_n)}{\displaystyle\int_{-\infty}^{\infty} \exp\left\{\frac{\left[-\frac{k_A}{2}(\Delta x_n)^2 + \frac{g_A}{3}(\Delta x_n)^3\right]}{kT}\right\} d(\Delta x_n)}$$

$$= \frac{\displaystyle\int_{-\infty}^{\infty} x \exp\left(-ax^2 + bx^3\right) dx}{\displaystyle\int_{-\infty}^{\infty} \exp\left(-ax^2 + bx^3\right) dx},$$

with

$$a = \frac{k_A}{2kT}, \qquad b = \frac{g_A}{3kT}, \qquad x = \Delta x_n.$$

Rearranging terms gives

$$\overline{\Delta x_n} = \frac{\displaystyle\int_{-\infty}^{\infty} x \exp\left(-ax^2\right) \exp\left(bx^3\right) dx}{\displaystyle\int_{-\infty}^{\infty} \exp\left(-ax^2\right) \exp\left(bx^3\right) dx}.$$

Assuming that

$$bx^3 \ll 1$$

we can expand $\exp(bx^3)$ using the binomial expansion as

$$\exp(bx^3) \simeq 1 + bx^3.$$

Therefore,

$$\overline{\Delta x_n} \simeq \frac{\displaystyle\int_{-\infty}^{\infty} x \exp(-ax^2)[1 + bx^3]\,dx}{\displaystyle\int_{-\infty}^{\infty} \exp(-ax^2)[1 + bx^3]\,dx}$$

$$= \frac{\displaystyle\int_{-\infty}^{\infty} x \exp(-ax^2)\,dx + \int_{-\infty}^{\infty} bx^4 \exp(-ax^2)\,dx}{\displaystyle\int_{-\infty}^{\infty} \exp(-ax^2)\,dx + \int_{-\infty}^{\infty} bx^3 \exp(-ax^2)\,dx}.$$

These integrals can be readily evaluated as

$$\int_{-\infty}^{\infty} \exp(-ax^2)\,dx = \left(\frac{\pi}{a}\right)^{1/2}$$

$$\int_{-\infty}^{\infty} x \exp(-ax^2)\,dx = \int_{-\infty}^{\infty} bx^3 \exp(-ax^2)\,dx = 0$$

$$\int_{-\infty}^{\infty} bx^4 \exp(-ax^2)\,dx = \frac{3}{4}\left(\frac{\pi}{a}\right)^{1/2}\frac{b}{a^2}.$$

Finally

$$\overline{\Delta x_n} = \frac{\dfrac{3}{4}\left(\dfrac{\pi}{a}\right)^{1/2}\dfrac{b}{a^2}}{\left(\dfrac{\pi}{a}\right)^{1/2}}$$

$$= \frac{3}{4}\frac{b}{a^2}$$

$$= \frac{3}{4}\frac{g_A}{3kT}\left(\frac{2kT}{k_A}\right)^2$$

$$= \frac{kTg_A}{k_A^2}.$$

That is, the average displacement of the atoms increases in proportion to the temperature. The solid expands with increasing temperature.

The microscopic mean displacement $\overline{\Delta x_n}$ may be related to a macroscopic parameter, the **coefficient of linear expansion**, in the following manner. The macroscopic change in length ΔL is given by

$$\Delta L = \left(\frac{L}{a}\right)\overline{\Delta x_n},$$

where L is the length of the bar at $0°K$ and a is the interatomic spacing. The coefficient of linear expansion is defined by the relation

$$\Delta L = L\alpha T.$$

Equating the two expressions for ΔL yields

Table 22.1 Some typical values of the coefficient of linear expansion

Substance	$\alpha \times 10^4$ $(°K)^{-1}$
Aluminum	0.23
Copper	0.167
Lead	0.287
Silver	0.19
Invar	0.009

$$\frac{L}{a}\overline{\Delta x_n} = L\alpha T$$

or

$$\alpha = \frac{\overline{\Delta x_n}}{aT}$$

$$= \frac{kg_A}{ak_A^2}.$$

Some typical values of α are listed in Table 22.1.

PROBLEMS

1. In Section 22.3 it was stated that the necessary conditions for the existence of traveling wave solutions of the equations of motion of the two kinds of atoms for a one-dimensional array are

$$(2k_A - \omega^2 M)\xi = (2k_A \cos ka)\eta$$
$$(2k_A \cos ka)\xi = (2k_A - \omega^2 m)\eta.$$

Verify these conditions.

2. Compare the dispersion relation for a one-dimensional solid made of Ag and Br atoms with that for a one-dimensional solid made of Li and H atoms.

3. Given that the Fermi energy of the conduction electrons in copper is 7.04 eV, calculate the Fermi temperature and Fermi speed of the conduction electrons.

4. Estimate the Einstein characteristic frequency of oscillation for the atoms in copper. Use the atomic constant found in the example in Section 21.6.

5. At very low temperature, the molar specific heat of rock salt varies with the temperature according to Debye's T^3 law; therefore,

$$C_v = k\left(\frac{T}{\theta_D}\right)^3,$$

where $k = 464 \, \text{cal} \cdot (\text{g-mole})^{-1} \cdot (°\text{K})^{-1}$ and $\theta_D = 281°\text{K}$. Calculate the heat required to raise 2 g-mole of rock salt from $10°\text{K}$ to $50°\text{K}$.

6. The characteristic Einstein temperature for diamond is $1320°\text{K}$. Calculate C_v from the Einstein equation for $T = 300°\text{K}$ and compare your answer with the experimental value of $6.10 \, \text{J} \cdot (\text{g-mole})^{-1} \cdot (°\text{K})^{-1}$.

7. Does the Dulong and Petit prediction of C_v depend on the assumption that all atomic oscillators have the same frequency? Why?

8. For measurements of C_v made on metals in the region $T \ll \theta_D$, the data can be represented by an equation of the form

$$C_v = \alpha T^3 + \gamma T.$$

Results for metallic silver are shown in Fig. 22.14.

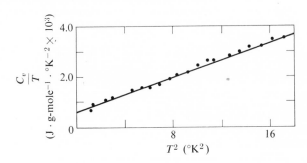

Fig. 22.14

(a) Deduce α and γ.
(b) Obtain a value of θ_D.
(c) Compare the electronic contribution that you would estimate for room temperature with the Dulong and Petit value.

*9. The specific heat of metallic silver as a function of temperature is shown in Fig. 22.15. Estimate the Einstein characteristic frequency of silver.

*10. Experimentally observed values of the specific heat of aluminum are shown in Fig. 22.16. Estimate the Einstein characteristic frequency.

11. The thermal conductivity of aluminum at $20°\text{C}$ is $2.3 \times 10^2 \, \text{J} \cdot \text{m}^{-1} \cdot \text{sec}^{-1} \cdot (°\text{C})^{-1}$. Taking the Fermi speed of the electrons as $10^6 \, \text{m} \cdot \text{sec}^{-1}$ and $\gamma = 1.46 \times 10^{-3} \, \text{J} \cdot (\text{g-mole})^{-1} \cdot (°\text{K})^{-2}$, deduce the electron mean free path.

*12. In the kinetic theory treatment of the heat transfer by the conduction electrons in a metal in Section

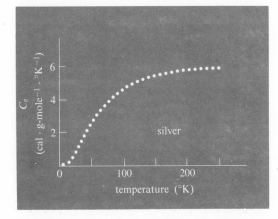

Fig. 22.15

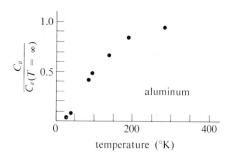

Fig. 22.16

22.6, it was stated that the number of electrons traveling in directions making angles between θ and $\theta + d\theta$ with the z axis is

$$\frac{N}{2} \sin \theta \, d\theta.$$

Verify this statement.

13. The thermal resistivity of a certain copper sample when multiplied by the absolute temperature and plotted as a function of the cube of the temperature yields a linear relation. The intercept on the ordinate axis is measured as $0.20 \text{ W}^{-1} \cdot \text{cm} \cdot (^\circ\text{K})^2$ and the slope of the graph is $1.66 \times 10^5 \text{ W}^{-1} \cdot \text{cm} \cdot (^\circ\text{K})^{-1}$. Plot a graph of the thermal conductivity of the sample as a function of temperature for temperatures from 1°K to 60°K.

14. (a) The period of an aluminum pendulum is 2.0016 sec at 0.0°C. Find its period at 25.0°C.
    (b) A clock controlled by this pendulum keeps correct time at 0.0°C. How many seconds a day will it gain or lose at 25.0°C?

15. A 1-kg mass is suspended from a copper wire of diameter 5.0×10^{-4} m. The temperature of the wire is then changed by an amount ΔT to bring the length of the wire back to its original value. Calculate ΔT.

16. Note the functional dependence of each of the following macroscopic parameters on the atomic force constant k_A: Young's modulus, the speed of sound, the Einstein frequency, the expansion coefficient. Predict values for each of these parameters for copper if k_A were suddenly doubled.

17. The coefficient of linear expansion of a metal rod can be measured experimentally by using a micrometer to measure changes of length ΔL for corresponding temperature changes ΔT. This method was used in a student laboratory to measure the expansion coefficient of an aluminum rod 0.5 m in length. Either steam or cold water was passed through the apparatus to alter the temperature of the rod. Some of the results obtained are given in Table 22.2. Use these data to calculate the expansion coefficient for aluminum. Comment on the accuracy of your result.

Table 22.2

Water temperature (°C)	Steam temperature (°C)	ΔL (m)
9	99	1.04×10^{-3}
7	99	1.08×10^{-3}

18. Calculate g_A for aluminum and then evaluate the ratio

$$\frac{\left(\dfrac{g_A}{3}\right)(\Delta x_n)^3}{\left(\dfrac{k_A}{2}\right)(\Delta x_n)^2}$$

at room temperature. Comment on the result.

19. In the derivation of the expression for the coefficient of linear expansion it was assumed that $bx^3 \ll 1$. Assuming the value of k_A for copper deduced in Section 21.6, show how well this approximation is satisfied at $T = 100°K$.

23 Thermodynamics

23.1 INTRODUCTION

In the preceding chapters we have examined the behavior of nature beginning with a detailed study of the motion of a single particle, proceeding to a discussion of systems of two particles and then a few particles, and finally ending with a look at the treatment of systems composed of large numbers of particles by the methods of kinetic theory, statistical physics, and quantum physics. The complexity of the statistical physics approach is due primarily to the large numbers involved; systems of a size large enough to make themselves readily observable to our senses are composed of countless numbers of atoms. For a complete description of the behavior of objects in our macroscopic world, we must take account of the atomistic quantized nature of our physical world. However, it is possible to describe correctly a great number of macroscopic phenomena even while ignoring the atomic description of matter. Long before men had clear ideas about atoms and quantum effects, the behavior of macroscopic objects had been studied in detail, and laws describing their behavior were formulated on the basis of these observations. In this chapter we shall study thermodynamics, which is concerned with the understanding and interpretation of the properties of matter insofar as they are affected by exchanges of energy. The laws of thermodynamics were arrived at as a consequence of observation and generalization; they have been found to be correct in all cases. However, classical thermodynamics makes no attempt to explain why the laws have their particular form. In this discussion of thermodynamics we shall attempt to relate some of these observational laws to the particle nature of matter.

23.2 TEMPERATURE AND THE ZEROTH LAW OF THERMODYNAMICS

We shall deal extensively with **isolated systems**, that is, systems separated from their surroundings by a wall that does not permit the passage of energy. Such an insulating wall can be approached quite easily in practice; a double-walled vacuum flask with silvered walls (Dewar vessel) is a good example. If a system is not isolated from its surroundings it is said to be in **thermal contact** with its surroundings. When a physical system is isolated it is found that it tends toward, and eventually reaches, a state in which no further change is noticeable, no matter how long we wait. The system is then said to be in **thermal equilibrium**. (The adjective "thermal" is usually omitted when discussing equilibrium.)

If two systems that are separately in equilibrium are brought into thermal contact, changes take place in each system until the composite system attains a new state of equilibrium; the two separate systems are said to be **in equilibrium with one another**. Examples are legion: the cooling down of a hot liquid to room temperature or the warming up of a cool liquid to room temperature are two common examples. The **zeroth law of thermodynamics** is a statement concerning equilibrium conditions for more than two systems. It describes such a common phenomenon that it was not thought necessary to formulate it as a law until other basic laws had been discovered, hence the name "zeroth" law. It may be stated in the following way.

If, of three bodies A, B, and C, A and B are separately in equilibrium with C, then A and B are in equilibrium with one another.

We operate under this law every time we take a temperature reading. We assume that if the reading of the thermometer is the same when it is placed in two different liquids, say, then those two liquids would be in thermal equilibrium if they were in thermal contact and we would say that they have the same **temperature**. In order to examine more closely the concept of temperature we shall discuss a simple experiment with two masses of fluid (either a liquid or a gas).

It is a fact of experience that any property of a fluid may be expressed in terms of the pressure P and the volume V of the fluid. For example, in Section 17.2 we developed the relation

$$PV = \frac{2U}{3}$$

for an ideal gas, which gives the dependence of the total energy U on P and V. If we now take two masses of fluid, a standard mass A and a test mass B, and keep A in a fixed state (P_A and V_A constant), we may vary P_B and V_B of the test mass in such a way as to keep A and B always in equilibrium with one another. Since P_A and V_A are fixed we may do this, for example, by choosing a new value for P_B and then adjusting V_B so that A and B remain in equilibrium with each other. As P_B is varied, V_B changes, and a graph of P_B as a function of V_B is as shown in Fig. 23.1. The curved line in Fig. 23.1 is called an **isotherm**.

Suppose now we had chosen instead of A another standard mass A' in equilibrium with A. Then for any state of B corresponding to a point on the isotherm we should have A in equilibrium

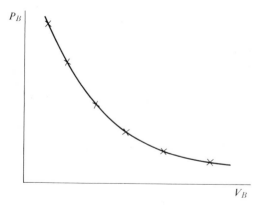

Fig. 23.1. An isotherm of a fluid.

with both A' and B. Therefore, according to the zeroth law, A' and B are also in equilibrium, and the isotherm of B determined from A' would pass through this point. In fact, the isotherms determined by using A or A' are the same and depend only on the fluid comprising system B.

By varying the state of the fluid mass A, a whole family of isotherms may be determined for the fluid mass B, as shown in Fig. 23.2. We now

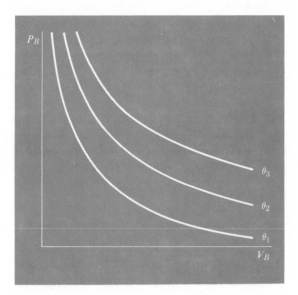

Fig. 23.2. Isotherms of a fluid for different empirical temperatures.

label the isotherms θ_1, θ_2, θ_3, etc., in some systematic way. It is obvious that there must be some relation between P, V, and θ for a fluid, which we may write in the general form

$$f(P, V) = \theta.$$

The quantity θ is called the **empirical temperature**, and the relation between P, V, and θ is the **equation of state of the fluid**. (The relation between empirical temperature θ and the absolute temperature T that we have used in previous chapters will be discussed in Section 23.4.)

The ideal gas is the simplest fluid to consider. Boyle's law (see Section 17.3) says that the isotherms must be given by the relation $PV = $ constant. The constant in Boyle's law must be a function of θ to satisfy the general equation of state given above. It is found experimentally that the required relation is

$$PV = nR\theta,$$

where R is the gas constant introduced in Section 17.3 and n is the number of kilogram-moles of the gas.

The temperature θ is called the empirical temperature because it is chosen on the basis of experiment rather than from fundamental considerations. Several types of thermometers were mentioned in Section 17.3, and two types of empirical temperature scales were introduced—the centigrade (Celsius) scale and the (ideal) gas scale. The (ideal) gas scale is based upon the fact that all gases tend to behave as ideal gases as the pressure approaches zero, so that all constant-volume gas thermometers at the same temperatures give the same reading as the pressure of the gas in them approaches zero. This temperature scale is based on the relation

$$PV = nR\theta,$$

which describes the behavior of an ideal gas.

There is one temperature scale, the **Kelvin or absolute temperature scale**, which is not empirical in nature but is derived from fundamental considerations; we shall discuss absolute temperature in Section 23.4 and show the relation between the empirical temperature as defined by the ideal gas scale and the absolute temperature.

23.3 THE FIRST LAW OF THERMODYNAMICS

The **first law of thermodynamics** is just a statement of energy conservation. In Section 17.6 we introduced the concept of internal energy for a gas as a product of the mean energy of the molecules in the gas and the total number of molecules. We also studied the behavior of a gas in a cylinder fitted with a movable frictionless piston as in Fig. 23.3 and showed that when heat is added to the gas, the gas does an amount of work

$$dW = P\,dV$$

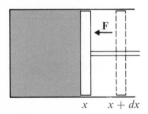

Fig. 23.3. A gas in a cylinder must do work in moving the piston against an external force.

in pushing out the piston against the external force F upon it (Section 17.6). It is also observed that the internal energy of the gas increases by an amount dU. Applying conservation of energy, we then wrote the general expression

$$dQ = dU + dW$$

or

$$dU = dQ - P\,dV$$

for an ideal gas. These equations are a mathematical statement of the **first law of thermodynamics**, which can be stated simply: energy is conserved if heat is taken into account. The equivalence of heat and mechanical work has been known for some time; you have only to rub your hands together vigorously to demonstrate that such a connection exists.

There are some important restrictions upon the use of the equation $dQ = dU + dW$ that should be stated now. The internal energy of a fluid may be changed in many ways, such as by doing mechanical work on it with, say, a paddlewheel, by putting heating coils into it and passing current through them, by introducing a hot or cold object into it, etc. In some cases only work is done to change the internal energy of the fluid, or only heat is added to change the internal energy, or as is common, a combination of work done and heat added changes the internal energy. The conclusion is that for any given change dU in internal energy, there are many combinations of dQ and dW that will effect the change dU. What this means is that the equation $dQ = dU + dW$ cannot necessarily be integrated, because dQ and dW are not uniquely determined. However, in the case of a gas in a cylinder fitted with a frictionless piston we showed that $dW = P\,dV$. Since P and V are parameters that define the equation of state of the gas, we can consider dW and hence dQ to be uniquely defined, and in this case $dU = dQ - P\,dV$ can be integrated. The necessary condition is that the piston be frictionless so that no energy is lost when it moves. This is an example of a system that can undergo a **reversible process**. A reversible process is an idealized one that may be exactly reversed by an infinitesimal change in the external conditions. The internal energy U, the pressure P, and the volume V are known as **state functions**, since their values for a particular state of a system depend only on the state and not upon how the system arrived in that state.

Example. A quantity of 8 J of heat is added to a gas in a cylinder fitted with a piston. (a) If the internal energy of the gas increases by 4 J, what amount of work is done by the gas? (b) If the pressure in the fluid remains constant and equal to atmospheric pressure during the change of state, what change in volume will occur?

Solution
(a) We write

$$\Delta U = Q - W$$

for a finite change in the state of the system. Therefore,

$$W = Q - \Delta U$$

$$= 8 - 4$$

$$= 4\,\text{J}.$$

The work done by the gas is 4 J and is a positive quantity.
(b) Since $dW = P\,dV$ for a gas,

$$W = \int_{V_1}^{V_2} P\,dV$$

$$= P \int_{V_1}^{V_2} dV$$

$$= P(V_2 - V_1)$$

if P remains constant. Therefore,

$$V_2 - V_1 = \frac{W}{P}\,\text{m}^3.$$

Atmospheric pressure is $1.0 \times 10^5\,\text{N}\cdot\text{m}^{-2}$, and W is 4 J from (a), so that

$$V_2 - V_1 = \frac{4}{1.0 \times 10^5} = 4 \times 10^{-5}\,\text{m}^3.$$

The volume change is $4.0 \times 10^{-5}\,\text{m}^3$ or $40\,\text{cm}^3$.

Historically, the first law usually was stated in a negative way, as follows: it is impossible to construct a machine that operates in a cycle and does a net amount of work on its surroundings without obtaining energy from some external source. This form of the first law states the impossibility of building a **perpetual motion machine of the first kind**, one that violates the principle of conservation of energy.

23.4   THE SECOND LAW OF THERMODYNAMICS

We have already put a rather severe restriction on the possible changes of state that may occur in a system through the first law; only those that conserve energy are permitted. However, there are many possible processes that do conserve energy but do not occur in practice. The **second law of thermodynamics** expresses an essential asymmetry of nature. For example, gases always spontaneously mix and never unmix; gases always seep through an opening from a region of high pressure to a region of low pressure; heat always flows from a hot object to a cold object; rocks always weather and crumble; living organisms always grow old.

In Section 20.4 we introduced the entropy that is another state function of a system. The entropy change dS that occurs when an amount of heat dQ is added to a system that undergoes a reversible change and whose temperature is T is defined as

$$dS = \frac{dQ}{T}.$$

Moreover, the entropy of an isolated system is always observed to increase in any real process involving changes of state within the isolated system; for an idealized reversible process the entropy remains constant. The second law of thermodynamics may be stated in the following manner.

Only those processes can occur in an isolated system for which the entropy of the system increases or remains constant.

Historically, long before the second law was formulated, it was known that mechanical energy could be converted completely into heat at will (by means of friction, for example) but that only part of a given amount of heat could be converted into mechanical energy (by doing work) even in machines with very little friction. The possible efficiencies of machines, such as steam engines, for the conversion of heat into work were studied by Sadi Carnot long before the second law was formulated. For these reasons early statements of the second law were phrased in terms of the impossibility of certain machines. One form due to Rudolf Clausius says the following.

It is impossible to construct a device that, operating in a cycle, will produce no effect other than the transfer of heat from a cooler to a hotter body.

A later form due to Lord Kelvin and Max Planck states:

It is impossible to construct an engine that operates in a cycle and produces no effect other than the extraction of heat from a reservoir and the performance of an equal amount of work.

A machine such as that described in the Kelvin–Planck statement of the second law would have an efficiency of 100% for the conversion of heat into work. Note, however, that it would not violate the first law. A machine that violates the second law is called a **perpetual motion machine of the second kind**.

There is another statement of the second law, not obviously related to any of the other statements, which is due to Carathéodory. Before stating it, however, we must introduce the **adiabatic process**, which is a change of state during which no heat is gained or lost from the system. The formulation of Carathéodory is:

In the neighborhood of any equilibrium state of a system there are states that are inaccessible by an adiabatic process.

As the state of a fluid is changed adiabatically, the pressure and volume change continuously, and an **adiabatic line** is traced out in a plot of P versus V as in Fig. 23.4. An isotherm of the fluid is drawn for comparison. It is a fact of experience that adiabatics (lines of constant heat) always have steeper slopes than isotherms (lines of constant temperature). Suppose we consider a point (P_1, V_1) in Fig. 23.4. The second law tells us, upon application to the fluid in question, what other states characterized by other values of P and V can or cannot be reached by processes taking place in the fluid. For example, the statement due to Carathéodory

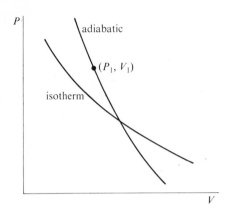

Fig. 23.4. An isothermal line and an adiabatic line for a fluid.

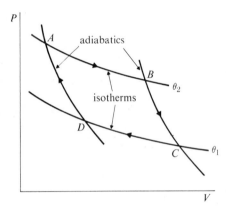

Fig. 23.5. The Carnot cycle.

tells us, after a bit of development a little too involved for this discussion, that only those states whose coordinates P and V lie on the adiabatic line through (P_1, V_1) can be reached by an adiabatic process and that there is only one adiabatic line passing through (P_1, V_1).

In discussions of the consequences of the second law it is useful to make use of an idealized process known as the **Carnot cycle**, which is the basis of a special engine devised by Carnot in his study of the efficiencies of machines. A Carnot cycle is presented diagrammatically in Fig. 23.5. The fluid is originally in a state corresponding to point A in Fig. 23.5. The state is then changed isothermally from A to B, adiabatically from B to C, isothermally from C to D, and adiabatically from D to A, the original state, which completes the cycle. During the isothermal transition A to B an amount of heat Q_2 is taken from a reservoir at temperature θ_2, whereas during the isothermal transition C to D an amount of heat $-Q_1$ is lost to a reservoir at temperature θ_1. By very general arguments based upon the first and second laws[1], it is possible to show that

$$-\frac{Q_1}{Q_2} = \frac{f(\theta)|_{\theta=\theta_1}}{f(\theta)|_{\theta=\theta_2}}$$

for the Carnot cycle, where $f(\theta)|_{\theta=\theta_1}$ and $f(\theta)|_{\theta=\theta_2}$ are related by some function to the constant temperatures θ_1 and θ_2.

At this point we introduce the **absolute temperature scale** by writing

$$T = f(\theta),$$

where T is the absolute temperature. Therefore, for any Carnot cycle we write

$$-\frac{Q_1}{Q_2} = \frac{T_1}{T_2},$$

where the minus sign merely indicates that heat is lost during one part of the cycle. This temperature scale is called absolute because it does

[1] A. B. Pippard, *Classical Thermodynamics* (New York: Cambridge University Press, 1961), Chapter 4.

not depend on the properties of a particular substance as other scales do. The unit of absolute temperature is the Kelvin degree (°K). There is still an arbitrary constant of proportionality available, since we have only a ratio of temperature at the moment.

It can be inferred from experiments with real gases that

$$\frac{dP}{dT} = \frac{P}{T}$$

for an ideal gas at a constant volume. Therefore, we can write

$$P = Tf(V),$$

where $f(V)$ is some function of the volume V of the system; that is, the pressure is always proportional to the absolute temperature, but the constant of proportionality depends upon the volume. However, we have seen in Section 17.3 that Boyle's law requires that $PV = \text{constant}$, or

$$P \propto \frac{1}{V}$$

for an ideal gas. Therefore, we must have $f(V)$ of the form

$$f(V) \propto \frac{1}{V}$$

or

$$f(V) = \frac{b}{V}$$

and

$$PV = bT,$$

where b is a constant. The ideal gas scale of temperature is defined by the equation

$$PV = R\theta,$$

so that $\theta/T = \text{constant}$. By choosing the same fixed points we can put $\theta = T$, and the equation of state becomes

$$PV = RT,$$

where R is the gas constant. Since we often work with the number of kilogram-moles (kg-mole) of gas, we often write

$$PV = nRT,$$

where R is now the gas constant per kilogram-mole of gas.

Entropy S is defined so that $dS = dQ/T$ or

$$dQ = T\,dS$$

for a reversible process. If we substitute for dQ in $dU = dQ - P\,dV$ (which is the statement of the first law for a simple fluid) for a reversible change, we have

$$dU = T\,dS - P\,dV,$$

where U, T, S, P, and V are all state functions. Therefore, dU, dS, and dV are all uniquely defined, and the equation may be integrated once the functional forms of the variables are known.

Example. One gram-mole of an ideal gas in a cylinder has a pressure of 5×10^3 $\text{N} \cdot \text{m}^{-2}$ and a volume of $0.5\,\text{m}^3$. The gas undergoes a reversible isothermal change to a final volume of $1.0\,\text{m}^3$ when 3 J of heat are added to the system.
(a) What is the final pressure and the temperature?
(b) What is the change in entropy of the gas?
(c) What is the change in internal energy of the gas?
(d) What is the work done by the gas?

Solution
(a) For an ideal gas $PV = nRT = RT$ for $n = 1$ g-mole. The isothermal process is indicated on a P-V diagram in Fig. 23.6. The gas, initially at pressure $P_1 = 5.0 \times 10^3\ \text{N} \cdot \text{m}^{-2}$ and volume $V_1 = 0.5\,\text{m}^3$, has a final volume of $V_2 = 1.0\,\text{m}^3$ and a final pressure of P_2. Initially we have

$$P_1 V_1 = RT,$$

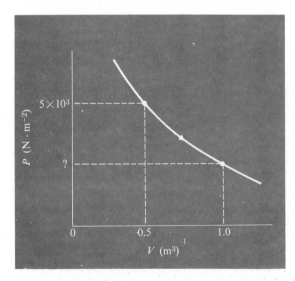

Fig. 23.6. An isothermal line for an ideal gas.

and finally we have

$$P_2 V_2 = RT.$$

Therefore,

$$P_1 V_1 = P_2 V_2$$

$$P_2 = \frac{P_1 V_1}{V_2}$$

$$= 5.0 \times 10^3 \left(\frac{0.5}{1.0}\right)$$

$$= 2.5 \times 10^3 \, \text{N} \cdot \text{m}^{-2}.$$

Therefore, the final pressure is $2.5 \times 10^3 \, \text{N} \cdot \text{m}^{-2}$. The temperature is

$$T = \frac{P_1 V_1}{R} = \frac{5.0 \times 10^3 \times 0.5}{8.32} = 300°\text{K}.$$

(b) Since $T = $ constant, the change in entropy of the gas is

$$\Delta S = \frac{\Delta Q}{T} = \frac{3}{300} = 10^{-2} \, \text{J} \cdot (°\text{K})^{-1}.$$

(c) For an ideal gas $PV = $ constant during an isothermal change. But $PV = 2U/3$ for an ideal gas. Therefore, the **internal energy U is constant during an isothermal change**.

(d) Since $dU = 0$, the first law becomes

$$dQ = P \, dV$$

or upon integration,

$$\Delta Q = \int P \, dV = \text{work done.}$$

Therefore, the gas does 3 J of work during the isothermal expansion.

In the last section we wrote the first law in the form

$$dQ = dU + dW$$

or

$$dU = dQ - dW$$

for differential changes in the state of a system and indicated that, in general, dQ and dW are not uniquely defined. For a *reversible* change in a fluid the first law becomes

$$dU = T \, dS - P \, dV,$$

where U, T, S, P, and V are all state variables. Since this last equation states a relation between quantities that are all state functions, we can state that it is applicable to *any* differential change, reversible or irreversible. It is *only for reversible changes*, however, that

$$dQ = T\,dS \quad \text{and} \quad dW = P\,dV,$$

and *only for irreversible changes* that

$$dQ \neq T\,dS \quad \text{and} \quad dW \neq P\,dV,$$

but we can still use the relation

$$dU = T\,dS - P\,dV$$

to determine the change in U during the irreversible change because U has a unique value for all states of the fluid.

These same considerations can be applied to systems other than fluids in which other types of work are involved. For example, if a solid metallic bar is held under a force of tension F that gives rise to an increase in length dl, the work done by the external force on the bar is $F\,dl$, and we may write

$$dU = dQ + dW$$
$$= T\,dS + F\,dl$$

for the change in internal energy of the bar. This equation may be applied even if the extension is not reversible. All that is required is that the stress be uniquely related to the strain. However, only for an elastic (reversible) process can we write

$$dQ = T\,dS \quad \text{and} \quad dW = F\,dl.$$

Example. The efficiency η of a heat engine is defined as

$$\eta = \frac{\text{work output}}{\text{heat input}}.$$

Develop an expression for the efficiency of a Carnot engine (that is, an engine operating in a Carnot cycle).

Solution. During a Carnot cycle (refer to Fig. 23.5), heat Q_2 is taken from a reservoir at temperature T_2, and a heat $-Q_1$ is ejected into a reservoir at temperature T_1. The net heat absorbed by the engine is

$$Q = Q_2 - (-Q_1)$$
$$= Q_2 + Q_1$$

(recall that Q_1 is negative due to our sign convention). Writing the first law in the form

$$\Delta U = Q - W$$

and observing that $\Delta U = 0$ in a cyclic process, we have

$$W = Q$$

$$= Q_1 + Q_2$$

for the work done during the Carnot cycle. The heat input to the engine is the heat Q_2 absorbed at temperature T. Therefore, the efficiency of the Carnot engine is

$$\eta = \frac{W}{Q_2}$$

$$= \frac{Q_2 + Q_1}{Q_2}$$

$$= 1 + \frac{Q_1}{Q_2}$$

or

$$\eta = 1 - \frac{T_1}{T_2},$$

since

$$\frac{Q_1}{Q_2} = -\frac{T_1}{T_2}.$$

Example. Show that no heat engine operating in cycles between two reservoirs at constant temperature can have a greater efficiency than a reversible (Carnot) engine operating between the same two reservoirs.

Solution. Suppose that the engine absorbs heat Q_2' at temperature T_2 and ejects heat $-Q_1'$ at temperature $T_1 < T_2$ while doing work W' and that the Carnot engine absorbs heat Q_2 at T_2, ejects heat $-Q_1$ at T_1, and does work W. Since both engines work in cycles,

$$\Delta U = \Delta U' = 0$$

and, from the first law,

$$W' = Q_2' - (-Q_1')$$

$$= Q_2' + Q_1'$$

$$W = Q_2 - (-Q_1)$$

$$= Q_2 + Q_1.$$

Let us assume that the engines are built or operated so that $W' = W$. Since the Carnot engine is reversible, we can reverse its mode of operation (making it a **refrigerator**), and it can be driven by the (supposedly) higher efficiency heat engine. Putting $W' = W$, we have

$$Q_2' + Q_1' = Q_2 + Q_1$$

$$Q_2 - Q_2' = Q_1' - Q_1$$

$$= -(Q_1 - Q_1').$$

Since we assume the heat engine to have higher efficiency than the Carnot engine but to produce the same work, we must have

$$Q_2 > Q_2' \quad \text{and} \quad Q_1 > Q_1',$$

since the heat engine will take in less heat at temperature T_2 and eject less heat at temperature T_1. However, the Carnot engine has been reversed so that it now absorbs heat at T_1 and ejects heat at T_2. Since

$$Q_2 - Q_2' = -(Q_1 - Q_1')$$

and

$$Q_2 > Q_2', \qquad Q_1 > Q_1',$$

the coupled engines are performing the sole function of absorbing heat $Q_1 - Q_1'$ at temperature T_1 and ejecting heat $Q_2 - Q_2'$ at temperature T_2 in violation of the second law of thermodynamics. Therefore, no heat engine operating in cycles between two reservoirs can have a greater efficiency than a Carnot engine operating between the same two reservoirs.

23.5   ABSOLUTE ZERO AND THE THIRD LAW OF THERMODYNAMICS

As a result of experiments with the liquefying and freezing of various gases, it has been discovered that temperatures far lower than those experienced in nature can be achieved in the laboratory. There does, however, appear to be a natural lower limit to the range of temperatures we can produce. As techniques for attaining low temperatures are perfected, the lower limit is approached more and more closely, but it has never been reached. In establishing the absolute scale of temperatures a decision was made to put $T = 0$ at this experimental lower limit of observable temperatures. Since the absolute temperature scale is a linear one, only one other fixed point is necessary in order to fix the slope of the line. The point chosen is that temperature at which a mixture of ice, water, and water vapor are in equilibrium with each other. This temperature is taken to be 273.16°K. The reasons for this particular value are historical and not of fundamental importance except insofar as this value allows for almost exactly 100 degrees between the freezing points and boiling points of water. One statement of the **third law of thermodynamics** is suggested by experiment.

The absolute zero of temperature can never be reached in any physical process.

An alternative statement of the third law arises from a discussion of entropy changes in reversible processes and can be stated in the following way.

As the temperature tends to zero, the magnitude of the entropy change in any reversible process tends to zero.

Room temperature on the absolute scale is about 300°K. Helium liquefies at 4.2°K and may be cooled further to about 1°K by reducing the pressure over the liquid[2] and letting it boil at a lower temperature. If the liquid helium is isolated from its surroundings, the heat necessary to produce helium vapor in boiling must come from the liquid whose temperature then falls. One might be tempted to ask whether that isn't just about low enough. For many materials the answer is yes, but for some systems even 1°K is a very high temperature in the sense that they are still in a state of almost complete disorder, whereas at 0°K all systems must be in their lowest energy states and therefore in a completely ordered state or, in a few cases, as ordered as it is possible for them to be ordered. In many cases, the properties of the system depend upon the logarithm of the temperature, and, on a log scale, a temperature of 0.003°K is as far below 1°K as 1°K is below 300°K. Therefore, there is much information of physical significance to be obtained by conducting experiments at temperatures below 1°K.

One way to obtain temperatures below 1°K is by a method known as **magnetic cooling**[3]. The basis of the method is the existence of substances called paramagnetic salts that may be appreciably magnetized only in large magnetic fields and at low temperatures. In a magnetic-cooling experiment a quantity of paramagnetic salt, together with any other material one wishes to investigate, is cooled to about 1°K by thermal contact with liquid helium. A large magnetic field is applied, causing the paramagnetic salt to become magnetized. During the magnetization process heat is evolved and is allowed to drain off into the liquid helium. Thermal contact with the helium is then broken, the magnetic field is removed and the magnetization tends to disappear. The demagnetization process requires energy that must be supplied by the paramagnetic salt itself, since it is now isolated from its surroundings. As a result the salt gives up heat energy, and its temperature falls to a low value of the order of 10^{-3} °K. By an extension of this technique, temperatures of the order of 10^{-5} °K have been reached.

23.6 ADIABATIC PROCESSES

An **adiabatic change** is one involving no heat gain or loss by the system; therefore the entropy remains constant. The equation of state of

[2] D. K. C. MacDonald, *Near Zero* (Garden City, N.Y.: Doubleday and Company, Inc., 1961).

[3] N. Kurti, "Nuclear Orientation and Nuclear Cooling," *Physics Today*, March, 1958.

a fluid is a relation involving the pressure P, the volume V, and the temperature T. It is also possible to write down a relation between P and V for an adiabatic change; all that is required is a knowledge of the variation of P and V at constant entropy. We shall derive this relation for an ideal gas after we review a few ideas about specific heats.

In Section 17.6 we determined values for the specific heat of an ideal gas kept at constant pressure or at constant volume using conservation of energy. The first law for a fluid written in differential form is

$$dU = dQ - P\,dV$$

or

$$dQ = dU + P\,dV.$$

Therefore, the molar specific heat at constant volume is

$$C_v = \frac{1}{n}\left(\frac{dQ}{dT}\right)_V$$

$$= \frac{1}{n}\left(\frac{dU}{dT}\right)_V,$$

where the subscript V outside the parentheses means that the quantity inside the parentheses is to be evaluated under conditions of constant volume. The specific heat at constant pressure is

$$C_p = \frac{1}{n}\left(\frac{dQ}{dT}\right)_P$$

$$= \frac{1}{n}\left(\frac{dU}{dT}\right)_P + \frac{P}{n}\left(\frac{dV}{dT}\right)_P,$$

where the quantities inside the parentheses are now evaluated at constant pressure. Alternatively, since we have $dQ = T\,dS$, we may write

$$C_v = \frac{T}{n}\left(\frac{dS}{dT}\right)_V, \qquad C_p = \frac{T}{n}\left(\frac{dS}{dT}\right)_P.$$

The specific heats of a fluid are related to the change in entropy with temperature of the fluid under conditions of constant pressure or constant volume.

In Section 17.6 we made extensive use of the **ratio of specific heats** γ defined as

$$\gamma = \frac{C_p}{C_v}$$

in our discussion of the behavior of real gases. We see now that γ may be written very generally for a fluid as

$$\gamma = \frac{C_p}{C_v} = \frac{\left(\dfrac{dS}{dT}\right)_P}{\left(\dfrac{dS}{dT}\right)_V}.$$

We can now determine the adiabatic equation for an ideal gas by noting that $dQ = 0$ for an adiabatic process, so that $dU = -P\,dV$ from the first law. Since $nC_v = dU/dT$ (omitting subscripts for simplicity), we have

$$\frac{dU}{dT} = -\frac{P\,dV}{dT} = nC_v$$

or

$$dT = \frac{-P}{nC_v}\,dV.$$

For an ideal gas, the equation of state is

$$PV = nRT;$$

therefore,

$$P\,dV + V\,dP = nR\,dT$$

$$= \frac{-P}{C_v}R\,dV.$$

But $R = C_p - C_v$ for an ideal gas (see Section 17.6). Therefore,

$$P\,dV + V\,dP = -P(C_p - C_v)\frac{dV}{C_v}$$

or

$$P\,dV + V\,dP = -P(\gamma - 1)\,dV$$

$$V\,dP + P\gamma\,dV = 0.$$

Dividing by PV yields

$$\frac{dP}{P} + \gamma\frac{dV}{V} = 0;$$

integrating gives

$$\ln P + \gamma \ln V = \text{constant}$$

or

$$PV^\gamma = \text{constant}.$$

This is the **adiabatic equation** for an ideal gas.

Example. One gram-mole of a monatomic ideal gas initially at a pressure of $5 \times 10^3 \, \text{N} \cdot \text{m}^{-2}$ in a volume of $0.5 \, \text{m}^3$ undergoes an adiabatic expansion to a volume of $1.0 \, \text{m}^3$.

(a) What is the final pressure?

(b) What are the initial and final temperatures?

Solution

(a) For a monatomic ideal gas $C_p = \frac{5}{2}R$ and $C_v = \frac{3}{2}R$ independent of temperature, so that $\gamma = 1.67$. Since $PV^\gamma = $ constant, we have

$$P_1 V_1^\gamma = P_2 V_2^\gamma$$

or

$$P_2 = P_1 \left(\frac{V_1}{V_2}\right)^\gamma$$

$$= 5 \times 10^3 \left(\frac{0.5}{1.0}\right)^{1.67}$$

$$= 1.57 \times 10^3 \, \text{N} \cdot \text{m}^{-2}.$$

(b) The gas is ideal, so that $PV = nRT$. Therefore,

$$T_1 = \frac{P_1 V_1}{nR}$$

$$= \frac{5 \times 10^3 (0.5)}{1 \times 8.32}$$

$$= 300°\text{K},$$

and

$$T_2 = \frac{P_2 V_2}{nR}$$

$$= \frac{1.57 \times 10^3 \times 1.0}{8.32}$$

$$= 189°\text{K}.$$

The temperature change is $121°\text{K}$. The adiabatic process AB is plotted in Fig. 23.7. As a comparison, isotherms for $T = 300°\text{K}$ and $T = 189°\text{K}$ are also shown. The isothermal process AC is the one calculated in the first example in Section 23.4. Note that the adiabatic line has a steeper slope than the isothermal lines.

23.7 PHASE CHANGES

We are all familiar with the fact that matter exists in various forms dependent upon the temperature. Water, for example, has a solid form (ice) and a gaseous form (steam or water vapor), as well as its usual liquid form. This behavior is not always as readily observed for a given material as it is for water. Plastics, for example, gradually soften as the temperature

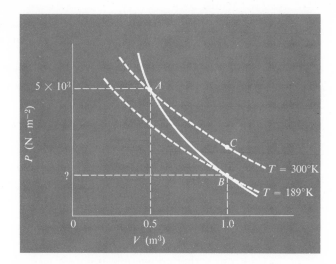

Fig. 23.7. An adiabatic line AB for an ideal gas. The dashed lines through A and B are isothermal lines for $T = 300°$K and $T = 189°$K, respectively.

rises, but it is difficult to measure any particular melting point for the material. Dirt and rock appear rather inert normally, but we know that the molten lava spewed forth by volcanoes is actually liquid rock. The gases in the air liquefy and eventually freeze as the temperature nears absolute zero, with the rather important exception (in that it does not freeze) of helium, which we will discuss later in this chapter.

This behavior is easy to picture from the molecular point of view. Figure 23.8 is the graph introduced in Section 17.2 showing the variation in the force between molecules in a gas. For low pressures and high temperatures, the molecules in a gas are, on the average, separated by distances much greater than R_0 and move about essentially independently of each other, being contained only by the walls of the vessel enclosing them. As the temperature decreases, the average energy of the molecules and therefore their average speed decreases, and the effect of attractive forces between molecules becomes more important. The effect is reinforced if the pressure increases or the volume decreases at the same time. The net result is that the molecules spend more time making collisions with each other until finally the attractive forces between them on the average are so large that the gas collapses into a much more compact

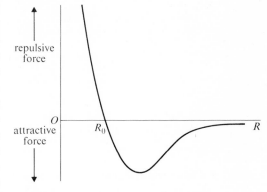

Fig. 23.8. The variation of the force between two molecules as a function of separation.

assembly of molecules that we call a liquid. The molecules are still free to move about in the liquid but cannot easily escape from the surface of the liquid because the combined force of their neighbors pulls them back.

As the temperature drops still further a point is finally reached when the available energy of motion is not sufficient to keep the individual molecules moving freely, and the liquid becomes a solid wherein the molecules are fixed into some sort of regular structure and are only free to vibrate about their mean positions in the structure. The transition from one form (solid, liquid, or vapor) to another is known as a **change in phase**. The preceding picture is entirely qualitative in nature; a detailed quantum mechanical analysis is necessary to predict the behavior of any particular system on the basis of its constituent molecules. From a thermodynamic point of view, however, changes of phase may be studied successfully even as we ignore the molecular nature of matter.

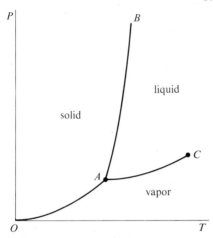

Fig. 23.9. A phase diagram typical of most simple substances.

Figure 23.9 is a **phase diagram** typical of most simple substances. The substance exhibits vapor, liquid, or solid properties depending upon the pressure and temperature. The point A is the **triple point** at which the solid, liquid, and vapor phases exist together in equilibrium. Along the line OA the solid and vapor phases are in equilibrium. If the system is at a pressure and temperature that correspond to a point on this line, then an increase in pressure would cause the vapor to solidify, and a decrease in pressure would cause the solid to vaporize. Also, an increase in temperature would cause the solid phase to vaporize, and a decrease in temperature would cause the vapor to solidify. The line AB is the **solid-liquid equilibrium line**, and the line AC is the **liquid-vapor equilibrium line**. The solid-liquid equilibrium line is observed experimentally to extend without end as P and T increase. The liquid-vapor line AC, however, terminates at a well-defined pressure and temperature known as the **critical pressure** and **critical temperature**, respectively. For combinations of pressure and temperature that take the system beyond this **critical point**, there is no sharp differentiation between the liquid and vapor phases. That is, it is possible to change the vapor

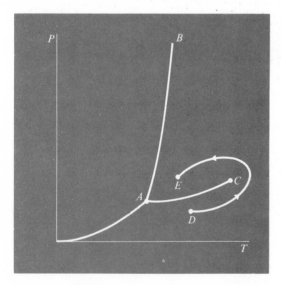

Fig. 23.10. The vapor phase may be changed to the liquid phase by a process which follows the path DE during which there is no abrupt change of phase.

into a liquid without going through an observable condensation by choosing a path for the process that does not intersect the line AC; such a path DE is shown in Fig. 23.10.

There are a few substances that have a phase diagram such as that shown in Fig. 23.11; water is an important example. The solid-liquid equilibrium line for these substances has a negative slope rather than the more normal positive slope of Fig. 23.9. The result is that water, for example, expands upon freezing, whereas most substances contract upon freezing.

Phase transitions may also be looked at on a P-V diagram. Figure 23.12 shows the isotherms for a typical substance in the region of liquid-vapor equilibrium. Considering the isotherm $ABCD$, we see that at D we have a vapor at low pressure and large volume. As P increases and V

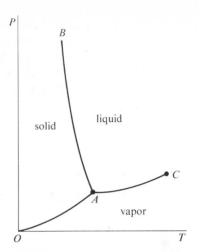

Fig. 23.11. A phase diagram for a substance such as water which expands on freezing.

falls the point C is approached. Between C and B, P remains constant as V decreases; the vapor is condensing to a liquid with a subsequent decrease in volume but no change in pressure. At B all the vapor has become liquid, and a further decrease in volume requires considerable increase in pressure. This behavior is what we normally observe. The isotherm

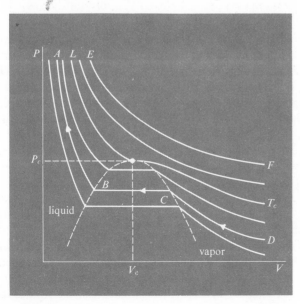

Fig. 23.12. Isotherms in the region of liquid-vapor equilibrium showing the critical isotherm L.

EF, however, has no region of constant pressure and therefore there is no liquid-vapor phase change observable along *EF*. The isotherm *L* marks the transition point between these two isotherms; it passes through the critical point of pressure P_c and volume V_c, at which point it has zero slope. The temperature corresponding to this isotherm is the critical temperature T_c.

There is an approximate empirical relation, namely

$$T_l = \tfrac{3}{5}T_c,$$

where T_l is the temperature at which a gas liquefies at normal atmospheric pressure. This is a useful equation in thermodynamic problems. The validity of this relation is illustrated in Table 23.1.

Table 23.1 Liquefaction temperature as deduced from critical temperatures

Gas	T_c (°K)	T_l (*from* T_c) (°K)	T_l (*experiment*) (°K)
Hydrogen (H_2)	33.3	20	20
Argon (Ar)	151	90	87
Methane (CH_4)	190.7	114	109
Ethylene (C_2H_4)	282.3	169	170

When a change of phase occurs, energy in the form of heat is either liberated or absorbed by the substance undergoing the change of phase. Heat is absorbed during the solid-to-liquid, liquid-to-vapor, and solid-to-vapor transformations, whereas heat is liberated during the reverse transformations. The reason for this behavior is easy to see. For the solid-to-liquid transformation energy is required to free the atoms of the solid from their fixed positions in the lattice. The pressure and temperature remain constant during the change of phase, for all heat energy coming in is used to separate the atoms, which results only in an increase in volume. In the reverse transformation from liquid to solid excess energy of motion is liberated in the form of heat when the atoms fall into their fixed positions in the lattice. The same sort of behavior holds true in the liquid-to-vapor and solid-to-vapor phase changes. The heat absorbed or released during a change of phase is known as a **latent heat of transformation** and has units of joules per kilogram or joules per kilogram-mole.

Plots of *P*, *V*, and *T* in three dimensions form *PVT* surfaces characteristic of the material in question. Figures 23.9, 23.10, and 23.12 are two-dimensional slices through such three-dimensional surfaces. Readers who

are interested in the appearance of such *PVT* surfaces should consult
F. W. Sears[4].

23.8 HELIUM

Helium is a unique substance in that it does not become a solid
even at 0°K unless it is subjected to very high pressure[5]. Naturally occurring helium consists entirely of ^{4}He (two protons and two neutrons in
the nucleus), with one part in 10^5 of ^{3}He (two protons and one neutron).
The phase diagram for ^{3}He is shown in Fig. 23.13. Note that there is no
triple point. ^{3}He will not solidify until pressures of the order of 30 atm
are applied at extremely low temperatures. The reason for this is essentially quantum-mechanical in nature and will be discussed further later
in this section.

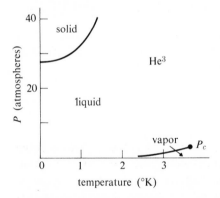

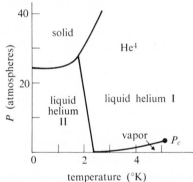

Fig. 23.13. The phase diagram of ^{3}He;
^{3}He remains liquid even at 0°K.

Fig. 23.14. The phase diagram for ^{4}He;
the liquid phase comes in two forms—an
ordinary fluid (Helium I) and a superfluid
form (Helium II).

The phase diagram for ^{4}He is shown in Fig. 23.14. It is very similar
to that for ^{3}He but has one remarkable difference: the liquid occurs in
two forms labeled I and II. Liquid helium I is not unusual in any sense,
but as the temperature of the liquid is reduced a sharp transition to the
helium II form occurs. Helium II has unique properties; it has such high
heat conductivity and such extremely low frictional effects that it is

[4] F. W. Sears, *Thermodynamics, The Kinetic Theory of Gases, and Statistical Mechanics*
(Reading, Mass.: Addison-Wesley Publishing Co., Inc., 1959), Chapter 6.

[5] See "Liquid Helium," a film available from the Instructional Media Center, Michigan
State University, East Lansing, Mich.

termed a **superfluid**. For example, it can flow with perfect ease through channels a few millionths of a centimeter wide.

As we outlined in the last section, when we cool a gas at constant pressure we should expect it to liquefy at some temperature and then to solidify at some lower temperature (normally greater than 50% of the temperature of liquefaction as measured in degrees Kelvin). As the solid is cooled toward 0°K we might expect that all motion would slow down, ceasing at 0°K. However, we know from the uncertainty principle (Section 18.5) that the product of the uncertainty Δx in the position of a particle and the uncertainty Δp in the momentum of the particle is given by

$$\Delta x \, \Delta p \geqslant \frac{h}{4\pi}$$

where h is Planck's constant. This relation means that even at 0°K the particle cannot be motionless, since neither Δp nor Δx can become zero, and the particle must have some energy (of vibration) even at 0°K. This is known as the **zero-point energy**.

Now helium atoms are very light and have very weak forces acting between them, since they are particularly stable configurations of protons and electrons (the reason for this is quantum-mechanical in nature and need not concern us here). This combination of very small mass and very weak attractive forces is not strong enough to bind the helium atoms together into a solid in the presence of the zero-point energy. In other words, if helium atoms try to coalesce into a solid structure at low pressures, the zero-point energy vibrations cause them to fly apart again. Only if some outside force is applied to the liquid helium by raising the pressure can a solid phase be formed. Since ^{3}He has only 75% of the mass of ^{4}He, we should expect that more pressure would be needed to overcome the zero-point vibrations; from Figs. 23.13 and 23.14 we see that this is so.

The explanation of the superfluid properties of helium II is thought to lie in the fact that ^{4}He is a boson (see Section 20.6), and, as the temperature falls below 2.182°K at normal pressure, the ^{4}He atoms start to fall into their lowest energy states, where they all have almost the same energy and *almost the same momentum*. This behavior is known as an **Einstein condensation**. At temperatures below 2.18°K, almost all the ^{4}He atoms are in their ground states, in which the atoms have almost the same momentum and the liquid exhibits a **cooperative behavior** and flows much more freely than a normal liquid.

An ordinary liquid flowing through a capillary tube encounters frictional effects in which energy is transferred to the liquid due to interactions between atoms in the liquid and atoms in the wall of the tube. Since energy is quantized this energy transfer takes place in terms of quanta or bundles of energy. If almost all the atoms in helium II are in their ground states, energy can be added only if a transition to a higher state occurs. Since the temperature is so low, there is very little thermal energy available, and it is difficult to transfer energy to the fluid. In other words, at very low temperatures it may well be impossible to produce energy changes in helium II that will allow energy transfer to take place, and if that is so, there just cannot be any friction. There are many other particles that are bosons (all even-mass-number nuclei, for example), but they are all solids at temperatures at which they would be likely to exhibit superfluidity. ^{4}He is unique in this respect.

23.9   SUPERCONDUCTIVITY

Certain metals, for example, aluminum, lead, mercury, niobium, tantalum, and tin, have the remarkable property of losing all trace of electrical resistance when cooled below a certain temperature characteristic of the metal. The disappearance of resistance, called **superconductivity**, is virtually instantaneous if the metal is of very high purity[6]. Lattice defects or chemical impurities in the metal will cause the resistance to disappear gradually over a temperature interval of up to one degree. The transition temperature is less than about 8°K for elementary metals but can be higher for compounds (about 10°K for a compound of niobium and tin). If a magnetic field is applied to the metal the transition temperature becomes lower as the strength of the field increases. This is illustrated in Fig. 23.15 for lead.

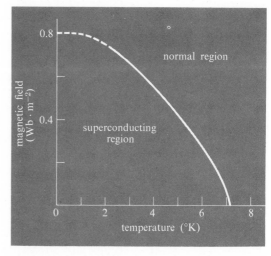

A superconducting sample behaves as if there is no magnetic field within it. At the transition temperature any magnetic field present is expelled. This is known as the **Meissner effect**. It is an essential property of the superconducting state.

Thermodynamically, superconductivity is of interest because it is an example of a phase change involving three independent variables (in this case, pressure, temperature, and magnetic field). At this time, however, we want only to discuss the reason for the phenomenon of superconductivity in terms of the atomic nature of matter.

Fig. 23.15. The transition temperature for lead decreases from 7.1°K as the magnetic field increases.

---

[6] See "An Introduction to Superconductivity," a film available from the Educational Development Center, Newton, Mass.

In an ordinary metal, **resistance** is offered to the flow of current (which is due to "drift" motions of conduction electrons superimposed on their random thermal motions, which causes a net flow of charge in the metal) due to thermal motions of the atoms in the lattice and to impurities and defects in the metal, all of which impede the drift of the electrons. As a metal is cooled down from room temperature, the resistance decreases, since thermal motions slow down. As $T \to 0$ we should expect the resistance to reach a minimum value determined by impurities and defects in the metal. This **residual resistance** is observed in some metals and is shown in Fig. 23.16(a). However, in other metals a **superconducting transition** takes place, as shown in Fig. 23.16(b).

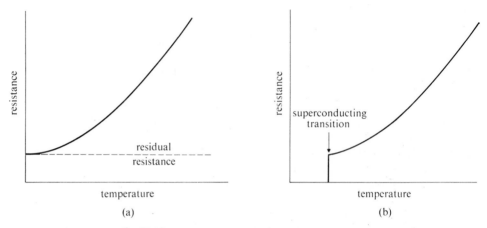

Fig. 23.16. (a) In some metals, the resistance approaches a residual value as the temperature is decreased. (b) In other metals, a superconducting transition takes place before a residual resistance is reached.

The phenomenon of superconductivity is connected with the zero-point vibrations of atoms discussed in the last section. Scattering of electrons by atoms undergoing zero-point vibrations does not give energy to the electrons, as there is none to spare. Instead, the interaction of electrons with the lattice atoms gives rise (in some quantum-mechanical way) to an electron-electron interaction such that pairs of electrons couple together to form a boson-like structure. Therefore, the electron pairs form a superfluid that is not disturbed by defects and thermal motions in the lattice, provided these other influences are not too strong. As the temperature rises, thermal motions become strong enough to overcome the electron-electron coupling, and the metal reverts to normal behavior. The coupling varies in strength for different metals, and those in which it is not strong enough do not become superconductors.

We saw in Sections 7.11 and 11.6 that a force is exerted on charged particles moving in a magnetic field. It should not be surprising then that applying a magnetic field to a superconductor can cause it to revert back to its normal state. At some critical value of the magnetic field the force exerted on the moving electrons will destroy the electron-electron coupling, and the change of state occurs.

We must stress that this picture of superconductivity is just that, an attempt to describe in simple terms a quantum-mechanical effect. As such it should not be regarded too seriously but should rather serve to illustrate the fact that nature is quantized and that, under the proper circumstances, the quantization gives rise to very surprising and remarkable effects.

PROBLEMS

1. Calculate the volume of 1 kg-mole of an ideal gas at standard temperature ($T = 273.16°K$) and standard pressure ($1.01325 \times 10^5 \ N \cdot m^{-2}$) using the value $R = 8.3143 \times 10^3 \ J \cdot (kg\text{-mole})^{-1} \cdot (°K)^{-1}$ for the gas constant. Calculate the number of molecules in this volume of gas using the value $k = 1.3805 \times 10^{-23}$ $J \cdot (°K)^{-1}$ for Boltzmann's constant.

2. An ideal gas is initially at temperature T_1, pressure P_1, and volume V_1. It then expands isothermally to a final pressure P_2 and final volume V_2. Develop an expression for the work done by the gas in terms of its volume.

3. The addition of heat to 1 kg-mole of nitrogen gas under conditions of constant pressure results in a change in temperature from $300°K$ to $310°K$. Treat nitrogen as an ideal gas.
   (a) Calculate the work done on the gas.
   (b) Calculate the change in internal energy of the gas.

4. What amount of work is required to reduce the volume of 10 kg of oxygen initially at a temperature of $300°K$ from $20 \ m^3$ to $10 \ m^3$ (assuming ideal-gas behavior)
   (a) at constant pressure?
   (b) at constant temperature?

5. One gram-mole of a monatomic ideal gas has an initial volume of $V_1 = 2 \times 10^{-3}$ m^3 and is at pressure $P_1 = 1.0 \times 10^5 \ N \cdot m^{-2}$. The pressure is then increased to $5.0 \times 10^5 \ N \cdot m^{-2}$, and the volume is held constant.
   (a) Calculate the work done on the gas.
   (b) Calculate the heat supplied to the gas.
   (c) Find the change in internal energy of the gas.

6. One gram-mole of a monatomic ideal gas initially of volume $V_1 = 2 \times 10^{-3} \ m^3$ at pressure $P_1 = 1.0 \times 10^5 \ N \cdot m^{-2}$ is expanded to a volume of $10^{-2} \ m^3$ at constant pressure.
   (a) Calculate the work done on the gas.

(b) Calculate the heat supplied to the gas.

(c) Find the change in internal energy of the gas.

7. One kilogram-mole of an ideal gas initially at a pressure of $1.01 \times 10^5 \, \text{N} \cdot \text{m}^{-2}$ and a temperature of 300°K expands reversibly against a piston until its volume is doubled. During the expansion the relation $P = bV$ is satisfied, where b is a constant. Calculate the work done by the gas.

8. One kilogram-mole of air at standard temperature and pressure is compressed isothermally to one-half its original volume. How much work is done on the gas?

9. Is it feasible to attempt to build an engine that will take in 10^8 J from its fuel supply, will reject 2.5×10^7 J in its exhaust, and will deliver 25 kWh (2.5×10^4 $\text{J} \cdot \text{sec}^{-1}$ for one hour) of mechanical work?

10. Calculate the change in entropy of 1 kg of ice when it freezes at 273°K. What can we say about the change in entropy of the "surroundings" of the block of ice (that is, of the material in thermal contact with the ice)?

11. What is the change in entropy of 1 kg-mole of water when it is heated from a temperature of 275°K to a temperature of 350°K?

12. One kilogram of water initially at 273°K is mixed with 1 kg of water initially at 373°K.

(a) What is the final temperature?

(b) What is the change in entropy?

*13. Show that the Clausius and Kelvin–Planck statements of the second law are equivalent. *Hint*: Consider a cyclic heat engine working between two reservoirs and violating the Kelvin–Planck statement and use it to drive a refrigerator working between the same two reservoirs.

14. The difference in specific heats C_p and C_v for any substance can be written in the form

$$C_p - C_v = \frac{\beta^2 T V}{K},$$

where

$$\beta = \frac{1}{V}\left(\frac{dV}{dT}\right)_P$$

is the **coefficient of volume expansion** and

$$K = -\frac{1}{V}\left(\frac{dV}{dP}\right)_T$$

is the **isothermal compressibility**. For copper at 300°K, $\beta = 4.9 \times 10^{-5} (°\text{K})^{-1}$, $K = 7.7 \times 10^{-12} \, \text{m}^2 \cdot \text{N}^{-1}$, and 1 kg-mole occupies a volume of $7.15 \times 10^{-3} \, \text{m}^3$. Calculate $C_p - C_v$ for copper at 300°K and compare the result

(a) with the value of C_v for copper (consult a physical table).

(b) with the value of $C_p - C_v$ for an ideal gas.

15. Under what conditions is the efficiency of a Carnot engine at a maximum?

16. A Carnot cycle is usually plotted on a P-versus-V graph (Fig. 23.5). Plot this cycle on an S-versus-T graph.

17. An ideal gas initially at a pressure $P_1 = 2.0 \times 10^5$ N·m^{-2}, volume $V_1 = 1$ m^3, and temperature $T_1 = 400°$K is compressed adiabatically to volume $V_2 = 0.3$ m^3. Find the final pressure and temperature, assuming the gas to be
    (a) monatomic.
    (b) diatomic.

18. An ideal gas, initially at temperature $T_1 = 300°$K, pressure $P_1 = 3.0 \times 10^5$ N·m^{-2}, and volume $V_1 = 8$ m^3, is expanded adiabatically to a volume of 24 m^3 according to the relation $PV^{1.4} =$ constant. Find the final temperature.

*19. Show that the work done in an adiabatic expansion of an ideal gas from pressure P_1 and volume V_1 to pressure P_2 and volume V_2 is

$$W = \frac{1}{1 - \gamma}(P_2 V_2 - P_1 V_1).$$

20. Water is one of the few substances that expands upon freezing. Supposing water to be a normal fluid that contracts upon freezing, describe some of the changes in our physical environment that would occur as a result.

21. Compare the amounts of heat required to melt 1 g of each of the following substances at normal atmospheric pressure (consult a physical table): aluminum, copper, lead, mercury, ice.

22. Compare the amounts of heat required to vaporize 1 g of each of the following liquids (consult a physical table): ethyl alcohol, chloroform, turpentine, water.

23. Calculate the increase of internal energy of 1 kg-mole of water when it is changed to steam at 100°C and atmospheric pressure.

APPENDICES

Some aspects of physical A
measurement

A.1   INTRODUCTION

Experimentation is fundamental to physics. In this appendix we shall try to give some insight into the manner in which physical measurements are obtained and the uncertainties to be associated with such measurements. To this end we shall discuss three simple experiments.

A.2   AN EXPERIMENTAL MEASUREMENT OF THE DENSITY OF COPPER

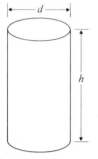

Fig. A.1. A copper cylinder.

To determine the density of copper from a particular sample, we must measure its volume and its mass. Let us assume that the sample is of cylindrical shape with a circular base of diameter d and height h, as shown in Fig. A.1. The volume of the sample is

$$V = \frac{\pi d^2 h}{4}.$$

To measure d and h, let us choose a simple steel rule with millimeter markings. The measurement of h is shown in Fig. A.2, where we see that $h = 4.4$ cm. We do not mean that h is exactly 4.4 cm but rather that it is closer to 4.4 cm than to 4.3 cm or 4.5 cm. That is, h lies somewhere in the range 4.35 to 4.45 cm. It is customary to write

$$h = 4.4 \pm 0.05 \text{ cm.}$$

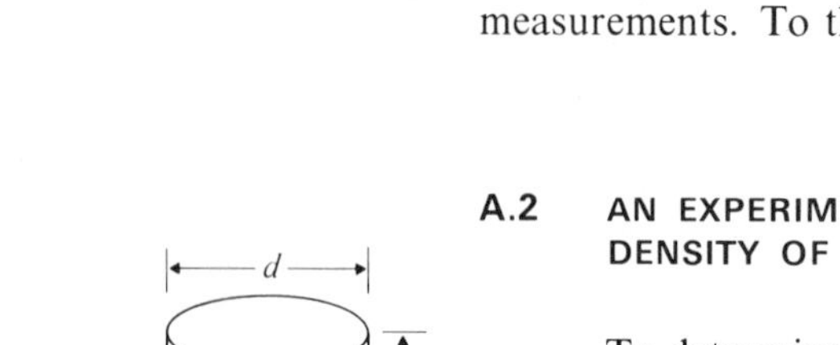

The 0.05 cm is the **absolute error** associated with the measurement of h using the particular steel rule selected. Someone else might estimate the reading to be

$$h = 4.43 \pm 0.02 \text{ cm.}$$

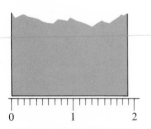

Fig. A.2. The height h is closer to 4.4 cm than to 4.3 cm or 4.5 cm.

This could be perfectly reasonable. An absolute error is determined both by the instrument used for the measurement and the skill of the experimenter. For convenience we shall always take one-half of the smallest subdivision of the measuring instrument as the absolute error.

The diameter d is measured to be

$$d = 1.9 \pm 0.05 \text{ cm,}$$

Fig. A.3. The diameter d is 1.9 ± 0.05 cm.

as shown in Fig. A.3. Note that in Fig. A.2 h is actually slightly greater than 4.4 cm, whereas in Fig. A.3 d is actually slightly less than 1.9 cm. The uncertainty introduced into our results because the ruler is only capable of measuring to a finite precision is an example of a general class

516

of errors called **random errors**. It is just as likely that a measured length will be greater or less than the actual length (the length that would be obtained if there were no measuring error).

What if we had chosen some other steel rule that also had millimeter markings? Would we necessarily obtain the same results? No, not necessarily. That is, although the random error inherent in the use of rulers capable of measuring to the nearest millimeter is the same, the calibrations of the rulers with respect to the standard meter may differ. Any deviation between a ruler used for a measurement and the standard meter will introduce a **systematic error** into the measurement.

For example, if each 1-mm interval of your ruler is in fact only 0.999 mm as compared to the standard meter, then any measurement made with the ruler will give too large a result. If each 1-mm interval is 1.001 mm as compared to the standard meter, then any measurement made with the ruler will give too small a result. This is the characteristic feature of a measuring instrument possessing a systematic error; it always gives results that are in error in the same direction. For our present example we shall assume for the moment that any systematic error present in our measuring instrument is negligible.

To measure the mass suppose that we select an equal-arm balance and a set of calibrated masses, the smallest of which is 0.01 g. The balance is shown in Fig. A.4. With nothing on either pan the pointer is at zero. With an 0.01-g mass on the right-hand pan the pointer moves to the right and reads $+4$. That is, four scale divisions correspond to 0.01 g. The sensitivity of the balance is therefore

$$\frac{0.01}{4} = 0.0025 \text{ g} \cdot \text{division}^{-1}.$$

Fig. A.4. An equal-arm balance.

Therefore, any measurement will have an associated absolute error of ± 0.0013 g. With the copper cylinder on the left-hand pan it requires 111.84 g on the right-hand pan to affect a balance with the pointer almost at -4. The measured mass of the copper cylinder is therefore

$$M = 111.84 + 4 \times 0.0025$$

$$= 111.850 \pm 0.0013 \text{ g}.$$

The set of calibrated masses that we used for the measurement is only a secondary standard that in turn is calibrated in terms of the standard kilogram. Deviations between the calibrated masses and the standard kilogram will result in a systematic error in our measurements of the mass of the copper cylinder. We shall assume that any systematic error present is negligible.

We are now ready to calculate the density of copper. First we calculate the volume of the cylindrical sample.

$$V = \frac{\pi d^2 h}{2}$$

$$= \frac{3.14 \times (1.9)^2 \times 4.4}{4}$$

$$= 12.(5)\,\text{cm}^3.$$

The figure "5" in brackets is really not significant, since the experimental measurements only contain two significant figures and arithmetical operations cannot improve upon the accuracy inherent in the experiment. The density of copper is then

$$\rho = \frac{111.850}{12.(5)}$$

$$= 8.9\,\text{g} \cdot \text{cm}^{-3},$$

where only two significant figures have been retained in the final answer.

How certain are we of this result? Clearly this is not the exact density of copper, since our measurements were not exact. There was an absolute error of 0.05 cm associated with the measurements of d and h and an absolute error of 0.0013 g associated with the measurement of M. First let us consider the error associated with d^2.

$$d^2 = (1.9 \pm 0.05)^2$$

$$\simeq (1.9)(1.9)\left[1 \pm \frac{2(0.05)}{1.9}\right]$$

neglecting the second order term $(0.05/1.9)^2$. Now 0.05 m is the absolute error in the measurement of d, which we shall designate Δd; 1.9 m is the value of d. The quantity

$$\frac{\Delta d}{d} = \frac{0.05}{1.9}$$

is called the **fractional error** associated with the measurement of d. We see that the fractional error associated with d^2 is $2\,\Delta d/d$, that is, twice that associated with d. The absolute error associated with d^2 is

$$\Delta(d^2) = 2\left(\frac{\Delta d}{d}\right)d^2$$

$$= 0.2\,\text{cm}^2.$$

In a similar manner, we see that

$$d^2 h = (1.9)(1.9)(4.4)\left\{1 \pm \left[2\left(\frac{0.05}{1.9}\right) + \left(\frac{0.05}{4.4}\right)\right]\right\}.$$

That is, the fractional error in the product of d^2 and h is just the sum of the fractional errors associated with d^2 and h considered individually. The absolute error associated with the product is the product of the fractional error associated with d^2h and the value of d^2h to two significant figures. That is

$$\Delta(d^2h) = d^2h\left[2\left(\frac{\Delta d}{d}\right) + \left(\frac{\Delta h}{h}\right)\right]$$

$$= 0.9 \text{ cm}^3.$$

It is not difficult to generalize the above discussion and thereby obtain the following rule.

When products of quantities are taken, the error associated with the product is obtained by adding the fractional errors associated with the individual quantities.

The fractional error so deduced gives the extreme width of the range of possible fractional errors for the product. It may be argued that the probability of so large an error is small[1].

The fractional error associated with the volume of the cylinder, $\Delta V/V$, is the same as that associated with the product d^2h, since $\pi/4$ is an exact number. The absolute error ΔV associated with V is therefore

$$\Delta V = \left(\frac{\Delta V}{V}\right)V$$

$$= 0.9 \text{ cm}^3.$$

To find the absolute error associated with our determination of the density ρ, we must consider the quotient

$$\frac{M}{V} = \frac{(111.850 \pm 0.0013)}{[12.(5) \pm 0.9]} \text{ g} \cdot \text{cm}^{-3}.$$

Using the binomial expansion we can expand the quantity $[12.(5) \pm 0.9]^{-1}$ and obtain

$$[12.(5) \pm 0.9]^{-1} \simeq [12.(5)]^{-1}\left[1 \mp \left(\frac{0.9}{12.5}\right)\right].$$

The quotient therefore becomes

$$\frac{M}{V} \simeq \frac{111.850}{12.5}\left[1 \pm \frac{0.0013}{111.850} \mp \frac{0.9}{12.(5)}\right],$$

---

[1] D. C. Baird, *Experimentation* (Englewood Cliffs, N.J.: Prentice-Hall, Inc., 1962), Chapter 3.

where 0.0013/111.850 is the fractional error associated with the measurement of M and 0.9/12.(5) is the fractional error associated with V. The possible fractional error in the quotient is the sum of these two fractional errors, which correspond to the worst possible cases in which the measured value of M exceeds the actual value, or vice versa.

This discussion may also be readily generalized, thereby giving us the following rule.

> When the quotients of quantities are taken, the error associated with the quotient is obtained by adding the fractional errors associated with the individual quantities.

This too is a rather pessimistic estimate of the fractional error of a quotient[2].

Hence, the fractional error associated with our density determination is

$$\frac{\Delta \rho}{\rho} = \frac{0.0013}{111.850} + \frac{0.9}{12.(5)}$$

$$= 0.00001 + 0.07 = 0.07.$$

Note that the fractional error associated with the mass determination is negligible compared with that associated with the volume determination. The absolute error associated with our density determination is

$$\Delta \rho = 0.07 \times 8.9 = 0.6 \, \text{g} \cdot \text{cm}^{-3},$$

and we write our final result as

$$\rho = 8.9 \pm 0.6 \, \text{g} \cdot \text{cm}^{-3}.$$

Percentage errors are also commonly used; they are defined as the fractional error times 100. For our example

$$\rho = 8.9 \, \text{g} \cdot \text{cm}^{-3} \pm 7\%.$$

If we wish to improve upon our experimental result then it is clear that we must choose a more sensitive device for measuring distances. Suppose we choose a micrometer such as the one shown in Fig. A.5. The calibrations along the barrel are in 0.5-mm intervals. The sleeve is marked off in 50 equal intervals; one complete rotation of the sleeve opens or closes the micrometer jaws by 0.5 mm. Therefore, rotating the sleeve through one division opens or closes the jaws of the micrometer by 0.5/50 = 0.01 mm. If readings with the micrometer are taken to the

---

[2] The corresponding rule for addition and subtraction is the following: for addition or subtraction absolute errors are added to give an upper limit to the absolute error in the sum or difference.

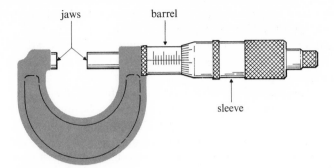

Fig. A.5. A micrometer is an instrument for making precise measurements. One complete rotation of the sleeve opens or closes the jaws by 0.5 mm.

nearest sleeve division, then the random error associated with a length measurement is 0.005 mm. If we close the jaws of the micrometer with nothing between them, the micrometer should read zero. If it does not, then a systematic error known as a **zero error** is present and must be taken into account every time the micrometer is used. For our micrometer the zero error is 0.02 mm, as shown in Fig. A.6; this must be subtracted from each of the measurements. We shall assume that the calibration on the barrel checks with the standard meter. Note that with this micrometer we can improve the accuracy of our length measurements by a factor of 100.

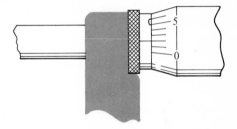

Fig. A.6. The zero error of this micrometer is 0.02 mm.

Six height and six diameter measurements were taken with this micrometer at different places on the copper cylinder. These are summarized in Table A.1. Corrections for the zero error have been made.

Table A.1   Length and diameter measurements of a copper cylinder

	Length (cm)	|Dev|* (cm)	Diameter (cm)	|Dev|* (cm)
	4.416	0.002	1.901	0.005
	4.428	0.006	1.887	0.009
	4.419	0.003	1.893	0.003
	4.423	0.001	1.902	0.006
	4.421	0.001	1.898	0.002
	4.426	0.004	1.896	0.000
Mean	4.422	0.003	1.896	0.004

* The symbol "|Dev|" means "the absolute value of the deviation from the mean."

A glance at the table shows that the variations between measurements are larger than the inherent limitation of the micrometer to measure distances. These variations are in fact a measurement of the deviations of the geometry of the copper sample from that of a true cylinder. If we approximate the actual sample by a cylinder, then the error introduced into our determination of the volume will not be that due to the inherent limitation of the micrometer but rather will depend upon the degree to which we succeed in approximating the actual sample by an equivalent cylinder. The simplest method of assigning an absolute error to the cylinder parameters resulting from averaging the height measurements to obtain $h = 4.422$ cm and averaging the diameter measurements to obtain $d = 1.896$ cm is the following. The absolute values of the differences between the mean value of the measurements and each individual measurement are calculated. These are included in Table A.1. These deviations are then averaged to obtain **mean deviations**, which we shall take as the absolute errors. That is, we shall take as the cylinder parameters

$$h = 4.422 \pm 0.003 \text{ cm}, \qquad d = 1.896 \pm 0.004 \text{ cm}.$$

The volume of the cylinder now becomes

$$V = \frac{\pi d^2 h}{4}$$

$$= \frac{3.14 \times (1.896)^2 \times 4.422}{4}$$

$$= 12.49 \text{ cm}^3,$$

and the density

$$\rho = \frac{M}{V}$$

$$= \frac{111.850}{12.49}$$

$$= 8.955 \text{ g} \cdot \text{cm}^{-3}.$$

The fractional error associated with our determination of the volume is

$$\frac{\Delta V}{V} = 2\left(\frac{0.004}{1.896}\right) + \frac{0.003}{4.422} = 0.004(9).$$

The fractional error associated with our mass determination is still negligible in comparison to the error in the volume determination, so that the fractional error associated with our density determination is

$$\frac{\Delta\rho}{\rho} = 0.004(9).$$

We may write as our final experimental value for the density of copper

$$\rho = 8.955 \pm 0.044 \,\text{g} \cdot \text{cm}^{-3}.$$

Assigning absolute errors to d and h using the mean deviation of the experimental results from the mean values is not the best method, although it is a very good approximation. A full discussion of more advanced mathematical techniques for treating experimental measurements is contained in the text by D. C. Baird[3], to which the reader is referred.

Finally we might note that, through all of the above considerations, it has been implicitly assumed that the sample was pure copper. If the impurity content of the sample is appreciable, then this assumption would lead to a further systematic error in the final result.

A.3   THE DETERMINATION OF A LOCAL VALUE OF THE ACCELERATION DUE TO GRAVITY

To determine the local value of the acceleration g due to gravity, we shall use a simple pendulum as shown in Fig. A.7. The experiment consists of measuring the length l and the time t for the pendulum to make ten complete oscillations. The length was measured from the point of suspension to the geometric center of the metal object by using an ordinary steel rule and was found to be

$$l = 94.2 \pm 0.3 \,\text{cm}.$$

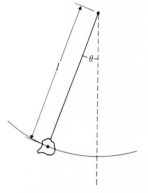

Fig. A.7. A simple pendulum may be used to determine the acceleration due to gravity.

A relatively large absolute error has been assigned because of the uncertainty in knowing exactly what is meant by the length of the pendulum. A stopwatch was used to time the oscillations of the pendulum, the time being read to the nearest 0.1 sec. The results appear in Table A.2. The fact that the results show a larger spread than can be explained by the inherent inaccuracy of the stopwatch may be attributed to the impossibility of exactly releasing the object in the same manner for each trial and to a variation in personal reaction times from trial to trial.

From the table we see that the period when the amplitude of oscillation is small is

$$T = 1.94 \pm 0.01 \,\text{sec},$$

where 0.01 sec is the mean deviation. We assume that our clock has been

[3] D. C. Baird, *Experimentation.*

Table A.2   Measurement of the period of oscillation of a pendulum

Small amplitude		Large amplitude	
Time for ten swings (*sec*)	*|Dev|** (*sec*)	*Time for five swings* (*sec*)	*|Dev|** (*sec*)
19.3	0.1	9.9	0.1
19.4	0.0	10.1	0.1
19.3	0.1	9.9	0.1
19.6	0.2	10.0	0.0
19.4	0.0	10.1	0.1
19.5	0.1	10.1	0.1
Mean 19.4	0.1	10.0	0.1
Period $= 1.94 \pm 0.01$ sec		Period $= 2.00 \pm 0.02$ sec	

* See Table A.1.

checked against a standard of time. Another set of measurements was taken for large-amplitude oscillations of the pendulum. By "small" and "large" we mean that the angle θ shown in Fig. A.7 is either of the order of 2° or greater than 45°, respectively. For large-amplitude oscillations the period of the pendulum has a value

$$T = 2.00 \pm 0.02 \text{ sec.}$$

The local acceleration due to gravity is given by (see Section 11.4)

$$g = \frac{4\pi^2 l}{T^2}.$$

This equation is correct only for small amplitudes of oscillation. This illustrates another type of systematic error, that due to approximations made in deriving physical equations. In this example, using T determined from large-amplitude oscillations would introduce an appreciable error into our determination of g.

The value of g determined by this experiment is

$$g = \frac{4 \times (3.14)^2 \times 94.2}{(1.94)^2} = 988 \text{ cm} \cdot \text{sec}^{-2}.$$

The fractional error associated with our determination of g is

$$\frac{\Delta g}{g} = \frac{0.3}{94.2} + 2 \times \frac{0.01}{1.94} = 0.013.$$

We may write as our final experimental result

$$g = 988 \pm 13 \text{ cm} \cdot \text{sec}^{-2}.$$

A.4   DECAY OF A SAMPLE OF RADIOACTIVE COBALT

In certain types of experiments random errors are introduced by fluctuations inherent in the process being studied. Radioactivity studies are a common example. The time at which a particular radioactive nucleus will spontaneously disintegrate is neither controllable nor predictable. To say it another way, if we were to prepare two radioactive nuclei, identical in every known respect, then in general the nuclei would disintegrate at different times, and we could not predict which one would disintegrate first. The decay of radioactive nuclei is governed by a probability law according to which one-half of a very large sample of nuclei spontaneously disintegrates within a characteristic time known as the **half-life**. The half-life parameter is well defined, even though the individual disintegrations occur randomly and unpredictably.

Let us consider an experiment involving a cobalt-60 (^{60}Co) sample, with a half-life of 5.2 years. The number of spontaneous disintegrations occurring in a 30 sec period was recorded by a geiger counter and an electronic counting system. The measurement was repeated 304 times; the results are plotted in Fig. A.8 with the numbers of counts recorded in the 30-sec counting interval plotted along the horizontal axis and the number of times each total count occurred plotted along the vertical axis.

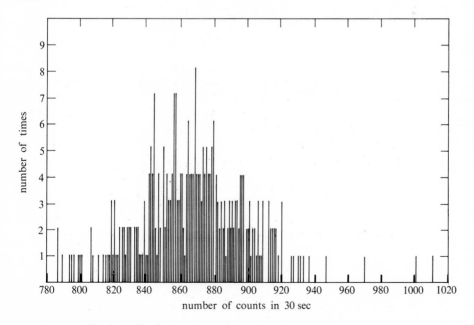

Fig. A.8. Distribution of counts from a radioactive source.

Although there are a considerable number of data points, it is seen that a given value is not repeated very many times, and the results of the various measurements are spread over a rather broad range. It is apparent from the graph that the majority of the results lie between 840 and 900 counts per 30-sec interval. However, many of the readings are outside this range, some considerably so. Another interesting fact is that if we were to ask how many times the result 849 or 881 occurs, the answer would be zero even though these numbers are in the thick of the data. A more meaningful plot of the data can be made by grouping the results in intervals of ten and plotting them as in Fig. A.9. Since the results are scattered over such a wide range, it is more meaningful to inquire whether a particular result is between 840 and 850 or between 880 or 890 without being concerned about precisely where in the range it occurs.

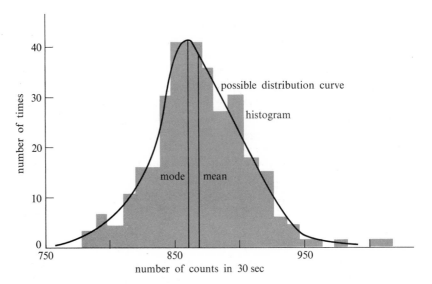

Fig. A.9. Properties of the experimental distribution.

Figure A.9 shows, for example, that more results lie between 850 and 870 than in any other comparable range, a fact not readily apparent from Fig. A.8. Figure A.9 is called a **histogram** of the results. By grouping we have produced a graph with much less fluctuation. The size of the group, one decade, was chosen because it was convenient, was large enough to smooth out the rapid fluctuations in the results, but was not so large that broad details of the fluctuations were obscured.

It is reasonable to expect that as the number of observations is increased the relative size of the fluctuations in Fig. A.8 would become smaller. In the limit of a very large number of observations the fluctuations would become negligible, and the broad features of the cobalt-60 decay would be apparent from Fig. A.8. Such a limiting curve is known as a **distribution curve** and, in this case, would be a distribution curve for the number of counts from the cobalt-60 source in 30-sec intervals. The histogram of Fig. A.9 gives some indication of the shape of the distribution curve; a possible distribution curve is shown superimposed on the histogram.

For many purposes it is desirable to have some representative value for the measurements that constitute the distribution. One such average quantity, known as the **mean value** $\bar{x}$, is found by taking the sum of the observed values and dividing by the total number of observations. That is,

$$\bar{x} = \frac{x_1 + x_2 + \cdots + x_{n-1} + x_n}{n},$$

where $x_1, x_2, \cdots, x_n$ are the experimental numbers of counts and n is the total number of trials. For the radioactivity example just discussed the mean is found to be 867.5 counts per 30 sec.

Another representative value often quoted is the **mode** or most probable number, which for the present example is just the number of counts per 30 sec corresponding to the peak of the curve. The mode is therefore the number of counts that is most likely to be obtained in a single 30-sec period. The mean and the mode for this distribution are not quite identical; they are shown in Fig. A.9.

From a study of radioactive decay it has been shown that the distribution curve is of a type called the **Poisson curve**, which is appropriate to experiments involving the counting of random events. The Poisson distribution is asymmetric (the mean and the mode have different numerical values) but tends to become symmetric (the mean and the mode coalesce) as the number of counts observed in a given period becomes larger and larger. For large enough numbers of counts per period, the Poisson distribution becomes indistinguishable from the **Gaussian distribution**, which has the form

$$y = C \exp\left[-h^2(x - \bar{x})^2\right],$$

where C and h are constants, $\bar{x}$ is the mean value, and y is the relative frequency with which the value x occurs.

The distribution is shown in Fig. A.10. When $x = \bar{x}$, $y = C$, so that C gives the height of the peak. When $x - \bar{x} = 1/h$, y has fallen to $1/e$ of its peak value; $1/h$ is a measure of the width of the distribution.

The Gaussian distribution may be mathematically analyzed, and an absolute error can be

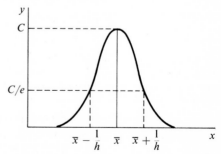

Fig. A.10. The normal or Gaussian distribution.

defined in terms of the parameter h (see Footnote 1 of this Appendix). For any type of experiment, if the experimenter can convince himself that the observed fluctuations are equally likely to increase or decrease the measured value, then he is justified in assuming that the measured values belong to a Gaussian distribution. For example, the micrometer measurements of the diameter and the length of the copper cylinder discussed earlier could have been treated in such a manner.

PROBLEMS

1. A stopwatch, graduated in 0.1-sec intervals, is used to measure time intervals of 0.5 sec, 5 sec, and 50 sec.
   (a) What is the absolute error associated with each of the measurements?
   (b) What is the fractional error in each case?
   (c) What is the percentage error in each case?

2. The stopwatch of Problem 1 was used to time the fall of a steel ball through a distance s measured to be 1.76 ± 0.01 m. The stopwatch reading t was 0.6 sec. From the measurements a value of the local acceleration due to gravity g was calculated from the relation

$$s = \tfrac{1}{2}gt^2.$$

   Determine the absolute, fractional, and percentage errors associated with the calculated value of g.

3. For any four measuring instruments that you wish, write down as many possible sources of systematic error as you can. Indicate briefly how you might discover that the instruments possessed the systematic errors that you mention.

4. The activities of two radioactive samples are found to be 1800 ± 42 counts per minute and $30{,}000 \pm 173$ counts per hour, respectively. Determine the combined counting rate of the two sources and the associated absolute error.

5. A simple pendulum experiment is used to measure the local acceleration g due to gravity. The period of the pendulum T is measured to 0.5% and the length l to 1%. What is the precision of the value of g calculated from these measurements?

6. Suggest methods and measuring instruments that you could use to make the following length measurements with a precision of 5%:
   (a) the thickness of a sheet of paper,
   (b) the height of a flagpole,
   (c) the horizontal distance that you have hit a golf ball.

7. Flip a coin ten times and record the number of heads and the number of tails obtained. Repeat the experiment a large number of times, say 50. Illustrate your results by means of a histogram.

8. Suggest a method by means of which a person standing near a highway might measure the speed of a passing car. Estimate the accuracy of the method that you have suggested.

9. Suggest a possible method for measuring the density of an irregularly-shaped piece of material.

10. A student measures the inside diameter of a cylindrical container to be $d_i = 5.73 \pm 0.01$ cm and the outside diameter to be $d_o = 5.76 \pm 0.01$ cm.
(a) Use this data to calculate the wall thickness of the cylinder.
(b) Compare the percentage errors associated with the measured diameters and with the calculated wall thickness.
(c) Suggest a more accurate method to obtain the wall thickness.

11. A surveyor marks off a distance of 20.0 ± 0.1 m from the base of a tree and from that point measures the angle between the horizontal and the top of the tree to be $31.0 \pm 0.5°$. Calculate the height of the tree and the associated absolute error.

12. Suppose you are given two spring balances with which to weigh an object you know to weigh approximately 100 g. The first balance measures 500 g for a full-scale deflection, with the scale divided into 250 divisions. The second balance measures 250 g full scale, with the scale divided into 100 divisions. Describe a balance that would permit a measurement of twice the precision of the more precise of the two balances described.

13. If an object with an initial speed v_0 is subjected to a constant acceleration a for a time t, its final speed is given by

$$v = v_0 + at.$$

In a particular experiment v_0 is measured to be $2 \text{ m} \cdot \text{sec}^{-1}$ with a precision of 2%, a is measured to be $10 \text{ m} \cdot \text{sec}^{-2}$ with a precision of 3%, and t is measured to be 5 sec with a precision of 2%. Determine the absolute error in the calculated value of v.

14. The diameter of a mechanical shaft is measured with a precision micrometer. Twenty readings give the following results (in m $\times 10^3$).

25.399	25.401	25.400	25.401
25.400	25.402	25.401	25.400
25.402	25.399	25.401	25.399
25.400	25.401	25.399	25.401
25.403	25.401	25.400	25.400

Draw a histogram of the distribution and calculate the mean and the mean deviation for this sample. Indicate in your diagram the position of the mean and a tentative shape for the distribution curve.

*15. A layman reports to have seen an unidentified flying object (UFO) traveling at about $300 \text{ km} \cdot \text{hr}^{-1}$ at an altitude of 2000 m. He also stated that the length of the object was about 60 m. When questioned about the accuracy of his determination he confidently stated that it was reliable to within $50 \text{ km} \cdot \text{hr}^{-1}$. Discuss the possibility of the average person making such a measurement.

16. A series of 1-min counts were made of the activity of a certain radioactive uranium source. The results were tabulated and then sorted into bins (or groups) 25 counts in width. The results are shown in Table A.3. The mean

Table A.3

Total number of counts	Number of results in bin	Total number of counts	Number of results in bin
9726–9750	1	10,026–10,050	10
9751–9800	0	10,051–10,075	8
9801–9825	1	10,076–10,100	5
9826–9850	1	10,101–10,125	5
9851–9875	0	10,126–10,150	2
9876–9900	6	10,151–10,175	2
9901–9925	4	10,176–10,200	3
9926–9950	6	10,201–10,250	0
9951–9975	8	10,251–10,275	2
9976–10,000	11	10,276–10,350	0
10,001–10,025	8	10,351–10,375	1

value $\bar{x}$ was found to be 10,013 counts per minute. Plot histograms of the readings using bins 25, 50, 100, and 200 counts in width. Which histogram conveys the most information about the characteristics of the distribution?

17. A rocket is launched straight outward from a nonrotating earth alone in the universe. The initial speed that the rocket requires to just escape from the influence of the earth's gravitational attraction is given by

$$v_0 = \left(\frac{2GM}{R}\right)^{1/2},$$

where G is the gravitational constant, M is the mass of the earth, and R is the radius of the earth. Given that $G = (6.670 \pm 0.005) \times 10^{-11}\,\mathrm{m^3 \cdot kg^{-1} \cdot sec^{-2}}$, $M = (5.980 \pm 0.005) \times 10^{24}\,\mathrm{kg}$, $R = (6.37 \pm 0.01) \times 10^6\,\mathrm{m}$, find v_0.

18. The local acceleration g due to gravity is to be determined by the method suggested in Problem 2 and using the same distance of fall s. Determine the absolute error needed in the time measurement t to yield a value of g precise to 5%.

*19. A particle moves subject to a velocity-dependent retarding force

$$F = -kv,$$

where k is a positive constant. If at $t = 0$ the speed of the particle is v_0, then the speed at a later time t is given by

$$v = v_0 \exp\left(-\frac{kt}{m}\right),$$

where m is the mass of the particle. Plot a graph to illustrate the variation of

v with t. If v, v_0, m, and t are measured to be 2.0 ± 0.1 m·sec^{-1}, 10.0 ± 0.1 m·sec^{-1}, 1.0 ± 0.1 kg, and 10.0 ± 0.1 sec, respectively, calculate k and the associated absolute error.

*20. The mass and volume of a copper sample are measured to be 17.51 ± 0.01 g and 1.96 ± 0.02 cm^3, respectively.
(a) Calculate the density of copper and the absolute error.
(b) The calculation in part (a) implicitly assumes that systematic errors are negligible. Estimate the maximum amount of an impurity of density 7.16 g·cm^{-3} permissible in the sample if this assumption is to be valid.

21. Choose one of the histograms of Problem 16 and plot on the same sheet of paper the normal curve that best represents the data. What are the values of h and C required to give the best fit? *Hint:* Take $\bar{x} = 10{,}013$ and estimate the value of x for which $y = C/2$. Use this value of x to determine h.

B Some useful mathematics

B.1   TRIGONOMETRY

The trigonometric functions $\sin\theta$, $\cos\theta$, and $\tan\theta$ may be defined with respect to a right-angled triangle as shown in Fig. B.1. That is,

$$\sin\theta = \frac{B}{C}, \qquad \cos\theta = \frac{A}{C}, \qquad \tan\theta = \frac{B}{A}.$$

These functions are related to one another by the identities

$$\sin^2\theta + \cos^2\theta = 1$$

$$\tan\theta = \frac{\sin\theta}{\cos\theta}.$$

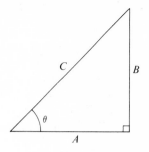

Fig. B.1. The functions $\sin\theta$, $\cos\theta$, and $\tan\theta$ may be defined as ratios of the sides of a right triangle as $\sin\theta = B/C$, $\cos\theta = A/C$, $\tan\theta = B/A$.

When two sides and the included angle of any triangle are known, we may calculate the remaining side from the relation (refer to Fig. B.1)

$$B^2 = A^2 + C^2 - 2AB\cos\theta;$$

this rule is known as the **law of cosines**. If two angles and the included side are known, the other two sides may be calculated from the **law of sines**,

$$\frac{A}{\sin a} = \frac{B}{\sin b} = \frac{C}{\sin c},$$

where a, b, and c are the angles opposite sides A, B, and C, respectively.

From a geometrical consideration of Fig. B.2 it can be shown that, if $\chi = \theta + \phi$ as shown in Fig. B.2a, then

$$\sin\chi = \sin(\theta + \phi)$$

$$= \sin\theta\cos\phi + \cos\theta\sin\phi$$

$$\cos\chi = \cos(\theta + \phi)$$

$$= \cos\theta\cos\phi - \sin\theta\sin\phi.$$

If $\chi = \theta - \phi$ as shown in Fig. B.2b, then

$$\sin\chi = \sin(\theta - \phi)$$

$$= \sin\theta\cos\phi - \cos\theta\sin\phi$$

$$\cos\chi = \cos(\theta - \phi)$$

$$= \cos\theta\cos\phi + \sin\theta\sin\phi.$$

For the special case of $\theta = \phi$,

$$\chi = \theta + \phi = 2\theta$$

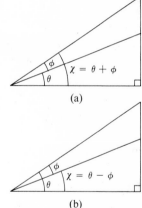

(a)

(b)

Fig. B.2. The sines and cosines of the sums and differences of angles can be deduced from geometrical considerations.

533

and

$$\sin 2\theta = 2 \sin \theta \cos \theta$$

$$\cos 2\theta = \cos^2 \theta - \sin^2 \theta$$

$$= 2 \cos^2 \theta - 1$$

$$= 1 - 2 \sin^2 \theta.$$

From these formulas the following useful relations can be derived.

$$\sin \theta + \sin \phi = 2 \sin \frac{\theta + \phi}{2} \cos \frac{\theta - \phi}{2}$$

$$\sin \theta - \sin \phi = 2 \cos \frac{\theta + \phi}{2} \sin \frac{\theta - \phi}{2}$$

$$\cos \theta + \cos \phi = 2 \cos \frac{\theta + \phi}{2} \cos \frac{\theta - \phi}{2}$$

$$\cos \theta - \cos \phi = -2 \sin \frac{\theta + \phi}{2} \sin \frac{\theta - \phi}{2}$$

$$2 \sin \theta \cos \phi = \sin (\theta + \phi) + \sin (\theta - \phi)$$

$$2 \cos \theta \sin \phi = \sin (\theta + \phi) - \sin (\theta - \phi)$$

$$2 \cos \theta \cos \phi = \cos (\theta + \phi) + \cos (\theta - \phi)$$

$$2 \sin \theta \sin \phi = -\cos (\theta + \phi) + \cos (\theta - \phi).$$

B.2   THE BINOMIAL THEOREM

It may be shown by direct multiplication that

$$(a + b)^n = a^n + na^{n-1}b + \frac{n(n-1)}{2!}a^{n-2}b^2 + \cdots + b^n,$$

where n is a *positive* integer. This is the **binomial theorem**, and the series on the right is said to be the **expansion** of $(a + b)^n$. If n is not a positive integer, the expansion has an infinite number of terms. The case for which $a = 1$ and $b = x$ is used in many places in the text. The binomial expansion of $(1 + x)^n$ is written

$$(1 + x)^n = 1 + nx + \frac{n(n-1)}{2!}x^2 + \frac{n(n-1)(n-2)}{3!}x^3 + \cdots.$$

This series is infinite unless n is a *positive* integer, in which case it terminates after $n + 1$ terms.

B.3   DIFFERENTIAL CALCULUS

Consider a function of one variable

$$y = f(x),$$

such as, for example, that plotted in Fig. B.3a. Differential calculus techniques allow us to determine from y the curve

$$y' = \frac{df(x)}{dx},$$

where y' is known as the **derivative** of $f(x)$ with respect to x. For the curve shown in Fig. B.3a the derivative is shown in Fig. B.3b. For a

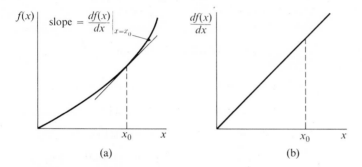

(a)   (b)

Fig. B.3. The derivative of a function $f(x)$ with respect to x evaluated at $x = x_0$ is the slope of the tangent drawn to $f(x)$ at x_0.

particular value of x, say x_0, $df(x)/dx$ evaluated at x_0 is just the slope of the tangent drawn to $f(x)$ at x_0. We represent this value symbolically by

$$\frac{df(x)}{dx}\bigg|_{x=x_0}$$

The derivatives of some common functions are listed in Table B.1.

For a function $h(x)$ that can be written as a product of two functions $f(x)$ and $g(x)$, the derivative of $h(x)$ with respect to x can be expressed in terms of the derivatives of $f(x)$ and $g(x)$ with respect to x. That is,

$$\frac{d}{dx}[h(x)] = \frac{d}{dx}[f(x)g(x)] = f(x)\frac{dg(x)}{dx} + g(x)\frac{df(x)}{dx}.$$

This is known as the **product rule**.

Table B.1   Derivatives of some common functions

$f(x)$	$\dfrac{df(x)}{dx}$
ax^n	nax^{n-1}
$e^{\pm ax}$	$\pm ae^{\pm ax}$
$\log_e ax = \ln(ax)$	$\dfrac{1}{x}$
$\sin ax$	$a\cos ax$
$\cos ax$	$-a\sin ax$

Example. Differentiate $x^2 \sin kx$ with respect to x using the product rule.

Solution. The derivative is

$$\frac{d}{dx}(x^2 \sin kx) = x^2 \frac{d}{dx}(\sin kx) + \sin kx \frac{d}{dx}(x^2)$$

$$= kx^2 \cos kx + 2x \sin kx.$$

If $y = f(t)$ and $x = g(t)$, then

$$\frac{dy}{dx} = \frac{dy}{dt}\frac{dt}{dx}.$$

This is known as the **chain rule**.

Example. If $y = A \sin t$ and $x = A \cos t$, calculate dy/dx directly, using the chain rule.

Solution. Squaring and adding these equations gives

$$x^2 + y^2 = A^2$$

Therefore,

$$2x\,dx = -2y\,dy$$

and

$$\frac{dy}{dx} = -\frac{x}{y}.$$

Using the chain rule

$$\frac{dy}{dt} = A\cos t, \qquad \frac{dx}{dt} = -A\sin t$$

Therefore,

$$\frac{dy}{dx} = \frac{dy}{dt}\frac{dt}{dx} = -\frac{A\cos t}{A\sin t} = -\frac{x}{y}.$$

Consider a function of two variables

$$z = f(x, y).$$

The differential change dz is given in terms of the differential changes dx and dy by

$$dz = \frac{\partial f(x, y)}{\partial x} dx + \frac{\partial f(x, y)}{\partial y} dy,$$

where $\partial f(x, y)/\partial x$ is the **partial derivative** of z with respect to x for y a constant and $\partial f(x, y)/\partial y$ is the partial derivative of z with respect to y for x a constant.

Example. For

$$z = 3x^2 y,$$

write an expression for dz.

Solution

$$dz = \frac{\partial}{\partial x}(3x^2 y)\, dx + \frac{\partial}{\partial y}(3x^2 y)\, dy$$

$$= 6xy\, dx + 3x^2\, dy$$

$$= 3x(2y\, dx + x\, dy).$$

B.4   INTEGRAL CALCULUS

Let us consider a function of one variable

$$y = f(x),$$

as for example that plotted in Fig. B.4a. Integral calculus techniques allow

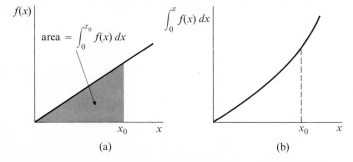

Fig. B.4. The definite integral of a function $f(x)$ with respect to x evaluated at $x = x_0$ is the shaded area shown in (a).

us to determine from y the curve

$$y = \int_a^x f(x)\,dx,$$

where y is known as the **definite integral** of $f(x)$ with respect to x. For the curve shown in Fig. B.4a and for $a = 0$, the definite integral is shown in Fig. B.4b. For a particular value of x, say x_0, $\int_a^{x=x_0} f(x)\,dx$ evaluated at x_0 is just the area under the curve $f(x)$ lying between $x = a$ and $x = x_0$. The definite integrals of some common functions are listed in Table B.2.

Table B.2   Definite integrals of some common functions

$f(x)$	$\int_a^x f(x)\,dx$
$Ax^n, \quad n \neq -1$	$\dfrac{A(x^{n+1} - a^{n+1})}{n+1}$
$\dfrac{A}{x}$	$A \ln\left(\dfrac{x}{a}\right)$
$e^{\pm Ax}$	$\pm \dfrac{e^{\pm A(x-a)}}{A}$
$\sin Ax$	$-\dfrac{(\cos Ax - \cos Aa)}{A}$
$\cos Ax$	$\dfrac{(\sin Ax - \sin Aa)}{A}$

Let us consider a function of two variables

$$z = h(x, y).$$

Integral calculus techniques allow us to determine from z the function

$$z = \int_a^x \int_b^y h(x, y)\,dx\,dy.$$

For the special case of $x = x_0$ and $y = y_0$, where x_0 and y_0 are constants and

$$h(x, y) = f(x)g(y),$$

we obtain

$$z = \left(\int_a^{x_0} f(x)\,dx\right)\left(\int_b^{y_0} g(y)\,dy\right).$$

That is, z is just the product of two one-dimensional integrals.

Example. If

$$z = 3x^2 y$$

and

$$0 \leqslant x \leqslant 4$$
$$1 \leqslant y \leqslant 3,$$

evaluate

$$\int_0^4 \int_1^3 3x^2 y \, dx \, dy.$$

Solution. The required integral is the product

$$3 \left(\int_0^4 x^2 \, dx \right) \left(\int_1^3 y \, dy \right)$$

$$= 3 \left(\frac{4^3}{3} \right) \left(\frac{3^2 - 1^2}{2} \right)$$

$$= 3 \times \frac{64}{3} \times \frac{8}{2}$$

$$= 256.$$

The **indefinite integral** is also widely used and is denoted by

$$Y(x) = \int f(x) \, dx.$$

The indefinite integral does not have limits between which x can vary as does the definite integral. The indefinite integral $Y(x)$ of $f(x)$ is just that function of x that when differentiated with respect to x yields $f(x)$. It is arbitrary with respect to an additive constant (called the constant of integration), since the constant will disappear upon differentiation. The indefinite integrals of some common functions are listed in Table B.3.

Example. Determine the indefinite integral and the definite integral between the limits $x = 2$ and $x = 4$ of

$$f(x) = 3x^2 + 2x.$$

Solution. The indefinite integral is

$$Y(x) = \int f(x) \, dx$$

$$= \int (3x^2 + 2x) \, dx$$

$$= \int 3x^2 \, dx + \int 2x \, dx$$

$$= x^3 + x^2 + \text{constant}.$$

Table B.3 Indefinite integrals of some common functions

$f(x)$	$\int f(x)\,dx$
$Ax^n, \quad n \neq -1$	$\dfrac{Ax^{n+1}}{n+1}$
$\dfrac{A}{x}$	$A \ln x$
$e^{\pm Ax}$	$\dfrac{e^{\pm Ax}}{A}$
$\sin(Ax)$	$-\dfrac{\cos(Ax)}{A}$
$\cos(Ax)$	$\dfrac{\sin(Ax)}{A}$

The definite integral is

$$Y = \int_2^4 (3x^2 + 2x)\,dx$$

$$= \int_2^4 3x^2\,dx + \int_2^4 2x\,dx$$

$$= 3\left(\frac{4^3}{3} - \frac{2^3}{3}\right) + 2\left(\frac{4^2}{2} - \frac{2^2}{2}\right)$$

$$= 68.$$

If we evaluate the indefinite integral at $x = 4$ and $x = 2$, we obtain

$$Y(4) = 64 + 16 + \text{constant}$$

$$= 80 + \text{constant}$$

$$Y(2) = 8 + 4 + \text{constant}$$

$$= 12 + \text{constant}.$$

Subtracting,

$$Y(4) - Y(2) = 68.$$

The definite and indefinite integrals are related by the equation

$$Y(x) - Y(a) = Y$$

$$= \int_a^x f(x)\,dx.$$

C

Physical constants, astronomical data, and conversion factors

C.1 PHYSICAL CONSTANTS

Constant	Symbol	Value	
Speed of light	c	2.9979×10^8	$m \cdot sec^{-1}$
Electron rest mass	m_e	9.1091×10^{-31}	kg
Proton rest mass	m_p	1.6725×10^{-27}	kg
Neutron rest mass	m_n	1.6748×10^{-27}	kg
Planck's constant	h	6.6257×10^{-34}	$J \cdot sec$
Boltzmann's constant	k	1.3805×10^{-23}	$J \cdot (°K)^{-1}$
Electron charge	e	-1.6021×10^{-19}	C
Avogadro's number	N_0	6.0225×10^{26}	$(kg\text{-}mole)^{-1}$
Gas constant	R	8314.3	$J \cdot (kg\text{-}mole)^{-1} \cdot (°K)^{-1}$
Volume of ideal gas at 0°C and 1 atm	V_0	22.414	$m^3 \cdot (kg\text{-}mole)^{-1}$
Gravitational constant	G	6.670×10^{-11}	$N \cdot m^2 \cdot kg^{-2}$

C.2 ASTRONOMICAL DATA

Constant	Symbol	Value	
Acceleration due to gravity at sea level at the equator	g	9.78	$m \cdot sec^{-2}$
Mass of earth	M_e	5.98×10^{24}	kg
Mass of sun	M_s	1.99×10^{30}	kg
Mass of moon	M_m	7.36×10^{22}	kg
Mean radius of earth	R_e	6.37×10^6	m
Mean radius of sun	R_s	6.96×10^8	m
Mean radius of moon	R_m	1.74×10^6	m
Mean distance earth to sun		1.50×10^{11}	m
Mean distance earth to moon		3.84×10^8	m

C.3 CONVERSION FACTORS

Force

$$1 \text{ newton (N)} = 10^5 \text{ dynes (dyn)}$$

Pressure

$$1 \text{ atmosphere (atm)} = 1.013 \times 10^5 \text{ N} \cdot m^{-2}$$

$$1 \text{ N} \cdot m^{-2} = 9.265 \times 10^{-6} \text{ atm}$$

Temperature

$$0°K = -273.15°C$$

$$0°C = 273.15°K$$

Mass and Energy

$$
\begin{aligned}
1 \text{ amu} &= 1.660 \times 10^{-27} \text{ kg} \\
1 \text{ amu} &= 1.492 \times 10^{-10} \text{ J} \\
1 \text{ amu} &= 931.0 \quad\quad\quad \text{MeV} \\
1 \text{ kg} &= 6.025 \times 10^{26} \quad \text{amu} \\
1 \text{ kg} &= 8.989 \times 10^{16} \quad \text{J} \\
1 \text{ kg} &= 5.611 \times 10^{35} \quad \text{eV} \\
1 \text{ eV} &= 1.602 \times 10^{-19} \text{ J} \\
1 \text{ eV} &= 1.074 \times 10^{-9} \quad \text{amu} \\
1 \text{ eV} &= 1.782 \times 10^{-36} \text{ kg} \\
1 \text{ J} &= 6.704 \times 10^{9} \quad\ \text{amu} \\
1 \text{ J} &= 6.242 \times 10^{18} \quad \text{eV} \\
1 \text{ J} &= 1.112 \times 10^{17} \quad \text{kg} \\
1 \text{ keV} &= 10^{3} \quad\quad\quad\quad \text{eV} \\
1 \text{ MeV} &= 10^{6} \quad\quad\quad\quad \text{eV} \\
1 \text{ GeV} &= 10^{9} \quad\quad\quad\quad \text{eV}
\end{aligned}
$$

Answers to odd-numbered problems

CHAPTER 2

3. 280 mi

13. (a) 3; (b) 5.66; (c) 2

17. 79°

19. (a) $-10\mathbf{i} - 3\mathbf{j} - 11\mathbf{k}$; (b) $10\mathbf{i} + 3\mathbf{j} + 11\mathbf{k}$

CHAPTER 3

3. (a) $0 \leqslant t < 2$ sec, $2 < t \leqslant 4$ sec; (b) 4 m; (c) 1.6 m $\cdot$ sec^{-2}, 2.3 m $\cdot$ sec^{-2}

15. (a) 6.3 m $\cdot$ sec^{-1}, 11.5 m $\cdot$ sec^{-1}; (b) 1.2 m $\cdot$ sec^{-2}; (c) 39 m

17. 100 km $\cdot$ hr^{-1}

19. 3.2 units; $v = 1.2t^{4/3}$

CHAPTER 4

1. (a) 35 m $\cdot$ sec^{-1}; (b) 25 m $\cdot$ sec^{-1}; (c) 30.4 m $\cdot$ sec^{-1}

3. 2 km $\cdot$ hr^{-1}

7. (a) $t = 0$, $x' = 1$ m, $y' = 2$ m, $z' = 3$ m
      $t = 2$, $x' = -4.8 \times 10^8$ m, $y' = 2$ m, $z' = 3$ m
      $t = 10$, $x' = -2.4 \times 10^9$ m, $y' = 2$ m, $z' = 3$ m
      $t = 100$, $x' = -2.4 \times 10^{10}$ m, $y' = 2$ m, $z' = 3$ m
   (b) $t = 0$, $x' = 1.667$ m, $y' = 2$ m, $z' = 3$ m
      $t = 2$, $x' = -8.0 \times 10^8$ m, $y' = 2$ m, $z' = 3$ m
      $t = 10$, $x' = -4.0 \times 10^9$ m, $y' = 2$ m, $z' = 3$ m
      $t = 100$, $x' = -4.0 \times 10^{10}$ m, $y' = 2$ m, $z' = 3$ m

9. (a) $x_2 = 2.71$ m, $x_1 = 2.00$ m   (b) $x_2 = 1.18$ m, $x_1 = 0.87$ m
     $y_2 = 1.71$ m, $y_1 = 1.00$ m $y_2 = 1.71$ m, $y_1 = 1.00$ m
     $\phi = 45°$ with x, x' direction $\phi = 66°\ 24'$ with x, x' direction

11. $(1, 0, 0, 1/c)$

CHAPTER 5

1. $0.95c$

3. The reply will be received

5. (a) $1.4c$ and 0; (b) $0.6c$ and $0.8c$; (c) $1.2c$ and $0.57c$, 25° to the x axis

7. 0.6 m and 1.0 m; 0.44 m and 1.0 m; 0.14 m and 1.0 m; 0.014 m and 1 m

9. 1.4 m

13. (a) $1.3(5) \times 10^{-9}$ sec; (b) 6.8×10^{-9} sec

15. $2.1(4) \times 10^{-6}$ sec

17. $0.87c$

19. $0.999999969c$

21. (a) 2.0×10^9 m; (b) 0; (c) 2.0×10^9 m, but x_1 and x_2 reversed from (a)

CHAPTER 6

1. 3.24 m, 3.24 m $\cdot$ sec^{-1}, 2.16 m $\cdot$ sec^{-2}

3. (a) $A[1 + (Bt - 1)\exp(-Bt)]$, $AB\exp(-Bt)[2 - Bt]$; (b) $v \to A, a \to 0$

7. 980 m $\cdot$ sec^{-1} making an angle of $53'$ with the vertical

9. 24 rad $\cdot$ sec^{-1} **11.** 0.55 rev $\cdot$ sec^{-1}

13. 2.13×10^3 m $\cdot$ sec^{-2}, 50 rad $\cdot$ sec^{-1}

17. $x = vt - (D/2)\sin(2vt/D)$, $y = (D/2)[1 - \cos(2vt/D)]$
$v_x = v[1 - \cos(2vt/D)]$, $v_y = v\sin(2vt/D)$
$a_x = (2v^2/D)\sin(2vt/D)$, $a_y = (2v^2/D)\cos(2vt/D)$

19. (a) 2.16×10^5 rev; (b) 186 miles

21. (a) 2.09 rad $\cdot$ sec^{-2}; (b) 20.9 m $\cdot$ sec^{-1}; (c) 437 m $\cdot$ sec^{-2}; (d) 2.09 m $\cdot$ sec^{-2}

23. (a) 4 rad $\cdot$ sec^{-2}, 16 rad $\cdot$ sec^{-1}, 0.354 m $\cdot$ sec^{-1} in direction opposite to that of railway car, $73°\,6'$; (b) $24°\,36'$

CHAPTER 7

1. 1.67×10^{-15} N   **5.** 41.8 m $\cdot$ sec^{-1}

7. (a) 106 ft $\cdot$ sec^{-1}; (b) 87.0 ft; (c) 351 ft

9. within 5 m   **11.** $45°$

13. 14.1 m $\cdot$ sec^{-1}

15. safe since ball is too high for catcher to catch

19. (a) $2mv/\Delta t$; (b) $2mv\cos\theta/\Delta t$ **21.** 3.0×10^{-3} sec

23. 2.4×10^3 m $\cdot$ sec^{-1}

25. (a) 49.0 kg $\cdot$ sec^{-1}; (b) 196 kg $\cdot$ sec^{-1} **27.** 0.9 unit

31. (a) $9.7(6)$ N, $9.7(6)$ m $\cdot$ sec^{-2}, 1.6×10^{-23} m $\cdot$ sec^{-2};   (b) $9.7(6) \times 10^3$ N, $9.7(6)$ m $\cdot$ sec^{-2}, 1.6×10^{-20} m $\cdot$ sec^{-2}; (c) $9.7(6) \times 10^6$ N, $9.7(6)$ m $\cdot$ sec^{-2}, 1.6×10^{-17} m $\cdot$ sec^{-2}

33. 2.7×10^{33} J   **35.** 2.8×10^9 J

37. 2.41×10^{-43} **39.** 3.1×10^{17} N, 8.9×10^{-6}

41. 9×10^{-2} N, 7.2×10^{-2} J, -9×10^{-2} J, -1.8×10^{-2} J

43. 5.66×10^{-2} sec gained

47. (a) $-2GmM/(r^2 + x^2)^{1/2}$; (b) $2GmMr/(r^2 + x^2)^{3/2}$; (c) $r = 0.707x$

CHAPTER 8

1. (a) 1.1×10^{14} kg $\cdot$ m^2 $\cdot$ sec^{-1}; (b) 9.6×10^{10} N $\cdot$ m, 9.6×10^{11} N $\cdot$ m $\cdot$ sec

5. circular path of radius 3.7 cm; no work is done on the proton; 3.2×10^{-6} N $\cdot$ m

7. $1.3 \times 10^7 \, \text{m} \cdot \text{sec}^{-1}$ **11.** (a) $0.125 \, \text{kg} \cdot \text{m}^2$; (b) $0.073 \, \text{kg} \cdot \text{m}^2$

13. (a) $343 \, \text{g} \cdot \text{cm}^2$
    (b) $280 \, \text{g} \cdot \text{cm}^2$
    (c) $210 \, \text{g} \cdot \text{cm}^2$

CHAPTER 9

1. (a) $1.00 \times 10^{-23} \, \text{kg} \cdot \text{m} \cdot \text{sec}^{-1}$; (b) $1.00 \times 10^{-17} \, \text{N}$

3. $784 \, \text{N}$

5. (a) $0.04 \, \text{m} \cdot \text{sec}^{-1}$; (b) $44.7 \, \text{m} \cdot \text{sec}^{-1}$ (*note:* if summation is approximated by an integration, we get 44.6)

7. $0, 15 \, \text{m} \cdot \text{sec}^{-1}$

9. (a) $0.4 \, \text{m} \cdot \text{sec}^{-1}$; (b) $0.1 \, \text{J}$; (c) $0.04 \, \text{J}$

11. (a) 0.86; (b) $2.64 m_1 \, \text{J}$

15. $v(\alpha\text{-particle}) = 1.85 \times 10^7 \, \text{m} \cdot \text{sec}^{-1}$; $v(^{12}\text{C nucleus}) = 6.77 \times 10^6 \, \text{m} \cdot \text{sec}^{-1}$; $\phi = 52.0°$; $K(\alpha) = 1.14 \times 10^{-12} \, \text{J}$; $K(^{12}\text{C}) = 0.46 \times 10^{-12} \, \text{J}$

CHAPTER 10

3. a spiral in the plane of rotation

7. $3.6 \, \text{N}$ at an angle of $33° \, 42'$ with the radius

9. Toronto $\lambda = 43° \, 40'$; $1.78 \times 10^{-2} \, \text{m} \cdot \text{sec}^{-2}$; 0.18%

11. $4.2 \times 10^3 \, \text{m} \cdot \text{sec}^{-2}$ toward the north

17. $0.36 \, \text{cm}$

CHAPTER 11

1. (a) $197 \, \text{m} \cdot \text{sec}^{-2}$, $6.28 \, \text{m} \cdot \text{sec}^{-1}$; (b) $0.027 \, \text{sec}$

3. $0.052 \, \text{m}$ **5.** (b) $0.01 \, \text{Hz}$

7. experimental period $= 1.192 \pm 0.007 \, \text{sec}$; calculated period $= 1.09(9) \, \text{sec}$

9. (a) $1.04 \, \text{sec}$; (b) $0.058 \, \text{m}$; (c) a small amount of kinetic energy is lost since this is an elastic collision

11. 1.414 **13.** (a) $49.7 \, \text{min}$; (b) $88.9 \, \text{min}$

17. (a) $T = 2\pi(l \cos \theta/g)^{1/2}$

19. $(6.25 \pm 0.024) \times 10^{-11} \, \text{N} \cdot \text{m}^2 \cdot \text{kg}^{-2}$; neglects systematic error due to other interactions between balls

21. $1.4 \times 10^{-2} \, \text{sec}$

CHAPTER 12

1. 2.28×10^{-31} kg; 6.10×10^{-31} kg; 1.18×10^{-30} kg; 5.55×10^{-30} kg

3. 0.548 MeV   **5.** 1.02 MeV

7. 21.2 m

9. 1.32×10^{17} kg, assuming energy radiated uniformly from the sun in all directions

11. $(0.378 \pm 0.001)c$ **15.** 7.07 MeV

17. 15.6 MeV   **19.** 0.335 m

21. 180.11 MeV

CHAPTER 13

1. 89%

7. (a) l-values: Mercury: 34.9 million miles; Venus: 67.2 million miles; Earth: 92.9 million miles; Mars: 140.4 million miles

9. (a) $r_{min} = 6.5 \times 10^6$ m; $r_{max} = 9.7(5) \times 10^6$ m;
    (b) $J = 5.6 \times 10^{13}$ kg $\cdot$ m^2 $\cdot$ sec^{-1}; $E = -2.4 \times 10^{10}$ J;
    (c) r_{max}: $K = 1.6 \times 10^{10}$ J; $U = -4.0 \times 10^{10}$ J
    r_{min}: $K = 3.7 \times 10^{10}$ J; $U = -6.1 \times 10^{10}$ J

11. 0.97

13. (a) $1.2(8) \times 10^{30}$ kg; (b) 5.8×10^{11} m

15. 6.25×10^3 sec

19. (a) 6.8×10^3 m $\cdot$ sec^{-1}, 9.1×10^3 m $\cdot$ sec^{-1}; (b) 312 kg

21. A_2: $v = 86$ m $\cdot$ sec^{-1}; A_3: $v = 35$ m $\cdot$ sec^{-1}; energy is conserved

23. 1.985×10^{30} kg

25. (a) (i) 1.17 (ii) 0.13 $\times$ density of earth (iii) 3.2 $\times$ escape velocity from earth

CHAPTER 14

1. (a) 5×10^{-11} m; 6.7×10^{-11} m; 9.5×10^{-11} m; 9.72×10^{-11} m;
    9.88×10^{-11} m; 9.92×10^{-11} m; all distances measured from H atom
    (b) 0.5 a.m.u.; 0.67 a.m.u.; 0.95 a.m.u.; 0.973 a.m.u.; 0.988 a.m.u.; 0.993 a.m.u.

3. (a) midway between C atoms;
    (b) (i) 5.74×10^{-47} kg $\cdot$ m^2 (ii) 2.80×10^{-46} kg $\cdot$ m^2 (iii) 3.37×10^{-46} kg $\cdot$ m^2

5. (a) at center of cube; (b) 0.064×10^{-10} m from C along CX$_4$;
    (c) 0.070×10^{-10} m from C along perpendicular to line X$_3$X$_4$;
    (d) 0.057×10^{-10} m from C along line X$_1$C on side opposite to X$_1$;
    (e) at center of cube

7. 6.56×10^{11} Hz   **9.** 2.0 g

11. 2.1×10^{-24} kg $\cdot$ m^2 $\cdot$ sec^{-2}; 1.5×10^{22} rad $\cdot$ sec^{-2}

13. 1.41   **15.** (a) 0.16 m; (b) 4.5 rev $\cdot$ sec^{-1}

17. $E_{rot} = 3J^2/16mr^2$, where r is the C–Cl distance

21. (e), (c)   **23.** a linear molecule

25. 3, 3, 2, 1

27. (b) first three nondegenerate; last two doubly degenerate

CHAPTER 15

3. (a) 3.0×10^{-2} m; (b) $3\pi/2$ rad; (c) 1.3π rad; (d) 1.11 mm

5. 0   **7.** $\pi/3$ rad

9. 228 Hz   **11.** 5 m·sec^{-1}

13. $[M^{-1}L^3T^{-2}]$ **15.** $A \propto r^{-1/2}$; $I \propto r^{-1}$

17. 0.37   **19.** (a) 1.088; (b) 0.912

21. (a) 13 m·sec^{-1}; (b) 550 Hz   **23.** 2.3×10^5 m·sec^{-1}

25. 5 Hz or less   **27.** $v'/v_0 = v + u_m/v - u_s$

CHAPTER 16

5. -0.25 m

7. adjacent maxima cannot be separated (resolved) by the eye

11. $v \propto (\text{tension})^{1/2}$

13. frequency $= (60n \pm 5)$Hz with n an integer

15. 30 kHz; 5.56×10^{-2}, 2.50×10^{-4}

19. (b) $v = v_0 \sin(ka/2)/(ka/2)$; $v_g = v_0 \cos(ka/2)$

21. $10^5 - 10^6$ Hz   **23.** $\cos^3 t = \frac{3}{4}\cos t + \frac{1}{4}\cos 3t$

CHAPTER 17

3. 0.2 N

5. (a) 2.62×10^{11} kg; (b) 3.49×10^{21} m^{-2}·sec^{-1}; (c) 9.26×10^{-20} N·m^{-2}

7. $T' = T + mv_0^2/3k$ **9.** methane only

11. (a) 0.8 kg·m^{-3}; (b) 1.96×10^{-2} N   **13.** (a) 3.53×10^7; (b) 3.05×10^{-5} m

21. 1.00×10^{-19} J; 4830°K

25. W.D. $= 1.4 \times 10^4$ J; $\Delta U = 3.7 \times 10^4$ J

27. (a) 0.56×10^{-10} m from C atom toward N atom; (b) 1.91×10^{-46} kg·m^2; (c) 8; (e) 2.08×10^4 J·(k-mole)$^{-1}$(°K)$^{-1}$

CHAPTER 18

3. $E_\lambda \Delta\lambda = (8\pi hc/\lambda^5)[\Delta\lambda/\exp(hc/\lambda kT) - 1]$

5. 8.50×10^{-13} m   **7.** 0.51 eV; no

9. energy: $\qquad h\nu = h\nu' + m_e(\gamma - 1)c^2$

momentum: $\quad \dfrac{h\nu}{c} = \dfrac{h\nu'}{c}\cos\theta + \gamma m_e v \cos\phi$

$$0 = \frac{h\nu'}{c}\sin\theta - \gamma m_e v \sin\phi$$

11. 3.29×10^{15} sec^{-1}

13. energy levels differ by a factor Z^2 from those of an H atom if charge on nucleus is Ze

15. 4.06×10^{-14} m; 1.24×10^{-15} m; 1240 MeV

17. 3.29×10^{-8} eV   **21.** ~ 0.1 J $\cdot$ sec

23. (a) 4.60×10^{-48} kg $\cdot$ m^2;

(b) 0 J, 2.42×10^{-21} J, 7.26×10^{-21} J, 1.35×10^{-20} J, 2.42×10^{-20} J;

(c) 4.37×10^{-20} J, 1.31×10^{-19} J, 2.19×10^{-19} J, 3.06×10^{-19} J, 3.94×10^{-19} J

CHAPTER 19

1. 0.23   **3.** 0.031

5. 60; 30; 24   **7.** 1.57×10^{-12}

9. (a) 40; (b) 36

11. $n = 0.589 N^{1/2}$; $n/N = 5.9 \times 10^{-12}, 5.9 \times 10^{-6}, 5.9 \times 10^{-4}, 5.9 \times 10^{-3}, \simeq 0.19$

15. $N(10^{-3}\,\text{eV})/N(10^{-1}\,\text{eV}) = 0.0217$ **17.** 2.57×10^{4}°K

19. hydrogen: one state; nitrogen: two states; oxygen: three states; chlorine: seven states; iodine: 16 states

21. $Z = [\exp(h\nu/2kT) - \exp(-h\nu/2kT)]^{-1}$

25. (a) $w(n_1, n_2, \cdots, n_k) = \dfrac{N!}{n_1! n_2! \cdots n_k!}(f_1)^{n_1}(f_2)^{n_2}\cdots(f_k)^{n_k}$

(b) $n_i = Nf_i$

CHAPTER 20

1. for temperature T, maximum occurs at momentum p; for temperature $5T$, maximum occurs at momentum $2.24p = (5)^{1/2}p$

3. 9.7 cm   **5.** 5×10^{10} neutrons $\cdot$ m^{-3}

7. $S(n = 0) = 0$; $S(n = N/100) = -2.1 \times 10^{-4}$ Nk

11. hydrogen: I $= 4.71 \times 10^{-48}$ kg $\cdot$ m^2, $r = 7.5 \times 10^{-11}$ m; hydrogen chloride: $I = 2.63 \times 10^{-47}$ kg $\cdot$ m^2, $r = 1.28 \times 10^{-10}$ m; carbon monoxide: $I = 1.45 \times 10^{-46}$ kg$\cdot$m^2, $r = 1.13 \times 10^{-10}$ m; oxygen: $I = 1.93 \times 10^{-46}$ kg$\cdot$m^2; $r = 1.20 \times 10^{-10}$ m

13. $C_v = \frac{3}{2}k + \varepsilon[dn(T)/dT]$, $n(T)$—fraction of molecules in higher energy internal state

CHAPTER 21

1. identity; clockwise rotation through 120°; clockwise rotation through 240°; reflection in the three planes, perpendicular to the plane of the triangle, which pass through the center of the triangle as well as one of the atoms

3. identity; clockwise rotations through 120° or 240° about each of four axes through the center of the cube and one of the atoms; rotations through 180°,

about each of three axes through the center of the cube and perpendicular to a
cube face; clockwise rotations through 90° and 270° about each of the preceding
three axes followed by reflection in a plane perpendicular to the axis; reflection
in each of the six planes through the center of the cube and containing two of the
atoms

5. $8.97 \times 10^3 \, \text{kg} \cdot \text{m}^{-3}$

7. (a) a^3, a^3, a^3; (b) $a^{-3}, 2a^{-3}, 4a^{-3}$; (c) $a, (3)^{1/2}a/2, (2)^{1/2}a/2$; (d) $(2)^{1/2}a, a, a$;
(e) 6, 8, 12; (f) 12, 6, 6

9. $r = 0.207a$ (sc); $r = 0.147a$ (fcc); $r = 0.097a$ (bcc) where a is the length of the
cube edge

11. $4.2 \times 10^{-3} \, \text{keV}$ **13.** $(001), (010), (100), (\bar{1}00), (0\bar{1}0), (00\bar{1})$

15. 12.1×10^{18} (100); 8.56×10^{18} (110); 14.0×10^{18} (111)

17. $6.75 \times 10^{10} \, \text{N} \cdot \text{m}^{-2}$ **19.** $18 \, \text{N} \cdot \text{m}^{-1}$

21. $-5 \times 10^{-5} \, \text{m}$

23. (a) $7.0 \times 10^{10} \, \text{N} \cdot \text{m}^{-2}$; (b) aluminum; (c) $5.1 \times 10^3 \, \text{m} \cdot \text{sec}^{-1}$

CHAPTER 22

3. $8.16 \times 10^{4}°\text{K}$; $1.57 \times 10^6 \, \text{m} \cdot \text{sec}^{-1}$ **5.** 65.2 cal

7. no   **9.** $\theta_E \simeq 180°\text{K}$

11. $1.61 \times 10^{-3} \, \text{m}$ **15.** 25.5°K

17. $\alpha = (0.233 \pm 0.002) \times 10^{-4} \, (°\text{K})^{-1}$

CHAPTER 23

1. 22.414.m³; 6.0227×10^{26} molecules   **3.** (a) $8.31 \times 10^4 \, \text{J}$; (b) $2.08 \times 10^5 \, \text{J}$

5. (a) no work done; (b) $1.2 \times 10^3 \, \text{J}$; (c) $1.2 \times 10^3 \, \text{J}$

7. $21.4b \, \text{J}$

9. The proposed operation of this engine violates the first law

11. $1.8 \times 10^5 \, \text{J}$ **15.** $T_1 = 0$ or $T_2 = \infty$

17. (a) $1.5 \times 10^6 \, \text{N} \cdot \text{m}^{-2}$; 900°K; (b) $9.4 \times 10^5 \, \text{N} \cdot \text{m}^{-2}$; 560°K

21. aluminum: 92.4 cal; copper: 43.0 cal; lead: 5 cal; mercury: 3 cal; ice: 79.67 cal

23. $4.07 \times 10^5 \, \text{J}$

APPENDIX A

1. (a) 0.05 sec for each; (b) 0.1, 0.01, 0.001; (c) 10%, 1%, 0.1%

5. 2%   **11.** $12.02 \pm 0.30 \, \text{m}$

13. $2.5 \, \text{m} \cdot \text{sec}^{-1}$ **17.** $(1.119 \pm 0.002) \times 10^4 \, \text{m} \cdot \text{sec}^{-1}$

19. $(0.161 \pm 0.024) \, \text{kg} \cdot \text{sec}^{-1}$

Index